1st International Conference on Functional and Engineering Materials (FEM 2019)

AA007451

IOP Conference Series: Materials Science and Engineering
Volume 743

Lodz, Poland
16 - 18 October 2019

ISBN: 978-1-7138-3498-4
ISSN: 1757-8981

Printed from e-media with permission by:

Curran Associates, Inc.
57 Morehouse Lane
Red Hook, NY 12571

Some format issues inherent in the e-media version may also appear in this print version.

This work is licensed under a Creative Commons Attribution 3.0 International Licence.
Licence details: http://creativecommons.org/licenses/by/3.0/.

No changes have been made to the content of these proceedings. There may be changes to pagination and minor adjustments for aesthetics.

Printed with permission by Curran Associates, Inc. (2021)

For permission requests, please contact the Institute of Physics at the address below.

Institute of Physics
Dirac House, Temple Back
Bristol BS1 6BE UK

Phone: 44 1 17 929 7481
Fax: 44 1 17 920 0979

techtracking@iop.org

Additional copies of this publication are available from:

Curran Associates, Inc.
57 Morehouse Lane
Red Hook, NY 12571 USA
Phone: 845-758-0400
Fax: 845-758-2633
Email: curran@proceedings.com
Web: www.proceedings.com

TABLE OF CONTENTS

SUPPRESSION OF $(CU,NI)_6SN_5$ INTERMETALLIC COMPOUND IN SN-0.7CU-0.05NI+1WT.TIO_2 SOLDER PASTE COMPOSITE SUBJECTED TO ISOTHERMAL AGING 1
R Mohd Said, N Saud, MAA Mohd Salleh, M N Derman

IN VITRO BIOACTIVITY STUDY OF THERMOPLASTIC STARCH/BENTONITE/HYDROXYAPATITE COMPOSITES FOR BIOMEDICAL APPLICATIONS 8
Nurul Husna Zakaria, Noorhafiza Muhammad, Mohd Mustafa Al Bakri Abdullah, S N F M Noor

EFFECT OF HALOGEN FREE-FLAME RETARDANTS ON THE MECHANICAL AND MORPHOLOGICAL PROPERTIES OF KENAF/POLYURETHANE FOAM COMPOSITES 16
Nur Suhaili Mohd Soberi, R Rozyanty, Firuz Zainuddin, Hazim Mohamad Amini

CORROSION RESISTANCE OF MICRO-TEXTURED SURFACE MODIFIED ALUMINA-TITANIA COATING 22
J. A. Wahab, M. J. Ghazali, M. N. Derman

COMPRESSIVE STRENGTH AND THERMAL CONDUCTIVITY OF METAKAOLIN GEOPOLYMERS WITH ANISOTROPIC INSULATIONS 30
Nur Ain Jaya, Liew Yun-Ming, Mohd Mustafa Al Bakri Abdullah, Kamarudin Hussin, Heah Cheng Yong, Ridho Bayuaji, Muhammad Faheem Mohd Tahir

EFFECT OF SOLID-TO-LIQUID RATIO ON THIN FLY ASH GEOPOLYMER 36
Ng Yong-Sing, Liew Yun-Ming, Mohd Mustafa Al Bakri Abdullah, Ng Hui-Teng, Kamarudin Hussin, Heah Cheng Yong, Nurul Aida Mohd Mortar, Andrei Victor Sandu

THE EFFECT OF POLYETHYLENE GLYCOL (PEG) ON TIO2 THIN FILMS VIA SOL-GEL METHOD 43
D S Che Halin, K A Razak, N S Mohamad Sukeri, A Azani, M M A B Abdullah, M A A M Salleh, N Mahmed, M M Ramli, AW Azhari, V Chobpattana

INFLUENCE OF DIFFERENT MATERIALS ON THE MECHANICAL ASPECTS IN THE DESIGN OF CYCLONE GASIFIER 49
C. Y. Khor, M. A. M. Nawi, Muhammad Ikman Ishak, Z. X. Wei, M. U. Rosli

EXPERIMENTAL AND COMPUTATIONAL EVALUATIONS OF THE MECHANICAL STRESSES OF BANANA TRUNK FIBRE-REINFORCED EPOXY RESIN COMPOSITE IN COFFEE TABLE APPLICATION 54
Muhammad Ikman Ishak, M U Rosli, C Y Khor, C N Ismail, M A M Nawi

A NOVEL GREEN TECHNOLOGY KAFFIR LIME EXTRACT AS LIZARD REPELLENT 59
M. A. M. Nawi, Muhammad Ikman Ishak, M. U. Rosli, M. A. R. M. Ayub, C. Y. Khor, Siti Nor Azreen Ahmad Termizi, Noorhidayah Ishak

SIMULATION-BASED OPTIMIZATION OF PLASTIC INJECTION MOLDING PARAMETER FOR AUTOMOTIVE CAR WHEEL FABRICATION USING RESPONSE SURFACE METHODOLOGY (RSM) 65
M. U. Rosli, C. Y. Khor, M. A. M. Nawi, Haris Ismail, Muhammad Ikman Ishak

PARAFFIN AS A PHASE CHANGE MATERIAL IN CONCRETE FOR ENHANCING THERMAL ENERGY STORAGE 71
Rafiza Abd Razak, Adeline Choo Khang Zhe, Mohd Mustafa Al Bakri Abdullah, Zarina Yahya, Subaer Junaidi, Khairunnisa Muhamad, Nurul Aida Mohd Mortar

SEAWATER EXPOSURE EFFECT ON FLY ASH BASED GEOPOLYMER CONCRETE WITH INCLUSION OF STEEL FIBER 77
Zarina Yahya, Mohd Mustafa Al Bakri Abdullah, Leow Yee Jing, Long-Yuan Li, Rafiza Abd Razak

SYNTHESIS AND CHARACTERIZATION OF FLY ASH BASED GEOPOLYMER CERAMICS: EFFECT OF NAOH CONCENTRATION 86
Romisuhani Ahmad, Wan Mastura Wan Ibrahim, Mohd Mustafa Al Bakri Abdullah, Andrei Victor Sandu, Nurul Aida Mohd Mortar, Noratikah Hashim, Warid Wazien Ahmad Zailani

PROPERTIES OF BLENDED ALKALINE SYSTEM GEOPOLYMER- A REVIEW 93
Fakhryna Hannanee Ahmad Zaidi, Romisuhani Ahmad, Zarina Yahya, Muhammad Faheem Mohd Tahir, Wan Mastura Wan Ibrahim, Ahmad Syauqi Saufi

SPECTROPHOTOMETRIC ANALYSIS ASSAY FOR THE MEASUREMENT OF ARSENIC USING NANOCOMPOSITE OF SILICA AND GRAPHENE 97
M. N. A. Uda, Uda Hashim, C. B. Subash Gopinath, M. N. Afnan Uda, N. A. Parmin, Hussain Adam, Nur Hulwani Ibrahim

ANALYSIS ON GOLD NANOROD INTERDIGITATED ELECRTRODE SENSOR USING SIMPLE PHOTOLITHOGRAPIC PROCESS 103
Hussaini Adam, C. B. Subash Gopinath, M. N. A. Uda, Uda Hashim, Tijjani Adam, M. N. Afnan Uda

NEW DEVELOPMENT QUANTIFICATION METHODS FOR SALT IODINE AND URINARY IODINE USING MICROFLUIDICS BASED NANOTECHNOLOGY 109
Nur Hulwani Ibrahim, M. N. Salimi, M. N. A. Uda, N. A. Parmin, Uda Hashim, M. N. Afnan Uda

DEVELOPMENT OF VOLTAGE AMPLIFIER ELECTRONIC READER FOR MULTIPLEX DETECTION OF TWO ELECTRODE ELECTRICAL BIOSENSORS 115
M. N. Afnan Uda, Asral Bahari Jambek, U. Hashim, M. N. A. Uda

DEVELOPMENT OF INTERNET OF THINGS (IOT) BASED ELECTRONIC READER FOR MEDICAL DIAGNOSTIC SYSTEM 124
M. N. Afnan Uda, Asral Bahari Jambek, U Hashim, M. N. A. Uda, M. A. F. Bahrin

NANO-MICRO-MILI CURRENT TO MILI VOLTAGE AMPLIFIER FOR AMPEROMETRIC ELECTRICAL BIOSENSORS 133
M. N. Afnan Uda, N. A. Parmin, Asral Bahari Jambek, U. Hashim, M. N. A. Uda, S. N. A. Shaharuddin, Hussaini Adam

ELECTRICAL DNA BIOSENSOR USING ALUMINIUM INTERDIGITATED ELECTRODE FOR SALMONELLA DETECTION 141
M. N. Afnan Uda, Asral Bahari Jambek, U. Hashim, M. N. A. Uda

DEVELOPMENT OF GEOPOLYMER CERAMIC AS A POTENTIAL REINFORCING MATERIAL IN SOLDER ALLOY: SHORT REVIEW 148
Nadiah 'Izzati Zulkifli, Mohd Mustafa Al Bakri Abdullah, Mohd Arif Anuar Mohd Salleh, Romisuhani Ahmad, Andrei Victor Sandu, Nurul Aida Mohd Mortar

A REVIEW OF MORPHOLOGY ANALYSIS ON DOLOMITE AS AN ADDITIVE MATERIAL IN GEOPOLYMER 155
Ahmad Syauqi Sauffi, Wan Mastura Wan Ibrahim, Romisuhani Ahmad, Nurul Aida Mohd Mortar, Fakhryna Ahmad Zaidi, Warid Wazien Ahmad Zailani

NEW HIGH STRENGTH WATER RETAINING INTERLOCKING PAVERS BLOCK FOR HIGH MECHANICAL PERFORMING PAVEMENT AND REDUCING RUNOFF 161
M A H Abdullah, N A Rashid, A L Abdul Rani, M F Omar

INFLUENCE OF VARIOUS SAND GRADATION ON MECHANICAL PROPERTIES AND CONCRETE QUALITY OF HIGH PERFORMANCE CONCRETE (HPC) 168
K Muhamad, N F Isa, N H Hashim, Z Tusimin, N Yahya, N A Naamandadin, R A Razak, P Y Tham

IMPROVING INDOOR ENVIRONMENTAL QUALITY (IEQ) AND COMFORTABILITY OF LECTURE ROOMS: CASE STUDY – UNICITI ALAM CAMPUS, PADANG BESAR, MALAYSIA 174
Nurul Akmam Naamandadin, Siti Nur Aishah Mohd Noor, Norrazman Zaiha Zainol, Arita Hanim Awang, Syakir Amir Abdul Rahman, Khairunnisa Mohamad, Muhammad Naim Ahmad

MANUFACTURING OF CAST IN SITU ULTRA HIGH PERFORMANCE FIBRE REINFORCED CONCRETE (UHPFRC) - WORKABILITY, TENSILE AND COMPRESSIVE STRENGTH 179
M. Z. A. Mohd Zahid, B. H. Abu Bakar, F. Mohamed Nazri, A. Ayob, R. Abdul Razak

MECHANICAL AND PHYSICAL PROPERTIES OF BOTTOM ASH/FLY ASH GEOPOLYMER FOR PAVEMENT BRICK APPLICATION 189
Wan Mastura Wan Ibrahim, Mohd Mustafa Al Bakri Abdullah, Romisuhani Ahmad, Muhammad Faheem Mohd Tahir, Nurul Aida Mohd Mortar, Muhammad Amirul Aziem Noor Azli

OPTIMIZATION OF PRETREATMENT AND ENZYMATIC HYDROLYSIS OF SPENT COFFEE GROUND FOR THE PRODUCTION OF FERMENTABLE SUGAR 197
Lai Shan Jin, Midhat Nabil Salimi, Syazni Zainul Kamal

EVALUATION AND OPTIMIZATION OF GENOMIC DNA EXTRACTION FROM FOOD SAMPLE FOR MICROFLUIDIC PURPOSE 207
M. N. Afnan Uda, N. A. Parmin, Asral Bahari Jambek, U. Hashim, M. N. A. Uda, S. N. A. Shaharuddin

ALUMINIUM INTERDIGITATED ELECTRODE BASED BIOSENSOR FOR SPECIFIC SSDNA TARGET LISTERIA DETECTION 214
M. N. Afnan Uda, Asral Bahari Jambek, U. Hashim, M. N. A. Uda, M. A. F. Bahrin

MECHANICAL AND RHEOLOGICAL PROPERTIES OF CONCRETE WITH CERAMIC TILE WASTE AS PARTIAL REPLACEMENT OF FINE AGGREGATE 221
N Yahya, S M S Razak, M K Othman, S N A M Noor, K Muhamad, M K M Jaih

A CONCEPTUAL PAPER ON APPLICATION OF ENERGY EFFICIENCY MEASURES FOR REDUCING CARBON FOOTPRINT IN ZINC PHOSPHATE COATING PROCESS OF AUTOMOTIVE INDUSTRIES 227
Abuzaar Mohd Zulkifli, Fazdliel Aswad Ibrahim, Norrazman Zaiha Zainol, Mardhiah Farhana Omar, Lim Chun Keat

PARTIALLY REPLACEMENT OF CEMENT BY SAWDUST AND FLY ASH IN LIGHTWEIGHT FOAM CONCRETE 235
M F Omar, M A H Abdullah, N A Rashid, A L Abdul Rani

PRODUCTION OF IRON OXIDE NANOPARTICLES BY CO-PRECIPITATION METHOD WITH OPTIMIZATION STUDIES OF PROCESSING TEMPERATURE, PH AND STIRRING RATE 241
Beh Hui Hui, Midhat Nabil Salimi

GLUCOSE PRODUCTION FROM SUGARCANE BAGASSE BY TWO STAGES CHEMICAL PRETREATMENT & HYDROLYSIS 254
Nurul Syiffa Husna Che Roslan, Midhat Nabil Salimi

FABRICATION OF ELECTROLYTE-GATE NANOCRYSTALLINE DIAMOND- BASED FIELD EFFECT TRANSISTOR (NCD-EGFET) FOR HIV-1 TAT PROTEIN DETECTION 263
Nurul Atiqah Ahmad, Ruslinda A Rahim, Nur Syakimah Ismail, Bohuslav Rezek

MECHANICAL PROPERTIES OF COCONUT SHELL REINFORCED UNSATURATED POLYESTER COMPOSITES WITH THE PRESENCE OF TITANIUM DIOXIDE AS ADDITIONAL FILLER 268
Nurzarihah Shahidan, Mohamad Bashree Abu Bakar, Mazlan Mohamed, Azwan Mat Lazim, Ayub Ahmed Janvekar

APPLICATION OF IONIC NANO PARTICLES IN FISH PROCESSING WASTEWATER BY USING BIOCHEMICAL TREATMENT: A SHORT REVIEW 275
Mohamad Anuar Kamaruddin, Lim Ewe Teik, Muhd Nazmi Ismail

FOOD WASTE QUANTIFICATION AND CHARACTERIZATION AS A MEASURE TOWARDS EFFECTIVE FOOD WASTE MANAGEMENT IN UNIVERSITY 281
Mohamad Anuar Kamaruddin, Nurul Nadia Jantira, Rasyidah Alrozi

STUDY OF THE UTILIZATION OF CARBIDE WASTE AND FLY ASH ON GEOPOLYMER PASTE 290
Ridho Bayuaji, N S Putri, Liew Yun Ming, Nurul Aida Mohd Mortar

ANALYSIS OF THE CONNECTION OF A NICKEL-CHROMIUM ALLOY WITH DENTAL CERAMICS AFTER ALUMINUM OXIDE ABRASIVE BLASTING 296
W Czepułkowska, E Wołowiec-Korecka

INFLUENCE DIFFERENT AMOUNT OF CELLULOSE ON THE MECHANICAL STRENGTH OF DENTAL ACRYLIC RESIN 302
Joanna Taczała, Chang Fu, Jacek Sawicki, Joanna Pietrasik

IMPACT OF NETWORK EXPANSIONS ON ENERGY LOSSES IN COMPRESSED AIR 310
Karolina Perz, Aleksandra Rewolińska

AUTOMATIC SYSTEMS SUPPORTING RESEARCH ON MODERN MATERIALS AND INDUSTRIAL BONDING TECHNOLOGIES 317
R Wojciechowski, M Bąkała, A Rylski

NUMERICAL SIMULATION OF PHASE TRANSFORMATION DURING GAS QUENCHING AFTER LOW PRESSURE CARBURIZING 326
Krzysztof Krupanek, Jacek Sawicki, Victoria Buzalski

INFLUENCE OF QUENCHING TEMPERATURE OF THE INDIVIDUAL QUENCHING METHOD ON THE GEOMETRICAL DIMENSIONS OF THE ELEMENTS 335
E Wołowiec-Korecka, P Zgórniak, A Brewka, K Krupanek

THE INFLUENCE OF GRINDING WHEEL TYPE ON MICROHARDNESS AND RESIDUAL STRESSES IN VACUUM-CARBURISED 20MNCR5 STEEL USING THE SINGLE-PIECE FLOW METHOD 341
P Zgórniak, E Wołowiec-Korecka, B Januszewicz, M Sikora, R Rosik

THE INFLUENCE OF GRINDING CONDITIONS ON SURFACE ROUGHNESS PARAMETERS OF 20MNCR5 STEEL VACUUM-CARBURISED BY SINGLE-PIECE FLOW METHOD 348
M Sikora, P Zgórniak, E Wołowiec-Korecka

FRICTIONAL BEHAVIOUR OF TIN COATINGS GROWN BY ATOMIC LAYER DEPOSITION 355
Wojciech Grodzicki, Lukasz Kolodziejczyk

COMPARISON OF MECHANICAL BEHAVIOUR OF MICROSTRUCTURES OF 2024 ALUMINIUM ALLOY CONTAINING PRECIPITATES OF DIFFERENT MORPHOLOGIES 362
Anna Staszczyk, Jacek Sawicki

Author Index

IOP Conf. Series: Materials Science and Engineering **743** (2020) 011001 doi:10.1088/1757-899X/743/1/011001

Organizer

Lodz University of Technology

INSTITUTE
OF MATERIALS SCIENCE
AND ENGINEERING

FACULTY OF MECHANICAL ENGINEERING · TUL · WYDZIAŁ MECHANICZNY
PŁ
M
1945

Partner

UNIVERSITI
MALAYSIA
PERLIS
UniMAP

www.fem.p.lodz.pl

Content from this work may be used under the terms of the Creative Commons Attribution 3.0 licence. Any further distribution of this work must maintain attribution to the author(s) and the title of the work, journal citation and DOI.
Published under licence by IOP Publishing Ltd

IOP Conf. Series: Materials Science and Engineering **743** (2020) 011001 doi:10.1088/1757-899X/743/1/011001

Preface

1st International Conference "FEM 2019" on Functional and Engineering Materials was organized in the frame of cooperation between two universities: the Lodz University of Technology in Poland and the Perlis University in Malaysia, under the honorary auspices of the Rector of the LUT Professor Sławomir Wiak, D.Sc, Ph.D., Eng.

Science of functional and engineering materials constitutes a common base for emergence and advancement of new materials, as well as nanomaterials from metal alloys, ceramics, glasses or polymers for smart materials, biomaterials or composites of thereof used in every domain of engineering and other fields of human activity.

On the other hand, three weeks after Climate Change Summit at UN Headquarters in New York, USA, against worsening of the world climate, scientist all over the world and the international community have the obligation to take measures for protecting the natural environment and the atmosphere of the Earth.

It follows from that the contemporary scientists and engineers are obliged not only to make the best choice from enormously great number of existing functional and engineering materials from the point of view of their durability and reliability, but also to consider the raw material consumption, operating costs as well as the carbon footprint of the designed products.

The basic objectives of the Conference were:

- to provide a forum for researchers, educators, students and industrial partners to share and exchange ideas and research findings;
- to create networks and stimulate potential collaboration between researchers and practitioners from different countries working in the same fields of research.

The Chairman and the Organizing Committee thank the keynote lecturers and the invited speakers as well as all the participants for preparing their manuscripts and contribution to the success of the Conference.

Prof. Bogdan Wendler
Chairman of Scientific Committee

IOP Conf. Series: Materials Science and Engineering **743** (2020) 011001 doi:10.1088/1757-899X/743/1/011001

1st International Conference on Functional and Engineering Materials - FEM 2019

October 16-18, 2019, Lodz, Poland

Organizer

Institute of Materials Science and Engineering, Lodz University of Technology, Poland

Scientific Committee:

Honorary Chairman

Prof. Piotr Kula *(Lodz University of Technology, Poland)*

Chairman

Prof. Bogdan Wendler *(Lodz University of Technology, Poland)*

Vice-Chairman

Prof. Mustafa Mohd Al. Bakri *(UniMAP, Malaysia)*

Secretary

Assoc. Prof. Jacek Sawicki *(Lodz University of Technology, Poland)*

Members

Prof. Long Yuan-Li	*University of Plymouth, GREAT BRITAIN*
Prof. Dr Che Mohd Ruzaidi Ghazali	*Universiti Malaysia Terengganu, MALAYSIA*
Assoc. Prof. Dr. Andrei Victor Sandu	*Gheorghe Asachi Technical University of Iiasi, ROMANIA*
Prof. Subaer, M.Phil, Ph.D., D. Sc.	*Universitas Negeri Makassar, INDONESIA*
Assoc. Prof. Dr. Faiz Uddin Ahmed Shaikh	*Curtin University, AUSTRALIA*
Dr. Mukhallad Mohammad Mawlood Al Mashhadani	*Istanbul Gelisim University, TURKEY*
Assoc. Prof. Dr. Nattapong Damrongwiriyanupap	*University of Phayao, THAILAND*
Assoc. Prof. Dr. Hrvoje Kušić	*University of Zagreb, CROATIA*
Prof. Valentin N. Tikhenko	*Odessa National Polytechnical University, UKRAINE*
Prof. Thomas G. Mathia	*National Research Centre Ecole Centrale de Lyon, FRANCE*
Prof. Jolanta E. Klemberg-Sapieha	*Montreal University of Technology Montreal, CANADA*
Prof. Eng. Jindrich Musil	*University of West Bohemia Pilsen, CZECHIA*
Prof. Vadim V. Silberschmidt	*Loughborough University Loughborough, GREAT BRITAIN*
Prof. Joachim Roesler	*Technical University of Braunschweig, GERMANY*
Prof. Andrei Stanishevskiy	*University of Alabama Birmingham, USA*
D.Sc. Eng. Karel Saksl	*Slovak Academy of Sciences, Kosice, SLOVAK REPUBLIC*
Prof. Veli-Tapani Kuokkala	*Tampere University of Technology, FINLAND*
Prof. Juan Pedro Espinos	*University of Sevilla, SPAIN*
Prof. Neville Baker	*Strathclyde University, SCOTLAND*
Prof. Emma Angelini	*Politecnico di Torino, ITALY*
Assoc. Prof. Matteo Pavese	*Politecnico di Torino, ITALY*
Prof. Victor R. Weber	*Novgorod State University, RUSSIA*
Prof. Piotr Niedzielski	*Lodz University of Technology, POLAND*
Assoc. Prof. Łukasz Kaczmarek	*Lodz University of Technology, POLAND*
Assoc. Prof. Hieronim Szymanowski	*Lodz University of Technology, POLAND*
Assoc. Prof. Michał Szota	*Czestochowa University of Technology, POLAND*
Assoc. Prof. Tomasz Tański	*Silesian University of Technology, POLAND*

IOP Conf. Series: Materials Science and Engineering **743** (2020) 011001 doi:10.1088/1757-899X/743/1/011001

1st International Conference on Functional and Engineering Materials - FEM 2019

October 16-18, 2019, Lodz, Poland

Organizing Committee:

Chairman

Adam Rylski (*Lodz University of Technology, Poland)*

Secretary

Norbert Kępczak (*Lodz University of Technology, Poland)*

Members

Sebastian Miszczak (*Lodz University of Technology, Poland)*
Paulina Byczkowska (*Lodz University of Technology, Poland)*
Mariola Grzelak (*Lodz University of Technology, Poland)*
Norazian Mohamed Noor (*UniMAP, Malaysia)*
Mohd Arif Anuar Mohd Salleh (*UniMAP, Malaysia)*

FEM 2019 Editors:

Dr. Adam Rylski
Institute of Materials Science and Engineering
Lodz University of Technology, POLAND.
Email: adam.rylski@p.lodz.pl

Assoc. Prof. Dr. Jacek Sawicki
Institute of Materials Science and Engineering
Lodz University of Technology, POLAND.
Email: jacek.sawicki@p.lodz.pl

Assoc. Prof. Dr. Mohd Mustafa Al-Bakri Abdullah
Center of Excellence Geopolymer & Green Technology (CEGeoGTech),
School Of Materials Engineering Universiti Malaysia Perlis, MALAYSIA.
Email: mustafa_albakri@unimap.edu.my

IOP Conf. Series: Materials Science and Engineering **743** (2020) 011002 doi:10.1088/1757-899X/743/1/011002

Peer review statement

All papers published in this volume of *IOP Conference Series: Materials Science and Engineering* have been peer reviewed through processes administered by the proceedings Editors. Reviews were conducted by expert referees to the professional and scientific standards expected of a proceedings journal published by IOP Publishing.

Content from this work may be used under the terms of the Creative Commons Attribution 3.0 licence. Any further distribution of this work must maintain attribution to the author(s) and the title of the work, journal citation and DOI.
Published under licence by IOP Publishing Ltd

IOP Conf. Series: Materials Science and Engineering **743** (2020) 012001 doi:10.1088/1757-899X/743/1/012001

Suppression of $(Cu,Ni)_6Sn_5$ Intermetallic Compound in Sn-0.7Cu-0.05Ni+1wt.TiO_2 Solder Paste Composite Subjected to Isothermal Aging

R Mohd Said[1], N Saud[1], MAA Mohd Salleh[1] and M N Derman[1]

[1]Centre of Excellence Geopolymer and Green Technology, School of Materials Engineering, Universiti Malaysia Perlis (UniMAP), Taman Muhibbah, 02600, Jejawi, Arau Perlis, Malaysia.

Email: rita@unimap.edu.my

Abstract. This paper investigated the intermetallic compound (IMC) layer that formed in Sn-0.7Cu-0.05Ni+1wt.% TiO_2 (SCNT) composite solder paste added with reinforcement (TiO_2) particles. Besides, the growth of the IMC layer during subsequent aging at temperature of 75 °C, 125 °C and 150 °C also being studied. Scanning electron microscopy was used to observe the IMC growth and to measure the thickness of IMC layer. The interfacial IMC layer has been suppressed whereas the activation energy value of the composite solder paste was high (37.35 kJ/mol) which in turn improved the thermal stability of the IMC layer. Results also show the IMC formed at bulk solder microstructure of SCNT solder paste composite was refined. The presence of TiO_2 particles has become the obstacle for the Cu atom diffusion from the substrate to the solder and Sn atom from the solder, thus, successful for the suppression of the IMC layer.

1. Introduction

Restriction on the use of lead (Pb) containing solder paste in surface mount assemblies has promoted the development of lead-free solder paste. The Sn-based lead-free solder pastes such as tin silver (Sn-Ag), tin copper (Sn-Cu), and tin silver copper (Sn-Ag-Cu) have been identified to replace Pb containing solder in electronic assemblies [1]. In the electronic packaging industries, solder paste was used as primarily an interconnect material for surface mount assemblies. As the joining materials, solders paste to be used in electronic industries should provide necessary electrical, mechanical, and thermal continuity [2]. Therefore, the reliability and quality of solder paste are crucial to ensure proper functioning of the solder joint during surface mount assemblies.

Intermetallic compounds (IMCs) that formed and growth play an important role in electronic package. The formation of IMC occurred during soldering a solder on the substrate. When tin (Sn) in the molten solder reacts with the copper (Cu) from the substrate, the interfacial IMC formed on the interface between the solder and contacted substrate. Generally, the reaction between molten solder and substrate produces IMC layers, such as Cu_6Sn_5, Cu_3Sn, Ni_3Sn_4 and $(Cu,Ni)_6Sn_5$ [3]. Excessive growth of IMC can jeopardize the solder joint performance which in turn influence the reliability of electronic devices. In order to improve the performance of Pb-free solders paste, variety reinforcements were incorporated into Pb-free solders paste to obtain composite solders paste.

Research conducted by Chellvarajoo et al. using diamond nanoparticles reinforcement found a significant improvement in Sn-3Ag-0.5Cu (SAC305) solder paste [4]. Microstructure and mechanical properties of as-reflowed Sn-0.7Ag-0.5Cu-2.5Bi-0.05Ni (SAC-BN) composite solder paste were

Content from this work may be used under the terms of the Creative Commons Attribution 3.0 licence. Any further distribution of this work must maintain attribution to the author(s) and the title of the work, journal citation and DOI.
Published under licence by IOP Publishing Ltd

studied by Liu et al. The bulk microstructure of composite solder paste reveals that the grain size of the β- Sn phase has decreased which causes the increase of the hardness and elastic modulus [5]. Tang et al. have studied the effects of adding TiO_2 particles on the properties of Sn-3.0Ag-0.5Cu-xTiO_2 lead-free solder. The microstructure becomes more uniform with a decrease in Ag_3Sn grains size and the spacing between the Ag_3Sn grains has also decreased relatively. Other than that, an increase in micro-hardness and tensile strength was also reported. This is due to the strengthening effect attributed to the reduction of the spacing between Ag_3Sn grains [6].

In this study, the effect of TiO_2 particles added into Sn-0.7Cu-0.05Ni solder paste on microstructure evolution was investigated during isothermal aging. This study also aims at exploring the kinetic of IMC growth in Sn-0.7Cu-0.05Ni+1wt.% TiO_2 solder paste composite.

2. Experimental

Sn-0.7Cu-0.05Ni (SCN) solder paste supplied by Nihon Superior Co. Ltd, Japan was used in this studied. The solder joints were prepared by printing the solder paste onto Cu-OSP surface finish FR4 substrate using stainless steel with a thickness of 0.15 mm and reflowed in reflow oven followed the standard lead-free solder reflow profile. Then, solder joints were isothermally aged at various temperatures at 75, 125 and 150 °C for 24, 240 and 480 hours in a Memmert Oven to study the growth kinetic of the interfacial intermetallic compound.

The reflowed samples were cross sectioned and undergo metallographic processes samples preparation for microstructure observation. The IMC layer formed at solder/substrate was observed using the optical microscope (Eclipe L300N) and Scanning Electron Microscope (SEM). For each specimen, at least five SEM images were taken at different positions. The thickness of interfacial IMC growth was measured using the J-Image software. The average IMC thickness was determined by dividing IMC area (A) by its total length (L) as shown in Figure 1.

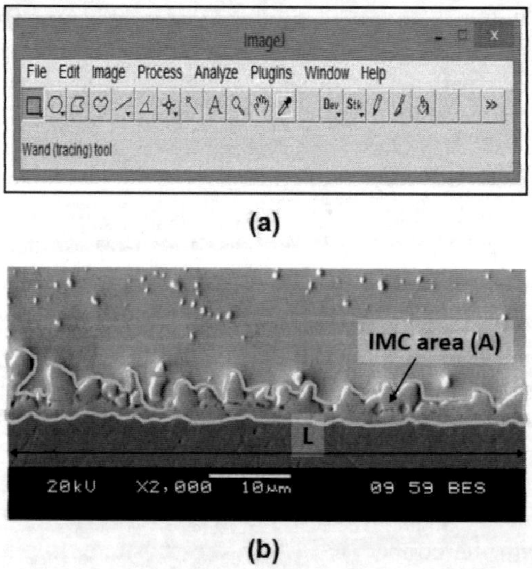

Figure 1. (a) J-Image software and (b) Illustration of the area of IMC layer and length of the measured area.

3. Results and discussion

3.1. Bulk Microstructure

Figure 2 (a and c) shows the SEM image for the bulk solder microstructure after reflowed for Sn-0.7Cu-0.05Ni (SCN) and Sn-0.7Cu-0.05Ni-1wt.%TiO_2 (SCNT) solder paste composite. Figure 2 (b and d) shows the SEM morphology of all solders after deep etched. The microstructure of SCN solder consists of β-Sn and small $(Cu,Ni)_6Sn_5$ IMC particles and SCNT solder possess finer $(Cu,Ni)_6Sn_5$ IMC particles. In the case of SCNT solder paste composite, refinement of $(Cu,Ni)_6Sn_5$ IMC is due to the heterogeneous nucleation site that formed when TiO_2 particles added into SCNT solder paste [7]. During soldering, the areas in which the area containing reinforcement particles has become the heterogeneous nucleation sites for $(Cu,Ni)_6Sn_5$ IMC. With the addition of TiO_2 reinforcement particles, the surface energy of $(Cu,Ni)_6Sn_5$ grains has decreased resulting in a decrease of the growth energy of $(Cu,Ni)_6Sn_5$ grains. According to heterogeneous nucleation theory, the $(Cu,Ni)_6Sn_5$ will nucleate on the reinforcement surface in order to reduce the thermodynamic barrier for the nucleation process. When the reinforcement particles were added into the solder, it will lead the particles to precipitate on the top of the substrate thus produce nucleation site. The increase in the nucleation of Cu_6Sn_5 grains consequently leads to the refinement of the grains [8].

Figure 3 presenting the SEM image for bulk solder microstructure evolution of SCN solder paste after aging for varying aging temperatures (75, 125 and 150 °C) at 24 h, 240 h and 480 h. The bulk solder microstructure of SCNT solder pastes composite subjected to isothermal aging for 75 °C, 125 °C, and 150 °C at 24 h, 240 h and 480 h were presented in Figure 4. After increasing the aging time, the IMC particles have become larger and coarsened. The results reveal that the aging temperature and time significantly influenced the microstructure evolution by increasing the β-Sn area. The plausible reason is that the growth of $(Cu,Ni)_6Sn_5$ IMC is governed by coarsening process. During isothermal aging, coarsening and the growth of $(Cu,Ni)_6Sn_5$ IMC was promoted due to the diffusion of Cu atoms from the substrate and Sn atoms from the solder. High temperature aging such as 150 °C provided enough energy for the reaction of Cu and Sn elements to reach each another.

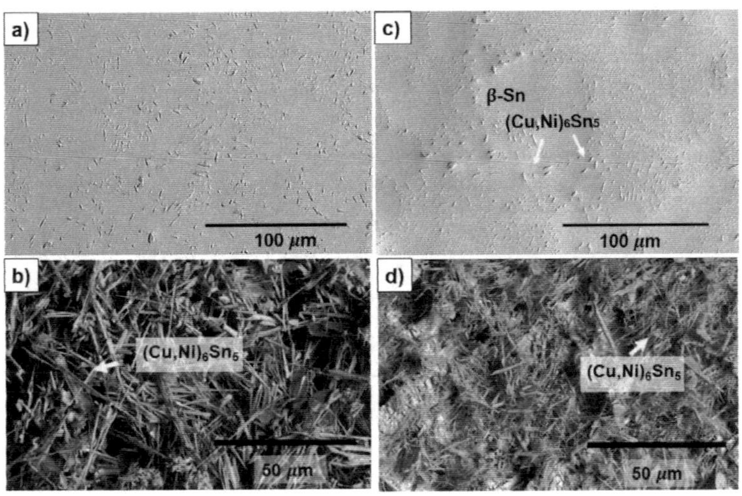

Figure 2. SEM morphology of (a-b) SCN and (c-d) SCNT solder after reflowed and deep etching.

Furthermore, the number of Cu-Sn IMC particles in the aged solders joints has reduced as shown in Figure 3 and 4 (i). This is due to the reduction in smaller IMC particles and the larger IMC particles increased in size. It is likely the IMCs undergo phase coarsening with exposure to prolong aging time and aging temperature. During coarsening, smaller IMC particles are combined with larger IMC particles. The coarsening process occurs in accordance with Thompson-Freundlich solubility relationship, according to which concentration gradient exists between adjacent particles in a matrix [9].

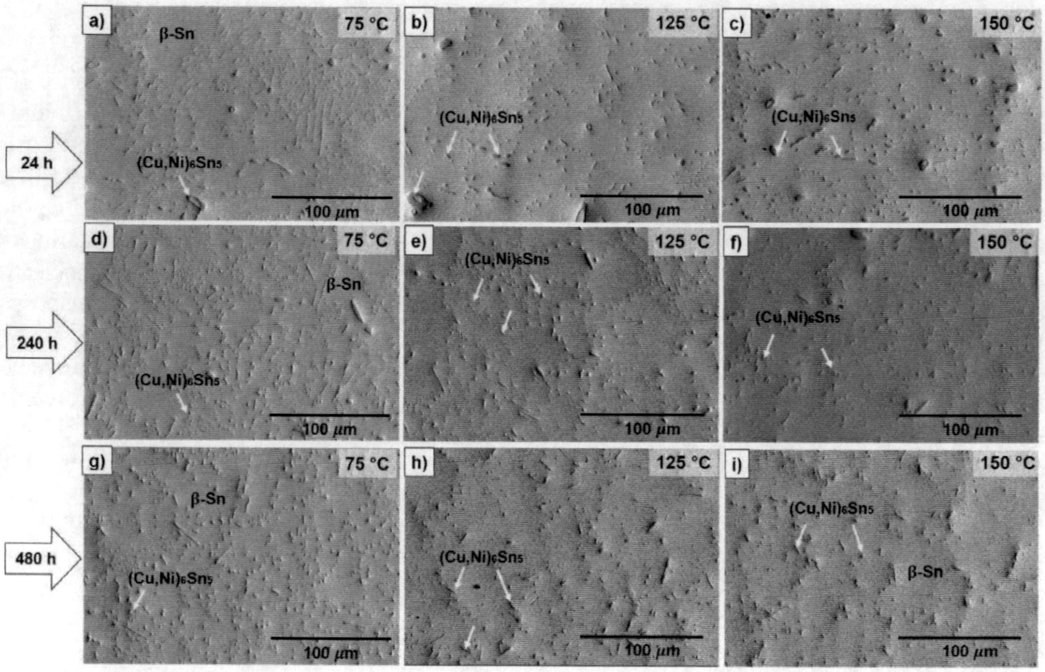

Figure 3. Bulk solder microstructure evolution of SCN during isothermal aging various aging temperature and time.

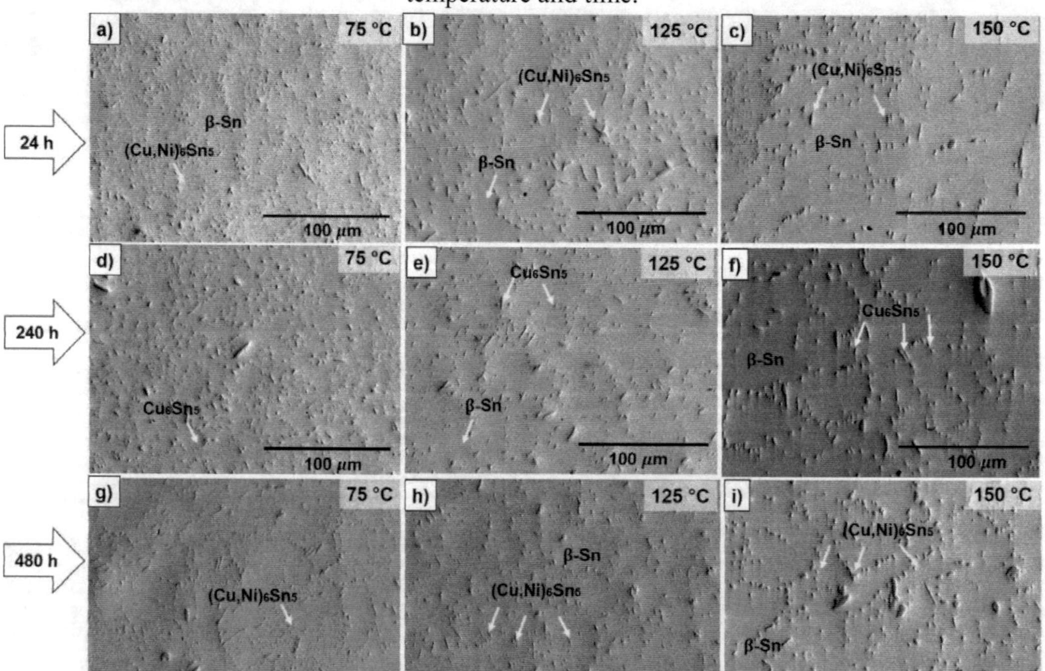

Figure 4. Bulk solder microstructure evolution of SCNT during isothermal aging various aging temperature and time.

According to Thompson-Freundlich solubility relationship, due to the concentration gradient that exists in the matrix, it will cause the particles to diffuse in the direction of the largest particles and away from the smallest particles. As a result, phase coarsening causes small particles to combine with large particles. The small particles dissolve to compensate for the decrease in the concentration at their interface as the flow of particles continue. This dissolution and growth process continue until the system is finally phase separated [9]. After aged to 480 hours for high temperatures such as 125 °C and 150 °C, the total amount of (Cu, Ni)$_6$Sn$_5$ IMC has reduced and the size of IMC was getting larger. In addition, it can be seen that the β-Sn area has increased.

3.2. Intermetallic compound growth
Figure 5 show the interfacial intermetallic compound (IMC) evolutions of SCN and SCNT solder joint subjected to reflow and subsequent isothermal aging at different temperature for 240 h. After isothermal aging, morphology of the IMC layer changed from scallop type to the combination of layer and scallop type of (Cu,Ni)$_6$Sn$_5$ intermetallic was clearly observed in the solder joint. Moreover, it can be seen that the IMC of aged solder grew thicker. However, IMC thickness for SCNT composite seems to be reduced as compared to SCN solder paste.

Generally, Cu atoms from the Cu substrate can diffuse into the (Cu,Ni)$_6$Sn$_5$ solder interface to react with Sn atoms, causing (Cu,Ni)$_6$Sn$_5$ IMC to grow towards the solder and attributed to an increase in the IMC thickness during isothermal aging. It is possible that TiO$_2$ particles in intimate contact with IMC interfacial layer stabilize the Cu dissolution path which in turn prevent the IMC from growing further during isothermal aging. The main mechanism on the growth of IMC layer in solder joints when TiO$_2$ was added is the grain boundary pinning [10]. During isothermal aging, the diffusion of the constituent atomic species through the IMC layer to reaction sites located at the solder/Cu$_6$Sn$_5$ and Cu$_3$Sn/Cu interface led to the growth of IMC layer of solder joints.

Normally, the diffusion of elements through grain boundaries is more rapid through the grain in order to maintain the continuous interfacial reaction. The existence of TiO$_2$ particles on the Cu$_6$Sn$_5$ grain boundary might be significantly helpful in suppressing the IMC growth during aging. The TiO$_2$ particles will obstruct the grain boundaries diffusion of Sn and Cu atom via the IMC layer and hence inhibit the growth of the IMC [11]. A study by Chen et al. (2016) shows that addition TiC particles notably suppressed the growth of interfacial layer of Sn-3.0Ag-0.5Cu under isothermal aging condition [12].

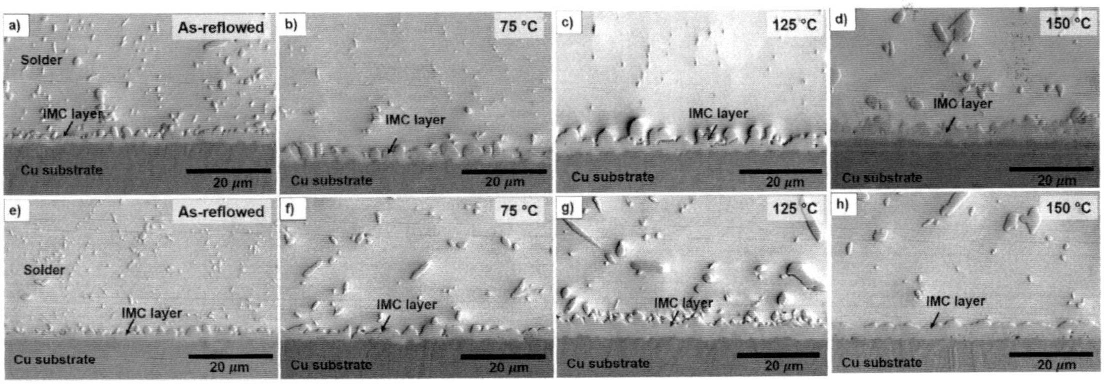

Figure 5. SEM micrograph of intermetallic compound formation of Sn-Cu based solder paste of (a - d) SCN and (e-h) SCNT aged for 240 h.

3.3. Interfacial Intermetallic Compound Layer Growth Kinetic
The activation energy for Sn-0.7Cu-0.05Ni (SCN) and Sn-0.7Cu-0.05Ni-1wt.%TiO$_2$ (SCNT) solder paste composite depicted in Figure 6. Results reveal that the activation energies of IMC growth are 34.12 kJ/mol and 37.35 kJ/mol for SCN and SCNT solder paste respectively. It was found that the activation energy of SCNT composite solder is higher compared to SCN solder paste. Generally, the activation energy of a reaction is known as the minimum energy required for the reaction to occur. The higher the activation energy of the solder, the higher the thermodynamic stability of the reaction. The presence of

Ni had stabilized the $(Cu,Ni)_6Sn_5$ IMC after reflowed and during subsequent aging. Apart from Ni, TiO_2 particles additions revealed that it can suppress the growth of IMC in SCNT solder paste composite. During aging, the presence of TiO_2 particles can prevent excessive IMC growth by restricting the diffusion activity of Cu atom from the substrate and Sn atom from the solder to react to each another. This finding indicates that the IMC growth of TiO_2 containing solder can be suppressed and could improve the solder joint reliability.

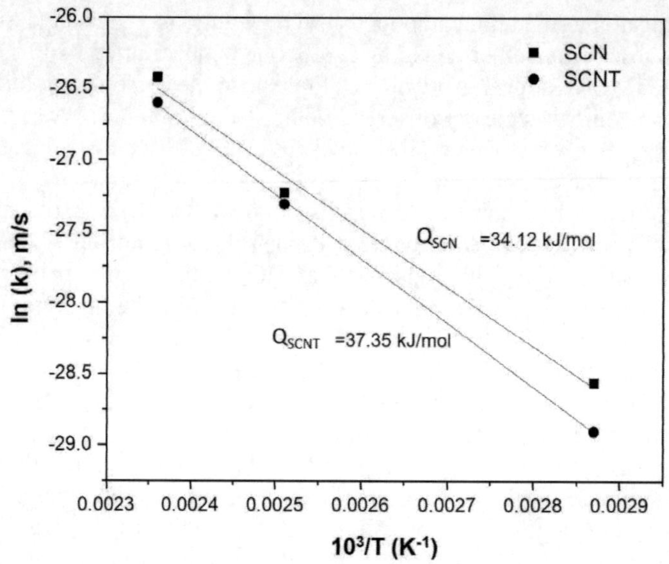

Figure 6. Arrhenius plot of ln k versus 1000/T for the solid-state IMC growth kinetics in SCN and SCNT.

4. Conclusion

The effect of TiO_2 particles addition on the microstructure evolution and interfacial intermetallic layer growth kinetic of Sn-0.7Cu-0.05Ni+1wt.%TiO_2 (SCNT) solder paste composite was examined through this work. The conclusion that can be drawn from this research are:

a) The $(Cu,Ni)_6Sn_5$ IMC in bulk solder area of SCNT solder paste composite are finer after introducing the TiO_2 particles.
b) The interfacial $(Cu,Ni)_6Sn_5$ IMC of SCNT solder paste was thinner during soldering and after isothermal aging.
c) The thermodynamically stable SCNT solder joint composite with high activation energy of 37.35 kJ/mol was obtained.

Acknowledgment

The authors gratefully acknowledge Nihon Superior Co. Ltd, Japan for supporting the research effort through materials.

References

[1] Tsao L C 2011 *J Alloy Compd* **509** 8441– 8448.
[2] Amin N A A M, Shnawah D A, Said S M, Sabri M F M, and Arof H 2014 *J Alloy Compd* **599** 114–120.
[3] Rabiatul Adawiyah M A and Saliza Azlina O 2018 *J Alloy Compd* **740** 958 966.
[4] Chellvarajoo S, Abdullah M Z, and Khor C Y 2015 *Mater Des* **82** 206-215.
[5] Liu H. Fu, Zhang F S H, Kong H, and Xin T 2016 *J. of Mater. Process Tech* **238** 290–296.

[6] Tang Y, Li G Y, and Pan Y C 2014 *Mater Des* **55** 574-582.
[7] Mohd Salleh M A A, Bakri M M A, Zan@Hazizi M H, Somidin F, Alui N F M, and Ahmad Z A 2012 *Mater Sci Eng A.* **556** 633–637.
[8] Tang Y, Li G Y, and Pan Y C 2013 *J Alloy Compd* **554** 195–203.
[9] Choubey A 2007 *Microstructural Changes Under Isothermal Aging and Their Influence on Thermal Fatigue Reliability for Tin-Lead and Lead-Free Solder Joints, Including Microstructural Changes Under Isothermal Aging in Mixed Solder Joints*, in *Department of Mechanical Engineering*. University of Maryland, College Park.
[10] Li, Cheng L X, Li G Y, Huang J H, and Tang Y 2017 *J Alloy Compd* **697** 104-113
[11] Li, Bi X D, Chen Q, and Shi, X Q 2011 *J Electro Mater* **40**(2) 165-175.
[12] Chen G, Peng H, Silberschmidt V V, Chan Y C, Liu C, and Wu F 2016 *J Alloy Compd* **685** 680 - 689.

In Vitro Bioactivity Study of Thermoplastic Starch/Bentonite/Hydroxyapatite Composites for Biomedical Applications

Nurul Husna Zakaria[1, 2, a*], Noorhafiza Muhammad[1, 2, b], Mohd Mustafa Al Bakri Abdullah[2, c] and S N F M Noor[3, d]

[1]School of Manufacturing Engineering, Pauh Putra Main Campus, Universiti Malaysia Perlis, Jalan Arau-Changlun, 02600, Arau, Perlis, Malaysia
[2]Center of Excellence Geopolymer & Green Technology, School of Materials Engineering, Universiti Malaysia Perlis, Kompleks Pusat Pengajian Jejawi II, Taman Muhibbah, 02600 Arau, Perlis, Malaysia
[3]Craniofacial and Biomaterials Sciences Cluster, Advanced Medical and Dental Institute, Universiti Sains Malaysia, 13200 Kepala Batas, Bertam, Pulau Pinang, Malaysia

Email: [a*] nurulhusna.zakaria@gmail.com, [b] noorhafiza@unimap.edu.my, [c] mustafa_albakri@unimap.edu.my, [d] fazliah@usm.my

Abstract. Potato starch composites film incorporating hybrid bentonite and hydroxyapatite (HA) were investigated as novel biomaterial for potential biomedical applications. The selected potato starch/bentonite composites were prepared using different ratios of HA (1, 5 and 10 wt%) through solution casting technique. The *in vitro* bioactivity of potato starch composites film and neat potato starch film to be used as a control were evaluated through immersion in simulated body fluid (SBF) at 37°C for 14 days. The composites films were also subjected to characterization such as morphology, pH evaluation and biodegradability analyses following immersion in SBF. The obtained results showed that the incorporation of HA into starch/bentonite composites improves the bioactivity of the starch hybrid composites by formation of apatite layer on the surface. Increasing HA content in the composite is directly proportional to the quantity of the apatite formed on the scaffolds indicating active surface interactions between the composites and SBF. Furthermore, increasing HA content contributes also to lowering biodegradability of the composite decreases.

Keywords: Bioactivity, biodegradability, potato starch, hybrid composites, simulated body fluid.

1. Introduction

Presently, disease and health problems are the main issue in society that need constant demand for cost effective and innovative biomaterials for biomedical applications. A critical requirement for biomaterial is biocompatibility, which is the ability of a material to function and adapt with an appropriate host response in a specific application. In addition to biocompatibility, several other important properties must be considered for biomaterial which are biodegradable, suitable

Content from this work may be used under the terms of the Creative Commons Attribution 3.0 licence. Any further distribution of this work must maintain attribution to the author(s) and the title of the work, journal citation and DOI.
Published under licence by IOP Publishing Ltd

permeability and processability for intended application [1]. Thus, research and development of quality biomaterials are therefore vital in order to fulfil the requirements of diverse medical industry.

Polymers are one of the biomaterials which received greater attention in biomedical applications such as coronary stent. This is due to their versatility properties, particularly easy to be shaped, good mechanical characteristics and their biocompatibility is unmatched by other biomaterials, such as metals [2]. Recently, natural polymers such as starch has been introduced as a promising biodegradable material for biomedical applications [3]. Starch has been famously used in food, pharmaceutical and biomedical applications because of its biocompatibility, total degradability without toxic residues, low cost, wide availability, renewability and has thermoplastic behavior [3, 4]. In addition, starch has shown great potential in biomedical fields including tissue engineering scaffolds, bone cements, drug delivery systems and stent [5, 6]. However, to perform as an ideal implantable material, further improvement in starch mechanical properties and biocompatibility is needed. The enhancement in starch performance as implantable material can be achieved by adding small amount of nanofillers as reinforcing fillers to form composites.

Nanoclay such as bentonite is the most frequently employed nanofiller for polymers because they are abundant, cheap and own tailorable surface chemistry [7], is used commonly as nanofiller for starch. Besides, bentonite addition has been proven to provide enhancement in tensile strength, toughness and thermal stability to the host starch [8, 9]. While numerous literature in this area is focused on improving mechanical properties, however bioactivity constraints are not adequately addressed. Therefore, researchers began to employ hybrid fillers in order to obtain composites material with greater combination properties.

Hybrid composites is an alternative method allowing a third phase materials to be integrated into starch based composites system for achieving synergistic effects in mechanical and bioactivity properties compared to single filler. Various types of reinforcement fillers is being used in hybrid composites such as nano hydroxyapatite (HA), titanium dioxide and carbon nanotubes. HA has been used widely in biomedical applications, particularly for contact with bone tissue due to its resemblance with mineralized bone. HA exists in natural or synthetic form and possesses favorable properties such as biocompatibility and osteoconductivity [10, 11]. Owing to these properties, HA is capable in promoting fast cell regeneration without intermediate connective tissues.

A common characteristic of HA is the formation of a carbonated apatite layer on their surface when they are in contact with a physiological fluid, which is very similar to the apatite in bone [12]. Kokubo and Takadama confirmed that the *in vivo* apatite formation can be simulated *in vitro* using an artificially prepared solution known as simulated body fluid (SBF) [13]. It was initially developed by Kokubo for initial prediction about the bioactivity and dissolution behavior of bioactive materials by examining their apatite-forming ability in the fluid. Thus, this study aims to highlight enhancement of bioactivity of the starch/bentonite composites with the addition of HA through *in vitro* bioactivity study. The composites are being subjected to characterization such as morphology, pH evaluation and biodegradability analyses following immersion in SBF.

2. Materials and Method

2.1 Materials

Commercially available potato starch, $(C_6H_{10}O_5)n$ (HmbG Chemicals) was used as matrix material. Bentonite and hydroxyapatite powder purchased from (Multifilla Sdn. Bhd and Sigma Aldrich, respectively) were used as nanofillers. Glycerol, $(C_3H_8O_3)$ obtained from (HmbG Chemicals) was used as plasticizer.

2.2 Preparation of Thermoplastic Potato Starch Composites Film

The hybrid composites film (potato starch/glycerol/bentonite/hydroxyapatite) were prepared according to the formulation design in Table 1. Potato starch hybrid composite films were prepared by mixing two different solutions which is bentonite/hydroxyapatite and starch solutions. Specified amount of dried bentonite and HA were first dispersed at room temperature in 40 ml deionized water with the following mixing modes:

I. Mechanical mixing with 5 cm conventional impeller at 1000 rpm for 30 minutes, using a RW 20 mixer (IKA,USA) in a 100 ml beaker.
II. Sonication mixing for 30 minutes, perform on a Bandelin HD 2200 ultrasonic homogenizer with the high frequency power at a frequency of 20 kHz and 6 mm probe diameter.

Meanwhile, 5 g of potato starch was dispersed in 100 ml deionized water with (30 wt% relative to dry strach) of glycerol content. Subsequently, the hybrid nanofillers suspension were added to the aqueous dispersion of starch and the suspension was then heated to 85 °C for 30 minutes under constant stirring. Afterwards, the mixture was poured homogeneously onto 14 cm diameter petri dish. The mixture were then dried at 45 °C in an oven with air force circulation overnight. The dry films were removed from the petri dish and stored in a desiccator before being tested to ensure the stabilization of films. The films were then subjected to *in vitro* bioactivity test using simulated body fluid (SBF) and characterized through morphology (SEM), pH evaluation, thermal stability and biodegradability analyses.

Table 1. Formulation design of potato starch hybrid composites film.

SAMPLE MIXTURE	RATIO	ACRONYM	AMOUNT			
			POTATO STARCH	GLY	BENT	HA
Potato starch/Gly/ Bentonite /HA	5:30:15:1	SBHA1	5 g	30 wt%	15 wt%	1 wt%
Potato starch/Gly/ Bentonite /HA	5:30:15:5	SBHA5	5 g	30 wt%	15 wt%	5 wt%
Potato starch/Gly/ Bentonite /HA	5:30:15:10	SBHA10	5 g	30 wt%	15 wt%	10 wt%

2.3 In Vitro Bioactivity Test

The bioactivity of the composites was analyzed using Simulated Body Fluid (SBF). The SBF with ion concentrations nearly equal to those of human blood plasma was prepared according to Kokubo method [13]. The composites films were incubated in the SBF solution by using the film to liquid ratio of 10 mg/ml (w/v). The film samples were soaked with SBF solution in a 50 ml conical sterile polypropylene centrifuge tubes. The film samples were incubated in the incubator shaker IKA KS4000, USA at 37 °C with 100 rpm for 1 day, 4, 7 and 14 days. At each interval, the film samples were removed from the SBF solution and rinsed with deionized water and dried at room temperature for 24 hours. Morphology, pH level and biodegradability analyses were performed on the film samples.

2.4 Morphology Analysis

Scanning electron microscopy (SEM) model JEOL JSM-6460LA was used to characterize the formation of phosphate layer and apatite on starch hybrid composites film following *in vitro* bioactivity test. Before examining using SEM, the samples were first sputter-coated with platinum to avoid charging effect during characterization.

2.5 pH Evaluation

At designated time points of 1 day, 4, 7 and 14 days, pH of the SBF solution containing the potato starch/glycerol film and potato starch hybrid composites films for *in vitro* bioactivity test was evaluated by using pH meter (Fisher Scientific Accumet AB150). The pH meter was calibrated with buffer solution at pH of 4, 7 and 10 prior to pH evaluation of the samples.

2.6 Biodegradability Analysis

The potato starch/glycerol film and potato starch hybrid composites films were cut into the same weight of 250 mg (Wi) and immersed in 25 mL SBF maintained at 37°C. After 14 days, films were taken out, washed with deionized water and dried at room temperature and weighed (Wd). The biodegradability of the films were assessed by weight loss percentage of each specimen calculated using the equation 1:

$$\text{Weight loss } (\%) = (Wi - Wd)/Wi \times 100 \quad (1)$$

Where:

Wi = Initial weight of the film
Wd = Dried weight of the film

3. Results and Discussion

3.1 Morphological Analysis

Figure 1 (a, c, e and g) shows SEM micrographs of the surface of S-control, SBHA1, SBHA5 and SBHA10 films after 1 day of immersion in SBF solution. Observation from the SEM demonstrated that smaller cauliflower-like spherical crystals is formed on the surface of the films especially for potato starch films containing bentonite and HA, while no spherical crystal formation was observed on the surface of S-control. The presence of the spherical crystals noticeably increased proportional to the HA content. After 7 days immersion, the larger cauliflower-like spherical crystals that were connected to each other had grown and formed on the surface of all SBHA films as shown in Figure 1 (b, d, f and h). Furthermore, the spherical crystals grew more rapidly and almost covering on the surface of SBHA5 and SBHA10.

These observations indicated that the potato starch containing bentonite and HA nanocomposites films showed better bioactivity than neat potato starch film. This finding is in accordance with a study which reported that the addition of HA in the polymer composites could improve the bioactivity of the composites [14]. The S-control film did not show formation of spherical crystals and after 7 days immersion some voids were noticed on the surface of the films which may be due to degradation process. On the other hand, the cauliflower-like spherical crystals formed on the surface of the SBHA films, were attributed to the formation of calcium-phosphate (Ca-P) layer as reported by other researchers [15, 16]. Besides, it was also reported that the composite containing larger amount of HA had greater ability to induce the formation of minerals *in vitro*.

3.2 pH Evaluation

The pH changes of S-control, SHA1, SBHA5 and SBHA10 films in SBF solution with different immersion time (days 1, 4, 7 and 14) is shown in Figure 2. The pH behavior of the SBF solution for all films showed similar trends throughout 14 days immersion period. The starting pH of SBF was set to 7.4 which is similar to human blood plasma condition. Observation from the graph shows that, at the first 24 hours of immersion the pH of all films was rapidly increased to pH value of 7.6 and followed by significant decrease during 4 days of immersion in SBF solution. The pH value of S-control and SBHA1 were slowly decreasing until day 14 and stabilized at pH value of 7.3. Meanwhile, the pH of SBHA5 and SBHA10 exhibited a slightly increase at day 7 and then slowly decreased until day 14, which posted a pH value of 7.4.

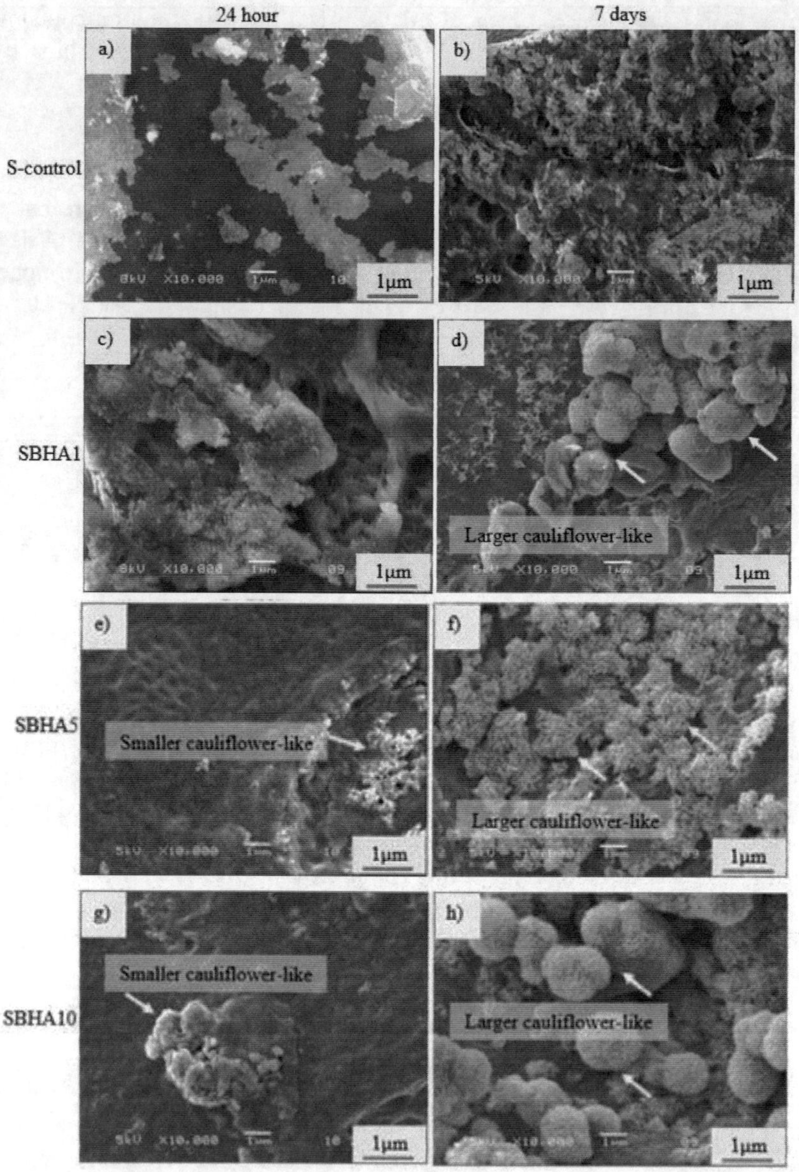

Figure 1. SEM micrographs of (a,b) S-control, (c,d) SBHA1 (e,f) SBHA5 and (g,h) SBHA10 after immersed in SBF solution for 1 and 7 days.

The pH values of the SBF solution upon immersion of the nanocomposites films showed rapid increase corresponding with the dissolution behavior of HA in SBF solution which maybe contributed by the formation of Ca-P. Fast release of calcium (Ca) ions from the nanocomposites films containing bentonite and HA coupled with hydrogen ions (OH) from SBF solution may cause high concentrations of Ca, thus resulted in a pH increase [17]. Meanwhile, the pH of SBF solution significantly dropped at day 4 that might be the results of radical hydroxyl release from the films surface. A recent study described that Ca-P compounds precipitated spontaneously when the pH value of the physiological solution was over 7.4. During the prolonged immersion period, Ca and phosphorus (P) ions are consumed largely due to the formation of abundant apatite and simultaneously, OH- is consumed correspondingly. This will cause a decline in the SBF pH value [18].

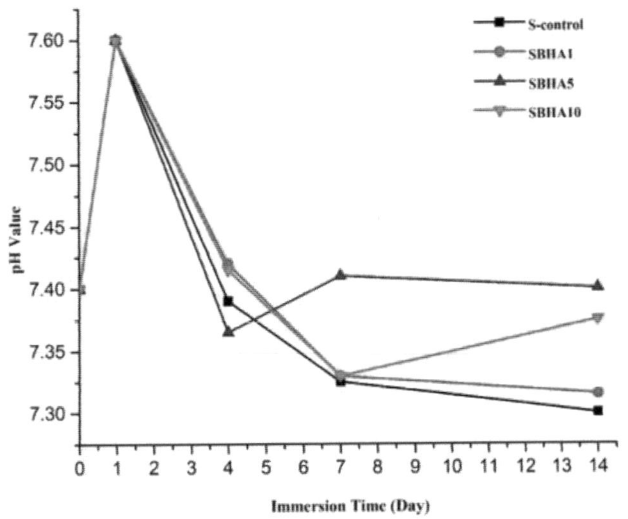

Figure 2. pH behavior of SBF solution upon immersion of nanocomposites films.

3.3 Biodegradability Analysis

The *in vitro* biodegradability analysis of starch/bentonite/HA hybrid composites film at various HA content (SBHA1, SBHA5 and SBHA10) compared with neat starch (S-control) were performed through weight loss measurement in SBF solution at 37°C for 14 days. Figure 3 shows the percentage of weight loss for all composites and neat starch film. Observation from the graph shows that all samples experienced weight loss after 14 days exposure in SBF fluid. Besides, the results showed a decreasing trend in weight loss with increasing HA content. Neat starch (S-control) exhibits the highest weight loss of 33.52 %, followed by SBHA1, SBHA5 and SBHA10 with 31.74 % and 30.82 % respectively.

The presented results revealed that the incorporation of HA in starch matrix affected the biodegradation properties of the composites film. It was found that increases HA content lead to decreased SBHA composites film weight loss compared to S-control. This reason is attributed to the low degradation rate of HA. Furthermore, the decreased weight of SBHA composites film is also correlated to the significant increase in apatite layer formation on the surface of the SBHA composites film. This phenomenon was proven by morphological analysis (section 4.7.1) in which spherical crystals grew more rapidly on the SBHA5 and SBHA10 surface. This finding is in accordance with the previous reported research [19].

Meanwhile, S-control film displayed the highest weight loss percentage. This is mainly due to hydrophilic properties of the potato starch which contributed to the highest weight loss during exposure in SBF fluid. Besides, glycerol utilized in preparation of S-control film also possess hydrophilic properties. Although glycerol is known to be a good plasticizer for potato starch, however the fact that glycerol domains for being hydrophilic really do affects the degradation rate of the starch when in contact with SBF fluid. This finding was in line with R. Shi et al. [20], which reported that the degradation of starch is initiated by the separation of glycerol among the starch chains out of the sample during degradation in SBF. The cleavage of the linkages of glycerol resulted in the destruction of crystallinity of starch sample during degradation.

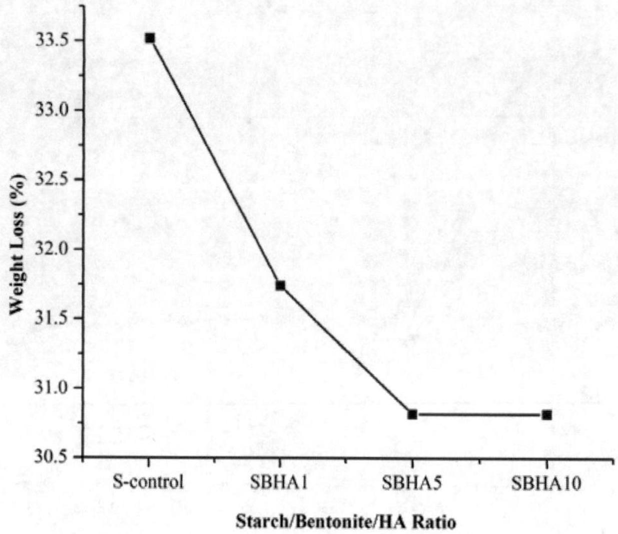

Figure 3. Percentage of weight loss of neat potato starch and starch hybrid composites after 14 days of SBF immersion.

4. Conclusion

The effect of hybridized bentonite and HA with various HA content in potato starch composites films with 1, 5 and 10 wt% on the bioactivity of potato starch films were examined through immersion in SBF solution. The result is supported by surface morphological, pH evaluation and biodegradability analyses. The current finding shows that the increasing HA incorporation into starch/bentonite composites improves the bioactivity of the starch hybrid composites indicating its potential to be used in biomedical applications.

Acknowledgements

The authors gratefully acknowledge School of Manufacturing Engineering, UniMAP for their expertise and support. The authors would also like to thank the funding support from the Fundamental Research Grant Scheme (FRGS-9003-00584) under Ministry of Education Malaysia (MOE).

References

[1] R. Song, M. Murphy, C. Li, K. Ting, C. Soo, and Z. Zheng 2018 *Drug Des. Devel. Ther.* 12 3117–3145.
[2] R. Boni, A. Ali, A. Shavandi, and A. N. Clarkson 2018 *J. Biomed. Sci.* 25(90) 1–21.
[3] F. Torres and D. Arce 2015 73–94.
[4] M. L. Sanyang, S. M. Sapuan, M. Jawaid, M. R. Ishak, and J. Sahari 2015 *Polymers (Basel).* 7 1106–1124.
[5] A. Beilvert, F. Chaubet, L. Chaunier, S. Guilois, G. Pavon-djavid, and D. Letourneur 2014 *Carbohydr. Polym.* 99 242–248.
[6] D. Velasquez, G. Pavon-djavid, L. Chaunier, A. Meddahi-pellé, and D. Lourdin 2015 *Carbohydr. Polym.* 124 180–187.
[7] A. F. Osman *et al. J. Appl. Polym. Sci.* 44640 1–15.
[8] A. Amalia, A. Fauzi, A. F. Osman, M. Aidil, A. Abdullah, and S. Mandal 2019 *J. Vinyl Addit. Technol.* 1–16.
[9] L. Lendvai, Á. Kmetty, and J. Karger-Kocsis 2017 *Mater. Sci. Forum* 885 129–134.
[10] F. Y. Carmen M.O. Müller, João Borges Laurindo 2011 *Ind. Crops Prod.* 33(3) 605–610.
[11] W. Gao, H. Dong, H. Hou, and H. Zhang *Carbohydr. Polym.* 88(1) 321–328.

[12] S. A. S. N. Fadzli, S. Roslinda, F. Zainuddin, and H. Ismail *2016 UKM FST Postgrad. Colloq.* 040033, 1–5.
[13] T. Kokubo and H. Takadama 2006 *Biomaterials* 27(15) 2907–15.
[14] L. Kong, Y. Gao, G. Lu, Y. Gong, N. Zhao, and X. Zhang 2006 *Eur. Polym.*42 3171–3179.
[15] I. B. Leonor, A. Ito, K. Onuma, N. Kanzaki, and R. L. Reis 2003 *Biomaterials* 24 579–585.
[16] J. Ni and M. Wang 2002 *Mater. Sci. Eng. C* 20 101–109.
[17] S. Ibrahim *et al. Saudi J. Biol. Sci.* 2016 23 S56–S63.
[18] F. Xin, C. Jian, Z. Jian-peng, W. Qian, Z. Zhong-cheng, and R. Jian-ming 2009 *Trans. Nonferrous Met. Soc. China* 19 347–352.
[19] S. N. F. Syed Adam, A. F. Osman, and R. Shamsudin 2018 *Sains Malaysiana* 47(6) 1303–1310.
[20] R. Shi *et al.* 2006 *Polym. Degrad. Stab.* 2006 91 3289–3300.

Effect of Halogen Free-flame Retardants on The Mechanical and Morphological Properties of Kenaf/Polyurethane Foam Composites

Nur Suhaili Mohd Soberi[1], R Rozyanty[1], Firuz Zainuddin[2] and Hazim Mohamad Amini[3]

[1]Center of Excellence Geopolymer and Green Technology (CEGeoGTech), School of Materials Engineering, Universiti Malaysia Perlis, 01000 Kangar, Perlis, Malaysia.
[2]School of Materials Engineering, Kompleks Pusat Pengajian Jejawi 2, Taman Muhibbah, Universiti Malaysia Perlis, 02600, Jejawi, Arau, Perlis, Malaysia.
[3]Faculty of Earth Science, Universiti Malaysia Kelantan, Jeli, Kelantan, Malaysia

Email: rozyanty@unimap.edu.my

Abstract. Kenaf/polyurethane foam (PUF) was synthesized form PUF and kenaf core fiber with expandable graphite (EG) and aluminum hydroxide (ATH) at varying amounts. In the production of polymer foam composites, the processing methods determine the properties of the final products. In this work, the potential of high energy milled on EG has reduced the EG platelet size to improve its dispersion in PUF matrix. From the results obtained, PU/KF/6ATH/10EG shows the highest mechanical properties, i.e. modulus and compression strength, while PU/KF/2ATH/5EG shows the lowest mechanical properties. Scanning Electron Microscope (SEM) shows the effect of adding kenaf fiber, EG and ATH in PUF to the foam cell structure and size.

1. Introduction

Nowadays, studies on natural fiber/ PUF composites are widely exploited among the researcher. PUF have been commercially used in variety of applications, especially for rigid PUF that widely applied as thermal insulating materials [1] due to low thermal conductivity and good mechanical properties. PUF was produced from the complex generation of cellular structure that is made up of solid and gaseous phases [2]. In some cases, more than one solid component is added as fillers or additives in order to improve the physical, mechanical or thermal properties of the PUF. Generally, PUF are produced from the reaction between polyol and isocyanate via addition polymerization [3]. Presently, most of the polyol used in the production of PUF was synthesized from petroleum sources.

However, depletion of petroleum sources and high cost of petroleum based materials has induced to the production of natural based polyol. Based on several researches done, palm oil was extensively used as the polyol in the production of PUF due to its price and production efficiency compare to any other commercial oils [4].

The addition of natural fibers as the filler in PUF has been extensively used due to their characteristics such as economical and environmental friendly [5]. Moreover, addition of kenaf (Hibiscus cannabinus, L.) offers various advantages such as enhancing on mechanical and thermal properties of PUF. In addition, the existence of lignin in kenaf fiber can provide free OH groups which can contribute to polarization between the fiber and matrix [6]. It is well known that the addition of

Content from this work may be used under the terms of the Creative Commons Attribution 3.0 licence. Any further distribution of this work must maintain attribution to the author(s) and the title of the work, journal citation and DOI.
Published under licence by IOP Publishing Ltd

filler or additives in PUF leads to increases in foam density and decreases the foam cell size [7]. In general, the filler should be uniformly dispersed in the matrix to obtain excellent mechanical properties [7] [8].

Natural fiber/ PUF composite is well known with its favorable properties such as light weight, low production cost [9] and excellent hardness. However, the poor fire resistance of PUF caused by its porosity, low density and the nature characteristics of organic cellular materials itself has restricted its application especially as fire resistance materials [10]. For this purpose, flame retardant was added into the PUF to minimize and improve flame retardancy of PUF. In recent years, there have been worldwide demands on halogen free flame retardants such as intumescent system and mineral flame retardants, as they are quite effective in flame protection.

Recently, aluminum hydroxide (ATH) was commonly used as mineral flame retardants, which can contribute to low production of smoke released. Besides non-toxic, ATII is low cost [11], odorless and easy to handle. However, high load of ATH is required to achieve good flame retardancy properties, which sacrifices the mechanical performances of the polymer. Thus, ATH was combined with expandable graphite (EG) to improve the flammability properties and lower the ATH loading in polymer matrix which improved the mechanical properties of the polymer [12]. EG is a graphite intercalation compound in which sulfuric acid (H_2SO_4) is inserted between the carbon layers of the graphite [13]. Extensive studies proved that incorporation of EG into PUF able to improve its flame retardancy properties. However, the inclusion of EG could disrupt the mechanical properties of the PUF composites. This is due to the bigger size of EG has deteriorate the cell structure of PUF as EG locates on the cell strut which leads to weaken the mechanical properties of PUF [14]. Therefore, EG particles have been pulverized to achieve fine size to obtain smaller particle size and thus improve the dispersion quality of EG in PUF [13].

Therefore, in this study, rigid PUF was prepared by replacing petroleum based polyol with palm oil based polyol. Herein, we reported the effect of kenaf fiber with halogen-free flame retardants on the mechanical properties and morphology of PUF.

2. Experimental

2.1 Materials

Polyether polyol with amine catalyst (Maskimifoam 9935B/35) is palm oil based polyol while isocyanate (Maskiminate 80) contains the mixture of Polymeric MDI with a concentration of 55% and Methylene Diphenyldiisocyanate with 45% concentration. Both materials were supplied by Maskimi Polyol Sdn. Bhd. Additives used for polymerization of PUF are aluminum hydroxide (ATH) and expandable graphite (EG). Both of the additives were manufactured by Sigma-Aldrich. Kenaf core fiber (*Hibiscus cannabinus L*) was supplied from Lembaga Kenaf Tembakau Negara (LKTN).

2.2 Preparation of filler

Kenaf core fiber was dried naturally under the sunlight to remove its moisture content. Then, the fiber was ground using a grinder and sieved with 63 μm sieve.

2.3 Preparation of additives

EG was pulverized in a high energy mill (planetary type), with 450 rpm speed for 1 hour and 30 minutes to obtain small particulate size of EG. The EG content was varied from 5, 10 and 15 wt%. While, ATH content was varied at 2, 4 and 6 wt%.

2.4 Preparation of composites

Kenaf/PUF composites were prepared according to the formulation in Table.1. The palm oil based polyol was stirred at 2000 rpm with a high speed mechanical stirrer and allowed to degass for 2 minutes. Then, kenaf fiber was added at a fixed amount which is 1 wt%. While, the amount of EG and ATH were varied as shown in the Table.1. All the particles were added and wetted in the polyol and was stirred continuously until a homogenous mixture was obtained. Lastly, the mixture was reacted with isocyanate and continuously stirred. The mixture was cast into a mold, covered and allowed to cure. The cured foam was removed from the mold after 24 hours. Then, the samples were placed in the oven for 2 hours at 80°C for post cure.

Table 1. Formulation for polyurethane foam compsites based on ratio 1:1.1.

Kenaf (wt%)	Expandable graphite (EG) (wt%)	Aluminum hydroxide (ATH) (wt%) 1st series	2nd series	3rd series	4th series	Polyol (g)	Isocyanate (g)
1	5	0	2	4	6	23.66	26.03
1	10	0	2	4	6	23.66	26.03
1	15	0	2	4	6	23.66	26.03

2.5 Mechanical test

Compression test for rigid cellular plastic was conducted under ASTM D1621-10 by using Instron 4206 machine. The cure PUF composites were cut by using a band saw machine with dimensions of 50 mm x 50 mm x 50 mm (length x width x thickness). Compression test was performed at room temperature with a constant crosshead speed at 10mm/min and the load was applied until the foam was compressed to approximately 80% of its original thickness (height). The samples were compressed in the direction parallel to foam rise. The strength and modulus of five specimens for a sample were measured and averaged.

2.6 Morphology

Scanning electron microscope (SEM) was conducted using JEOL JSM 6460 model. This analysis was done to determine the effect of additives and filler loading on foam cell size and distribution. It also reveals information about the sample surface morphology and mechanical properties. The samples were cut into 5 mm x 5 mm x 5 mm (length x width x thickness) and were coated with a very thin layer of platinum by using Auto Fined Coater to attain observation on the samples. Observation was made at 100 x magnifications.

3. Results and discussion

Figure 1 and Figure 2 represent the compression strength and modulus of PUF composites with varying amount of ATH. Generally, the achievement of good mechanical properties in filled PUF system was determined by the distribution and dispersion of filler in the polymer matrix [15]. Based on the results obtained, incorporation of EG into PUF matrix generates to a significant increase in the compression strength and modulus of the pure PUF. The improvement in mechanical properties of PU/15EG was due to the large surface area provided by the pulverized EG. In addition, EG is located on the foam cell struts as shown in Figure 3 (b) which can contribute to increases the strength and modulus of PUF composites.

Addition of ATH into PUF composites was drastically reduced the compression strength and modulus. This is due to disruption in the PUF system that was caused by rigidity of ATH instead the presence of kenaf and EG particles. However, the compression strength and modulus was increased at 4 wt% of ATH and further increases at 6 wt% of ATH. But, it decreased slightly at 15 wt% of EG for both compositions. Thus, it can be concluded that ATH did not able to show excellent mechanical properties at low amount. Whereas, it tends to show significant increases as EG was incorporated in the PUF composites. Based on the results obtained, it is observed that, PU/KF/6ATH/10EG gives the highest value for compression strength and modulus, while the optimum amount of EG incorporated in PUF is10 wt%. This is because during the processing of PUF composites, the addition of EG, ATH and kenaf fiber has resulted in great viscosity increases, especially for high EG and ATH content. This will lead to produce poor dispersion of filler and additives in the PUF matrix.

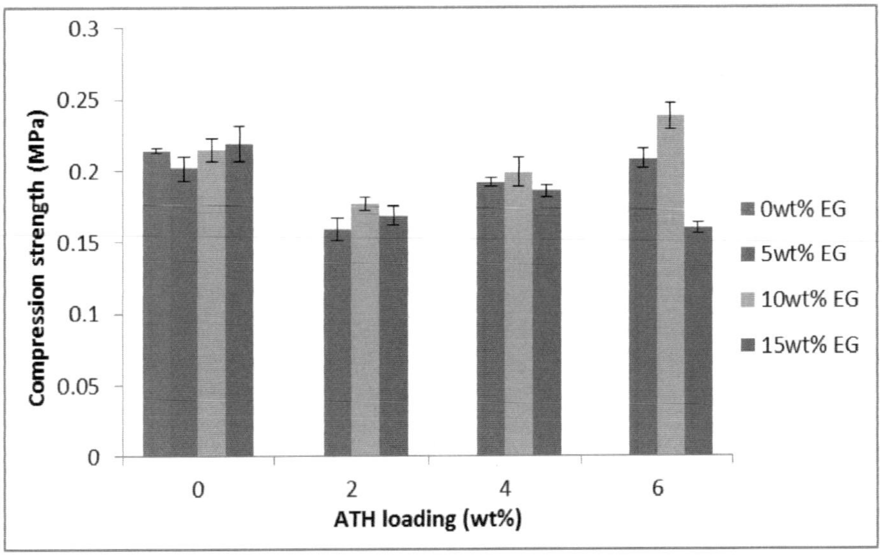

Figure 1. Compression strength of PUF composites.

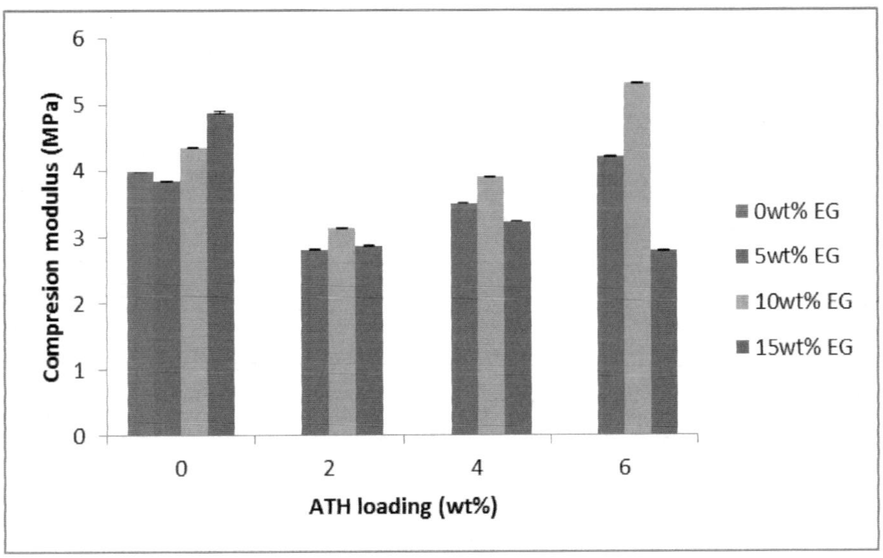

Figure 2. Compression modulus of PUF composites.

SEM was conducted in order to study on the morphology and cell structure of PUF composites. Figure 3 displays the SEM micrograph of rigid PUF composites. Based on Figure 3 (a), it can be seen that the cell shape of the pure PUF is approximately spherical. Besides that, the pure PUF cell size and cell distribution are nearly uniform with closed cell foam structure [6]. Figure 3 (b) represents PUF with 15 wt% of EG particles. According to Shi and his co-workers study [14], EG particles do not locate in the cell struts but between the cell walls due to their bigger size. Thus, in this work, EG particles have been pulverized to improve the distribution of EG in PUF matrix. From the image, it can be seen that EG particles was randomly distributed at the PUF cell struts which resulted to higher mechanical strength and modulus of the PUF composites. As shown in Figure 3 (c), the ruptured cell wall of PUF composites was mainly due to the rigidity of ATH particles. While, Figure 3 (d) shows the PUF structure becomes smaller and regular as 6 wt% of ATH and 10 wt% of EG was added into the PUF matrix. Generally, foam with regular and smaller cell size will possess better mechanical performance as proven in Figure 1 and Figure 2. Thus, the combination of EG and ATH could give synergistic effects in mechanical properties of PUF composites.

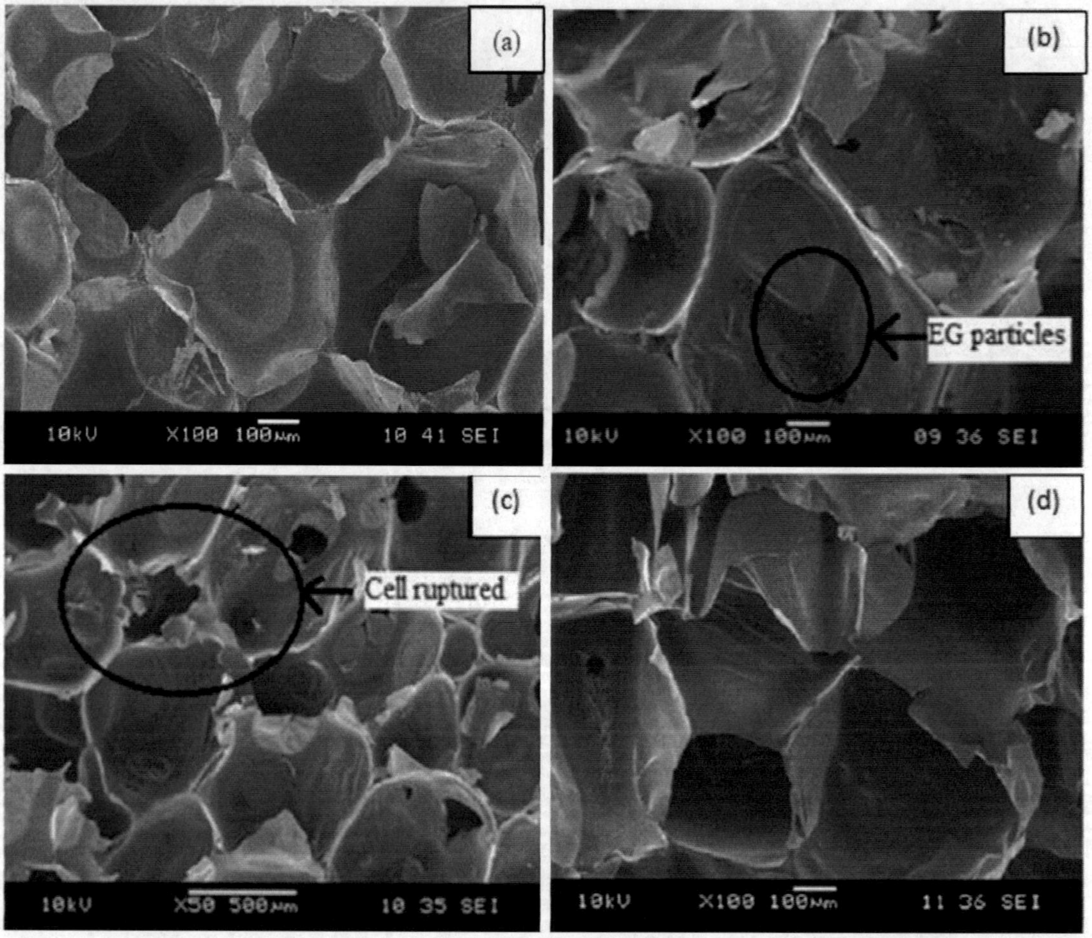

Figure 3. SEM micrograph of (a) PUF (b) PU/KF/15EG (c) PU/KF/2ATH/5EG (d) PU/KF/6ATH/10EG.

4. Conclusion

The incorporation of kenaf fiber, EG and ATH into palm oil based polyurethane foam has been carried out. Compression test shows that PU/KF/5EG shows the lowest compression strength and modulus while PU/KF/6ATH/10EG shows the highest. The incorporation of ATH at lower amount in the PUF has deteriorate the cell structure, thus lowering the compression strength and modulus of the composites. However, the compression strength and modulus was increased at higher ATH loading.

Addition of EG at 10 wt% shows good mechanical performance of the PUF composites. In conclusion, the combination of EG and ATH could produce synergistic effect on the mechanical properties of PUF composites.

Acknowledgement
The authors would like to thank University Malaysia Perlis (UniMAP) for facilities, Lembaga Kenaf Tembakau Negara (LKTN) for raw materials and the Ministry of Higher Education Malaysia for financial support under Fundamental Research Grant Scheme 9003-00483.

References
[1] K.H. Cheo, D.S. Lee, W.J. Seo, and W.N. Kim: submitted to Polymer Journal (2004).
[2] P. Kekicheff, , D. Mouhamadou, G. Dalongeville, C. Gauthier and T. Roland. 2016: submitted to Journal of Macromolecular Symposia.
[3] P. Li, Y. Tao, and Q.S. Sheldon: submitted to Bioresources (2014).
[4] K.H. Badri, Z.Othman, and S.H. Ahmad: submitted to Journal of Materials Science (2004).
[5] N.R.A. Norzali, K.H. Badri and Nuawi, M.Z: submitted to Sains Malaysiana (2011).
[6] N.S.M. Soberi, R. Rahman, & F. Zainuddin (2017). Mater. Sci. Forum Vol. 888 (2017), p. 188–192.
[7] G.K. Latinwo, D.S. Aribike, A.A Susu and S.A Kareem: submitted to Journal of Nature and Science (2010).
[8] V. Yakushin, L. Belkova, and I. Sevastyanova (2012). Paper presented at the *ECCM15-15TH European Conference On Composite Materials*,Venice, Italy.
[9] A.B. Frances, and M.V.N. Banon: submitted to Journal of Materials and Science Engineering (2014).
[10] A. Wolska, J. Ryszkowska, M. Grozdzikiewicz and J. Ryszkowska (2012). Paper presented at *ECCM15 - 15TH European Conference On Composite Materials*, Venice, Italy.
[11] J. Jin, Q. Dong, Z. Shu, W. Wang, and K. He: submitted to Journal of Procedia Engineering (2014).
[12] Z. Xu, L. Yan, L and L. Chen: submitted to Procedia Engineering (2016).
[13] L. Shi, Z. Li, W. Yang, M. Yang, Q. Zhou, Q and R. Huang: submitted to Journal of Powder Technology (2006).
[14] L. Ye, X. Meng, X. Ji, Z. Li and J. Tang: submitted to Journal of Polymer Degradation and Stability (2009).
[15] M. Thirumal, D. Khastgir, N.K. Singha, B.S. Manjunath and Y.P. Naik: submitted to Journal of Applied Polymer Science (2008).

Corrosion Resistance of Micro-Textured Surface Modified Alumina-Titania Coating

J. A. Wahab[1], M. J. Ghazali[2] and M. N. Derman[1]

[1]Center of Excellence Geopolymer and Green Technology (CEGeoGTech), School of Materials Engineering, Universiti Malaysia Perlis, Kompleks Pusat Pengajian Jejawi 2, Taman Muhibah, 02600 Jejawi, Arau, Perlis, Malaysia.
[2]Department of Mechanical and Materials Engineering, Faculty of Engineering & Built Environment, Universiti Kebangsaan Malaysia, 43600 UKM Bangi, Selangor, Malaysia

Email: juyana@unimap.edu.my

Abstract. Effect of micro-texture on the corrosion resistance of alumina-titania coated mild steel was investigated. The micro-texture was fabricated on the coating surface via laser surface texturing technique. Tafel extrapolation and immersion test was conducted to measure the corrosion resistance and corrosion mechanism of the coating in 3.5% NaCl solution. The results indicated that the micro-texture contributes to a significant improvement of corrosion resistance due to the formation small volume of air trapped in the micro-grooves, which resist the penetration of corrosive ions and reduce the area of solid – liquid interface. The WCA indicated that the textured surface had low wettability. The SEM analysis showed the occurrence of uniform corrosion. The analysis of EDS revealed that there was formation of corrosion product at the coating-substrate interface. In short, the resistance towards corrosion was increased up to 73% indicating that the resistivity of the coating against corrosion was improved by engraving the micro-texture on its surface.

1. Introduction

Development of new materials has affected ceramic film and coating technology. The relatively high hardness and inertness of ceramic coating make this type of coating materials gain interest for protection of substrate materials against corrosion, oxidation and wear especially in marine environment. Oil and gas industry, shipping industry and the marine structures suffer from corrosion due to the exposure of aggressive ions in seawater. Combinations of wear, oxidation and corrosion problems contribute to failure factors that would eventually cause total shutdown, economic loss and hazardous effects [1]. Previous studies summarized that the corrosion problem is not fully solved only by applying a layer of coating on material surface. For these reasons, an improvement on corrosion protection method is needed to overcome these problems. A material with well-defined surface textures have been prepared in order to utilizes the beneficial effects of surface texturing technology. Generally, the main purpose of producing the texture is specifically to improve the roughness of substrate surface to micro- to nano-scale. The micro- or nano- scale textures on surface will affect the wettability characteristic (hydrophobicity) which can be described by two main models as proposed by Wenzel and Cassie and Baxter [2]. Such textured material which has low surface energy characteristic can be prepared either by depositing the coating material on a textured substrate [3] or directly structuring the surface by several technologies such as chemical etching [4], mechanical machining [5], lithography [6] and laser surface texturing [7].

Content from this work may be used under the terms of the Creative Commons Attribution 3.0 licence. Any further distribution of this work must maintain attribution to the author(s) and the title of the work, journal citation and DOI.
Published under licence by IOP Publishing Ltd

In corrosion applications, plasma sprayed Al_2O_3 - 13%TiO_2 coating can reduce the corrosion rate due to its excellent properties; which are essential for corrosion applications [8]. The properties of Al_2O_3 - 13%TiO_2 coating, such as high hardness, low density, low thermal expansion and high resistance to oxidation and wear, make it widely used as a corrosion resistant coating, particularly in textile, machinery and printing industries [9, 10]. In marine environment, it has been a serious problem for many marine industrial structures such as hydraulic turbines, pumps and pipelines conveying solid particles [8]. Materials part or structures such the ships hulls, cooling-heat exchanger and structure in power plants are directly contact with corrosive ions (Cl-). They suffer from damaging effects induced by corrosion reaction and thus enhancing the material failures. The basic concept of corrosion protection by superhydrophobic surface is the micro-or nano-textured surface creates an interface layer with water by retaining air on the surface. The created interfaces that minimized the water molecules to contact the metal surface can prevent corrosive ion from adsorb and attack the surface materials.

The performance of Al_2O_3 - 13%TiO_2 coating, in these applications, could be made superior by applying some kind of texture on its surface. However, such experiments involving textured surfaces in mitigating corrosion problems have not been reported yet. In this study, the effect of micro-groove on the corrosion behaviour of plasma sprayed Al_2O_3 - 13%TiO_2 coating will be investigated. The micro-groove texture will be engraved on the Al_2O_3 - 13%TiO_2 surface using a short-pulsed laser system in a laser surface texturing process. Therefore, the performance of a laser textured Al_2O_3 - 13%TiO_2 coating in corrosion applications could be improved; thus enhancing the performance and lifespan of material components; especially in a harsh marine environment.

2. Experimental

2.1. Preparation of coating

To prepare the coating, commercial feedstock of Al_2O_3 - 13%TiO_2 (Metco 130SF), with a fine particle distribution, was deposited onto mild steel substrates using an atmospheric plasma spraying (APS) technique. The feedstock used had a fine particle distribution of -45+5 μm. A Praxair SG-100 plasma spray system, with a plasma torch mounted on a programmable robot was used in the deposition system. The surfaces of the substrates were sandblasted with alumina particles (mesh 24) as a pre-treatment to obtain good coating adhesion prior to the deposition process. A uniform coating thickness of 200–250 μm was prepared with a set of fixed spraying parameters of plasma power, 40 kW, primary gas pressure, 80 psi, secondary gas pressure, 60 psi, carrier gas pressure, 40 psi, powder feed rate, 3 rpm, and a stand-off distance of 100 mm.

2.2. Laser surface texturing

Table 1. Laser surface texturing parameters.

Parameter	Value
Model/system	Time Bandwidth, Duetto
Wavelength (nm)	1640
Pulse duration (ps)	10.3
Scanning speed	200 mm/s for 100 passes
Laser power (W)	2
Repetition rate (KHz)	300

A micro-groove texture was developed on the coating surface via a laser surface texturing technique. The Time Bandwidth, Duetto laser manufacturing system was used to produce the open-pore type texture, with a fixed distance of 150 μm and the depth of the groove was approximately 50 μm. A laser beam, with a diameter of approximately 5 mm, was introduced to the coating surface using a galvanometric scanner with a telecentric f-theta flat field lens. A galvanometer scanner was used to control the laser beam via a computer system. The features of the texture (Figure 1) were controlled via laser processing with a set of fixed parameters as shown in Table 1.

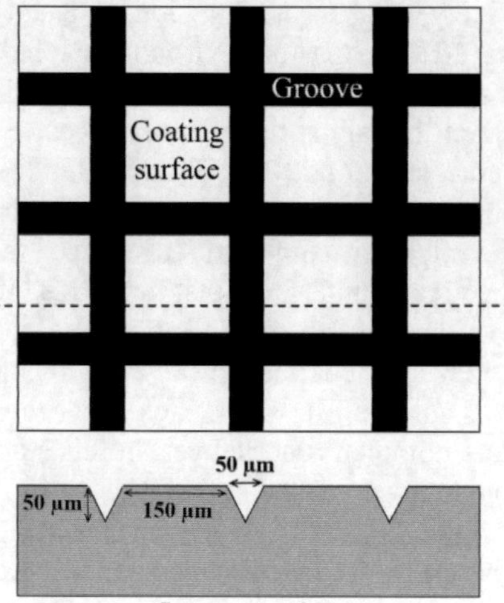

Figure 1. Schematic diagram of the texture feature with cross sectional view.

2.3. Characterisation and testing of the coating
The microstructural characteristics of the textured and non-textured coatings were observed using scanning electron microscope model FEI Nova NanoSEM 450. The phase compositions of the coatings were examined through X-ray diffraction (XRD) analysis, model Bruker D2 Phaser, Germany, using Cu-Kα radiation (1.5410Å).

2.4. Corrosion testing
The electrochemical properties of the coatings were studied using Tafel extrapolation method. The test was performed in a three-electrode cell with platinum rod as the counter electrode, a saturated calomel electrode (SCE) as the reference electrode and a coating specimen as the working electrode. 3.5% NaCl solution was used as the electrolyte, which was maintain at room temperature. Corrosion testing system of Metrohm Autolab with Nova V1.1 software was used to determine the corrosion behaviour of the sample. The Tafel curve was generated by a scan rate of 0.5mV/s and a potential range of ±250 mV. The corrosion mechanism of the coating sample was evaluated by conducting immersion test in a solution of 3.5% NaCl with a constant pH value of 7. The pH of the NaCl solution was controlled by using NaOH and HCl solution. Before the immersion test, the uncoated area of the sample was mount by using epoxy resin to avoid contact with the solution. All the specimens were placed in a water bath at temperature of 27±3 °C. After 12 weeks of immersion test, the specimens were cleaned with ethanol in ultrasonic cleaner and then rinsed with distilled water. The SEM was used to analyse the morphology of the specimens cross sectional area. The EDX was used to examine the corrosion product element.

3. Results and discussion
3.1. Morphological examination
Figure 2 shows the surface morphology of both non-textured and textured coatings. From the image, it shows that laser processing has the capability to develop a groove texture at a micro size scale with high precision geometry on a ceramic coating surface. For the textured coating, no cracks or other damage formed on the surface after the laser surface texturing process.

IOP Conf. Series: Materials Science and Engineering **743** (2020) 012004 doi:10.1088/1757-899X/743/1/012004

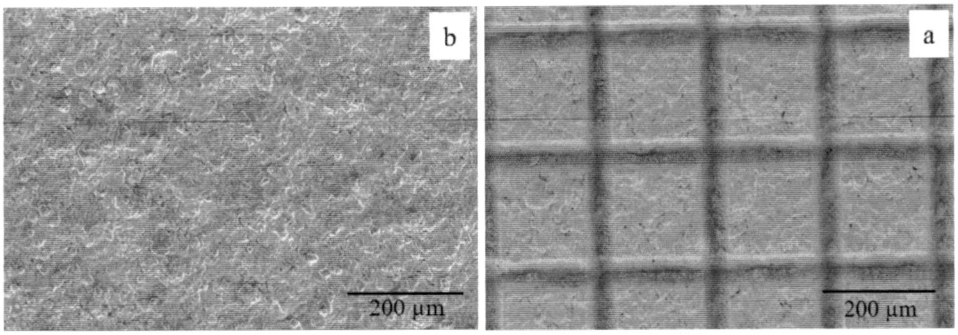

Figure 2. Surface morphology of the coating, a) non-textured and b) textured.

3.2. Corrosion behaviour

The effectiveness of grooves on coating surface in mitigating corrosion has been tested in this study by examine the electrochemical behaviour via Tafel extrapolation technique. The corrosion rate was determined by extracting the values of corrosion potential (E_{cor}) and corrosion current density (i_{cor}) from the Tafel plot. As can be seen in Figure 3, the corrosion potential of textured coating was more electropositive which was approaching a positive value. Meanwhile, for non-textured coating, the corrosion potential was more electronegative. The corrosion current density of non-textured coating also higher than that of textured coating, which indicated that the system of coating-substrate was less resistance to corrosion. It is important to mention that Al_2O_3 - 13%TiO_2 coating is an insulator, so that any current flowing in the coating system was due to the penetration of electrolyte through the coating. The non-textured coating exhibited a high corrosion resistance as compared to the non-textured coating. The calculated corrosion rate for both of textured and non-textured coating was around 1.63×10^{-3} mm/year and 6.06×10^{-3} mm/year, respectively. These results were found to be in agreement with Boinovich et al. [11] research, where they reported that the textured surface showed a reduction in the value of corrosion current density. Similar observation was discussed by the other researchers [12, 13], which stated that the reduction of corrosion current density value in the Tafel test was due to the changes in the wettability characteristic of the textured surface. The wettability characteristic of the surface can affect the corrosion performance of a material. Based on the investigation by Zhang and friends [14], the low wettability characteristic of the surface led to the high corrosion resistance properties. In their research, the superhydrophobic surface of steel contributed to large fraction of trapped air on surface, which was effectively prevents the direct contact between water droplet and the surface. Therefore, the corrosion reaction was decelerated and resulted in the reduction of corrosion rate of the steel.

Figure 4(a) and (b) displays the cross sectional area and EDS analysis of the coating after 12 weeks of immersion in NaCl solution. The cross sectional images of the corroded coatings revealed significant signs of degradation in the coating-substrate interface. The occurrence of uniform corrosion can be observed. The corrosion product formed at the coating-substrate interface indicated that corrosion reaction was occurred between the substrate and the NaCl solution. Further corrosion reaction at the coating-substrate interface will result in the delamination of the coating [15]. Based on the EDS analysis in Figure 4(c), the corrosion product consisted of Fe and O elements. This result showed that the corrosion product formed was attributed to the reaction between NaCl solution and the substrate. There was also Al element, which was originated from the coating compound.

IOP Conf. Series: Materials Science and Engineering **743** (2020) 012004 doi:10.1088/1757-899X/743/1/012004

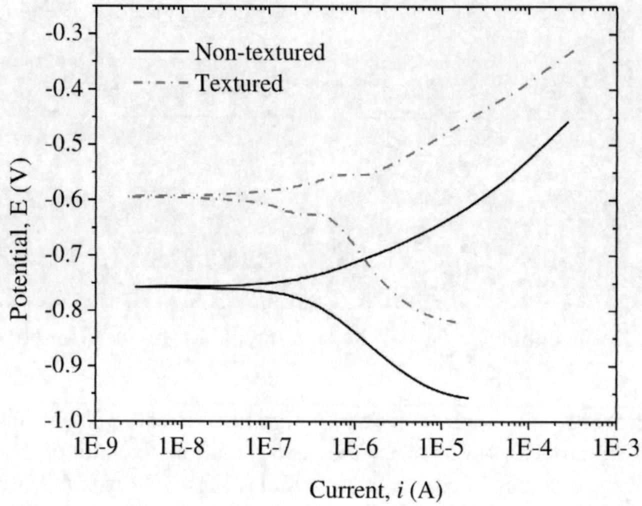

Figure 3. Tafel curve for the non-textured and textured coating.

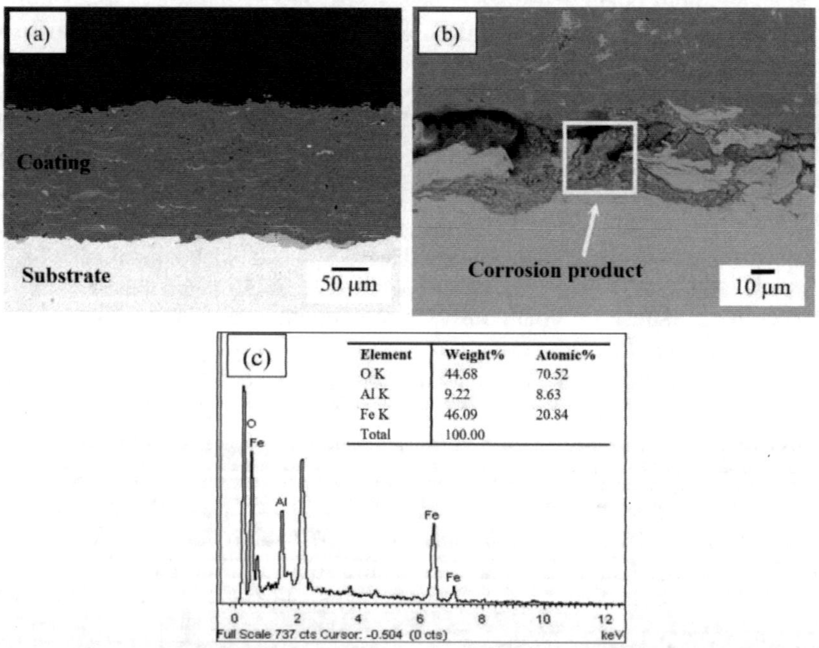

Figure 4. SEM image of cross sectional area of coating a) before immersion, b) after immersion and c) EDS analysis on corrosion product after immersion in 3.5% NaCl.

Based on the analysis of XRD pattern in Figure 5, there was corrosion product existed on the coating surface. The main corrosion product formed was consisted of iron oxide (Fe_2O_3) compound as confirmed by EDS. This result was supported by Tian and co-workers [16], which stated that the existence of Fe_2O_3 compound on the coating surface was due to the corrosion reaction that occurred at the interface of coating-substrate. Thus, it was confirmed that the coating layer consisted of porosity, which allowed the Fe_2O_3 compound moves out to reach at the coating surface.

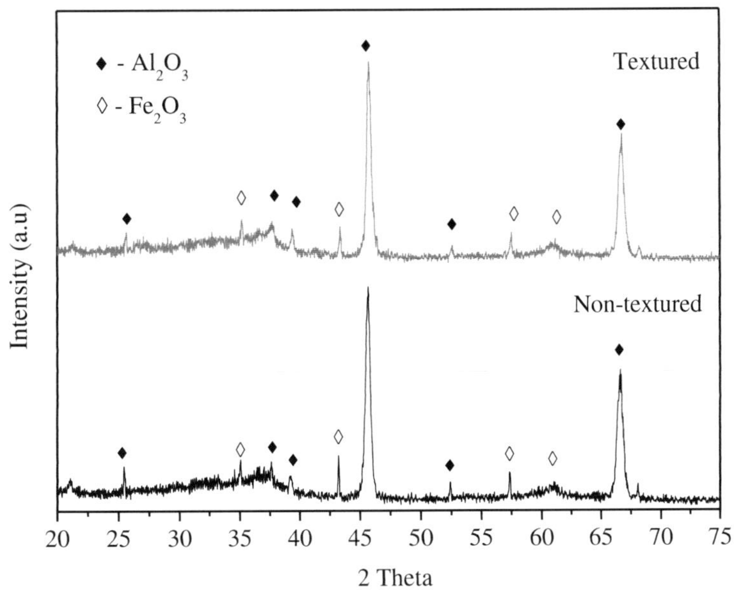

Figure 5. XRD pattern of non-textured and textured coatings after immersion in NaCl solution.

3.3. Corrosion mechanism

Through this study, it was clearly noted that introducing micro-grooves onto the surface enhanced the corrosion resistance of Al_2O_3 - 13%TiO_2 coating. From the results, it was showed the calculated corrosion rate revealed that the textured coating has high corrosion resistance as compared to that of non-textured coating. However, for both coating types, the electrolyte can penetrate through the coating layer and induce corrosion reaction at the interface of coating-substrate after 12 weeks of immersion in NaCl. This was occurred due to the existence of porosity in the coating system [15]. For the textured coating, the improvement in the corrosion resistance was due to the changes in wettability properties of the textured surface. As shown in Fig. 6, the contact angle of textured coating was around 97.1° ± 1.25 meanwhile for the non-textured coating, the contact angle was in the range of 74.5° ± 0.14. The surface of textured coating showed a hydrophobic characteristic, meanwhile the non-textured coating showed a hydrophilic characteristic. The hydrophobicity of the textured surface was resulted to the increment of the trapping air pockets in the grooves on textured coating surface. The trapped air pockets reduce the fractional area of electrolyte-coating interface and thus prevent direct contact of the NaCl solution with the coating surface [11], which led to the slow corrosion reaction. Therefore, the rate of corrosion reaction was reduced [14]. The investigation by Jagdheesh [17] revealed that the contact angle of alumina surface was increased around 153° ± 3 after laser surface texturing process. Similar trend was observed by [18], which stated that the laser surface texturing process was resulted to an increase in contact angle of zirconia surface up to 130°. The increment of surface contact angle (≥90°) value resulted to the high hydrophobicity characteristic of the surface [13]. High hydrophobicity of the surface has high capability to trap more air pockets on the surface [19], which can prevent the penetration of electrolytes into the substrate and thus, contributes to the higher corrosion resistance.

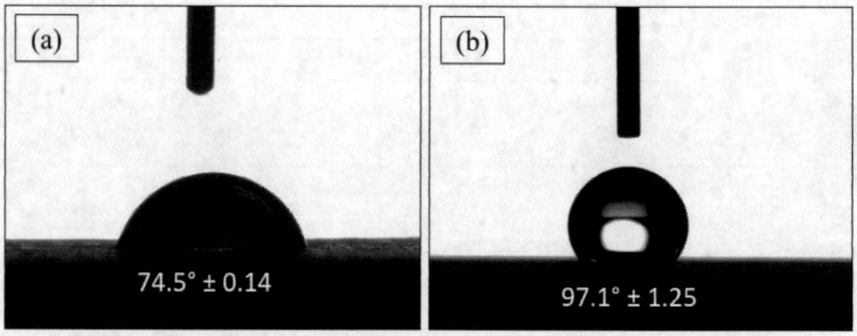

Figure 6. Contact angle of a) non-textured and b) textured coating.

4. Conclusion

Laser surface texturing is a convenient technique for developing surface texture on a ceramic coating with zero or minimal surface damage. In this study, a micro-groove texture, on the surface of plasma sprayed Al_2O_3 - 13%TiO_2 coating, was successfully fabricated via a laser surface texturing technique. The performance of the ceramic coating was found to be improved, particularly for tribological and corrosion applications. In the case of corrosion, the occurrence of uniform corrosion was detected and the resistance towards corrosion was increased up to 73%. Therefore, in this study, the resistivity of the coating against corrosion was improved by using the micro-groove texture on its surface. This would be extremely beneficial to several industries, particularly the marine industry.

Acknowledgement

The author acknowledges Universiti Kebangsaan Malaysia (UKM) for the meaningful supports and co-operations.

References

[1] M. A. Zavareh, A. A. D. M. Sarhan, B. B. A. Razak, and W. J. Basirun, "Plasma thermal spray of ceramic oxide coating on carbon steel with enhanced wear and corrosion resistance for oil and gas applications," Ceramics International, vol. 40, pp. 14267-14277, 2014.

[2] N. Valipour M, F. C. Birjandi, and J. Sargolzaei, "Super-non-wettable surfaces: A review," Colloids and Surfaces A: Physicochemical and Engineering Aspects, vol. 448, pp. 93-106, 2014.

[3] J. Koskinen, U. Tapper, P. Andersson, S. Varjus, J. Kolehmainen, S. Tervakangas, et al., "Friction reduction by texturing of DLC coatings sliding against steel under oil lubrication," Surface and Coatings Technology, vol. 204, pp. 3794-3797, 2010.

[4] A. Nagaoka, K. i. Yokoyama, and J. i. Sakai, "Evaluation of hydrogen absorption behaviour during acid etching for surface modification of commercial pure Ti, Ti–6Al–4V and Ni–Ti superelastic alloys," Corrosion Science, vol. 52, pp. 1130-1138, 2010.

[5] W. Grzesik, K. Żak, and P. Kiszka, "Comparison of Surface Textures Generated in Hard Turning and Grinding Operations," Procedia CIRP, vol. 13, pp. 84-89, 2014.

[6] U. Pettersson and S. Jacobson, "Influence of surface texture on boundary lubricated sliding contacts," Tribology International, vol. 36, pp. 857-864, 2003.

[7] D. Braun, C. Greiner, J. Schneider, and P. Gumbsch, "Efficiency of laser surface texturing in the reduction of friction under mixed lubrication," Tribology International, vol. 77, pp. 142-147, 2014.

[8] V.P. Singh, A. Sil, R. Jayaganthan, A study on sliding and erosive wear behaviour of atmospheric plasma sprayed conventional and nanostructured alumina coatings, Materials & Design 32(2) (2011) 584-591.

[9] N.H.N. Yusoff, M.J. Ghazali, M.C. Isa, A.R. Daud, A. Muchtar, S.M. Forghani, Optimization of plasma spray parameters on the mechanical properties of agglomerated Al2O3–13%TiO2 coated mild steel, Materials & Design 39 (2012) 504-508.

[10] S. Sinha, Thermal model for nanosecond laser ablation of alumina, Ceramics International 41(5) (2015) 6596-6603.

[11] U. Trdan, M. Hočevar, P. Gregorčič, Transition from superhydrophilic to superhydrophobic state of laser textured stainless steel surface and its effect on corrosion resistance, Corrosion Science 123 (2017) 21-26.

[12] L. B. Boinovich, A. M. Emelyanenko, A. D. Modestov, A. G. Domantovsky, K. A. Emelyanenko, Synergistic Effect of Superhydrophobicity and Oxidized Layers on Corrosion Resistance of Aluminum Alloy Surface Textured by Nanosecond Laser Treatment, ACS Applied Materials & Interfaces 7(34) (2015) 19500-19508.

[13] L. R. De Lara, R. Jagdheesh, J. L. Ocaña, Corrosion resistance of laser patterned ultrahydrophobic aluminium surface, Materials Letters 184 (2016) 100-103.

[14] H. Zhang, J. Yang, B. Chen, C. Liu, M. Zhang, C. Li, Fabrication of superhydrophobic textured steel surface for anti-corrosion and tribological properties, Applied Surface Science 359 (2015) 905-910.

[15] D. Thirumalaikumarasamy, K. Shanmugam, V. Balasubramanian, Corrosion performance of atmospheric plasma sprayed alumina coatings on AZ31B magnesium alloy under immersion environment, Journal of Asian Ceramic Societies 2(4) (2014) 403-415.

[16] W. Tian, Y. Wang, T. Zhang, Y. Yang, Sliding wear and electrochemical corrosion behavior of plasma sprayed nanocomposite Al_2O_3–13%TiO_2 coatings, Materials Chemistry and Physics 118(1) (2009) 37-45.

[17] R. Jagdheesh, Fabrication of a Superhydrophobic Al2O3 Surface Using Picosecond Laser Pulses, Langmuir 30(40) (2014) 12067-12073.

[18] B. S. Yilbas, Laser texturing of zirconia surface with presence of TiC and B4C: Surface hydrophobicity, metallurgical, and mechanical characteristics, Ceramics International 40(10) (2014) 16159-16167.

[19] M. Gharabaghi, S. Aghazadeh, A review of the role of wetting and spreading phenomena on the flotation practice, Current Opinion in Colloid & Interface Science 19 (4) (2014) 266-282.

IOP Conf. Series: Materials Science and Engineering **743** (2020) 012005 doi:10.1088/1757-899X/743/1/012005

Compressive strength and thermal conductivity of metakaolin geopolymers with anisotropic insulations

Nur Ain Jaya[1], Liew Yun-Ming[1], Mohd Mustafa Al Bakri Abdullah[1], Kamarudin Hussin[1,2], Heah Cheng Yong[1,2], Ridho Bayuaji[3] and Muhammad Faheem Mohd Tahir[1]

[1]Center of Excellence Geopolymer & Green Technology (CEGeoGTech), School of Materials Engineering, Universiti Malaysia Perlis (UniMAP), Perlis, Malaysia
[2]Faculty of Engineering Technology, Universiti Malaysia Perlis (UniMAP), Perlis, Malaysia
[3]Department of Civil Infrastructure Engineering, Faculty of Vocations, Institut Teknologi Sepuluh Nopember,Surabaya, Indonesia

Email: ymliew@unimap.edu.my

Abstract. This research investigated the properties of thermally insulating geopolymer prepared using waste filler (fibreboard and rubber) to act as anisotropic pore/insulation. The geopolymer matrix was synthesised using metakaolin and an alkaline solution consists of sodium hydroxide solution and sodium silicate mixture. Geopolymers with varying content (0, 3, 5 and 7 layers) of coin-shaped fibreboard and expanded polystyrene are produced to examine the anisotropic insulation effect on the material characteristics. The compressive strength and thermal conductivity were determined experimentally. From the results, it is proved that the use of anisotropic insulations can improve the thermal conductivity and minimizing the reduction of compressive strength. Geopolymer incorporated with fibreboard had better performance in terms of strength while geopolymer incorporated with rubber had better thermal conductivity.

1. Introduction

Building and construction is considered as the biggest contributor to the worldwide energy consumption and greenhouse gas emission and many efforts has been done to improve energy efficiency. In light to this matter, insulating concrete has been widely used to reduce energy consumption. Insulating concrete is a material designed to reduce thermal conductivity by incorporating insulating components, such as lightweight aggregates and entrained pores. The presence of pores and the volume ratio helps lowering the thermal conductivity of concrete. Geopolymers have been used to make lightweight concretes with aim of producing more environmentally friendly construction materials compared to the OPC-based concretes [1-5]. Geopolymers are class of environmentally friendly and sustainable inorganic aluminosilicate, firstly introduced by Professor Joseph Davidovits in 1978 [6-11].

Generally, the thermal conductivity decreases as the pore volume increases, while the compressive strength has a direct proportion to the pore volume. Several studies have been done to overcome the contrary behavior between these properties using OPC-based concrete [12,13]. Cabrillac et al. [14] studied the mechanical behavior of an anisotropic porous materials made up of ellipsoidal pores. The study showed that at an equal porosity, a material made up of ellipsoidal pores could be four times as

Content from this work may be used under the terms of the Creative Commons Attribution 3.0 licence. Any further distribution of this work must maintain attribution to the author(s) and the title of the work, journal citation and DOI.
Published under licence by IOP Publishing Ltd

rigid on a mechanical level in a preferential direction as compared to the material with spherical pores. Chung et al. [15] compared the effect of isotropic and anisotropic pores on materials properties of insulating concrete and discovered that the strength in parallel direction to the flat surface of the ellipsoidal pores is greater compared to the strength of isotropic pores. The thermal conductivity of the anisotropic pores is lower when measured in direction perpendicular to the flat surface of the ellipsoidal pores compared to the isotropic pores. From the study, it is proved that when the anisotropic pore is appropriately arranged in a specific direction, it can be used to reduce the thermal conductivity by minimizing strength loss. Zake-Tiluga et al. [16] also observed similar results when investigated anisotropic behavior of compressive strength in porous ceramic.

Most of the study on anisotropic pores were focused on OPC-based concrete and ceramics. There are limited knowledge of the pore anisotropy using geopolymer. Hence, the objective of this study is to develop thermally insulating materials with mechanical properties suitable for use in structural applications. In this work, rubber and corrugated fibreboard were incorporated in metakaolin geopolymer as anisotropic pores. The influence of directional factor on the compressive strength and thermal conductivity is assessed.

2. Experimental

2.1. Material

Metakaolin was used as the Si-Al source material in this work. The metakaolin was obtained by calcining the kaolin at 900°C for 6h in the furnace. The chemical composition of metakaolin determined by X-ray fluoroscene (XRF) spectrometer is shown in Table 1. The activator solution was a mixture of sodium hydroxide (NaOH) and liquid sodium silicate (Na_2SiO_3). The NaOH powder has 99% purity with a density of 2.13 g/cm^3 while the liquid Na_2SiO_3 contains 30.1% SiO_2, 9.4 Na_2O and 60.5% H_2O with a density of 2.4 g/cm^3. Corrugated fibreboard (from disposed carton box) and Insulflex (Lan Hoe, Malaysia) made of nitrile rubber were cut into coin-shaped. These insulation materials are denoted here as paper and rubber, respectively.

Table 1. Chemical composition of metakaolin as determined by XRF analysis.

Chemical	Wt%
SiO_2	55.7
Al_2O_3	38.6
Fe_2O_3	2.03
TiO_2	0.78
CuO	0.03
ZrO_2	0.04
K_2O	2.43
MnO_2	0.04

2.2. Preparation of Metakaolin Geopolymer Incorporated with Anisotropic Insulations

Geopolymer paste were prepared by mechanically mixing metakaolin and alkali activator solution at a ratio of 0.8. The alkali activator was obtained from mixture of 10M NaOH solution (prepared by dissolving sodium hydroxide pellets) and liquid sodium silicate at a ratio of 1.0. The geopolymer paste were set into moulds with dimensions of 100 x 100 x 100 mm. To produce regular arrangement of the insulations, a layer with specific thickness is spread in the mould. Then, the coin-shaped insulations are uniformly distributed as shown in Figure 1 and varied in 3, 5 and 7 layers. The thickness between each insulation layer depends on the total number of layers in the sample. Reference sample also was made without incorporation of coin-shaped insulations. The moulded samples were sealed with a thin film to prevent moisture loss. The samples were left at room temperature for 24 h before demoulding. Subsequently, the demoulded samples were cured in the oven at 80°C for 24 h. The samples were then kept at room temperature for 28 days prior to compressive strength test and thermal conductivity.

IOP Conf. Series: Materials Science and Engineering **743** (2020) 012005 doi:10.1088/1757-899X/743/1/012005

Figure 1. Geopolymer sample with anisotropic insulation materials (top view).

2.3. Test, Measurement and Characterization

The bulk density of the MkGPs was obtained as the ratio between the mass of the sample and its geometrical volume. The compressive strength was measured in two directions which were parallel and perpendicular (Figure 2) to the flat surface of the coin-shaped insulation and tested in accordance to ASTMC109 by using Instron machine series 5569 Mechanical Tester. Room-temperature thermal conductivity was measured in two directions which were parallel and perpendicular to the flat surface of the coin-shaped insulation using a Hot-Disk thermal constant analyser (Hot-Disk AB Uppsala, Sweden). At least 3 measurements were performed to ensure reproducibility.

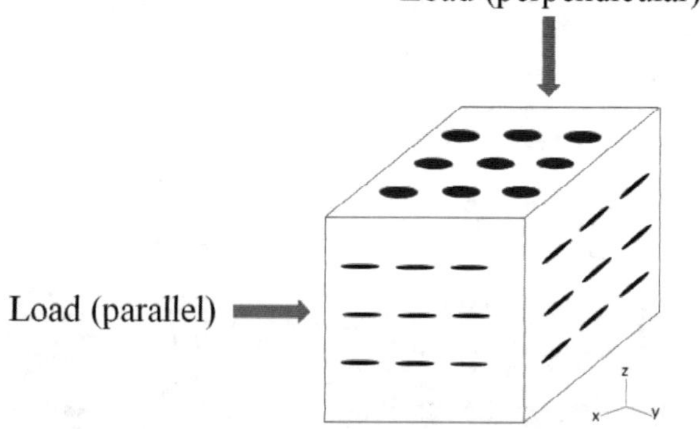

Figure 2. Schematic diagram showing the direction of load for compressive strength and thermal conductivity measurement.

3. Results and Discussions

Figure 3a presents the compressive strength of geopolymer with 0, 3, 5 and 7 layers of paper and rubber measured in directions parallel and perpendicular to the flat coin-shaped insulations. The compressive strength of the geopolymer decreased upon incorporation of the coin-shaped insulation, when measured in both parallel and perpendicular directions. However, the decrement of strength was more significant in parallel direction (38%) compared to perpendicular direction (86%) regardless of the type of insulation materials. According to Cabrillac et al. [14], when load was applied in the direction parallel to the flat surface of the insulation, the maximum stress is uniformly distributed over the whole sample. However, maximum stress can be found along the edge of anisotropic pores when the load is in perpendicular direction. It is showed that the orientation of anisotropic pores can be used to minimize the strength loss of the geopolymer. Besides, the strength decreases with the increasing layers of anisotropic pores, regardless of the directional factor. This is due to the decrease of the weight of the sample which leads to low density sample having low compressive strength (Table 2). Density has a direct relationship with the compressive strength of geopolymer.

Figure 3b shows the thermal conductivity of geopolymer with 0, 3, 5 and 7 layers of paper and rubber measured in directions parallel and perpendicular to the flat coin-shaped insulations. The reference sample has thermal conductivity of 0.213 W/mK. The thermal conductivity decreases upon incorporation of the coin-shaped insulation, when measured in both parallel (20%) and perpendicular (30%) directions, regardless of the type of insulation materials. However, larger reduction can be observed when measured perpendicular to the flat surface of the coin-shaped insulation. This is because less heat transfers through the perpendicular direction due to the disruption of heat flow by the flat surface of the pores [17].

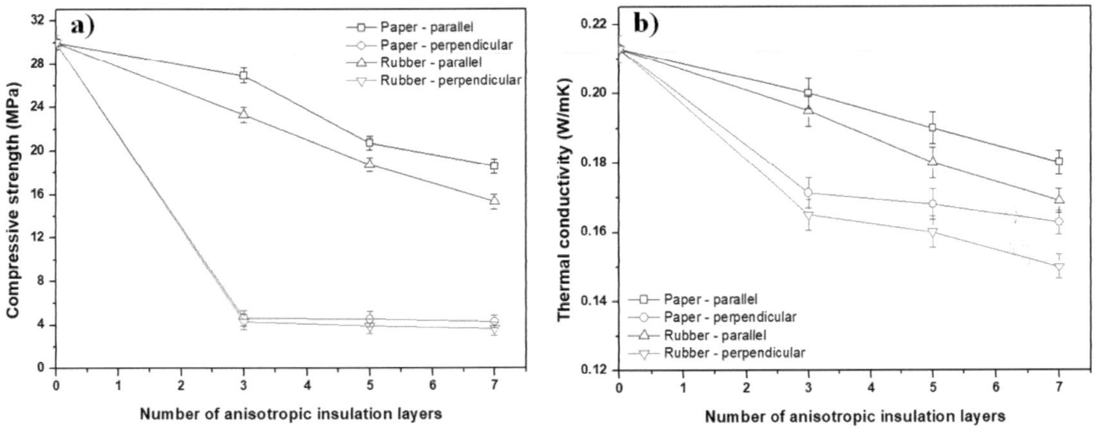

Figure 3. The compressive strength and thermal conductivity of geopolymers with different insulations materials for the parallel and perpendicular directions.

Table 2. Density of the metakaolin geopolymers.

Samples	**Density (g/cm^3)**			
	0	**3**	**5**	**7**
Reference	1.630	-	-	-
Paper	-	1.613	1.572	1.554
Rubber	-	1.601	1.565	1.540

Figure 4(a-d) showed the microstructure of rubber and paper and also the geopolymer incorporated with paper and rubber. The porous nature of rubber is the most effective to reduce the thermal conductivity while paper is the most beneficial insulation material for larger strength due to fibrous-like and larger strength. The thermal conductivity results suggest that the use of anisotropic insulations can improve the insulation effect of the material. The formation of microcracks at the interface, as shown in Figure 4c was due to the poor adhesion between geopolymer matrix and rubber which causes large strength loss. However, the good adhesion between paper and the geopolymer matrix shown by the proper embedment of paper to the geopolymer (Figure 4d) attributed to the smaller strength loss.

The geopolymer showed a very promising thermal properties compared to other thermally insulating building materials produced using lightweight aggregates and entrained air. For example, metakaolin geopolymer foams has a thermal conductivity of 0.091 W/mK at a strength of 0.3 MPa [18]. Lightweight geopolymer composites has a thermal conductivity of 0.12 W/mK at a strength of 2 MPa [3], however are not designed for load bearing purposes. This highlights the excellent properties of geopolymer obtained in this study with a thermal conductivity of 0.150 W/mK produced for when geopolymers with rubber as anisotropic insulation, measured in perpendicular direction with compressive strength of 16.34 MPa when measured in parallel direction. The properties of the geopolymer obtains in this study are within the range for structural and insulating lightweight materials based on ACI committee [19].

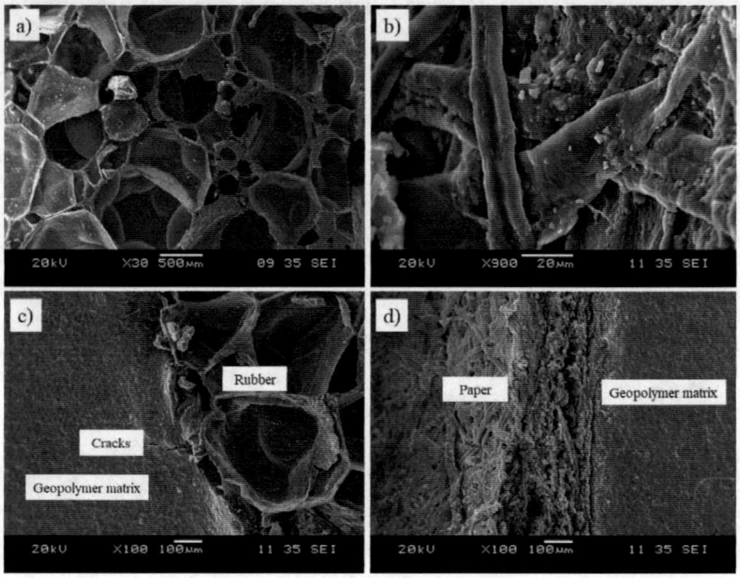

Figure 4. SEM images of a) rubber, b) paper, c) geopolymer incorporated with rubber and d) geopolymer incorporated with paper.

4. Conclusions

From the research, we can draw several conclusions which are:

I. The compressive strength is higher when measured parallel to the flat surface of the coin-shaped insulation, while thermal conductivity is smaller when measured in perpendicular direction to the flat surface of the coin-shaped insulation.
II. The compressive strength and thermal conductivity were improved compared to the geopolymer foams with isotropic pores.
III. Increasing the number of insulation layers decrease the thermal conductivity and compressive strength regardless of the insulation materials type.
IV. Rubber is the most effective to reduce the thermal conductivity due to the porous nature, while paper is the most beneficial insulation material for higher strength due to fibrous-like and larger strength.

Acknowledgement

The authors wish to deliver their gratitude to the funding support from "Partnership for Research in Geopolymer Concrete" (H2020-MSCA-RISE-2015-689857-PRIGeoC) sponsored by the European Union and Fundamental Research Grant Scheme (FRGS/1/2015/TK05/UNIMAP/02/2) by Ministry of Higher Education, Malaysia.

References

[1] Bai C, Li H, Bernardo E and Colombo P 2019 *Ceram Intl* **45** 7196-7202.

[2] Wongsa A, Zaetang Y, Sata V and Chindaprasirt P 2016 *Constr Build Mater* **111** 637-643.
[3] Colangelo F, Roviello G, Ricciotti L, Ferrandiz-Mas V, Messina F, Ferone C, Tarallo O, Cioffi R and Cheeseman C R 2018 *Cement Concrete Comp* **86** 266-272.
[4] Hamzah H N, Al Bakri Abdullah M M, Yong H C, Zainol M R R A and Hussin K 2015 *Key Eng Mater* **660** 298-304.
[5] Hamzah H N, Al Bakri Abdullah M M, Yong H C, Zainol M R R A and Hussin K 2015 *Mater Science Forum* **841** 59-64.
[6] Davidovits J 1991 *T Therm Anal* **37** 1633-1656.
[7] Abdullah M M A B, Hussin K, Bnhussain M, Ismail K N, Yahya Z, Razak R A 2012 *International Journal of Molecular Sciences* ***13*(6)** 7186–7198.
[8] Heah C Y, Kamarudin H, Bakri A M M Al, Luqman M, Nizar I K 2011 *Australian Journal of Basic and Applied Sciences* **5(7)** 1026–1035.
[9] Liew Y M, Heah C Y, Li L yuan, Jaya N A, Abdullah M M A B, Tan S J, Hussin K 2017 *Construction and Building Materials* ***156*** 9–18.
[10] Al Bakri Abdullah M M, Jamaludin L, Kamarudin H, Binhussain M, Ruzaidi Ghazali C M, Ahmad, M I 2013 *Advanced Materials Research* ***686*** 227–233 .
[11] Cheng-Yong H, Yun-Ming L, Abdullah M M A B, Hussin K 2017 *Scientific Reports*, **7 (November 2016)** 1–11.
[12] Shahedan N F, Abdullah M M A B, Mahmed N, Kusbiantoro A, Binhussain M, Zailan S N 2017 *AIP Conference Proceedings 1835* **1** 020046.
[13] Shahedan N F, Abdullah M M A B, Mahmed N, Kusbiantoro A, Bouissi A 2018 *AIP Conference Proceedings* **2030** 20294.
[14] Cabrillac R, Bruno F, Anna-lise B, Helene D and Sophie O 2006 *Const Build Mater* **20** 286-295
[15] Chung S Y, Elrahman M A and Stephen D 2016 *Energ Buildings* **125** 122-129.
[16] Tiluga Z, Svinka I R and Svinka V 2014 *Key Eng Mater* **604** 153-156.
[17] Chung S Y, Elrahman M A and Stephan D 2017 *Int J Concr Struct M* **11** 573-584
[18] Bai C and Colombo P 2017 *J Mater Res* **32** 3251-3259.
[19] Committee, A.C.I., *Specifications for Structural Concrete for Buildings (ACI 301-72).* ACI Journal Proceedings, 1972. **69.**

IOP Conf. Series: Materials Science and Engineering **743** (2020) 012006 doi:10.1088/1757-899X/743/1/012006

Effect of Solid-to-Liquid Ratio on Thin Fly Ash Geopolymer

Ng Yong-Sing[1], Liew Yun-Ming[1], Mohd Mustafa Al Bakri Abdullah[1], Ng Hui-Teng[1], Kamarudin Hussin[1,2], Heah Cheng Yong[1,2], Nurul Aida Mohd Mortar[1] and Andrei Victor Sandu[3]

[1]Center of Excellence Geopolymer and Green Technology (CEGeoGTech), School of Materials Engineering, Universiti Malaysia Perlis (UniMAP), Perlis, Malaysia
[2]Faculty of Technology Engineering, Universiti Malaysia Perlis (UniMAP), Perlis, Malaysia.
[3]Gheorghe Asachi Technical University of lasi, Faculty of Materials Science and Engineering, lasi, Romania

Email: ymliew@unimap.edu.my

Abstract. The present work studies the effect of solid-to-liquid (S/L) ratio on the properties of thin fly ash-based geopolymer. The fly ash geopolymers with dimension of 160 mm × 40 mm × 10 mm were synthesised by using various S/L ratios (1.5, 2.0, 2.5, 3.0 and 3.2). The alkali activator was prepared by mixing 10M sodium hydroxide (NaOH) solution and sodium silicate (Na_2SiO_3) with the Na_2SiO_3/NaOH ratio of 2.5. The samples were cured at 60°C for 6 hours. The performance of fly ash geopolymers was evaluated by testing the flexural strength after 28 days. Results showed that the S/L ratio had an effect on flexural strength. The optimum flexural strength of 5.12 MPa was achieved by the fly ash geopolymer with S/L ratio of 2.5. However, the flexural strength dropped with higher S/L ratio as the workability decreases. However, further experimental lab work should be carried out as there is less knowledge in the study on the flexural strength of thin fly ash geopolymer.

1. Introduction

Geopolymer is an inorganic aluminosilicate polymer synthesised by the alkali activation of the solid aluminosilicate base materials with the mixture of an alkali metal hydroxide and silicate solution [1, 2]. The types of solid aluminosilicate base materials could be divided into natural materials such as such as kaolin and metakaolin [3] or industrial by-products such as fly ash and slag [4]. Geopolymers is getting widespread attraction recently as it is sustainable with low carbon footprint [5, 6] and exhibits a wide range of properties such as high mechanical strength, low shrinkage, low water absorption, fast setting time and low thermal conductivity [7-13].

Geopolymerisation involves several processes including the dissolution of solid aluminosilicate oxide in alkali activator, followed by the diffusion of dissolved Al and Si species to an interparticle space, the gelation process of polymerisation between silicate solution and Al and Si species, and eventually the hardening that occurs in the gel phase [14]. The geopolymerisation process is affected by various factors such as raw materials, molarity of alkali solution used, alkali activator ratio, curing regime and solids-to-liquids ratio.

The solids-to-liquids (S/L) ratio indicated the aluminosilicate raw materials-to-alkaline activator solution ratio. The S/L ratio is crucial for homogenous mixing as it controls the quantity of solids and

Content from this work may be used under the terms of the Creative Commons Attribution 3.0 licence. Any further distribution of this work must maintain attribution to the author(s) and the title of the work, journal citation and DOI.
Published under licence by IOP Publishing Ltd

liquid. These compositional amount would directly influence the workability, dissolution, setting and most importantly, the strength of the geopolymers [15, 16].

The rate of geopolymerisation is affected by initial solids content. Increasing the S/L ratio leads to the larger formation of alkaline aluminosilicate as the reactant species are increasing dissolved [17]. Yao et al. [18] reported that the geopolymerisation reaction would be slower and take longer time with lower S/L ratio as large amount of liquid content decreased the interfacial connection between the precursor materials. In contrast, Fernandez-Jimenez and Palomo [19] reported a lower workability for high S/L ratio fly ash geopolymers.

Besides, Rahim et al. [20] reported that the mechanical strength of fly ash geopolymers increased with the increasing S/L ratio. However, the strength dropped once the S/L ratio exceed its optimum ratio of 4. This decrement in strength was associated with the faster setting time resulting in faster hardening of gel phase and more rapidly bonding of the geopolymeric structure when S/L ratio increased [21]. Thus, the optimum solid-to-liquids ratio for geopolymer formation is significant in determining the properties of geopolymer. In this study, the influence of solids-to-liquids ratio on the thin fly ash geopolymer was investigated. The study focused on the relationship of this parameters between the physical properties and flexural strength of thin geopolymers.

2. Methodology

2.1. Materials

Fly ash collected from Cement Industries of Malaysia Berhad (CIMA) Perlis Plant, Malaysia was used as the source of aluminosilicate. Table 1 shows the chemical composition of the fly ash obtained through XRF analysis. The fly ash contains total composition of SiO_2 and Al_2O_3 of 84.30%. According to ASTM C618, the fly ash used was classified as Class F fly ash as the CaO content is 3.89% which is less than 20% The SEM micrograph of fly ash particle is revealed in Figure 1. The morphology of fly ash particles sieved with siever 68μm was spherical in shape with smooth surfaces. The alkaline activator used was a mixture of liquid sodium silicate (Na_2SiO_3) and sodium hydroxide (NaOH). The technical grade Na_2SiO_3 has chemical composition of 30.1% SiO_2, 9.4% Na_2O and 60.5% H_2O with modulus SiO_2/Na_2O of 3.2, specific gravity at 20°C = 1.4 and viscosity at 20°C = 0.4 Pa·s. The NaOH pellets used was classified as caustic soda pellets by the brand of HmbG® Chemicals supplied by Sigma-Alrich, Germany with assay of 97.0%.

Table 1. Chemical composition of fly ash.

Compound	SiO_2	Al_2O_3	Fe_2O_3	CaO	TiO_2	K_2O	Others
Mass (%)	56.30	28.00	6.86	3.89	2.17	1.49	1.29

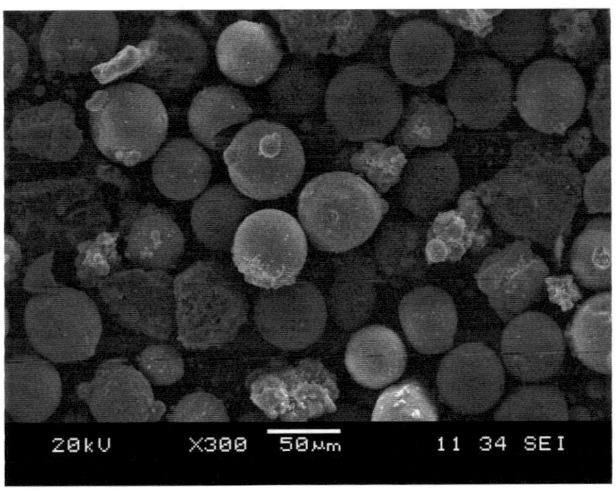

Figure 1. SEM micrograph of fly ash particles.

IOP Conf. Series: Materials Science and Engineering **743** (2020) 012006 doi:10.1088/1757-899X/743/1/012006

2.2. Formation of Thin Geopolymer

The alkaline activator was mixed with the Na_2SiO_3/NaOH ratio of 2.5 and the concentration of NaOH solution was fixed at 12M. Then, the fly ash was mixed well with the alkaline activator at a S/L ratio ranging from 1.5 – 3.2. S/L ratio of 3.2 was set as the upper boundary of this experimental work as the mixing process beyond that was undesirable. The fresh paste was then rapidly moulded and compacted into a 160 × 40 × 10 mm moulds. The moulded geopolymer samples were cured at 60°C in oven for 6 hours and then left at room temperature for another 24 hours. The cured samples were kept at room temperature for 28 days before testing.

2.3. Testing and Analysis

The bulk density measurement was obtained by measuring the mass and dimension of the samples. The water absorption was measured by the wet mass (immersed in water for 24 hours) and dry mass (heated 100°C in oven for 24 hours) while apparent porosity was measured by the wet mass, dry mass and suspended mass (after immersed in water for 24 hours) according to ASTM C642. The flexural test was carried out using Instron Machine Series 5569 Mechanical Tester according to the ASTM C348 with a span length of 110 mm at a crosshead speed of 1 mm/min.

3. Results and Discussion

3.1. Physical Observation

Figure 2 shows the physical image of fly ash geopolymers synthesised with different S/L ratio. The fly ash geopolymers were grey in colour and the surface condition was comparatively similar. However, the fly ash geopolymer mixed using S/L ratio of 1.5 bended in a curvature shape as shown in Figure 3 after 28 days. This is due to the higher liquid content which lead to the formation of less viscous slurry during mixing process [22]. The high liquid content in the system creates tension force on the surface of samples and causes the occurrence of curvature in sample.

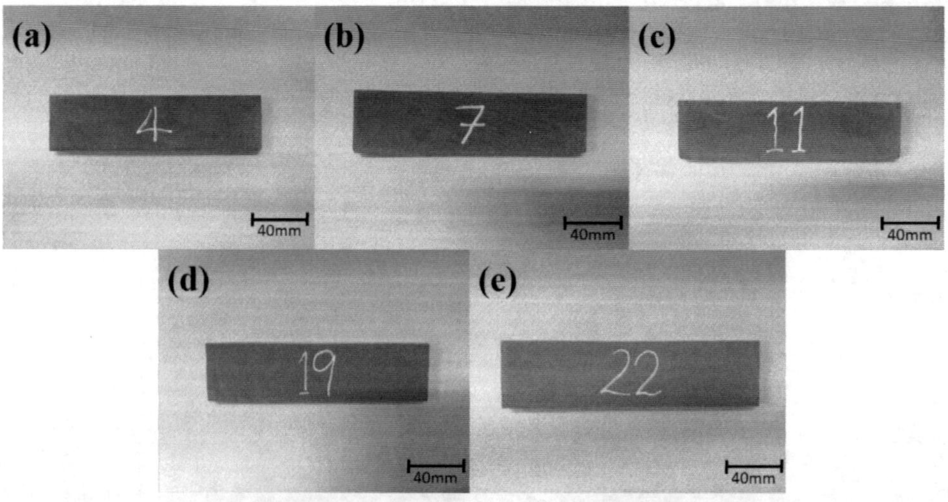

Figure 2. Physical images of fly ash geopolymers prepared with S/L ratio of (a) 1.5, (b) 2.0, (c) 2.5, (d) 3.0 and (e) 3.2.

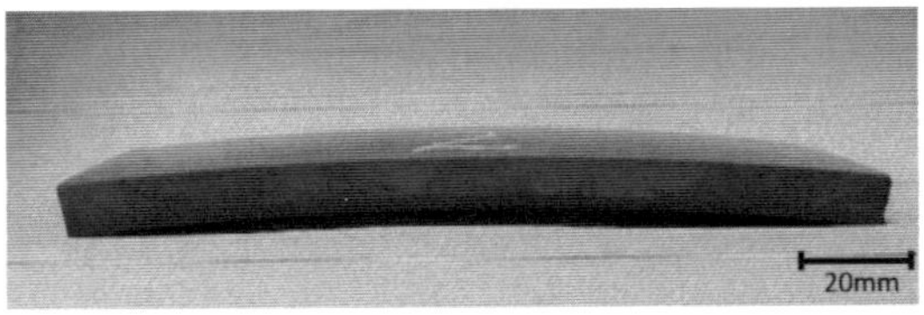

Figure 3. Bending of fly ash geopolymer synthesised with S/L ratio of 1.5.

3.2. Bulk Density, Apparent Porosity and Water Absorption

Figure 4 shows the bulk density of fly ash geopolymers with different S/L ratios. It was observed that the bulk density of fly ash geopolymers increased from ratio 1.5 to 2.5, but followed by a decrement with S/L ratio of 3.0 and 3.2. The highest bulk density obtained was 2020 kg/m^3 with S/L ratio of 2.5 whereas the fly ash geopolymers mixed with S/L ratio of 2.0 had the lowest bulk density of 1868 kg/m^3. However, when the solid content was in excess (S/L ratio of 3.0 and 3.2), the bulk density of fly ash geopolymers reduced. This is because the liquid content is insufficient for geopolymerisation reaction as the flow of mixture is hindered [18]. The fly ash particles could not interact fully with the limited amount of liquid during geopolymerisation process, and hence reduced the bulk density and less reaction products were formed.

Rahim et al. [21] observed a similar trend in the bulk density of fly ash geopolymers cured with one day with S/L ratios of 3 to 5. The bulk density first increased the S/L ratio increased from 3 to 4 but then decreased when the S/L ratio reached 5. However, the density obtained was higher (ranges from ~2900 kg/m^3 to ~3600 kg/m^3. This difference in bulk density might be due to data collection at different curing day.

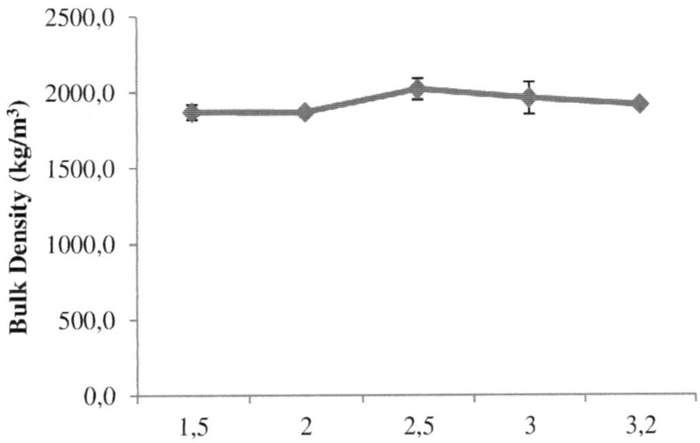

Figure 4. Bulk density of fly ash geopolymers with varying S/L ratios.

Figure 5 depicts the apparent porosity and water absorption of fly ash geopolymers with different S/L ratios. Generally, both the apparent porosity and water absorption shared the similar trend where the apparent porosity and water absorption decreased when S/L ratio increased from 1.5 to 2.5, but followed by an increment when the S/L ratio was beyond 2.5. This data was in parallel with the bulk density measurement obtained in Figure 4 as the lower percentage of porosity would contribute to the lower water absorption and more densified structure which lead to higher bulk density.

The fly ash geopolymers synthesised using S/L ratio of 2.5 had the lowest apparent porosity and water absorption of 19.9% and 10.6% respectively. Furthermore, fly ash geopolymers prepared with S/L ratio of 3.2 had the highest apparent porosity and water absorption of 22.2% and 11.5%. this is because the precursor materials could not fully react with alkaline activator when the S/L was too high. This led to the water retained in the geopolymer matrix and promoted the pore formation and hence increased the percentage of water absorption [24].

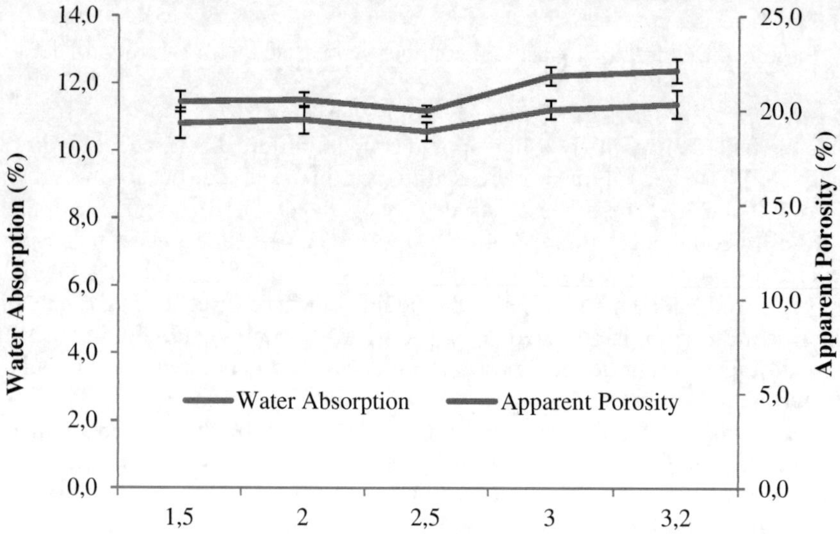

Figure 5. Water absorption and apparent porosity of fly ash geopolymers with varying S/L ratios.

3.3. Flexural Strength

Figure 6 depicts the flexural strength of fly ash geopolymers synthesised using different S/L ratios. It was observed that the flexural strength of fly ash geopolymers increased with increasing S/L ratios from 1.5 to 2.5, and then decreased as the S/L ratio increased until 3.0 and 3.2. The highest flexural strength was denoted by fly ash geopolymer synthesised using S/L ratio of 2.5 with the value of 5.12 MPa whereas the fly ash geopolymer prepared with S/L ratio of 3.2 had the lowest flexural strength of 3.61 MPa.

At lower S/L ratio, the intermolecular contact between the precursor materials with the activating solution was hindered and became limited as the volume of fluid medium was too large [25]. The rate for the aluminosilicate materials to be dissolved was decelerated and thus lower the flexural strength of fly ash geopolymers. As the S/L ratio was increased to 2.5, the solid content in system was increased. The probability for the contact between the precursor materials with activating solution was increased and thus increase the flexural strength of fly ash geopolymers.

However, the flexural strength of fly ash geopolymers dropped at high S/L ratios of 3.0 and 3.2. This is because the workability of fly ash geopolymers decreases when the S/L ratio is too high. The excess solid content makes the mixing becomes undesirable and faces the difficulty in compaction for moulding process [18]. Besides, the excess solid contents would retard the geopolymerisation reaction and consequently lower the flexural performance of fly ash geopolymers as the extent for the binder formation is hindered [23, 26].

Furthermore, the flexural results were in parallel with the bulk density measurement and apparent porosity results as shown in Figur and

Figure respectively. The lower the porosity content, the higher the bulk density, the higher the compressive strength. Some researchers evaluated that the S/L ratio could be influence the volume of voids and porosity in the slurry which directly affect the mechanical strength of geopolymer [27, 28]. In overall, the optimum S/L ratio for the fly ash geopolymers synthesis was 2.5, as it exhibited the highest flexural strength among the other specimens. However, further research activities should be implemented in order to enhance the performance of thin geopolymer and get better flexural properties.

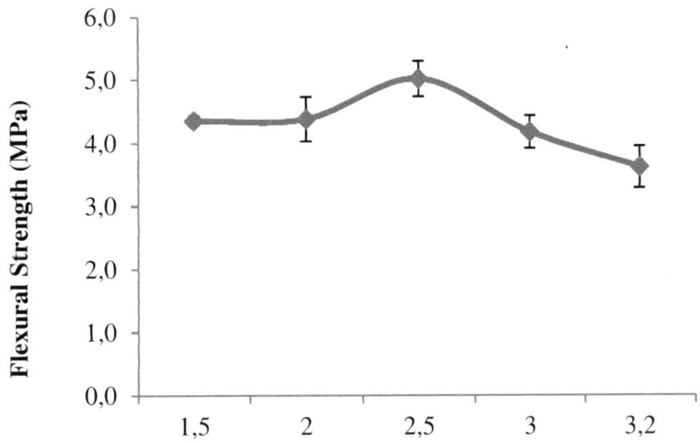

Figure 6. Flexural strength of fly ash geopolymers with varying S/L ratios.

4. Conclusion

This paper investigated the effect of various S/L ratios (1.5 to 3.2) on thin fly ash geopolymers. Physical and mechanical properties of thin geopolymers indicated with bulk density, apparent porosity and flexural strength reported that, ratio of solids-to-liquids plays an important role in enhancing the properties of thin fly ash geopolymer. The optimum S/L ratio of 2.5 produced thin fly ash geopolymer with the highest flexural strength of 5.12 MPa It was believed that the contact between the alkaline activator and the reacting materials were improved and thus produced denser geopolymer sample. The flexural strength results were complied with the density, porosity and water absorption values. However, further experimental lab work should be carried out to enhance the flexural strength of thin fly ash geopolymer.

Acknowledgements

The authors of the present work wish to acknowledge the funding support from "Partnership for Research in Geopolymer Concrete" (H2020-MSCA-RISE-2015-689857-PRIGeoC) sponsored by the European Union and Fundamental Research Grant Scheme (FRGS/1/2015/TK05/UNIMAP/02/2) sponsored by Ministry of Education Malaysia.

References

[1] J. Davidovits, J. Therm. Anal. Calorim. **37(8)**, 1633-1656 (1991).
[2] R.N. Thakur and S. Ghosh, J. Eng. Appl. Sci. (Asian Res. Publ. Netw.). **4**, 68-74 (2009).
[3] J. Davidovits, Spec. Publ. **144**, 383-398 (1994).
[4] D. Hardjito, S.E. Wallah, D.M. Sumajouw, and B.V.Rangan, ACI Mater. J. **101(6),** 467-472 (2004).
[5] M.B. Ogundiran and S. Kumar, Constr. Build. Mater. **125**, 450-457 (2016)
[6] H.N. Hamzah, M.M.A.B. Abdullah, C.Y. Heah, M.R.R.A. Zainol and K. Hussin, Key Eng. Mater. **600**, 298-304 (2015).
[7] E. Benhelal, G. Zahedi, E. Shamsaei, and A. Bahadori, J. Cleaner Prod. **51**, 142-161 (2013).
[8] J. Davidovits, *Geopolymer Chemistry and Applications* (Institut Géopolymère, Saint-Quentin, 2011).
[9] R. Ahmad, M. Mustafa, A. Bakri, K. Hussin, A. V. Sandu, M. Binhussain, and N. A. Jaya, *Rev.*

Adv. Mater. Sci. **44**, 26–32, (2016).

[10] N. A. Jaya, M. M. Al Bakri Abdullah, C. M. R. Ghazali, M. Hussain, K. Hussin, and R. Ahmad, *MATEC Web Conf.* **78**, (2016).

[11] N. A. Jaya, M. M. Al Bakri Abdullah, and R. Ahmad, *Mater. Sci. Forum* **803**, 81–87, (2015).

[12] A. Romisuhani, M. M. Albakri, H. Kamarudin, and S. V. Andrei, *IOP Conf. Ser. Mater. Sci. Eng.* **267(1)** (2017).

[13] Abdullah, M.M.A.B., Nordin, N., Tahir, M.F.M., Kadir, A.A., Sandu, A.V., *International Journal of Conservation Science* **7(3)**, (2016).

[14] Z. Yunsheng, S. Wei, L. Zongjin, Z. Xiangming, Eddie and C. Chungkong, Constr. Build. Mater. **22(3)**, 370-383 (2008).

[15] P. Chindaprasirt, T. Chareerat and V. Sirivivatnanon, Cem. Concr. Compos. **29(3)**, 224-229 (2007).

[16] H.N. Hamzah, M.M.A.B. Abdullah, C.Y. Heah, M.R.R.A. Zainol and M.N.M Nawi, MCRJ. **3**, 155-162 (2018).

[17] S. Alonso and A. Palomo, Mater. Lett. **47(1-2)**, 55-62 (2001).

[18] X. Yao, Z. Zhang, H. Zhu and Y. Chen,. Thermochim. Acta. **493(1-2)**, 49-54 (2009).

[19] A. Fernández-Jiménez and A. Palomo, Mater Construcc **57(287)**, 7-22 (2007).

[20] R. H. A. Rahim, K. A. Azizli, Z. Man, T. Rahmiati, and L. Ismail, Appl. Mech. Mater. **625**, 46-49 (2014).

[21] R. H. A. Rahim, K. A. Azizli, Z. Man, and M. F. Nuruddin, Appl. Mech. Mater. **699**, 15-19 (2015).

[22] A. Sathonsaowaphak, P. Chindaprasirt, and K. Pimraksa, J. Hazard. Mater. **168**, 44-50 (2009).

[23] D. Khale and R. Chaudhary, J. Mater. Sci. 42(3), 729-746 (2007).

[24] B. Krishnendu, B. Ankush, P. Surosrita, D. Dutta, and D. T. G. Biswas, IOSR-JMCE **15015**, 1-6 (2001).

[25] H.N. Hamzah, M.M.A.B. Abdullah, C.Y. Heah, Z. Rozainy and W.W. Zailani, *Correlation of the Na_2SiO_3 to NaOH Ratios and Solid to Liquid Ratios to the Kedah's Soil Strength.* **78**, 01071 (2016).

[26] J. L. Provis, C. Z. Yong, P. Duxson, and J. S. J. van Deventer, Colloids Surf. A. **336**, 57-63 (2009).

[27] D. L. Y. Kong and J. G. Sanjayan, Cement Concrete Comp. **30(10)**, 986-991 (2008).

[28] D. L. Y. Kong, J. G. Sanjayan and K. Sagoe-Crentsil, Cement Concrete Comp. **37(12)**, 1583-1589 (2007).

IOP Conf. Series: Materials Science and Engineering **743** (2020) 012007 doi:10.1088/1757-899X/743/1/012007

The Effect of Polyethylene Glycol (PEG) on TiO_2 Thin Films via Sol-Gel Method

D S Che Halin[1], K A Razak[1], N S Mohamad Sukeri[1], A Azani[1], M M A B Abdullah[1], M A A M Salleh[1], N Mahmed[1], M M Ramli[1,2], AW Azhari[3] and V Chobpattana[4]

[1]Center of Excellence Geopolymer & Green Technology (CEGeoGTech), School of Materials Engineering, Universiti Malaysia Perlis, (UniMAP), 02600 Jalan Kangar-Arau, Perlis.
[2]School of Microelectronic Engineering, Universiti Malaysia Perlis, Pauh Putra Campus, 02600, Arau, Perlis,Malaysia.
[3]Water Research Group (WAREG), School of Environmental Engineering, Universiti Malaysia Perlis 02600 Arau, Perlis, Malaysia.
[4]Department of Materials and Metallurgical Engineering, Faculty of Engineering, Rajamangala University of Technology Thanyaburi (RMUTT), Thailand.

Email: dewisuriyani@unimap.edu.my

Abstract. This research focus on the effect of polyethylene glycol (PEG) on TiO_2 thin film. Sol-gel method is the best method which tends to be used due to its simplicity, good chemical homogeinity and high purity of the product. Titanium (IV) isopropoxide, isopropanol and acetic acid are the three different chemical which being used to make sol solution. The samples was then will annealed at three different temperature which are 400 °C, 500 °C and 600 °C to observed the phase composition of TiO_2 added PEG thin film and pure TiO_2 by using x-ray diffraction (XRD) analysis and the structural surface by using scanning electron microscope (SEM) analysis. The XRD analysis show the anatase phase present for the sample of pure TiO_2 thin film and rutile phase present for the sample of TiO_2 added PEG thin film. The micrograph of SEM show that with the addition of PEG at high temperature will give the analysis of flaky large cracked which is not separated to each other on the surface coating. Meanwhile, pure TiO_2 give the result of irregular shape structure of the film.

1. Introduction

Titanium dioxide (TiO_2) has a variety of crystal structure which were anatase, brookite and rutile. The least stable of crystal structure is the distortion of brookite [1]. Titanium dioxide (TiO_2) is a nontoxic which has no harmful effects. Besides, TiO_2 gives a unique chemical, physical and electro-optic properties which it is become famous and well researched materials [2]. TiO_2 can be applied in many fields such as environmental including the self-cleaning surface, anti-fogging surface also energy fields. TiO_2 also widely used as photocatalyst. Based on research, TiO_2 has some limitations where high refractive index of TiO_2 will reduce the glass of transmittance. In addition, the self-cleaning efficiency will negatively affected because of the rapid reestablishment of hydrophobicity in the dark environments of TiO_2 films [3].

Content from this work may be used under the terms of the Creative Commons Attribution 3.0 licence. Any further distribution of this work must maintain attribution to the author(s) and the title of the work, journal citation and DOI.
Published under licence by IOP Publishing Ltd

IOP Conf. Series: Materials Science and Engineering **743** (2020) 012007 doi:10.1088/1757-899X/743/1/012007

Photocatalysis is a clean technology, efficient and attractive which has been used for pollutant abatement [4]. There are many method that can be used to prepare the TiO_2 film. In preparing the TiO_2 thin film, sol-gel process was considered to be used because of its own advantage in order to save the cost. It is an attractive technique because of its advantages and the preparation method is easier which just need less completing instrument and less time consuming. Sol-gel method also have been used for preparation of TiO_2 thin film where it can be utilized in many fields applying on superhydrophilic properties, photocatalytic properties and also antibacterial properties [5]. The phase of anatase will be produce at a low temperature because of the high activity in photocatalytic applications. Based on previous researchers, the photocatalytic decomposition of an aqueous acetic acid was affected by the morphology and crystal structure of TiO_2 coatings [6].

This research was focus on the effect of the additive Polyethylene Glycol (PEG) on TiO_2 thin films which play an important role in resulting the desired result. The TiO_2 thin film was fabricated by using spin coating of sol-gel method. Then, the thin film will characterized by using x-ray diffractometer (XRD) to observe the crystalline structure and the phase composition of TiO_2 film and Scanning Electron Microscopy (SEM) was used to determine the surface morphology.

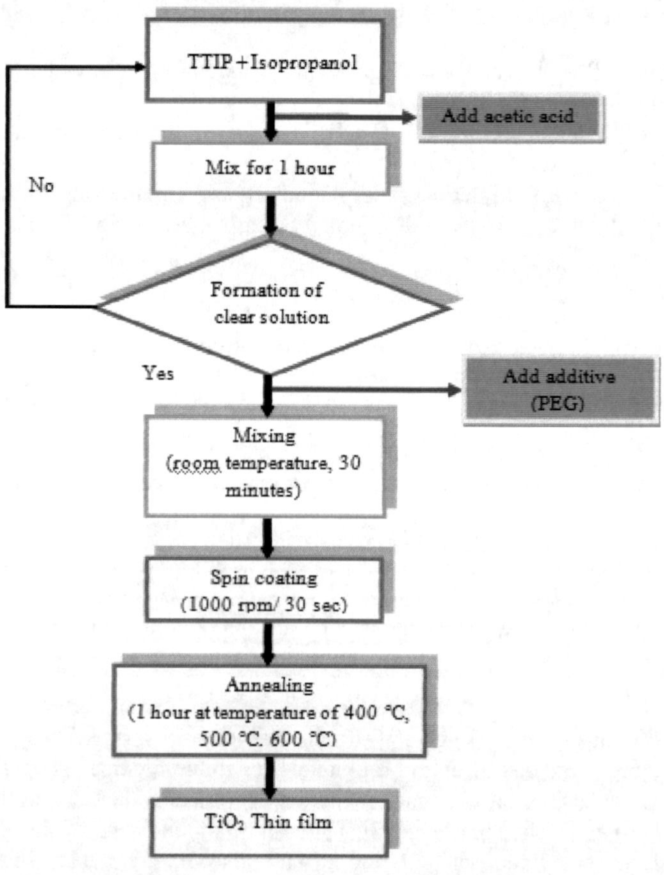

Figure 1. Flowchart preparation for TiO_2 thin film.

2. Methodology

The solutions of TiO_2 was prepared by using titanium (IV) isopropoxide (TTIP) as a source of TiO_2 and then, mixed together into the isopropanol by using a molar ratio mixture of 1:20 (TTIP:isopropanol). The solution was then stirred for 5 minutes. During stirring process, several drops of acetic acid was added into the mixing solutions which give the function of catalyst. The solutions were continuous mixing for about 1 hour by using magnetic stirrer. The clear and homogeneous solutions will be formed after 1 hour. Figure 1 show the flowchart preparation of TiO_2 thin film.

The final solution which was formed will be deposited on the silica wafer with a spin rate of 1000 rpm for about 30 seconds. The sample will dry in the oven for about 10 minutes at the temperature of 60°C. Then, the samples will annealed at three different temperatures which are 400 °C, 500 °C and 600 °C for about 1 hour. Then, the preparation of TiO_2 thin film with the 10 drops of PEG was prepared with the same steps as the previous one. PEG will be dropped into the solution of TiO_2 for about 10 drops after clear solution of TiO_2 was formed. Then, it will continuously stirred for 30 minutes at room temperature.

3. Results and discussion

Figure 2 show the XRD pattern of pure TiO_2 sample. It can be seen the diffraction pattern shows anatase phase and the diffraction peak was assigned at 61.20° and 69.12° which is corresponds to (2 1 3) and (1 1 6) as shown in Figure 2 (a) and (b). Meanwhile, Figure 2 (c) show the diffraction pattern at 25.8°, 61.20° and 69.12°. The diffraction pattern observed is labeled according to standard XRD pattern of JCPDS No. 21-1272. The phase corresponds to (1 0 1), (2 1 3) and (1 1 6) directions. Muthukumarasamy et al. (2010) was studied at high temperature of annealing process, the formation of anatase phase will be present where the film will oxidized and turn into oxide. This will affect the increasing of grain size. The formation of anatase phase and the grain growth were the results diffusion of titania species towards nucleated grain [7].

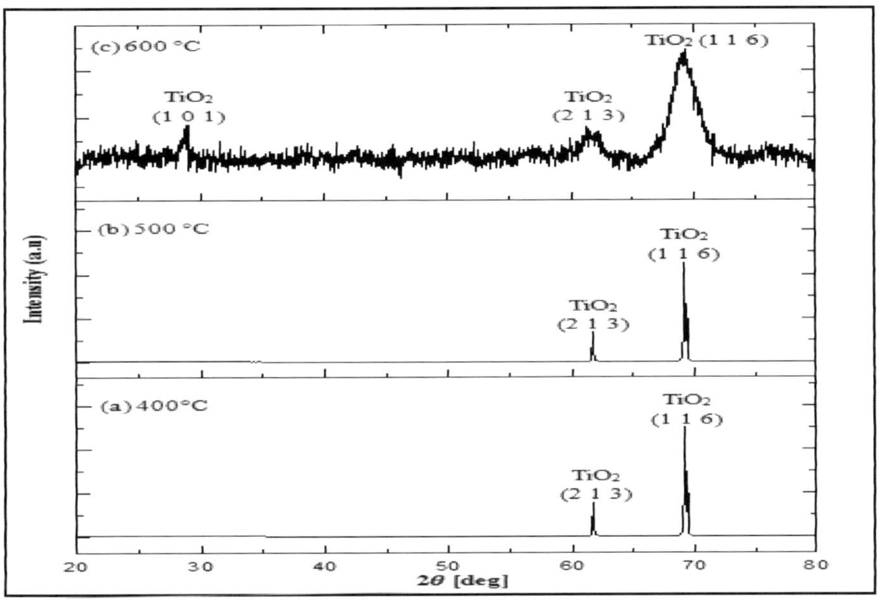

Figure 2. XRD pattern of pure TiO_2 film which annealed at 400 °C, 500 °C, 600 °C.

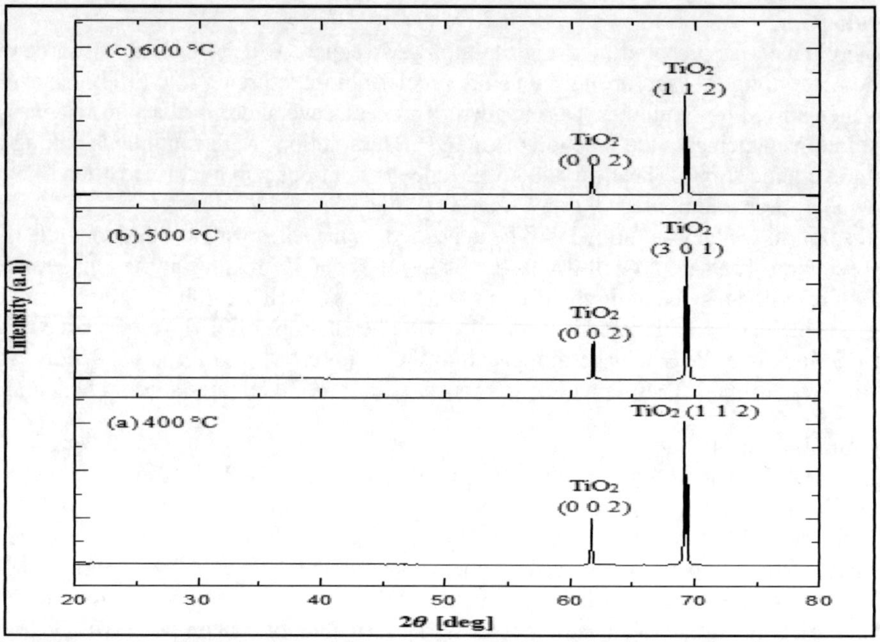

Figure 3. XRD pattern of TiO_2 added PEG film which annealed at 400 °C, 500 °C, 600 °C.

Figure 3 (a), (b) and (c) show the XRD pattern of 10 drops PEG which undergoes the annealing process at temperature of 400 °C, 500 °C and 600 °C. It is clearly seen that there is a rutile phase with a tetragonal structure existed. Based on Figure 3 (a) and (c), the diffraction peak at 61.799° and 69.927° was assigned to the diffraction pattern of rutile phase which corresponds to (0 0 2) and (1 1 2) direction. From Figure 3 (b) show the diffraction pattern 61.799° and 69.442° which have a plane of (0 0 2) and (3 0 1), respectively. The pattern observed with the agreement of XRD standard pattern of JCPDS No. 88-1172. Liu et al. (2005) was reported that the relatively large agglomerate tends to the formation of rutile phase [8]. Rutile is the most stable phase which has a large chemical stability, cheaper production cost also has high refractive index compared to anatase and brookite phase [9].

Figure 4 shows the SEM micrograph of pure TiO_2 thin film and TiO_2 added PEG thin film which annealed at 400 °C, 500°C, 600°C, respectively. The SEM micrograph in Figure 4a-400 revealed that there is a very tiny of irregular shape of structure on the crack surface. This might happen due to the low annealing temperature. This result is well agree with Mayabadi et al. (2014) [10]. Figure 4b-400 show there are many crack present with a large size of crack. Shuhui et al. (2014) was reported, the mismatch of thermal expansion coefficient between substrate and films, the evaporation of a solvent and also the precursors decomposition leads to formation of cracks [11].

Figure 4a-500 show the microstructure of pure TiO_2 is in irregular shape of crystal which was the solution is not added with any additive. Meanwhile, in Figure 4b-500 show the microstructure is crack free and exhibit a huge agglomeration. Regarding to Thomas et al. (2002), agglomeration occur when the combination of three set rate processes which involved the consolidation and growth, wetting and nucleation and attrition and breakage. The large agglomerate size is affected by the higher of PEG concentration was used [12]. From the Figure 4a-600 the microstructure of irregular shape start to merge with one another due to the increasing in annealing temperature, this can be proved by Bakri et al. (2017) [13]. Besides, the microstructure of the large flaky cracks which is not separated to each other can be seen in Figure 6b-600. The cracked film present may be due to the temperature increase from 400 °C to 600 °C. It is also can reduce the coating of lucidity and transparency and happen because of the surface tension between film and air during drying process [14-15]. Zhang et al. (2004) was studied that the addition of PEG will improve the surface roughness also the cracking-free films [16]. Hence, from the results of the crack obtained might be due to the insufficient of molar ratio of PEG to prevent the film from crack.

Figure 4. SEM image of; (a) pure TiO_2 thin film (b) TiO_2 added PEG thin film, which was annealed at 400 °C, 500 °C, 600 °C, respectively.

4. Conclusions

This research was focus on the effect of polyethylene glycol (PEG) on TiO_2 thin films via sol-gel method. There are three different annealing temperature were carried out for every samples which are 400 °C, 500 °C and 600 °C. The samples were then characterized by using X-Ray Diffraction (XRD) analysis and Scanning Electron Microscope (SEM) analysis. The different concentration of PEG and three different annealing temperature has been affected the XRD analysis and also the surface morphology of the thin film. Based on the XRD analysis, basically all of the samples with the PEG show there is a TiO_2-rutile phase existed. Meanwhile, the TiO_2-anatase phase exist for the samples of pure TiO_2. From the SEM micrograph, when the annealing temperature was increased from 400 °C to 600 °C, the films were non-uniform with an irregular shape and also have an agglomeration. At 600 °C, it can be clearly seen, there was no agglomeration but having a large flaky crack. At higher temperature, a small island will coalesce and channel each island.

Acknowledgements
The author would like to acknowledge the support from the Fundamental Research Grant Scheme (FRGS) under a grant number of FRGS/1/2017/TK07/UNIMAP/02/6 from the Ministry of Education Malaysia, TSTRG Research Grant from Tin Industry (Research and Development) Board under a grant number of 9002-00082 and Center of Excellence Geopolymer & Green Technology (CEGeoGTech), School of Materials, Engineering, Universiti Malaysia Perlis, UniMAP.

References
[1] Davis K A 1982 *J. Chem. Educ.* **59** 158.
[2] Guo B, Liu Z, Hong L, and Jiang H 2005 *Surf. Coatings Technol.* **198** 24–29.
[3] de Jesus M A M L, Neto J T da S ,Timò G, Paiva P R P, Dantas M S S, and Ferreira A de M 2015 *Appl. Adhes. Sci.* **3** 5.
[4] Arconada N 2009 *Appl. Catal. B Environ.* **86** 1–7.
[5] Yamashita H and Ichihashi Y 2016 *Nanostructured Photocataalysis: Advanced Functional Materials*, Switzerland: Springer International Publishing 178.
[6] Yu J, Zhao X, and Zhao Q 2000 *Thin Solid Films* **379** 7–14.
[7] Senthil T S, Muthukumarasamy N, Agilan S, Thambidurai M, and Balasundaraprabhu R 2010 *Mater. Sci. Eng. B Solid-State Mater. Adv. Technol.* **174** 102–104.
[8] Reyes-Coronado D, Rodríguez-Gattorno G, Espinosa-Pesqueira M E, Cab C, De Coss R, and Oskam G 2008 *IOP Publ. Nanotechnol. Nanotechnol.* **19** 145605–10.
[9] Hosono E, Fujihara S, Kakiuchi K, and Imai H 2004 *J. Am. Chem. Soc.* **126** 7790–7791.
[10] Mayabadi A H, Waman V S, Funde A M, Pathan H M, Jadkar S R 2014 *J. Phys. Chem. Solids* **75** 182–187.
[11] Yu S, Yao K, Shannigrahi S, Tay F and Hock E 2003 *Journal of Materials Research* **18** 737-741
[12] Abberger T, Seo A, and Schæfer T 2002 *Int. J. Pharm.* **249** 185–197.
[13] Bakri A S, Sahdan M Z, Adriyanto F, Raship N A, Said N D M, Abdullah S A, and Rahim M S 2017 AIP Conf. Proc. **1788** 030030-1–030030-8.
[14] Halin D S C, Abdullah M M A B, Mahmed N, Abdul Malek S N A, Vizureanu P, and Azhari A W 2017 *IOP Conf. Ser. Mater. Sci. Eng.* **209** 12002.
[15] Shokuhfar A, Alzamani M, Eghdam E, Karimi M, and Mastali S 2012 *Nanosci. Nanotechnol.* **2** 16–21.
[16] Zhang J, Au K, Zhu Z and O'Shea S 2004 *Opt. Mater. (Amst).* **26** 47–55.

IOP Conf. Series: Materials Science and Engineering **743** (2020) 012008 doi:10.1088/1757-899X/743/1/012008

Influence of Different Materials on the Mechanical Aspects in the Design of Cyclone Gasifier

C. Y. Khor[1], M. A. M. Nawi[1], Muhammad Ikman Ishak[1], Z. X. Wei[1] and M. U. Rosli[1]

[1]Simulation and Modelling Research Group (SimMReG), Faculty of Engineering Technology, Universiti Malaysia Perlis (UniMAP), Kampus UniCITI Alam, Sungai Chuchuh, Padang Besar, 02100 Perlis Indera Kayangan, Malaysia.

Email: cykhor@unimap.edu.my

Abstract. Cyclone gasifier is used as the energy conversion system for the biomass. The design and material selection for the cyclone gasifier is important, which affect the strength of the gasifier during the operation. This research aims to investigate the effect of the various materials to the mechanical aspects (i.e., stress, strain and displacement) of the cyclone gasifier. The mechanical aspects analysis is carried out by using the finite element (FE) based software. The stress, strain and displacement of the cyclone gasifier's structure was analyzed for various stainless steel materials (i.e., ferritic (FSS), AISI 316, 1023 CSS, 201 ASS and AISI 4130). The finite element analysis revealed the use of 201 ASS experienced highest stress (221 MPa). Lowest strain and displacement were found on 201 ASS and 1023 CSS, respectively. These research findings are expected to be used as the reference for the engineer in the material selection process for the design and fabrication of cyclone gasifier.

1. Introduction

The gasification system is an alternative way to convert the biomass or organic matter to the renewable energy. At the same time, it's also beneficial to the environment sustainable. In the energy conversion process, the gasifier may experience high temperature, mechanical and thermal stress on the gasifier's structure. Prior to the fabrication of the cyclone gasifier, the design and material selection are crucial to ensure the safety and functionality during the operation. A conventional trial-and error method is costly and time consuming in the design analysis process. Thus, simulation technique is an alternative way to tackle these problems. In the simulation modelling, various techniques like computational fluid dynamics [1,2], finite element analysis [3,4] and coupling of both finite element and finite volume method [5] can be applied to various engineering problems.

Recent years, the simulation techniques are widely applied in solving various engineering problems such as conventional and molded underfill process [6,7], integrated circuit (IC) encapsulation process [8], reflow soldering process [9,10], airflow analysis in the reflow oven [11] and fluid-structure interaction problems [12,13]. Besides, the simulation analysis also applicable to predict the structural problems in of the cyclone gasifier [14]. During the gasification process, the pressurized airflow enters the compartment and the structure may impose with a specific pressure. Cyclone gasifier is classified as the entrained-flow bed gasifier, which could produce low CH4 syngas and free of tar [15]. The unburned particles are separated from the syngas when passing through the cyclone gasifier. Thus, the cyclone gasifier saves the operating cost for the cleaning of the syngas. Moreover,

Content from this work may be used under the terms of the Creative Commons Attribution 3.0 licence. Any further distribution of this work must maintain attribution to the author(s) and the title of the work, journal citation and DOI.
Published under licence by IOP Publishing Ltd

IOP Conf. Series: Materials Science and Engineering **743** (2020) 012008 doi:10.1088/1757-899X/743/1/012008

the air flow characteristics in the gasifier were investigated via computational fluid dynamics on various types of gasifier [16-18].

The understanding of the mechanical aspect is crucial in the design process of the cyclone gasifier before the fabrication process. Many factors such as material selection, operating condition and physical design [14] may influence the reliability and functionality of the cyclone gasifier. Besides, the cyclone geometry is one of the significant factors to affect the internal airflow performance [19]. In the design process, the engineer needs the preliminary information as the input to design the cyclone gasifier. The preliminary information of the cyclone gasifier provides the reference to the engineer to select the most suitable material for the structure of cyclone gasifier. The current study is the extension of our previous work [14]. The finite element based [20-22] simulation is applied to explore the mechanical aspects [23] of the structure of cyclone gasifier when using different materials for its structure. The mechanical aspects which are stress, strain and displacement of each material are compared.

2. Methodology

This study was focused on the study of different materials towards the mechanical aspects of the cyclone gasifier. A similar cyclone gasifier model and finite element method were adapted from our previous work [14]. The 3D cyclone gasifier model was simulated in the finite element based software to study the stress, strain and displacement of the structure when different materials are applied. The cyclone's dimension and design are depicted in Figure 1.

i. Cyclone inlet height, D_i (0.210 m)
ii. Inlet width, D_i (0.210 m)
iii. Gas outlet (vortex finder) diameter, D_e (0.18 m)
iv. Length, S (1.90 m)
v. Barrel height, h (1.50 m)
vi. Total height, H (5.00 m)
vii. Cone-tip diameter, B (0.18 m)
viii. Body diameter, D (0.445 m)

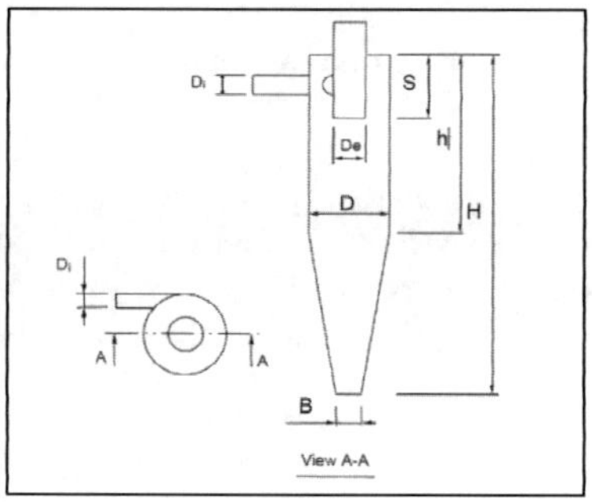

Figure 1. Dimension of the cyclone gasifier [19].

Five different materials were considered for the analysis, which are ferritic stainless steel (FSS), AISI 316 stainless steel (AISI 316), 1023 carbon stainless steel (CSS 1023), 201 annealed stainless steel (ASS 201) and AISI 4130 stainless steel (AISI 4130). The materials are applied to the structure of the cyclone gasifier as shown in Figure 1. During the simulation, the internal pressure was applied on the compartment of the cyclone (Figure2), which mimic the internal pressure during the operation. The stress, strain and displacement will be analyzed on the critical location for five different materials.

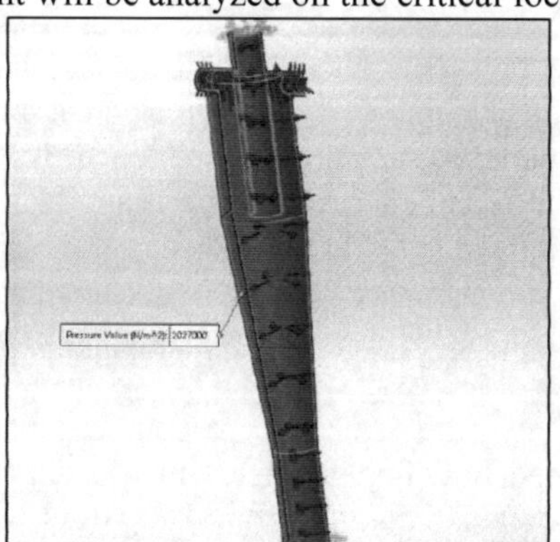

Figure 2. Pressure distribution in the cyclone gasifier [14].

3. Results and Discussion

Figures 3-5 depict the results from the mechanical aspects, which are maximum stress, strain and displacement of the cyclone's structure using various materials. The results demonstrate the variation of the mechanical aspects when different materials were applied to the cyclone structure. The use of CSS 1023 and AISI 4130 yields lowest stress that imposed to the cyclone, with only 217 MPa compared to other materials. However, 201 annealed stainless steel shows the highest stress (221 MPa) amongst the five types of material. In terms of strain, ASS201 has the lowest value of strain and AISI 4130 shows the highest strain during the operation. Moreover, the results revealed CSS1023 yields the lowest displacement, which means the cyclone's structure has better rigidity and resist to the deformation compared with AISI4130.

From the results, CSS1023 demonstrates the good mechanical aspects amongst the materials with lower stress and displacement. In addition, CSS1023 has the excellent machinability with reasonable strength. CSS1023 is easy to form and it is cheaper compared to the other steels. Thus, it is suitable for the fabrication of the cyclone gasifier. From the consideration of the mechanical aspects, CSS1023 is recommended for the fabrication of the cyclone gasifier. On the other hand, other approaches such as Analytical Hierarchy Process (AHP) [24, 25] and ELETRE methods can be also applied to select the material and process for the fabrication of the gasifier.

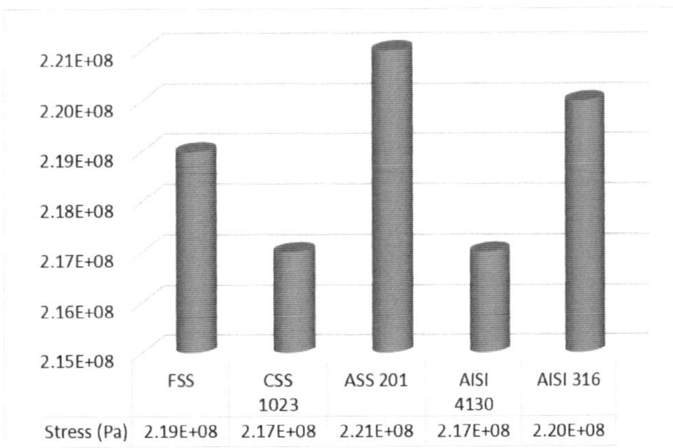

Figure 3. Maximum stress versus different materials.

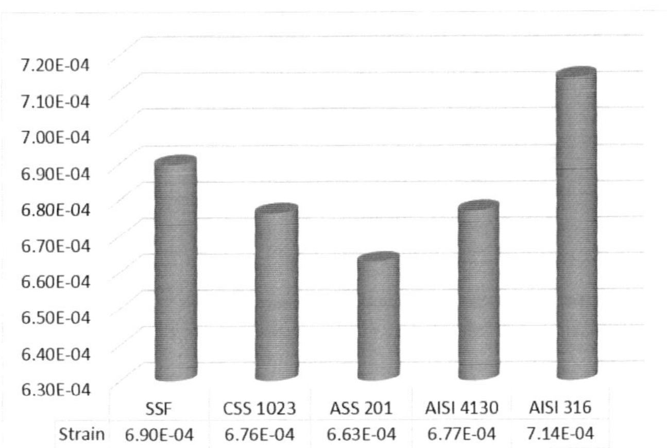

Figure 4. Maximum strain versus different materials.

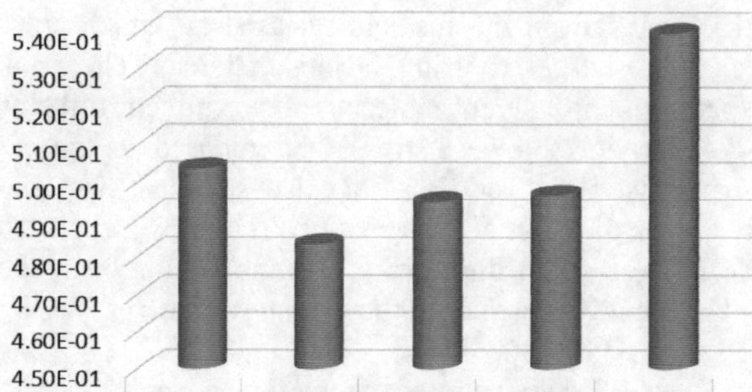

	SSF	CSS 1023	ASS 201	AISI 4130	AISI 316
Displacement (mm)	5.04E-01	4.83E-01	4.95E-01	4.96E-01	5.40E-01

Figure 5. Maximum displacement versus different materials.

4. Conclusion

The study of mechanical aspects of cyclone gasifier with different materials has been carried out in a FE based simulation. The results indicated that the selection of material significantly affects the mechanical aspects of the cyclone gasifier. Improper material selection will result in the higher stress concentration, strain and deformation, which lead to the safety issue when the gasifier operates. The top of the cyclone was a critical region and it can be minimized by increasing the physical design parameter of the cyclone. The simulation results revealed the use of CSS1023 resulted in low stress (217 MPa) and displacement (0.483 mm) and it is also machinability. With these characteristics, CSS1023 is suitable for the fabrication process of cyclone gasifier. The current results are expected to be used as the reference for the material selection and design of cyclone gasifier. Moreover, the application of simulation modelling technique provides better understanding of fundamental knowledge and also effectively reduces the time to market and cost in the fabrication process.

References

[1] Abdul Aziz M S, Abdullah M Z, Khor C Y, Iqbal A M and Che Ani F 2014 *International Journal of Heat and Mass Transfer* **72** 400-410.

[2] Khor C Y, Abdullah M Z, Majid M F M A, Mujeebu M A, Abduallah M K and Ariff Z M 2010 *Heat and Mass Transfer* **46** 1315-1325.

[3] Khor C Y and Abdullah M Z 2012 *Simulation Modelling Practice and Theory* **29**, 109-122.

[4] Ramdan D, Abdullah M Z, Khor C Y, Leong W C, Loh W K, Ooi C K and Ooi R C 2012 *IEEE Transactions on Components Packaging and Manufacturing Technology* **2**(11) 1786-1795.

[5] Ong E E S, Abdullah M Z, Khor C Y, Loh W K, Ooi C K and Chan R 2014 *Microelectronic Engineering* **113**(1) 40-49.

[6] Khor C Y and Abdullah M Z 2013 *Microelectronics Reliability* **53** 334-347.

[7] Khor C Y, Abdullah M Z, Lau C-S, Leong W C and Abdul Aziz M S 2014 *Microelectronics Reliability* **54** 796-807.

[8] Ong E E S, Abdullah M Z, Khor C Y, Loh W K, Ooi C K and Ooi R C 2012 *International Communications in Heat and Mass Transfer* **29**(10) 1616-1623.

[9] Lau C-S, Abdullah M Z and Khor C Y 2013 *Microelectronics International* **30**(3) 151-168.

[10] Lau C-S, Khor C Y, Soares D, Teixeira J C and Abdullah M Z 2016 *Soldering & Surface Mount Technology* **28** 41-62.

[11] Najib A M, Abdullah M Z, Khor C Y and Saad A A *2015 International Journal of Heat and Mass Transfer* **87**, 49-58.

[12] Leong W C, Abdullah M Z and Khor C Y 2012 *Microelectronics Reliability* **52** 744-756.

[13] Abdul Aziz M S, Abdullah M Z, Khor C Y, Fairuz Z M, Iqbal A M, Mazlan M and Rasat M S M 2014 *Advances in Mechanical Engineering* **2014** 275735.

[14] Wei Z X, Khor C Y, Rahim W M F W A, Razak N A, Ishak M I, Rosli M U, Jamalludin M R, Nawi M A M and Zakaria M S 2018 *AIP Conference Proceedings* **2030** 020047.
[15] Henrich E, Dahmen N and Niebel A 2015 *BioBoost Report* 1-38.
[16] Kumar U and Paul M C 2019 *Chemical Engineering Science* **195** 413-422.
[17] González W A and Pérez J F 2019 *Energy* **186** 115904.
[18] Luo H, Lin W, Song W, Li S, Dam-Johansen K and Wu H 2019 *Fuel Processing Technology* **195** 106146.
[19] Singh P, Couckuyt I, Elsayed K, Deschrijver D and Dhaene T 2016 *Applied Mathematical Modelling* **40** 4248-4259.
[20] Jizat N M, Mohamad S N and Ishak M I 2017 *AIP Conference Proceedings* **1885** 020126
[21] Ishak M I, Ahmad Shafi A, Abdul Kadir M R and Sulaiman E 2017 *Journal of Medical and Biological Engineering* **37**(3) 336-344
[22] Ishak M I, et al. 2017 *MATEC Web of Conferences* **97** 01040
[23] Tan J S, Khor C Y, Rahim W M F W A, Ishak M I, Rosli M U, Jamalludin M R, Zakaria M S, Nawi M A M, Abdul Aziz M S and Che Ani F 2017 *AIP Conference Proceedings* **1885**, 020063.
[24] Rosli M U, Jamalludin M R, Khor C Y, Ishak M I, Jahidi H, Wasir N Y, Faizal W M, Wan Draman W N A, Lailina N M, Ismail R I 2017 *MATEC Web Conf.* **97**, 01039.
[25] Luqman M, Rosli M U, Khor C Y, Zambree S, Jahidi H 2018 *IOP Conf. Series: Materials Science and Engineering*, **318**, 2018, 012058.

IOP Conf. Series: Materials Science and Engineering **743** (2020) 012009 doi:10.1088/1757-899X/743/1/012009

Experimental and computational evaluations of the mechanical stresses of banana trunk fibre-reinforced epoxy resin composite in coffee table application

Muhammad Ikman Ishak[1], M U Rosli[1], C Y Khor[1], C N Ismail[1] and M A M Nawi[1]

[1]Simulation and Modelling Research Group (SimMReG), Faculty of Engineering Technology, Universiti Malaysia Perlis, Level 1, Block S2, UniCITI Alam Campus, Sungai Chuchuh, 02100 Padang Besar, Perlis, Malaysia

Email: ikman@unimap.edu.my

Abstract. The banana plant including the trunk is normally disposed after it has produced fruits once in its lifetime and becoming unproductive. Herbicide injection and burning using kerosene are the examples of current disposing method. Those methods are unfavourable owing to bad effects to the environment. Therefore, the unproductive banana trunk through its fibre properties, has a high potential to be used as composite in saving the environment. Three tensile test specimens were prepared with two of them have different length of banana fibres – shorter and longer than 0.5 mm, whilst another specimen has no fibres. The elastic modulus, Poisson's ratio, and vertical load of 2 GPa, 0.3, and 981 N, respectively, were applied towards numerical models. The results exhibited that the composite with short banana fibre length promoted encouraging maximum and break stress values. Besides, the properties of the composite are well appropriate to be applied in furniture making as satisfactory stress value recorded in the computational analysis.

1. Introduction

Banana plant can be categorised as a gigantic herb which forms trunk with a height of 3 to 6 meters from an underground stem. The plant comprises several main parts such as fruits, leaves, and stem or trunk. The trunk or also recognised as pseudostem, developed by the firmly stuffed overlapping leaf sheaths. The leaves will emerge one after another that allowing the trunk to grow continuously until reaches its maximum height. As the banana fruits can be eaten, the trunk, on the other hand, is usually processed to make natural craft materials and garment [1]. Moreover, the advantages of banana trunk have also been expanded, to date, in medical treatment applications, as instances for blood pressure, digestion, detoxification, healing acidity, weight loss, and gastric problems.

The banana plant is only productive once in its lifetime before it dies. Commonly, the plant takes approximately 13 months to achieve maturity and the plucking process of the fruits is ready. When the banana plants die, they are normally disposed via several ways such as by burning using kerosene, cutting and chopping down using mechanical tools, and injecting herbicide. However, most of the methods are harmful to the environment due to the consequences of air pollution and toxicity. Thus, the unproductive banana plant especially its trunk can otherwise be used for related possible applications in regard to its fibre properties. Up to the present time, natural fibres are widely utilised in numerous appliances owing to less pollution, low density, low cost, light weight, and eco-friendly nature. Natural fibre can be described as substances that produced by animals or plants and it can be

Content from this work may be used under the terms of the Creative Commons Attribution 3.0 licence. Any further distribution of this work must maintain attribution to the author(s) and the title of the work, journal citation and DOI.
Published under licence by IOP Publishing Ltd

spun into thread, rope, or filament [2]. Banana trunk fibre possesses similar properties with that of natural bamboo, however, it has better spin ability and fineness.

Composite is defined as a material that developed from two or more basic materials (reinforcement and matrix) with different chemical and physical properties, to create a new material with different properties from the individual element. Among the examples of typical engineered composite materials are composite wood, reinforced concrete, reinforced plastics, metal matrix composite, and ceramic matrix composite [3]. The present study emphasises the use of generic organic matrix material which is polymer, to be evaluated.

This study, thus, places a major interest on the evaluation of the mechanical properties of the banana trunk fibre-reinforced polymer composite for furniture making application. To date, there is no study observed that considering the proposed type of composite in the making of furniture. Previous findings reported that the rattan being the typical material used as main parts of furniture since 1970s. The method of weaving reeds or other natural fibres is known as one of the prevalent techniques considered in fabricating related furniture elements. Furthermore, this study is basically an extension of our previous related investigation on the similar composite applied in the fabrication of furniture which focused on the effect of different weights of banana fibre [4]. The present evaluation, however, examined the influence of distinct sizes of the fibre – 0, < 0.5, and > 0.5 mm – on the mechanical stresses generated within the composite under experimental and numerical analyses.

2. Materials and methods

2.1. Preparation of banana trunk fibres

Generally, most of the materials and methods involved in this study were adopted from our past related investigation [4]. A banana trunk of an unproductive matured banana tree was collected from a farm. The trunk was later cleaned and washed with water before cut into some small parts using a cutter, with the thickness of 50 mm for each part. This study only consumes one part of the trunk and it was sliced into tiny fragments. The fibres were then dehydrated under the direct sunlight for 36 hours until they were absolutely dry or all moisture content was removed.

Through this study, the influence of banana trunk size (length) embedded in the polymer matrix on the mechanical properties was a major focus. Hence, it is important to note that the quantity of the fibre (weight) was set to be identical (0.9 g) in all specimens. Whilst, epoxy resin was selected as the material of the matrix which used to bind all the fibres together. In overall, there were three specimens made; 1) Specimen 1: Without fibre (pure epoxy resin matrix), 2) Specimen 2: 0.9 g fibre with the fibre size of less than 0.5 mm (14.6 ml epoxy resin matrix), and 3) Specimen 3: 0.9 g fibre with the fibre size of more than 0.5 mm (14.6 ml epoxy resin matrix). Two different manual sifters with the hole size of smaller and greater than 0.5 mm were utilised to strain out all the needed fibre size.

2.2. Preparation of specimen mould

Prior to the experimental testing of the specimens, there is a need to fabricate the specimen moulds first. Therefore, a three-dimensional (3-D) model of dog bone shape signifying the design and dimension of the specimen, was created using SolidWorks software. The 3-D model was then printed using a rapid prototyping machine and it has a length, width, and thickness of 150, 20, and 4 mm, respectively. The printed dog bone-shaped model was casted to produce the specimen mould by combining two types of material – silicon rubber and hardener. The ratio of silicon rubber to hardener considered is 25:1. Among the advantages of silicon rubber are good heat resistance and elasticity. After the mixture of both materials had been prepared, it was directly poured into a container. The specimen models were later put into the mixture solution and the mould was let harden. Figure 1 exhibits the hardened silicon rubber mould and the dried banana trunk fibres.

2.3. Preparation of tensile testing specimens

The testing specimens had been produced by mixing epoxy resin and hardener at a ratio of 2:1 to be as the composite matrix, and the mixture was poured into three separated containers. Two of the containers were inserted with different sizes of banana trunk fibre – smaller and greater than 0.5 mm,

IOP Conf. Series: Materials Science and Engineering **743** (2020) 012009 doi:10.1088/1757-899X/743/1/012009

respectively, and stirred evenly, while another container was left unmixed with any fibres. Afterward, the composite mixture was ran into the mould and dried at room temperature for approximately 24 hours. The mechanical properties of the composite namely maximum and break stresses (MPa) were determined through a tensile test. The tensile test machine was firstly calibrated to prevent any errors which may affect the result data. The speed of the stroke was fixed at 30 mm/min.

2.4. Computational analysis of furniture
A numerical examination on a furniture model was conducted to evaluate its performance in regard to the properties of the composite attained in the experimental analysis. Thus, a 3-D model of coffee table was developed using SolidWorks software and then transferred into Abaqus software for the finite element analysis (FEA). Finite element technique shows a vital role in solving mathematical modelling difficulties in numerous fields of technology, science, and industry [5-20].

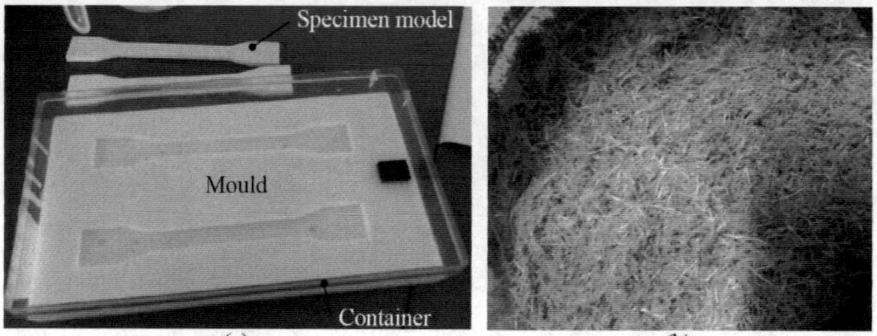

Figure 1. (a) The hardened silicon rubber mould. (b) The dried banana trunk fibres [4].

3. Results and discussions
3.1. Maximum and break stress magnitudes
Figure 2 illustrates the comparison of maximum and break stress magnitudes recorded among the specimens and the condition of the specimens after the tensile test conducted. It was found that the maximum stresses were considerably increased as the size of fibre decreased. The value of maximum stress was significantly lower in both Specimen 1 (no fibre) and Specimen 3 (> 0.5 mm fibre) than the one in Specimen 2 (< 0.5 mm fibre) with discrepancies of 12.64 and 11.97%, respectively. Similar situations were observed for the break stress outcomes in which Specimen 2 (26.1 MPa) still superior with about 1.18- and 1.01-times higher break stress magnitude as compared to Specimen 1 (22.1 MPa) and Specimen 3 (25.9 MPa), respectively. The specimens with fibre content irrespective of its size exhibited brittle-like fracture surface compared to the one without.

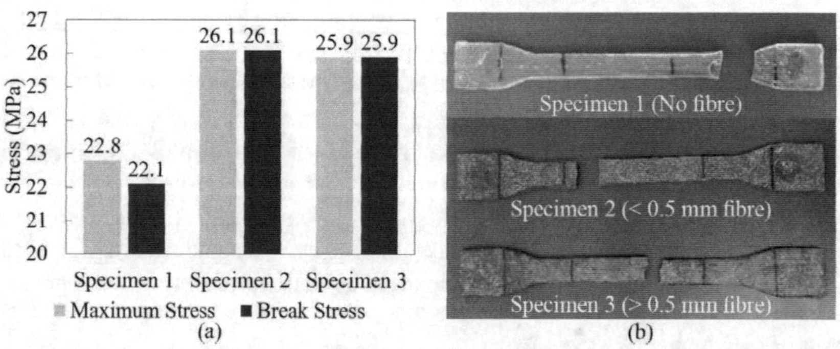

Figure 2. (a) The comparison of maximum and break stress values among all specimens. (b) The condition of specimens after the tensile test.

One of the promising clarifications for the greater stress values recorded in the specimen with decreased size of fibres is due to the high contact area of fibre-to-matrix. The decrease in fibre size has increased the total contact area of all fibres, accordingly. In comparison, it was important to note that the maximum and break stress values were similar for Specimen 2 and Specimen 3. Meanwhile, for Specimen 1, the break stress was slightly smaller than the maximum stress magnitude by nearly 3%. This could be due to the brittleness of the composite which has generated less strain hardening phase or plastic deformation before fracture. These were in accordance to the type of specimen fracture depicted by Specimen 2 and Specimen 3 wherein no necking or constriction seen to reveal the elasticity elongation under the application of tensile force.

3.2. *Maximum von Mises stress dispersion in furniture model*

The outcomes of FEA in terms of von Mises stress value and distribution within the coffee table model are exhibited in figure 3. From the tensile test results, it was reported that the Young's modulus, E of the tested composite is about 2 GPa, with the Poisson's ratio, v of 0.3 assumed. These material properties had been assigned for the table legs in the pre-processing settings of analysis. A vertical load of 981 N was onto the top surface of the table and all FEA models were meshed with four nodes solid tetrahedral elements.

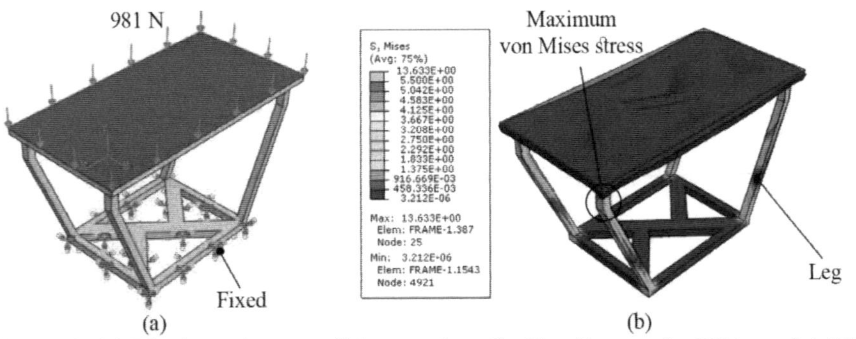

Figure 3. (a) The boundary conditions and applied loading on the FEA model [4]. (b) Von Mises stress plot for the table model that assigned with the composite properties.

The findings revealed that the load was uniformly dispersed all over the legs with excessive stress intensity found at the top regions. The highest von Mises stress level produced in the table was 13.6 MPa illustrated in grey colour spectrum scale. The magnitude appears significantly way lower than the normal range of ultimate tensile stress value of the natural fibre-reinforced polymer composite. It could thus be agreed that the properties of banana trunk fibre-reinforced epoxy resin composite are satisfactory in yielding promising stress value within the furniture.

4. Conclusions

The findings of the present study support the following conclusions. The mechanical properties of banana trunk fibre-reinforced epoxy resin composite was observed to be favourable in the application of furniture making owing to encouraging mechanical stress value obtained via numerical analysis. Also, the decrease in the size of fibre embedded within the polymer matrix was evident to have considerably improved the properties of the composite.

Acknowledgements

Appreciation is given to Faculty of Engineering Technology, Universiti Malaysia Perlis. The authors reported no conflicts of interest related to this study.

References

[1] Violette R 2016 *Ten Uses of Banana Trees* (New York: Dengarden).

[2] Rosli MU, et al. 2017 Analytical hierarchy process for natural fiber composites automotive armrest thermoset matrix selection *MATEC Web of Conferences* **97** 01039.
[3] John MJ and Thomas S 2008 Biofibers and biocomposites *Carbohydrate Polymers* **71**(3) 343-64.
[4] Ishak MI, Ismail CN, Khor CY, Rosli MU, Jamalludin MR, Hazwan MHM, Nawi MAM and Mohamad Syafiq AK 2019 Investigation on the mechanical properties of banana trunk fibre–reinforced polymer composites for furniture making application *IOP Conference Series: Materials Science and Engineering* **551** 012107.
[5] Omar A, Ishak MI, Harun MN, Sulaiman E and Abu Kasim NH 2012 Effects of different angulation placement of mini-implant in orthodontic *Applied Mechanics and Materials* **121-6** 1214-9.
[6] Ishak MI and Kadir MRA 2013 Treatment options for severely atrophic maxillae *SpringerBriefs in Applied Sciences and Technology* 9783642326028 9-26.
[7] Ishak MI, Abdul Kadir MR, Sulaiman E and Abu Kasim NH 2010 Effects of different zygomatic implant body surface roughness and implant length on stress distribution *2010 IEEE EMBS Conference on Biomedical Engineering and Sciences (IECBES)* **5742230** 210-5.
[8] Jizat NM, Mohamad SN and Ishak MI 2017 Resonator graphene microfluidic antenna (RGMA) for blood glucose detection *AIP Conference Proceedings* **1885** 020126.
[9] Ishak MI, et al. 2017 Conceptual design of automotive compressor for integrated portable air conditioning system *MATEC Web of Conferences* **97** 01040.
[10] Ishak MI and Abdul Kadir MR 2013 Biomechanics in dentistry: evaluation of different surgical approaches to treat atrophic maxilla patients *SpringerBriefs in Applied Sciences and Technology* 9783642326028.
[11] Ishak MI, Ahmad Shafi A, Abdul Kadir MR and Sulaiman E 2017 Effect of ferrule height and post length on mechanical stress and displacement of endodontically treated maxillary central incisor: a finite element analysis *Journal of Medical and Biological Engineering* **37**(3) 336-44.
[12] Ishak MI and Abdul Kadir MR 2013 Finite element modelling *SpringerBriefs in Applied Sciences and Technology* 9783642326028 37-52.
[13] Tan JS, et al. 2017 Influence of solder joint length to the mechanical aspect during the thermal stress analysis *AIP Conference Proceedings* **1885** 020063.
[14] Khor CY, Ishak MI, Rosli MU, Jamalludin MR, Zakaria MS, Yamin AFM, Abdul Aziz MS and Abdullah MZ 2017 Influence of material properties on the fluid-structure interaction aspects during molded underfill process *MATEC Web of Conferences* **97** 01059.
[15] Hanzah MR, Wan Abd Rahim WMF, Khor CY, Ishak MI, Rosli MU, Jamalludin MR, Zakaria MS and Nawi MAM 2017 New approach for quality control in manufacturing process *AIP Conference Proceedings* **1885** 020069.
[16] Ishak MI, Ahmad Shafi A, Rosli MU, Khor CY, Zakaria MS, Wan Abd Rahim WMF and Jamalludin MR 2017 Biomechanical evaluation of different abutment-implant connections – a nonlinear finite element analysis *AIP Conference Proceedings* **1885** 020064.
[17] Ramdan D, Abdullah MZ, Khor CY, Leong WC, Loh WK, Ooi CK and Ooi RC 2012 Fluid-structure interaction investigation in PBGA packaging *IEEE Transactions on Components Packaging Manufacturing Technology* **2**(11) 1786-95.
[18] Ong EES, Abdullah MZ, Khor CY, Loh WK, Ooi CK and Chan R 2014 Fluid-structure interaction analysis on the effect of chip stacking in a 3D integrated circuit package with through-silicon vias during plastic encapsulation *Microelectronic Engineering* **113**(1) 40-9.
[19] Tan JS, et al. 2017 Influence of solder joint length to the mechanical aspect during the thermal stress analysis *AIP Conference Proceedings* **1885** 020063.
[20] Luqman M, Rosli MU, Khor CY, Zambree S and Jahidi H 2018 Manufacturing process selection of composite bicycle's crank arm using analytical hierarchy process (AHP) *IOP Conference Series: Materials Science and Engineering* **318** 012058.

A Novel Green Technology Kaffir Lime Extract as Lizard Repellent

M. A. M. Nawi*[1,2,3], Muhammad Ikman Ishak[2,3], M. U. Rosli[3], M. A. R. M. Ayub[3], C. Y. Khor[2,3], Siti Nor Azreen Ahmad Termizi[2,4] and Noorhidayah Ishak[4]

[1]Energy and Materials Research Group (EMaReG), Faculty of Engineering Technology, Universiti Malaysia Perlis (UniMAP), 02100 Padang Besar, Perlis, Malaysia
[2]Simulation and Modelling Research Group (SimMReG), Faculty of Engineering Technology, Universiti Malaysia Perlis (UniMAP), 02100 Padang Besar, Perlis, Malaysia
[3]Department of Mechanical Engineering Technology, Faculty of Engineering Technology, Universiti Malaysia Perlis (UniMAP), 02100 Padang Besar, Perlis, Malaysia
[4]Department of Chemical Engineering Technology, Faculty of Engineering Technology, Universiti Malaysia Perlis (UniMAP), 02100 Padang Besar, Perlis, Malaysia

Email: * alhafiznawi@unimap.edu.my

Abstract. The ubiquity of lizard in homes and has caused huge problem to human. They are one of the very important agents to transmit Salmonella that cause food poisoning towards human being. Through this research the potential of Kaffir lime peel toward lizard was reported. The process to produce Kaffir lime extract is hydrodistillation extraction method. The repellency test of essential oil will be assess at different concentration (0%, 25%, 50%, 75% and 100%). The duration of observation will be test from 3 to 6 hour duration. The results has shown that the concentration up to 50% and above of essential oil from Kaffir lime peel extraction may reflect the lizards in a state of discomfort.

1. Introduction

Kaffir lime (*Citrus hystrix*) is a very common herb that was use as ingredients in the dishes in South East Asian country such Thailand, Laos, Malaysia, Vietnam and Indonesia. It has distinctive aroma which believe can aid to reduce stress. In fact, not many studies were reported on utilization of Kaffir lime peel as repellent. Kaffir lime consist of several part which is leaves, stem, and peels. For all of that oil extract obtained from peel is higher compare to oil obtained from leaves. Literally, fruit peels is consider as waste for user. Henceforth, this research will mainly focus on implement the citrus products and application of citrus by products [1]. Kaffir lime extract product have huge potential to be use as repellent which is applicable to lizard. Lizard have potential to mechanically carry and transmit salmonella bacteria to human being. Salmonellosis was a type of food poisoning caused by Salmonella bacteria .Salmonella disease can cause someone to feel irksome and if the person was infected during pregnancy the consequence can become worst [2] Through this study researcher will focusing on applying Kaffir lime extract as repellent towards common house gecko (*Hemidactylus frenatus*). Its commonly can be seen around the house and annoyingly this kind of lizard will leave its dropping. Generally, house geckos are insectivores and as for adults its going to eat spiders and other invertebrates.

Content from this work may be used under the terms of the Creative Commons Attribution 3.0 licence. Any further distribution of this work must maintain attribution to the author(s) and the title of the work, journal citation and DOI.
Published under licence by IOP Publishing Ltd

They are quite active at night and will produce vexatious sound that will disturb your sleep. This lizard has been considered as a pest in prevalent of country. Several concentration were studied to test the repellency effect of Kaffir lime extract toward lizard which is 0%, 25%, 50%, 75%, and 100%, v/v respectively. Research about repellents derived from Kaffir lime extract is quite limited at this time. Recently there was a study of repellents effect of Kaffir lime peel against cockroaches [3]. Until now, no studies have shown to directly focusing on application of Kaffir lime extract as lizard repellent. From, this research the Kaffir lime essential oil will be extract by using two methods which is Soxhlet extraction method and Steam distillation method which the aim is to observe which method is more effective. This extraction oil then will be test on lizard from *Hemidactylus frenatus* species under specific condition.

1.1. Research Motivation

One way to repel Lizard or other pest is either by using chemical insect repellent or hire pest Control Company. This chemical repellent and pesticide were used to control, repel or kill pest, for example lizard are considered as disturbance to human whether from its dropping or its presence that can scare children or their annoying sound that can disturb our sleep. Without consumer realise when they consume chemical repellent for a long period of time it will cause bad complication to their own health and even environment. Due to this issue they need to spend a lot of money to cure their health and even government need to add additional budget to solve environment issue that has been caused by this pesticide and repellent.

Toxic repellent spray not only kills pests but also effects the non-targets such as human being and pets. This repellent can be inhaled, ingested or absorbed through skin. Approximately 16 million people living in the United States of America (USA) are sensitive to pesticides. This is because their immune systems are already damaged by pesticides in their living environment. Studies to this illnesses such as asthma, thyroid disorder, and possible fatal illnesses such as cancer has been done by [4]. Furthermore, is not only effect human being but also effect the aquatic life which this repellents substance are highly toxic to fish and other aquatic life [4]. Consumer used toxic chemicals because they was unaware of alternative solution which we have safer repellent which derive from natural ingredients. Generally, this research was aim to provide alternative solution to consumer who are allergic to chemical repellents by developing a more safe and environmental-friendly lizard repellent that was derived from Kaffir lime extract.

2. Methodology

This topic includes detail process flow and methodology that will be apply in this research including the extraction method for kaffir lime peel as well as the methodology to test the repellent activity of kaffir lime extract against lizard (*Hemidactylus frenatus*). It is essential to have high understanding of the process from first process until the last process to avoid any unexpected error in research. Basically, from this research we going to clarify which is the best extraction method that act the best as lizard repellent. Furthermore, we also want to elucidate which concentration is the greatest to use for lizard repellent. For this current study, about six different concentration of kaffir lime extract will be tested which are 0%, 25%, 50%, 75%, and 100% respectively. In addition, this manuscript only describe one method of Kaffir lime extraction only, which steam distillation method.

Besides, to test repellency efficiency for the lizard same method from previous research [3] will be apply. Which we going to use 3 container to test the efficiency. In generating the best outcome of this research all the preparations are made according to the research order to get the optimum result. All the necessary information will be recorded carefully for further analysis that will be use to achieve research objective.

2.1. Material, chemical and equipment

This are the following material that will be used in this experiment from extraction process until efficiency test of repellent.

Table 1. Materials involved.

Material and chemical	Purpose
Kaffir lime peel	Starting material
Water	Medium for steam distillation process
Dichloromethane	Solvent
Acetone	Solvent
Heptane	Solvent
Anhydrous sodium sulphate	To dry the solution

Table 2. Equipment involved.

Equipment	Purpose
Steam distillation apparatus	Extract kaffir lime peel
PVC tubes	Link
Container	Channel for lizard path

2.2. Extraction of Kaffir lime peel

In this sub-section there was three important step that become the most important thing to extract the Kaffir lime peel. Analysis of Kaffir lime peel and repellency of oil extract against lizard. All of this thing will be explained in this sub-topic.

2.2.1. Hydrodistillation extraction.

As for the second extraction method, Kaffir lime essential oil will be extract by using steam distillation process. In general, steam distillation process consist of three main thing which are. Distillation column, a condenser and heating system that illustrated in Figure 1. About, 10 litre of filtered water will be fill inside the column and a packed bed contained of Kaffir lime peels is located at the upper compartment of the column. To heat the system immersion type of heater will be use to raise up the water temperature. Moreover, the water level of the tank will keep unchanged. In other to reduce loss of water during the process the water coolant will recycle until it reached desired water level Henceforth, it will reduce wasted water and also minimize the temperature difference between water tank and water flow-in. Steam will passes through Kaffir lime peel and transport small particles of extract to the condenser. Next, it will change again the liquid mixture that is actually a mixture of oil and water. The mixture will fully occupy the separator that separate between oil and water. The advantage of this automated steam distillation process is to provide proper temperature control to meet user requirement and selection of suitable temperature profiles for associated plant during extraction process with reasonable cost operation. The temperature used to for heating process was 80 celcius. The production yield will be obtain approximately 50 minutes after process started. Two layers will be form, the top layer corresponds to oil while bottom layer is water. The oil sample will be transfer into separating funnel. Petroleum ether will be add into the funnel to form immiscible solvent where the oil and dichloromethane solvent at the bottom layer and water at the upper layer. The oil at lower layer was taken and the liquid at the upper layer will be remove. Then, the evaporating process will be continue by dry up remaining water contained in oil layer.

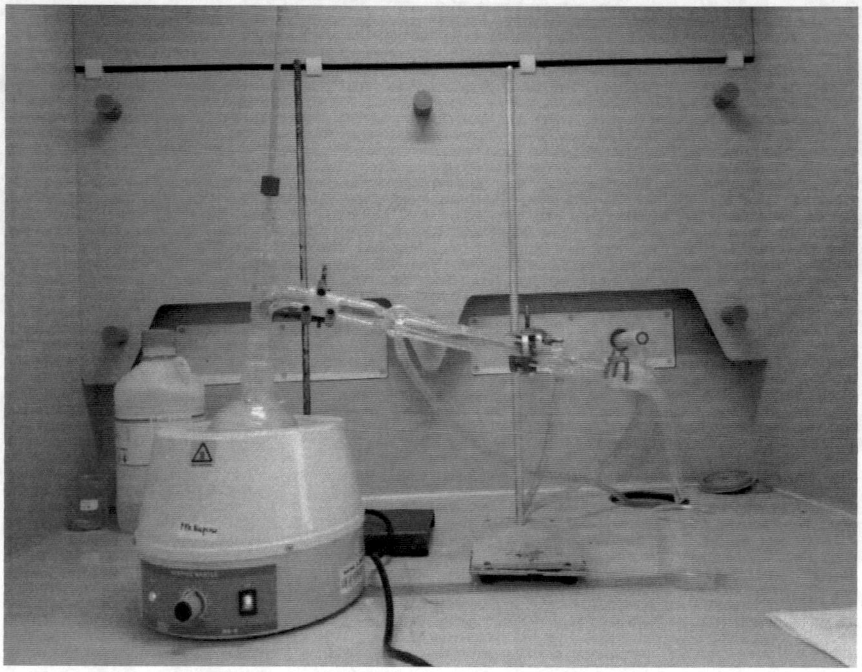

Figure 1. Hydrodistillation process setup.

2.2.2. Repellency activity.
The lizard use is from *Hemidactylus frenatus* species which will be collected around college area. Previous developments [3] in the fields of repellent have led to a different application. A required number of lizard will collect and keep in a container, which was perforated for aeration. Same method as [3] has been applied in the current study. These lizard were starved for 1 day for experiment. For the present study, an apparatus was improvised outside the laboratory with the following compartments: middle plastic chamber from which projects two PVC tubes (15 cm each) as shown in Figure 2. The end of each plastic tube was fitted to two containers and marked as A and B, respectively. The five (5) starved lizard has been introduced into the middle plastic chamber. Container A was taken as the treated side while container B as the control side. Biscuit powder will act as food for the lizard was coated with 1 mL of different concentrations of the essential oil extracts, i.e., 0-100% (*v/v*) oil/ethanol, and placed in container A, while biscuit powder in container B was not treated [3].

2.2.3. Percentage repellency.
The lizard located in the treated and control areas were carefully observed and counted 6 hours after treatment. The extent of repellency of essential oil extracted from the Kaffir lime peel was calculated using the following relationship [5]. From this formula the percentage repellency of Kaffir lime extract against lizard will be obtain. Henceforth, the results from both of this equation is really important because it was interrelated with the objective which is to produce an effective repellent from bio-based product.

$$PC = \left\lfloor 1 - \frac{NT}{NT+NC} \right\rfloor \times 100 \qquad (1)$$

NT : Number of insects trapped in the chemical treated test chamber
NC : Number of insects trapped in the control test chamber
PC : Percentage repellency (Percentage of lizard trapped in control test chamber

Figure 2. Repellency test setup.

3. Result and discussion

Figure 3 shows the increases of the Kaffir lime peel oil concentration in line with the values of the percentage repellency increased. Based on previous study [3] the main reason that this green substance have a strong repellent effect due to the main composition of the oil itself. This will make the lizard is in a state of discomfort. According to [6] the Kaffr lime peel oil has major citronellal component. The citronellal is also the main component in plant-based insect repellent. However, others researcher was claimed that the sense of smell stimulus has been one of the factor disrupt the insect behaviour. And more, the presence of volatile compounds having strong odour could block the tracheal respiration of the lizard leading to their death [6].

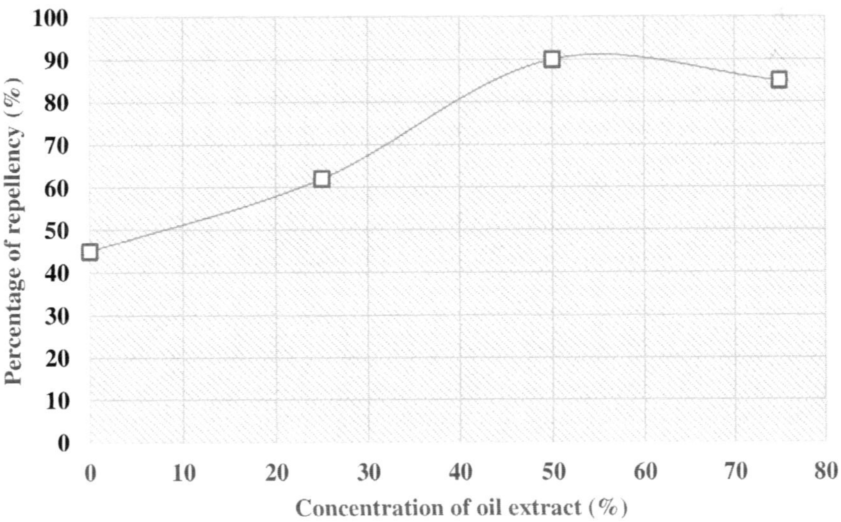

Figure 3. Percentage repellency of Lizard toward Kaffir lime peel oil.

4. Conclusion

As conclusion, the concentration of 50% and above of essential oil derived from Kaffr lime peel have given a direct impact on insect and that's what happened to the lizard in this study. Effect on the lizard due to the different concentration may be considered as novel findings in the potent botanical insecticides against the insect. The result in the current study of Kaffr lime peel would provide an information and knowledge about the effectiveness of insect repellent.

Acknowledgements

The authors acknowledge the facilities and the staff involvement, Sharifah Zati Hanani Syed Zuber from the Department of Chemical Engineering Technology, Faculty of Engineering Technology, Universiti Malaysia Perlis (UniMAP).

References

[1] Haroen U *et al* 2013 *Pak J Nutr* **12(8)** 730-735.
[2] Otokunefor T V et al 2003 *World J Microbiol Biotechnol* **19(5)** 545-548.
[3] Soplah S *et al* 2018 *Sustainable Technologies for the Management of Agricultural Wastes* 131-141.
[4] Roy D N *et al* 2017 *Environmental toxicology and pharmacology* **50** 91-102.
[5] Sakuma M and Fukami H 1985 *Applied Entomology and Zoology* **20(4)** 387-402.
[6] Nerio L S *et al* 2010 *Bioresource Technology* **101(1)** 372-378.

Simulation-Based Optimization of Plastic Injection Molding Parameter for Automotive Car Wheel Fabrication Using Response Surface Methodology (RSM)

M. U. Rosli[1], C. Y. Khor[1], M. A. M. Nawi[1], Haris Ismail[1], Muhammad Ikman Ishak[1]

[1]Simulation and Modelling Research Group (SimMReG), Faculty of Engineering Technology, Universiti Malaysia Perlis (UniMAP), Kampus UniCITI Alam, Sungai Chuchuh, Padang Besar, 02100 Perlis, Malaysia

Email: uzair@unimap.edu.my

Abstract. In the era of revolution industry 4.0, most companies can expect metal-to-plastic conversion to result in cost savings. The objectives of this study are to find the most appropriate material composition and the injection molding process parameter to produce a plastic car wheel as a substitute for metal and aluminum alloy. The Polyetherimide (PEI) with different compositions are simulated using MoldFlow software and optimized the quality of injection molding process parameter by using Central Composite Design (CCD) in Response Surface Methodology (RSM) as to get an optimal response and meets the car wheel requirement specification. The crucial factors are melting temperature, injection time and cooling time. The quadratic model fits the model response. As the results, the optimum values suggested by the software were melt temperature of 370.27°C, 1.8s of injection time and 2599.89s of cooling time for 0% of filler (pure PEI). With small differences error value between solution and simulation, 0.1% of shrinkage and 0.9% for warpage, the results were acceptable.

1. Introduction

In the automotive industry, there are many kinds of parts are available from the metal and plastic. Each of the parts has its own unique features and reason. The automotive industry keeps developing many technologies. It improves the performance, efficiency, and safety of the vehicles [1-5]. Most of the body parts use steel and aluminum alloy as their main or common material [6-10]. Apart from chassis to body part uses steel and metal because as for chassis; they can withstand a great impact if they got into a collision [11,12]. Aluminum alloy commonly used for the selected part in the vehicles to reduce its weight while maintaining its strength [13-15]. For the engine component, most of the manufacturers nowadays use the part from ceramics which is because they can withstand a higher temperature rather than steel [16]. The selection of plastic part will reduce the weight of the current part hence reducing the total weight of the car itself. While reducing the weight of the car, power efficiency will be higher because less power needed to move the car. Simulation based optimization is the most effective way to study the strength of plastic car wheel before the actual fabrication processes begin [17,18]. The injection molding parameters are to determine the most significant factor and accurate molding process parameter setting [19,20].

2. Methodology

2.1 Design

The design model of a car wheel is created in SolidWorks 2016 software. The diameter of the wheel is 14 inches or 35.56 cm and the width is 5 inches or 12.7 cm shown in Figure 1.

Content from this work may be used under the terms of the Creative Commons Attribution 3.0 licence. Any further distribution of this work must maintain attribution to the author(s) and the title of the work, journal citation and DOI.
Published under licence by IOP Publishing Ltd

Figure 1. Wheel design.

2.2 Numerical Factor

The processing parameters that are selected are melting temperature (A), molding temperature (B), and injection pressure (C). The factor selection is shown in Table 1. These parameters are selected because the value of three of them can be adjusted using Moldflow simulation. These factors will analyse the interaction with the response. Minimum and maximum level are based on the combination of the three materials. That means all factors of these three materials are in the range of level 1 to level 3.

Table 1. Level selection.

Column	Factor	Level 1	Level 2	Level 3
A	Melting temperature (°C)	366 °C	385 °C	404 °C
B	Injection time (s)	1.8 s	2.85 s	3.9 s
C	Cooling time (s)	1450 s	2025 s	2600 s

2.3 Categorical Factor

The chosen material is Polyetherimide (PEI). PEI is selected because of its characteristics are capable of high strength and at high temperature. There are three composites of PEI are chosen. PEI without filler, PEI with 10% carbon fiber high strength and PEI with 20% carbon fiber high strength.

2.4 Response Selection

Two responses have been selected which is volumetric shrinkage and warpage. High volumetric shrinkage can cause part warpage, sink marks, critical dimensions that are too small, and internal voids [18,19]. Excessive wall thickness and inadequate packing can both contribute to high volumetric shrinkage in a part. Warpage can cause the surface of the final product do not follow the designated shape [21,22]. Warpage resulted from molded-in residual stresses, which is affected by the shrinkage in the molded part.

3. Results and Discussion

The experiments were conducted to run the simulation using Moldflow software simulation. The experiment was conducted under difference melt temperature, injection time and cooling time to control the response control which is volumetric shrinkage (Y1) and warpage (Y2). The numerical run that generated from Design Expert software totalled of 60 times with 15 of them have same value whilst 45 of them have different value order of processing parameters.

3.1 Results of central composite design (CCD)

There are three minimum values of volumetric shrinkage from all 60 simulations, which is run number 2, 29 and 58. While the value of maximum volumetric shrinkage is at run number 22. For the minimum and maximum values of warpage, the minimum is at run number 19. While the values of maximum volumetric shrinkage are at run number 49 and 58.

Table 2. Result of CCD.

#	Parameters						#	Parameters					
	A	**B**	**C**	**D (%)**	**Y1**	**Y2**		**A**	**B**	**C**	**D (%)**	**Y1**	**Y2**
1	385	2.85	2025	20	12.91	2.68	31	385	1.8	2025	10	13.01	2.64
2	366	3.9	2600	10	11.96	3.27	32	385	2.85	2025	10	12.92	2.69
3	404	2.85	2025	20	13.74	2.16	33	385	2.85	2025	0	12.91	2.69
4	385	2.85	2025	10	12.92	2.69	34	366	1.8	1450	0	12.07	2.90
5	385	2.85	2025	20	12.91	2.68	35	404	2.85	2025	10	13.75	2.16
6	366	1.8	2600	20	12.06	2.89	36	404	1.8	1450	0	13.87	2.40
7	385	2.85	2025	20	12.91	2.68	37	366	3.9	1450	20	11.97	3.27
8	404	3.9	2600	0	13.68	2.33	38	366	1.8	1450	20	12.07	2.89
9	385	2.85	2600	10	12.91	2.68	39	385	2.85	2025	10	12.92	2.69
10	366	1.8	2600	10	12.06	2.89	40	366	1.8	2600	0	12.06	2.89
11	385	2.85	2025	10	12.92	2.69	41	385	2.85	2025	0	12.91	2.69
12	385	2.85	1450	10	12.92	2.69	42	385	2.85	2025	0	12.91	2.69
13	404	3.9	2600	20	13.68	2.32	43	404	3.9	1450	0	13.68	2.33
14	385	2.85	2600	0	12.91	2.68	44	366	2.85	2025	0	12	3.15
15	385	3.9	2025	20	12.85	2.65	45	404	3.9	1450	10	13.68	2.34
16	385	2.85	2025	0	12.91	2.69	46	385	3.9	2025	10	12.85	2.66
17	366	1.8	1450	10	12.07	2.89	47	385	2.85	2025	20	12.91	2.68
18	385	2.85	2025	10	12.92	2.69	48	385	2.85	2025	10	12.92	2.69
19	404	2.85	2025	0	13.74	2.15	49	366	3.9	1450	0	11.97	3.28
20	366	2.85	2025	10	12.01	3.15	50	385	2.85	1450	0	12.92	2.69
21	385	2.85	2025	0	12.91	2.69	51	404	3.9	1450	20	13.68	2.32
22	404	1.8	1450	10	13.88	2.41	52	385	3.9	2025	0	12.85	2.65
23	404	1.8	1450	20	13.87	2.40	53	385	1.8	2025	20	13.01	2.63
24	385	2.85	2025	20	12.91	2.68	54	385	1.8	2025	0	13.01	2.64
25	385	2.85	2025	20	12.91	2.68	55	366	3.9	2600	0	11.97	3.17
26	404	1.8	2600	10	13.85	2.40	56	385	2.85	2025	0	12.91	2.69
27	404	1.8	2600	0	13.85	2.38	57	404	1.8	2600	20	13.85	2.40
28	385	2.85	2600	20	12.91	2.69	58	366	3.9	1450	10	11.96	3.28
29	366	3.9	2600	20	11.96	3.27	59	366	2.85	2025	20	12	3.15
30	404	3.9	2600	10	13.68	2.33	60	385	2.85	1450	20	12.92	2.69

3.2 3D Surface Graph

In this study, the minimum value of the response volumetric shrinkage and warpage are the important goals that need to achieve. The 3D surface plot can display the maximum or minimum value need to get the most optimal results for shrinkage and warpage. The 3D surface response and contour plots of the warpage and shrinkage were plotted using Design Expert software to study the interactive relationship between each factor and responses. In the 3D surface response, the selection of two variables and a constant variable was determined according to the level of sensitivity towards the responses that depended on the perturbation plots. The significant factor for shrinkage for all the three materials is melting temperature and injection time. To find the most optimize results which are the lowest value of shrinkage, lower melt temperature and higher injection time were needed. In Figure 2 shows the most optimal results for shrinkage is at 366°C and 3.9s.

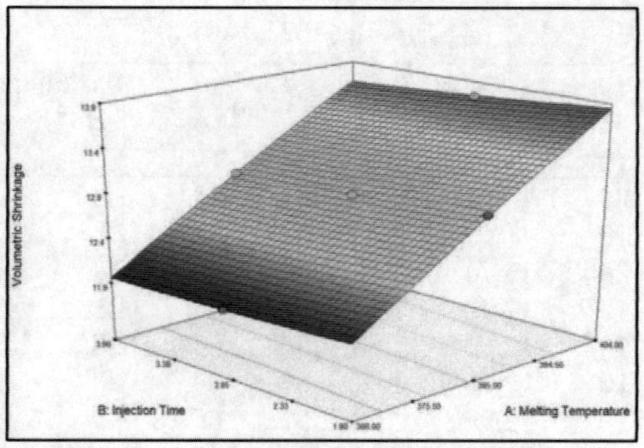

Figure 2. 3D surface graph for Volumetric Shrinkage.

As for warpage, the significant factor for all the three materials is also melting temperature and injection time. To find the lowest value of warpage, melting temperature needs to be high and the injection time likewise needs to be set to high setting. In Figure 3, it shows the lowest value of the warpage is when melting temperature at 404°C and injection time at 3.9s.

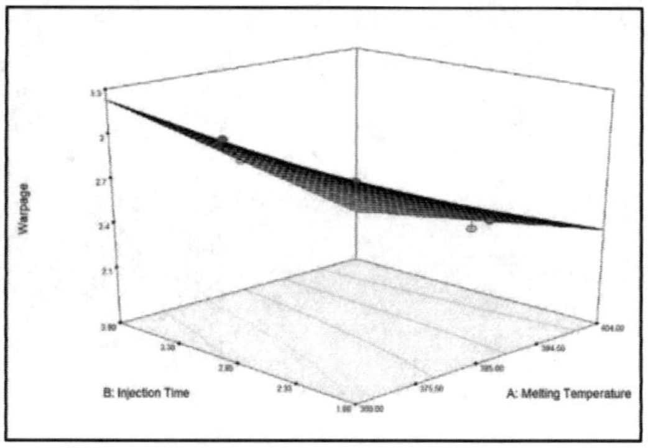

Figure 3. 3D surface graph for Warpage.

3.3 Optimization of design parameters

Table 3 shows the suggested optimized value suggest by Design Expert. The factors are put in range according to the design ranges. The shrinkage and warpage values of selected solutions are 12.27768% and 2.829566mm respectively. To minimize the responses, a simulation by Moldflow software need to be done by using the suggested parameters which are melting temperature = 370.27 c, injection time = 1.8 s mm and cooling time = 2599.89 s with 0% percentage of filler (pure PEI).

Table 3. Suggested optimized value by Design Expert.

A	B	C	D	Shrinkage	Warpage
370.27	1.8	2599.98	No filler	12.27768	2.829566

3.4 Validation of simulation result

The comparisons of responses between the predicted and simulation values are shown in Table 4.

Table 4. Result of error between predicted and simulation.

Response	Predicted	Simulation	% Difference
Volumetric Shrinkage	12.27768%	12.29%	0.1
Warpage	2.829566mm	2.803mm	0.94

The values of the percentage difference between predicted and simulation for volumetric shrinkage and warpage are only 0.1% and 0.9% respectively. The percentage difference between the two responses are small, therefore the result can be accepted. The result of volumetric shrinkage and warpage may not be lowest by individuals, but by combining two of the responses together, the result is the lowest compared to all 60 runs. In summary, we can conclude that the optimum values for each factor were determined by RSM.

4. Conclusions

In conclusion, the most significant factors and interaction relationship of plastic wheel parameter towards the strength of the plastic wheel have been evaluated. The optimum values suggested by DesignExpert software were melting temperature = 370.27 c, injection time = 1.8 s mm and cooling time = 2599.89 s with 0% percentage of filler (pure PEI). The values of the percentage difference between predicted and simulation for volumetric shrinkage and warpage are only 0.1% and 0.9% respectively. With the percentage of error value is small, thus the result is acceptable. In this case, cooling time does not affect the value of volumetric shrinkage and warpage.

Acknowledgments

The research was supported by the Faculty of Engineering Technology, Universiti Malaysia Perlis (UniMAP) for experiment facilities and financially supported under Short Term Grant (STG) through project number 9001-00503.

References

[1] Tian J, Chen M 2014 Waste Management, **34**(2), 458–467.
[2] Guo W, HuaL, MaoH, Meng Z 2012 Journal of Mechanical Science and Technology, **26**(4), 1133–1139.
[3] Rosli M U, Ariffin M K A, Sapuan S M, Sulaiman S 2013 International Journal of Materials, Mechanics and Manufacturing, **1**(1), 32-35.
[4] Rosli M U, Jamalludin M R, Khor C Y, Ishak M I, Jahidi H, Wasir N Y, Faizal W M, Wan Draman W N A, Lailina N M, Ismail R I 2017 MATEC Web Conf. 97, 01039.
[5] Rosli M U, Ariffin, M. K. A., Sapuan, S. M., & Sulaiman, S. 2013 International Journal of Engineering and Technology, **5**(3), 3158-3167.
[6] Technologies M, Republic S 2010 Automotive Materials. Institute of Production Technology, Sloval University of Technology Bratislava, 27–33.
[7] Robles Hernández F C, Cummings S, Kalay S, Stone D. 2011 Wear, **271**(1–2), 374–381.
[8] Bharvez Khan, Rosli M U, Jahidi H, Ishak M I, Zakaria M S, Jamalludin M R, Khor C Y, Faizal W M, Nawi M A M 2017 AIP Conference Proceedings 1885, 020074.
[9] Patil A, Patel A, Purohit R. 2017 Materials Today: Proceedings, **4**(2), 3807–3815.
[10] Tan J S, Khor C Y, Faizal W M, Ishak M I, Rosli M U, Jamalludin M R, Zakaria M S, Nawi M A M, Abdul Aziz M S and Che Ani F 2017 AIP Conference Proceedings 1885, 020063.
[11] Stearns J, Srivatsan T S, Prakash A, Lam P C 2004 Materials Science and Engineering A, **366**(2), 262–268.
[12] Shinde J, Kadam S, Pandit S 2017 International Research Journal of Engineering and Technology (IRJET), **4**(7), 2723–2725.
[13] Luqman M, Rosli M U, Khor C Y, Shayfull Zambree, Jahidi H 2018 IOP Conf. Series: Materials Science and Engineering, 318, 2018, 012058.
[14] Kale H N 2015 International Journal of Advance Research and Innovative Ideas in Education, (5), 241–243.
[15] Ishak M I, Khor C Y, Jamalludin M R, Rosli M U, Shahrin S, Norsyahadah Yeop Wasir, Zakaria M S, Yamin A F M, Noor Diyana Dahlan, Wan Nur A'tiqah Wan Draman 2017 MATEC Web Conf. 97, 01040.
[16] Oró E, de Jong E, Cabeza L F 2016 Journal of Energy Storage, **7**, 131–135.
[17] Rosli M U, Ishak M I, Jamalludin M R, Khor C Y, M.A.M. Nawi, A.K. Mohamad Syafiq 2019 IOP Conference Series: Materials Science and Engineering, **551**(1).
[18] Ibrahim M I F, Rosli M U, Ishak M I, Zakaria M S, Jamalludin M R, Khor C Y, Faizal W M, Nawi M A M, Shahrin S 2018 AIP Conference Proceedings, 2030, 020078.

[19] Khor C Y, Ishak M I, Rosli M U, Jamalludin M R, Zakaria M S, Yamin A F M, Abdul Aziz M S, Abdullah M Z MATEC Web Conf. 97, 01059.
[20] Doyle M M 1996 Doctoral dissertation, Dublin City University.
[21] Ishak MI, Ahmad Shafi A, Abdul Kadir MR and Sulaiman E 2017 Effect of ferrule height and post length on mechanical stress and displacement of endodontically treated maxillary central incisor: a finite element analysis Journal of Medical and Biological Engineering **37**(3) 336-344.
[22] Jizat NM, Mohamad SN and Ishak MI 2017 Resonator graphene microfluidic antenna (RGMA) for blood glucose detection AIP Conference Proceedings 1885 020126.

Paraffin as a Phase Change Material in Concrete for Enhancing Thermal Energy Storage

Rafiza Abd Razak[1,2], Adeline Choo Khang Zhe[1], Mohd Mustafa Al Bakri Abdullah[2], Zarina Yahya[1,2], Subaer Junaidi[3], Khairunnisa Muhamad[1] and Nurul Aida Mohd Mortar[2]

[1] Department of Civil Engineering Technology, Faculty of Engineering Technology, Universiti Malaysia Perlis (UniMAP), Perlis, Malaysia
[2] Center of Excellence Geopolymer and Green Technology (CEGeoGTech), School of Materials Engineering, University Malaysia Perlis (UniMAP), Perlis, Malaysia
[3]Geopolymers & Green Material Group, Physics Department, FMIPA, Universitas Negeri Makassar, Indonesia

Email: rafizarazak@unimap.edu.my

Abstract. Phase change material (PCM) for thermal energy storage (TES) is the material that can absorb energy during heating process as phase change takes place and release energy to environment during cooling process. Nowadays, energy consumption trends in construction building show a significant increase. It is important for reducing energy consumption in building by decreasing the energy demand and providing thermal human comfort inside the building. The objectives of this research are to determine the optimum percentage of PCM in concrete affected by time to achieve high thermal storage performance and evaluate the mechanical properties in different percentage of PCM concrete. Various percentage of 0%, 5%, 10% 15% and 20% of PCM concrete added with Ground Granular Blast Slag (GGBS) as filler were prepared and tested after 7 and 28 days of curing process. The result shows the thermal storage energy effective when integration with PCM in concrete. Besides, the compressive strength, water absorption and density measurement decrease with increasing amount of PCM. In conclusion, 5% of PCM with 30% GGBS added in concrete resulted in better thermal energy storage and compressive strength compared to other percentages of PCM concrete samples.

1. Introduction

Thermal energy storage (TES) is defined as the capable to store heat energy with the help of sensible heat storage or latent heat storage material. Latent heat materials are considered as phase change material (PCM) which is interchange solid and liquid state known as phase change. Therefore, PCM can act as insulator in concrete wall to prevent solar heat entering the building. When cement wall added with PCM material as a construction material in building, it will produce high performance of heat energy storage material. The thermal properties and heat transfer of this TES material will stimulate thermal energy storage in concrete. Phase change material (PCM) as a latent heat storage was use as application in building show had high potential can be used in passive cooling and heating strategies in Europe [1].

Moreover, it has been found that PCM helps indoor temperature to be reduced to maintain desirable temperature for a longer period of time [2]. Thus, PCM can be a useful way to improve the building thermal performance. When the temperature increases at a certain melting point, the PCM will start to absorb the heat from outside which it phases changed from solid state to liquid state. However, when

Content from this work may be used under the terms of the Creative Commons Attribution 3.0 licence. Any further distribution of this work must maintain attribution to the author(s) and the title of the work, journal citation and DOI.
Published under licence by IOP Publishing Ltd

the temperature decreases, the PCM will release energy away and turn back to solid state from liquid state. Paraffin is a type of organic PCM. Paraffin is a white or colourless soft solid, derivable from petroleum, coal or oil shale. In this research, paraffin is chosen as the phase change material to investigate the thermal storage capacity in building.

2. Experimental Procedures

2.1. Raw Materials

Paraffin is the hydrocarbons which having the general formula C_nH_{2n+2}, C being a carbon atom and H is a hydrogen atom. The mixture of solid straight-chain hydrocarbons ranging in melting point from about 48 °C to 66°C. The paraffin used with size not more than 1mm. Besides, the chemical composition of ground granulated blast furnace slag (GGBS) is identically to the Portland cement. It is produced from the blast furnaces. Replacing Portland cement with GGBS can increase the strength of concrete. 30% GGBS can achieved the highest strength in concrete compared to others percentage of GGBS [3].

2.2. Mix Design and Process

The grade of concrete is 35. The concrete mix design calculation is shown in appendix A. The proportion of coarse aggregates and fine aggregates are 65% and 35% of the total proportion of aggregates in concrete mixture. The water-cement ratio is 0.45. All the mix design for GGBS ratio was fixed at 30% of the total weight of cement replacement since the 30% of replacement is the limit and best percentage of replacement to reduce heat hydration and increase strength of concrete [3], [4]. Besides, there are 5 different percentages of paraffin added in proportion to dry materials concrete including 0%, 5%, 10%, 15%, and 20% [5].

For the concrete mixing process, the dry materials such as coarse aggregates, fine aggregates, GGBS, paraffin, and OPC cement were mixed together. Then, the water will be added into the dry mixtures. After the mixture reaching the proper consistency, the fresh concrete was poured by three layers into the cubic moulds and compact it by the rod with dropping the rod. After casting of specimens, the specimens were left inside the room at room temperature for 24 hours. Then, the specimens were immersed in the water tank which filled with water for 7 and 28 days.

2.3. Testing and Analyzing

The thermal energy storage (TES) test was according to EN ISO 9659-1:1993, the stimulation of change between day and night by periodical illumination is used in this experiment. The experiment will be examined the situation with two layered PCM concrete samples represent as the walls of the building materials. Besides, the compressive strength is determined according to BS 1881-116:1983. The compressive strength of concrete is used to test the capable of the concrete samples to resist the forces by compressing it when the weight is applied until concrete samples failure. Moreover, the density measurement test was determined according to BS EN12390-7. The density for 28th day after curing process of concrete specimen is determined. The density was calculated at 28th day of specimen taken out from the water. The density of different percentage of specimens are calculated and compared. The method for determination of water absorption was according to BS 1881-122. The different percentages of PCM in concrete with cubic size samples were obtained and tested at 28th day.

3. Results and Discussion

3.1. Thermal Energy Storage

Based on the result obtained in Figure 1 and Figure 2, it can be seen that the temperature quickly increases while the temperature between the concrete panels and the temperature inside the calorimetric chamber changes slowly. After turning off the lamp, the outside temperature decreases quickly. The temperature changes periodically by switching on and off the light source about every 10 minutes with a temperature wave is generated by about 4°C. Solidification and melting of PCMs in every cycle and night cooling is important to achieve full performance of the PCM storage [6].

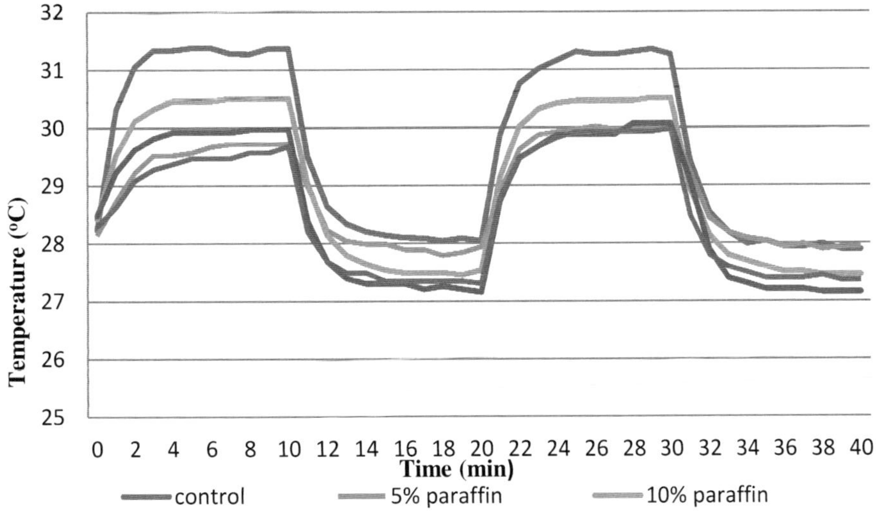

Figure 1. Outer surface concrete temperature versus time in different paraffin in concrete on 7 days.

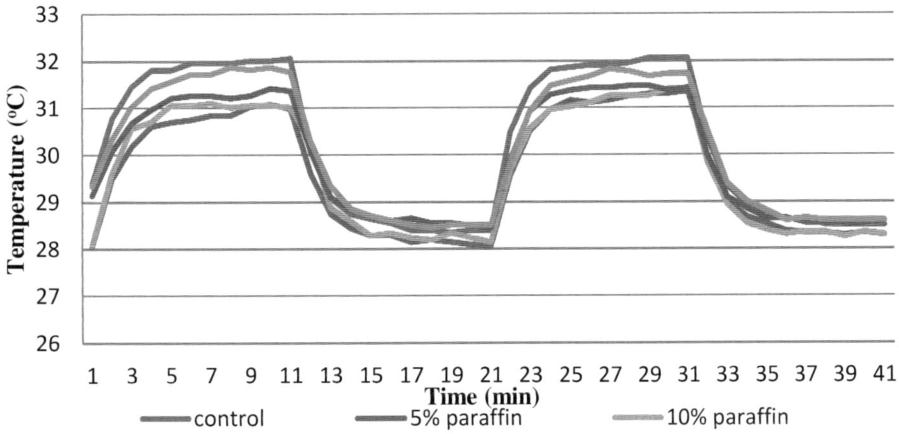

Figure 2. Outer surface concrete temperature versus time in different paraffin in concrete on 28 days.

The 5% paraffin in concrete had the lowest temperature obtained in peak temperature, that means highest level of latent heat storage of paraffin. Similarly, the peak temperature deceasing also mainly result from the sensible heat storage and also effected in the inside temperature decreasing. The 5% paraffin in concrete which was the best in heat reducing performance, showed the reduction of temperature by around 3°C compared to other percentage of paraffin concrete which is around 3.5°C.

Besides, the temperature of the 5% paraffin concrete did not change during 18-20 minute. It is believed that the heat transfer to the paraffin took place during this period. Then there was a temperature rise with significantly increase at about 20-22 minutes. The sudden increase in temperature indicated the completion of phase transition of PCM.

The results showed the peak temperature is lower when addition of paraffin into the concrete. It is due to the integration of paraffin to the concrete causing decrease in thermal conductivity. A reduction of thermal conductivity will affect the rate of heat to absorb and release. The reduction in thermal conductivity is due to increased air content in concrete and voids content when increasing the percentage paraffin in concrete [7].

3.2. Compressive Strength

From Figure 3, GGBS concrete had lower strength at the early ages, however long term strength will be increased [8], [9]. Both the trend lines at days 7th and 28th are decreasing from 0% of paraffin added in concrete until 20% of paraffin with constant 30% GGBS replace cement content. It showed that the compressive strength of the concrete samples decreased as the percentage of paraffin increased. This is due to the inhomogeneity of the mixture. Paraffin can be in solid and liquid phase which will not react with the concrete mixture [10] and more pores observe at high percentage of paraffin in concrete. It shows that the addition of paraffin does not contribute to the strength development, but paraffin can maintain a good thermal storage material.

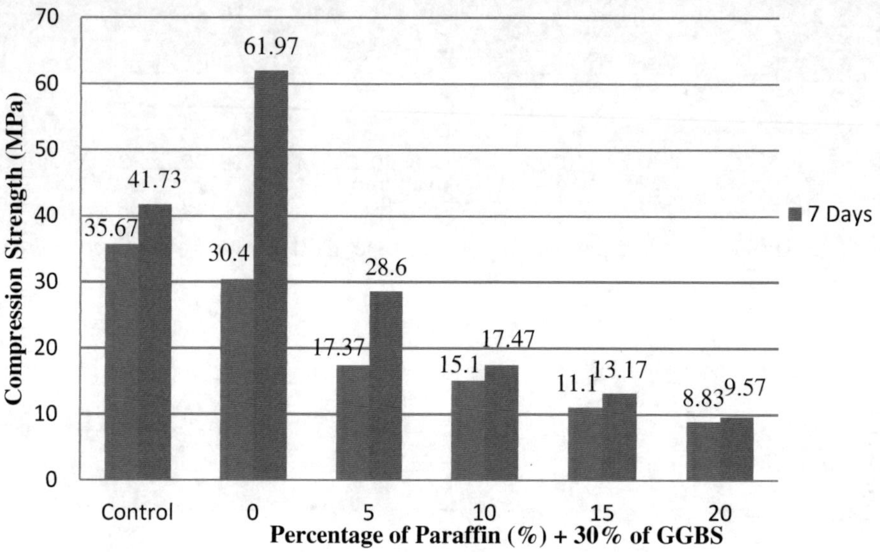

Figure 3. Compressive strength of concrete with addition paraffin at various percentage on 7 and 28 days.

3.3. Density

Figure 4 showed that the density of concrete samples decreasing when the percentage of the paraffin increasing in the concrete sample. It is because the paraffin in concrete containing high porosity values than the control concrete samples and containing air and voids in the concrete samples [11]. Besides, paraffin contains low specific gravity, thus density of concrete with high contain of paraffin were less density compared to normal concrete samples.

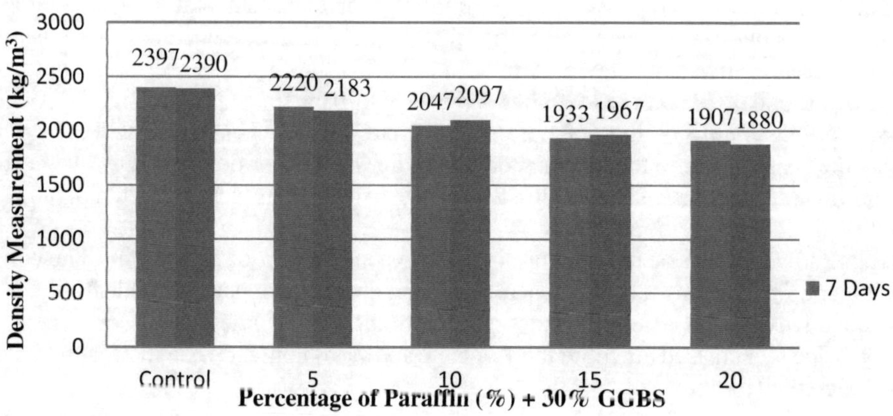

Figure 4. Comparison of the density measurement on different percentage of paraffin in concrete at 7 and 28 days after curing process.

Moreover, 20% of paraffin in concrete sample shows the density closely to be classified as lightweight concrete at 28th days. The density of unit weight concrete which determined in the dry state should not exceed 1840 kg/m³, and is usually between 1400 and 1800 kg/m³ [12]. Thus, the concrete sample contents more than 20% of paraffin can be used as internal wall for insulation purpose to achieve human thermal comfort without spending building energy use.

3.4. Water Absorption

According to the Figure 5, the trend line shows the percentage of water absorption is significantly decreasing until 10% paraffin in concrete. However, the water absorption value shows increasing when the percentage of paraffin in concrete is getting higher. The highest water absorption obtained is control concrete which is 1.79%. It observed that the paraffin concrete showed merely different when incorporation of paraffin in concrete more than 10%. It is because they undergoes the same transition of state of paraffin in concrete. The presence of paraffin in concrete leads to a decreasing in water absorption up to 79% when compared to control concrete due to partially or total occupation of the pores in concrete by the paraffin [13][14].

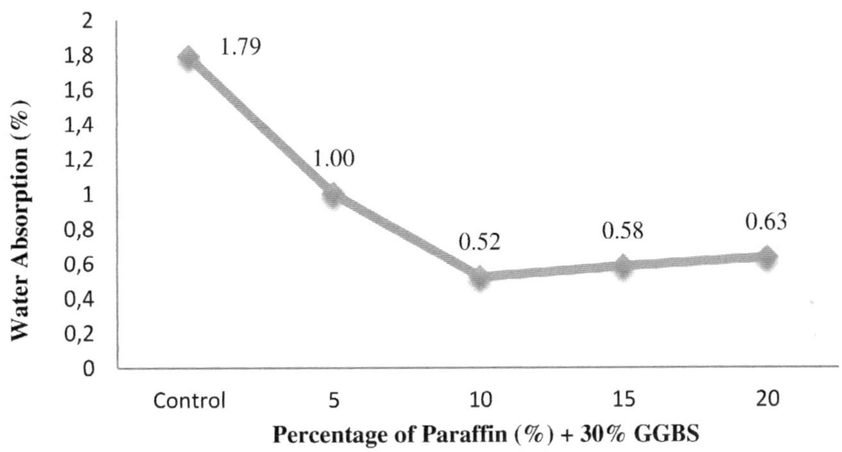

Figure 5. Comparison percentage of water absorption on different percentage of paraffin concrete on 28 days.

4. Conclusion

The various percentage of paraffin in concrete incorporating 30% GGBS in concrete samples were conducted to test the thermal energy storage, compression strength test, density measurement, and water absorption at 7 and 28 days. Based on the thermal results obtained, the 5% paraffin concrete achieve the highest thermal energy storage with lowest peak temperature obtained compared to other percentage paraffin. For the compressive strength test, the lower the percentage of paraffin in concrete, the higher the compression strength obtained. Moreover, the density is decreased when the percentage of paraffin in concrete is increased due to porosity of paraffin concrete increased. The water absorptions of the concrete were in the range of 0.5-2.0% which can be categorized as low. It can be concluded that 5% paraffin + 30% GGBS in concrete is the best percentage of paraffin in concrete. It is because it had highest thermal energy storage, high compression strength, optimized density and water absorption.

References

[1] Isa, M. H. M., Zhao, X. & Yoshino, H. 2010 *Sustainability* **2** 8.
[2] Sarbu, I. & Sebarchievici, C. 2018 *Sustainability (Switzerland)* **10** 1.

[3] Kumar Karri, S., Rao, G. V. R. & Raju, P. M. 2015 *Int. J. Civ. Eng.* **2** 10.
[4] Navarro, L. et al. 2016 *Renewable Energy* **85.**
[5] Norvell, C., Sailor, D. J. & Dusicka, P. 2013 *J. Green Build.* **8** 3.
[6] Cabeza, L. F., Castell, A., Barreneche, C., De Gracia, A. & Fernández, A. I. 2011 *Renewable and Sustainable Energy Reviews* **15** 3.
[7] Hunger, M., Entrop, A. G., Mandilaras, I., Brouwers, H. J. H. & Founti, M. 2009 *Cem. Concr. Compos.* **31** 10.
[8] D. Suresh, K. N. 2015 *IOSR J. Mech. Civ. Eng.* **12** 4.
[9] Fenollera, M., Míguez, J. L., Goicoechea, I., Lorenzo, J. & Álvarez, M. Á. 2013 *Materials* **6** 8.
[10] Razak R. et al. 2015 *Int. J. Molecular Sci.* **16** 5.
[11] Zalba, B., Marín, J. M., Cabeza, L. F. & Mehling, H. 2003 *Appl Thermal Eng.* **23** 3.
[12] Effting, C. et al. 2013 *Proceedings of the Twin Covilha International Conferences on Civil Engineering - Towards a Better Environment, CE 2013 and the Concrete Future, CF 2013*
[13] Deschner, F. et al. 2012 *Cem. Concr. Res.* **42** 10.
[14] Change, P., Pcms, M. & Šavija, B. 2018 *Materials* **11** 654.

IOP Conf. Series: Materials Science and Engineering **743** (2020) 012013 doi:10.1088/1757-899X/743/1/012013

Seawater Exposure Effect on Fly Ash based Geopolymer Concrete with Inclusion of Steel Fiber

Zarina Yahya[1,2,*], Mohd Mustafa Al Bakri Abdullah[1], Leow Yee Jing[2], Long-Yuan Li[3], and Rafiza Abd Razak[1,2]

[1]Center of Excellence Geopolymer and Green Technology, School of Materials Engineering, Universiti Malaysia Perlis (UniMAP), 01000, P.O. Box 77, D/A Pejabat Pos Besar, Kangar, Perlis, Malaysia
[2]Department of Civil Engineering Technology, Faculty of Engineering Technology, Universiti Malaysia Perlis (UniMAP), Sungai Chuchuh, 02100 Padang Besar, Perlis, Malaysia
[3]School of Engineering, University of Plymouth, Plymouth, PL4 8AA, UK

Email: * zarinayahya@unimap.edu.my

Abstract. Concrete is widely used in construction offshore such as concrete floating bridges and sea tank. This research is providing an alternative construction material to replace ordinary Portland cement (OPC) concrete known as geopolymer. The geopolymer concrete was produced by mixing fly ash with alkaline activator and 3% of steel fibre in order to improve the properties of fiber reinforced geopolymer concrete (FRGPC). The effects of aging period in term of strength, changes in weight and carbonation of FRGPC in seawater is investigated and compared with the fiber reinforced concrete (FROPC). The compressive strength obtained for FRGPC were higher than FROPC. The highest compressive strength obtained by FRGPC is 76.87 MPa at 28 days and 45.63 MPa at 28 days for FROPC concrete. The compressive strength was decreased as the period of immersing the concrete in seawater is increased. During the immersion process of both samples in seawater up to 120 days, the carbonation was not detected even though with the existence of steel fibres.

1. Introduction

Green concrete was developed rapidly in recent years in order to reduce carbon footprint especially carbon dioxide (CO_2) emission and limited landfill areas. By definition, green concrete can be defined as concrete that are produced using waste materials in order to decrease the natural resources depletion, environmental pollution as well as energy consumption [1]. Geopolymer has been introduced by Davidovits in 1978 where waste materials that rich in Silica (Si) and Aluminium (Al) was used as source materials. When these source materials activated with alkaline activator solution it possesses binder properties which is similar with ordinary Portland cement (OPC). The by-product materials such as silica fume, fly ash, rice husk ash, slag, red mud and natural minerals like kaolinite, albite, clays, feldspar and stilbite are the example of source materials [2-7]. However, slags, calcined clays and coal fly ashes are the common source materials used by the researchers for geopolymer [8]. Hardened binder is produced when the source materials were dissolved in an activating solution polymerizes into molecular chains and networks.

Geopolymer concrete plays an important role in producing a green environment. A number of 3.6 billion tons of cement have been produced in 2011 showed that the production of cement has increased

Content from this work may be used under the terms of the Creative Commons Attribution 3.0 licence. Any further distribution of this work must maintain attribution to the author(s) and the title of the work, journal citation and DOI.
Published under licence by IOP Publishing Ltd

year by year [9]. Cement production caused the liberation of CO_2 to surrounding due to burning of fossil fuels and the calcination of limestone. Therefore, the emission of CO_2 should be reduced by using the geopolymer concrete instead of OPC. This would help in turning down the global thermostat.

The inherent weakness of plain concrete is caused by the presence of micro-cracks in the concrete aggregate interface [10]. The addition of fibres in the mix can increase the strength of concrete. This is due to the loads transferred by the fibres at the internal micro-cracks. Therefore, adding the fibres which are short, discontinuous and discrete fine fibres of specific geometry dispersing randomly in the conventional concrete form a new composite material to a concrete. The function of fibres is like an aggregate but it has a totally different shape compared to aggregates which is rounded smooth. The fibres act as an interlocking agent assist to interlock and fill the voids between aggregates in the concrete. This will cause the workability of the concrete decreases as the mix is more cohesive and less prone to segregation [11]. Fibres can reduce the development of cracks and transform the properties of concrete with low tensile and impact resistance into a strong composite that has an excellent resistant to crack, good ductility and post-cracking behavior prior to failure. Based on the research by Hake et al. [12], the compressive strength of geopolymer concrete (GPC) increased when the percentage of steel fibre increases from 1% to 3%. Besides, compressive strength and flexural strength are enhanced when the steel fibres are added in alkali activated concrete. An optimum content of 3% of steel fibres is added and the optimum compressive strength is gained [11]. Conversely, Bernal et al. [13] investigated the effect of steel fibre on slag based geopolymer and the result showed reduction in compressive strength but enhanced the flexural strength and splitting tensile. The inclusion of steel fiber also improves the durability performance of alkali activated slag concrete in term of water absorption, capillary and water resistance penetration.

Concrete structures exposed to marine environment are highly vulnerable to corrosion due to chloride penetration and carbonation which compromise the durability of the concrete [14]. Moreover, the existence of various ions such as sulphates and magnesium in seawater contributed to the deterioration of concrete [15, 16]. Geopolymer exposed to seawater is subjected to high concentration of chlorides, sulphates and magnesium ions where it can cause several mechanisms such as crystallization of expansive salts, precipitation of insoluble composites or ionic attacks [17]. It was found that the concentration of ions in typical seawater is about 19,090 ppm of chloride and 2233 ppm of sulphate [18]. Previous researcher has found that during laboratories studies, the performance of geopolymer concrete (GPC) in term of chloride penetration especially chloride diffusion was relatively low compared to OPC concrete [19, 20]. Astutiningsih et al. [21], investigated the performance of fly ash and metakaolin based geopolymer concrete on seawater exposure for 7, 28, 56 and 90 days. It was found that the strength of fly ash geopolymer concrete decrease 20% during 56 days of exposure but did not decreased further until 90 days. Meanwhile, kaolin geopolymer concrete showed better performance due to the inexistence of calcium content. However, for controlled sample (OPC concrete) which consists of high amount of calcium showed major decrement in strength after exposure in seawater. Besides that, permeability also plays an important role in determining the durability of geopolymer in marine environment where denser geopolymer samples will be difficult for aggressive ions to penetrate through the pores. Frequently, previous research works more focused on one type of chemical exposure such as acid, sulphate and magnesium attack.

Vijai et al. [22] conducted a study on geopolymer concrete composite (GCC) that consists of 90% fly ash, 10% OPC and steel fibers. The steel fibers of 0.25% to 0.75% by volume in GCC enhanced the compressive strength, split tensile strength and flexural strength with 0.75% steel fiber contents contributed to optimum performance. The performance of fiber reinforced oil palm shell geopolymer concrete (FROPSGPC) prepared with mixture of slag, palm oil fuel ash (POFA) and 0.5% steel fibers was reported by Islam et al. [23]. The mechanical properties such as splitting tensile and flexural strength increased by about 19%-38% and 13%-44% compared to samples without addition of steel fibers.

This paper reports on the performance of fiber reinforced fly ash based geopolymer concrete (FRGPC) with 3% inclusion of wool steel fibers. The results are compared with fiber reinforced OPC concrete (FROPC), after both of samples were exposed in seawater up to 120 days. The performance of both types of concretes due chloride ingress from the seawater were measures by spraying phenolphthalein solution.

2. Experimental Methods

2.1. Materials

Fly ash was obtained from Cement Industries of Malaysia Berhad (CIMA), Perlis, Malaysia and it was categorized in class F according to ASTM C618-12 with the total content of silicon oxide, aluminum oxide and iron oxides more than 70%. This fly ash was used to produce fiber reinforced geopolymer concrete (FRGPC).

The Portland cement used to produce FROPC as a control concrete in this research is ASTM Type I Portland cement. Table 1 list the chemical composition of fly ash and OPC as determined by X-ray Fluorescence (XRF).

Table 1. Chemical composition of fly ash and OPC.

Composition	Fly Ash Weight (%)	OPC Weight (%)
SiO_2	55.9	20.99
Al_2O_3	27.8	6.19
CaO	3.95	65.96
Fe_2O_3	7.09	3.86
TiO_2	2.25	-
K_2O	1.55	0.60
SrO	0.37	-
ZrO_2	0.13	-
RuO_2	0.24	-

A technical grade of waterglass or sodium silicate solution was purchased from South Pacific Chemical Industries Sdn. Bhd. (SPCI), Malaysia. The waterglass consists of 30.1% SiO_2, 9.4% Na_2O and 60.5% H_2O (modulus SiO_2/Na_2O = 3.2), with the specific gravity of 1.4 g/cm^3 and viscosity of 0.4 Pas (at 20 °C).

Sodium hydroxide (NaOH) powder had 99% purity and was made in Taiwan under the brand name of Formosoda-P. The NaOH solution of 12 Molar is prepared by diluting NaOH powder with distilled water and allowed to cool down to room temperature before it was used. Activator solution was prepared by mixing waterglass with NaOH solution. The fly ash/alkaline activator ratio was fixed at 2.0 and sodium silicate/sodium hydroxide ratio was fixed at 2.5.

River sand was used as fine aggregates with size not more than 4.75mm. Coarse aggregates of maximum sizes 20mm with specific gravity 2.5 and water absorption of 0.17% were used.

High quality of low carbon steel with straight end is the steel fibres used in this research. The dimension of steel fibres with the diameter of 0.13mm and the length of 5mm.

2.2. Preparation of Samples

The percentage of aggregates used to produce FRGPC is 70% from the total concrete mixture. The proportion of coarse aggregates and fine aggregate are 60% and 40% of the total mass of aggregates used. The percentage of steel fibres in the FRGPC is 3% from the total weight of the concrete which is the optimum content percentage with the highest strength gained based on previous study by Meor et al. [11]. The mix proportion and quantity of fibre steels content in each mix is shown in Table 2.

Table 2. Details of mixture proportions for FRGPC and FROPC.

Mix No.	FRGPC	FROPC
Fly Ash (kg/m^3)	8.2	-
Cement (kg/m^3)	-	8.2
Coarse aggregates (kg/m^3)	16.1	16.1
Fine aggregates (kg/m^3)	10.8	10.8
NaOH solution (kg/m^3)	0.9	-
Na_2SiO_3 solution (kg/m^3)	2.4	-
Steel Fiber (kg/m^3)	1.2	1.2
Water (kg/m^3)	-	3.3

Fly ash and alkaline activator solution were mixed thoroughly using a mixer in a dry state. Then, coarse aggregates, fine aggregates and steel fibres were added into the mix until it is in homogeneous state. After the concrete had been thoroughly mixed, the fresh concrete was poured into the mold with dimension of 100mm×100mm×100mm. Each cube was casted in three layers. The fresh concrete was poured into the molds layer by layer for three layers and compacted it using the rod by dropping the rod to the fresh concrete for at least 25 times. Besides, at least 10 times was tapped on the side for each layer. The sample was vibrated by using vibrating table. All the samples were finished by using trowel to smooth the exposed surface of the concrete. The control samples (FROPC) also were casted with the same mix design as FRGPC.

After casting, the samples were preserved in the lab at room temperature for one day before being de-molded. Then, the samples were removed from the molds and cure at room temperature for 3 days. After 3 days of curing, the samples (FRGPC and FROPC) were situated in the water tank which filled with seawater and immersed for 28, 60, 90 and 120 days.

2.3. Testing

The compressive strength for FRGPC and FROPC samples were determined according to British Standard BS EN 12390-3 [24]. The compressive strength was measured when the concrete samples was crushed by using Universal Testing Machine (UTM). For each test, which are at 28th, 60th, 90th and 120th days, three concrete cubes were tested in order to obtain the average compressive strength values.

Carbonation depth in hardened concrete is determined by using phenolphthalein method with accordance to British Standard BS EN 14630 [25]. The test cube samples were taken out from the seawater tank at 28th, 60th, 90th and 120th day and split it by using concrete cutter. After the splitting of concrete sample was done, the freshly split concrete sample was cleaned and the phenolphthalein pH indicator was sprayed on the concrete sample. Then, the changes of colour on the concrete was observed. A purple colour is obtained on the concrete sample for the non-carbonated part of the samples as it is highly alkaline. While, no changes of colour is occurred for the carbonated part of the concrete samples as the low alkalinity of the concrete. The depth of the colourless phenolphthalein region was determined by measuring the depth of total six points which are three points perpendicular to the two edges of the split face of fresh concrete. The depth was measured immediately after spraying the phenolphthalein indicator. The average depth of the region of colourless surface was calculated.

3. Results and Discussion

3.1. Compressive Strength

Figure 1 showed the compressive strength of FRGPC and FROPC after being exposed to seawater for 28 days until 120 days. As it was expected, the compressive strength values for both samples demonstrated a linear decreasing trend as the exposure period increased. The highest compressive strength of 76.87MPa is obtained at 28 days whereas the lowest compressive strength obtained is

51.63MPa at 120 days for FRGPC samples. The trend for compressive strength of FROPC samples is almost identical with FRGPC. The highest compressive strength for FROPC is 45.63MPa at 28 days whereas the lowest compressive strength is 40.1MPa. The strength decrement for FROPC samples was almost constant after 60 days of exposure periods. Generally, the hydration process of OPC rely on lime (calcium) for strength properties where calcium silicate hydrate (CSH) and calcium hydroxide ($Ca(OH)_2$) as the main hydration product. However, due to the depletion of CSH in FROPC hence caused deterioration of strength.

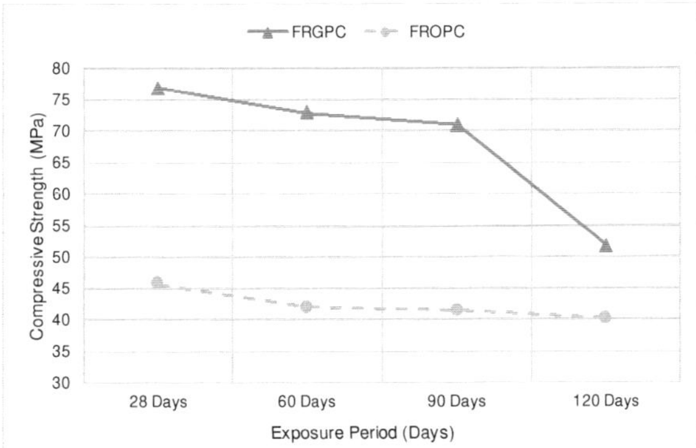

Figure 1. Compressive strength of FRGPC and FROPC.

The results obtained is accordance with the past research, which the compressive strength of geopolymer concrete is decreasing as the immersion time in seawater is increasing. In fact, the compressive strength of geopolymer concrete at 28 day was able to achieve highest compressive strength. This is due to low porosity of geopolymer concrete that resulting to lesser amount of sulfate ions can be ingress into geopolymer concrete [26]. Sulfate ions found in seawater will cause deterioration and loss in strength of concrete. Loss of cohesion of geopolymer concrete sample is caused when it is exposed to an external source of sulfate deteriorated by expansive reactions with magnesium hydroxide ($Mg(OH)_2$) and calcium sulfate ($CaSO_4$).

FRGPC has higher compressive strength than FROPC and this may be due to the lower calcium content in the fly ash that effect the compressive strength of concrete. While, FROPC sample contain more calcium content due to the presence of Calcium Silicate Hydrate gel (C-S-H). The presence of calcium compounds will affect the ability of concrete to resist the chemical agents such as chloride ion when exposed to seawater [27]. Ingression of sulphate ions or chloride ions is not allowed in geopolymer concrete as geopolymer concrete does not have transition zone [28].

3.2. Changes in Weight

Figure 2 displayed the change in weight for both types of samples which are subjected to continuous immersion in seawater. Variation in weight of concrete is calculated with difference in initial weight and weight after period of exposure to seawater. There was no reduction in the weight of all the samples, but weight of samples showed small increment due to the exposed liquid which is the seawater was absorbed by the concrete.

According to Figure 2, the lowest variation of change in weight for FRGPC concrete is 0.4% at 28 days. While, the highest variation of change in weight for FRGPC is 1.23% at 120 days. For FROPC which is the control concrete is same with FRGPC has the lowest variation of change in weight of 2.11% at 28 days. While, the highest variation of change in weight is 4.22% at 120 days for FROPC. The trend line pattern for FRGPC is almost identical to that FROPC which is increasing trend as the age in seawater is increasing. However, the trend line for FROPC is above that of FRGPC. Therefore, the test results show that the variation of change in weight of FROPC is generally higher than that of FRGPC.

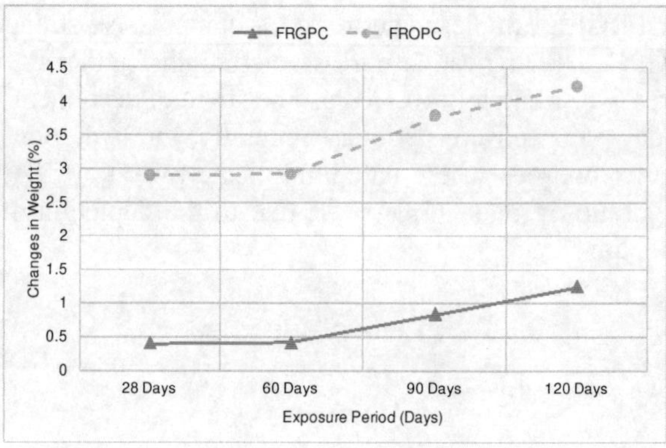

Figure 2. Changes in weight of FRGPC and FROPC.

The FRGPC performed a steady weight change at 28 days and 60 days and gradual increase for the following days. FRGPC has low weight change percentages compared to FROPC indicate that there is no sign of chloride accumulation in the samples [29]. A marginal microstructure alteration is occurred in the concrete sample as FRGPC concrete after 28 days showed a consistent weight change with time. In contrast, a gradual increase in the percentage of weight change with time is shown in FROPC concrete. This may be due to accumulation of chloride crystal in the concrete pores that was bound by the paste and this will increase the final weight of the concrete. There is less chloride accumulation in the FRGPC concrete compared to FROPC concrete [29]. Therefore, the change in weight of the FRGPC concrete is minor in the continuous immersion related to the increase of porosity. FRGPC concrete shows good resistance to acid and very less change in weight throughout the test. The percentage difference for changes in weight between FRGPC concrete and FROPC concrete is 70.62%.

3.3. Carbonation Depth

Table 3 depicts the change in colour on the freshly split surface of concrete between carbonated and non-carbonated region in both types of concrete which are FRGPC concrete and FROPC concrete using phenolphthalein indicator.

Based on the results obtained, there is no effect on carbonation for both FRGPC concrete and FROPC concrete. This is due to penetration depth is mostly found on outer layer of the samples when concrete is subjected to carbonation [30]. Dense microstructure of the sample concrete has slow rate of diffusion cause the carbonation is only occurred on the outer layer or surface of the concrete. Besides, immersion of concrete sample for 120 days might be too short for the carbonation process to occur. The strength of the concrete is affected by carbonation. Hence, the strength could be higher when the carbonation is not prominent. Furthermore, moisture content of concrete significantly affects the rate of carbonation [31]. The low moisture content of the concrete cause the carbon dioxide diffuses more rapidly. This is due to the diffusion if gas in air is rapid than in water. However, the diffusion of carbon dioxide is very low if the pore of concrete is fully filled with water when the concrete samples is fully immersed in seawater.

The carbonation of FRGPC concrete is different with FROPC concrete where there is no obvious boundary between the area of coloured region and colourless region compared to FROPC concrete. Change in colour on FROPC concrete is more obvious and rapid than geopolymer concrete after phenolphthalein indicator is sprayed on a freshly split concrete. Based on the research by Neville [32], occurrence of partial carbonation will cause unobvious carbonation 'front' on the freshly split concrete surface. The calcium hydroxide, $Ca(OH)_2$ and C–S–H gel are produced by FROPC concrete whereas FRGPC concrete produced $[M_z(AlO_2)_x(SiO_2)_y{\cdot}nMOH{\cdot}mH_2O]$ gel [33]. With the reason of $Ca(OH)_2$ and C-S-H are not present in FRGPC concrete, NaOH is the only constituent that can be carbonated.

Sodium carbonate, Na_2CO_3 is formed when NaOH reacts with CO_2 and releasing water [31]. The carbonation of NaOH is only partial carbonation while $Ca(OH)_2$ or C-S-H is a full carbonation. The overall colour of non-carbonated FROPC concrete was much darker than in the non-carbonated FRGPC concrete and take shorter time for changes in colour after it was sprayed with phenolphthalein indicator.

Table 3. The change in color between carbonated and non-carbonated region in FRGPC and FROPC.

Days	FRGPC	FROPC
28		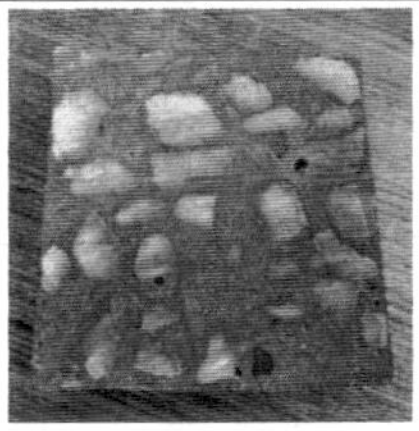
60		
90		
120		

4. Conclusion

The trend for compressive strength of FRGPC is in decreasing trend as the time of seawater immersion increased. The highest compressive strength obtained by FRGPC concrete is 76.87MPa at 28 days whereas the lowest compressive strength obtained is 51.63MPa. This is due to porosity of concrete that cause ingression of sulfate ions into concrete in seawater. The longer the immersion time in seawater, the higher the amount of sulfate ions can be ingress into geoplymer concrete.

FRGPC concrete showed the increasing trend for changes in weight. This is due to the chloride penetration that has been bound into the paste to increase the density of concrete when immersed in seawater.

There is no effect for both FRGPC and FROPC concrete on carbonation progress. This is due to the time exposure to seawater is too short for the carbonation progress. The change in colour after phenolphthalein solution is sprayed is lighter or not clear and slower for GPC concrete when compared to OPC concrete with the reason of partial carbonation is occurred for GPC concrete.

Acknowledgments
The authors would like to acknowledge the support received from The Academy Science of Malaysia via Newton Mobility Grants (NI170199).

References

[1] Jin R, and Chen Q 2013 *49th ASC Annual Int. Conf. Proceedings.*
[2] Abdullah M M A B, Hussin K, Bnhussain M, Ismail K N, Yahya Z, Razak R A 2012 *International Journal of Molecular Sciences* ***13*(6)** 7186–7198.
[3] Heah C Y, Kamarudin H, Bakri A M M Al, Luqman M, Nizar I K 2011 *Australian Journal of Basic and Applied Sciences* ***5*(7)** 1026–1035.
[4] Liew Y M, Heah C Y, Li L yuan, Jaya N A, Abdullah M M A B, Tan S J, Hussin K 2017 *Construction and Building Materials* ***156*** 9–18.
[5] Al Bakri Abdullah M M, Jamaludin L, Kamarudin H, Binhussain M, Ruzaidi Ghazali C M, Ahmad, M I 2013 *Advanced Materials Research* ***686*** 227–233.
[6] Heah C Y, Kamarudin H, Bakri A M M Al, Luqman M, Nizar I K 2011 *Australian Journal of Basic and Applied Sciences* ***5*(7)** 1026–1035.
[7] Abdullah, M.M.A.B., Nordin, N., Tahir, M.F.M., Kadir, A.A., Sandu, A.V., *International Journal of Conservation Science* **7(3)**, (2016).
[8] Francisco J 2013 Master of Science Thesis, Florida Atlantic University.
[9] Armstrong T 2012 *Int. Cem. Review.*
[10] Mahadik S, Kamane S and Lande A 2014 *International Journal of Advanced Structure and Geotechnical Engineering*, **3(04)** 388-392.
[11] Meor A F, Mustafa A M M, Khairul Nizar I, Ratnasamy M and Ramadhansyah P J 2016 *IOP Conference Series: Materials Science and Engineering* **133** 0120145 1-6.
[12] Hake S, Bade M, and Katkar C 2016 *International Journal of Engineering Research and Applications* **6(5)** 7-10.
[13] Bernal S, Gutierrez R B, Delvasto S, and Rodriguez E 2010 *Constr. and Build. Mater.* **24** 208-214.
[14] Suryavanshi A, and Swamy R N 1996 *Cem. and Conc. Res.* **26 (5)** 729 -741
[15] Reddy D V , Edouard J B, Sobhan K 2012 *J. of Mater. in Civil Eng.***25 (6)** 781-787
[16] Vera R, Villarroel M, Carvajal A M , Vera E, and Ortiz C 2009 *Materials Chemical Physical* **114 (1)** 467-474
[17] Reddy D V, Edouard J-B, Siobhan K, and Tipnis A 2011 *9th Latin American and Caribbean Conference for Engineering and Technology.*
[18] Lanthanum M, Cohen M, and Olek J 2006 *Cem. Conc. Res.* **36** 2132-2137.
[19] Kupwade-Patil K, and Allouche E N 2012 *J. of Mater. in Civil Eng.* **25 (10)** 1465 – 1476.
[20] Shaikh F U 2014 *Adv. in Conc. Constr.* **2 (2)** 109-123.
[21] Astutiningsih S, Nurjaya D W, Ashadi H W, and Swastika N 2010 *Advanced Science Technology* **69** 92-96.
[22] Vijai K, R. Kumutha R, and Vishnuram B G 2012 *Asian J. of Civil Eng. (Build. and Hous.)* **13 (3)** 381-389.
[23] Islam A, Alengaram U J, Jumaat M Z, Ghazali N, Yusoff S, and Bashar I 2017 *Constr. and Build. Mater.***152** 964-977.
[24] BS EN 12390-3 2009 *British Standards Institution*
[25] BS EN 14630 2006 *British Standards Institution.*
[26] Chindaprasirt P, and Chalee W 2014 *Constr. and Build. Mater.* **63** 303-310.

[27] Sotya A,Marta N, Wibowo A, and Niken S. 2010 *Advances in Science and Technology* **69** 92-96.
[28] Okoye F, Prakash S, and Singh N 2017 *J. of Cleaner Product* **149** 1062-1067.
[29] Olivia M 2013 *Hokkaido University Collection of Scholarly and Academic Papers: HUSCAP* 18.
[30] Ashraf W 2016 *Constr. and Build. Mater.* **120** 558-570.
[31] Adam A 2009 *Thesis.*
[32] Neville A, 1996 *Properties of concrete 4th edition New York: Wiley and Sons Inc.*
[33] Law D, Adam A, Molyneaux T , Patnaikuni I and Wardhono A 2014 *Mater. and Struc.* **48** 721-731.

IOP Conf. Series: Materials Science and Engineering **743** (2020) 012014 doi:10.1088/1757-899X/743/1/012014

Synthesis and Characterization of Fly ash based Geopolymer Ceramics: Effect of NaOH Concentration

Romisuhani Ahmad[1*], Wan Mastura Wan Ibrahim[1], Mohd Mustafa AlBakri Abdullah[2], Andrei Victor Sandu[3], Nurul Aida Mohd Mortar[2], Noratikah Hashim[1] and Warid Wazien Ahmad Zailani[4]

[1]Faculty of Engineering Technology, Universiti Malaysia Perlis (UniMAP), Perlis, Malaysia
[2]Center of Excellence Geopolymer and Green Technology, School of Materials Engineering, Universiti Malaysia Perlis (UniMAP), Perlis, Malaysia
[3]Gheorghe Asachi Technical University of lasi, Faculty of Materials Science and Engineering, lasi, Romania
[4]Faculty of Civil Engineering, Universiti Teknologi MARA, Shah Alam, Selangor

Email: romisuhani@unimap.edu.my

Abstract. Ceramic materials have demonstrated impressive performance under severe conditions in a wide number of areas including health, energy and environment, transportation and electronic applications. In order to meet the growing demand of specific properties of ceramics, the use of geopolymer method were introduced as an alternate way by providing new material to fabricate geopolymer as ceramic precursor and moving towards green technology. Geopolymer ceramic was produced using fly ash as source materials. Different molarities of sodium hydroxide solution in the range of 8 M to 14 M were used in the preparation of fly ash geopolymer ceramic. The ratio of fly ash/alkaline solution and alkaline activator were kept constant at 2.0 and 2.5 respectively. Physical and mechanical properties such as density, flexural strength and shrinkage were determined. The optimum flexural strength of 61.6 MPa is obtained at the sodium hydroxide concentration of 12 M with the density of 2.21 g/cm^3. The formation of almost fully crystalline phase with appearance of albite contribute to the flexural strength of fly ash geopolymer ceramic. Microstructure image result shows that a presence of pores due to sintering effect and smoother matrix which gives denser structure as supported by the excellent density.

1. Introduction

Ceramic are non-metallic inorganic material with high melting point and high hardness. Typical ceramics are thermal and electrical insulators with good strength in compression and good chemical stability. Ceramics play are critical role in a wide array of electronic, optical, magnetic and energy-related technologies [1]. Advanced ceramics play important role in providing thermal insulation and higher temperature properties. Several method are available for fabricating ceramic materials and fabrication method is important since it can affect the properties of product. More recently, geopolymers have been suggested as a low cost ceramic binder with application in some ceramic and high-technology applications [2]. Besides, the alternative way to produce ceramic materials using geopolymer since the amorphous to semi-crystalline behaviour of geopolymer will transform into

Content from this work may be used under the terms of the Creative Commons Attribution 3.0 licence. Any further distribution of this work must maintain attribution to the author(s) and the title of the work, journal citation and DOI.
Published under licence by IOP Publishing Ltd

crystalline ceramic phase upon heating. Furthermore, geopolymer can design the chemical composition of final product and directly converted into final structural ceramic part of interest.

The properties and characteristics of geopolymer depending on the raw materials and the processing conditions used. Davidovits (2002) was the first that introduced geopolymer technology. Geopolymer is an inorganic polymeric material that contain silicon (Si) and aluminum (Al) that formed with alkaline solution or alkaline silicate solution at higher temperature level [3]. Geopolymer shows numerous outstanding properties, for example high compressive strength, low shrinkage in addition to corrosive and fire resistance [4]. The strength of geopolymers is dependent on the nature of the source materials, with geopolymers synthesized from calcined source materials such as slag, fly ash and metakaolinite exhibiting a higher compressive strength than those from non-calcined materials, for example kaolinite and naturally occurring minerals. On the other hand, the abundant availability of fly ash worldwide creates opportunity to utilize this by-product of burning coal to manufacture ceramic [5-8].

The geopolymerization process need alkaline solution which consist of sodium hydroxide (NaOH) or potassium hydroxide (KOH) and glassy silicate that consist of sodium silicate or potassium silicate [9]. The most frequent alkaline solution used in geopolymerization are sodium hydroxide (NaOH) and sodium silicate (Na_2SiO_3) commonly used to be incorporated with source materials that in rich in silica and alumina due to it cost and availability [10]. In production of geopolymer material, the concentration of sodium hydroxide (NaOH) solution play important roles. The solubility of aluminosilicates in alkaline solution depends on the charge of aluminum and silicon as well as on hydroxide concentration. In this study, the fabrication of geopolymer ceramic is carried out by using fly ash and powder metallurgy method. Thus, this paper study the effect of sodium hydroxide (NaOH) solution on fly ash based geopolymer ceramic.

2. Experimental

2.1 *Material*

Fly ash is a fine grey powder that consist of spherical, glassy particles that are produced as a byproduct in coal fired power stations. The fly ash used in this study was obtained from Manjung power station, Lumut, Perak, Malaysia. The physical forms of fly ash used were in powder state with the mean particle size (X50) of 46.22 µm. The components of fly ash typically include SiO_2, Al_2O_3, CaO and Fe_2O_3, which exists in the form of amorphous and crystalline oxides or various minerals as shown in Table 1.0.

Table 1. Percentage of chemical composition of fly ash class type F.

Composition	SiO_2	Al_2O_3	Fe_2O_3	CaO	TiO_2	K_2O	SrO	SO_3	RuO_2	ZrO_2	LOI
Percentage (%)	55.9	27.8	7.093	3.95	2.25	1.55	0.371	0.33	0.24	0.134	0.382

The alkali activator used were a combination of sodium hydroxide (NaOH) and sodium silicate (Na_2SiO_3) solutions. The NaOH used is Formosoda-P brand name made in Taiwan supplied by Formosa Plastic Corporation, Taiwan while Na_2SiO_3 was supplied by South Pacific Chemicals Industries Sdn. Bhd. Malaysia with a chemical composition of 30.1 % SiO_2, 9.4 % Na_2O and 60.5 % H_2O as stated by the supplier upon purchased.

2.2 *Experimental method*

The alkaline activator solution was formed by mixing Na_2SiO_3 to NaOH solution at a ratio of 2.5 until clear solution was obtained for 24 hours. An alkaline activator with a combination of NaOH and Na_2SiO_3 was prepared just before mixing with fly ash to ensure the reactivity of the solution. In order to produce geopolymer paste, fly ash powder was mixed with the alkaline solution with 2.0 solid to

IOP Conf. Series: Materials Science and Engineering **743** (2020) 012014 doi:10.1088/1757-899X/743/1/012014

liquid ratio and stirred well until no clumps were visible. The mixture then mixed well for a few minutes and poured into high-density polyethylene (HDPE) mould. Then, the samples were put into oven at 60 °C for 24 hours for the curing process.

To obtained fine powder, the samples were sieved manually to a 150 micron. In order to produce geopolymer ceramic, the sieved powders were compacted using uniaxial isostatic pressing with a 12-mm-diameter cylindrical stainless steel die at 5 tonnes for 2 minutes. After being shaped into a pellet, the samples were sintered at a fixed temperature of 1000 °C with soaking time of 180 minutes and heating and cooling rate of 5 °C/minutes.

2.3 Testing

2.3.1 Flexural Strength

Flexural strength measurements of fly ash geopolymer ceramics were conducted on specimens (7 mm x 5 mm x 52 mm) using a three-point-bending fixture on an Instron- 500 tester machine with a span length of 30 mm at a crosshead speed of 0.5 mm/min. The three-point test configuration exposes only a very small portion of the specimen to the maximum stress. Minimum of five samples were subjected to flexural strength testing in accordance with ASTM C1161-02c.

2.3.2 Density

The theoretical density was measured using densitometer (DH-300X DahoMeter. A cylinder shaped sample with a diameter of 12 mm and a length of 5 mm were used in this testing.

2.3.3 Morphology Analysis

The morphology characterization in this study was done by scanning electron microscopy (SEM) using Hitachi TM3000. . Geopolymers are non-conductive materials, therefore samples were coated with platinum using Auto Fine Coater JEOL JFC 1600 model as a conductive layer to obtain better quality of microstructural images. Imaged in a scanning mode using conventional SEM techniques (magnification used were 500x and 1000x).

2.3.4 Phase Analysis

The phase analysis and crystallinity test was performed on fly ash, fly ash geopolymer and fly ash geopolymer ceramic using XRD 6000, Shimadzu diffractometer. The test was performed at operating conditions of 40 kV and 30 Ma using Cu-Kα radiation. In collecting data sets, the range diffraction were performed from 10 ° - 80 ° 2θ with a step size of 0.02 °at a rate of 2 °/ minutes. To analyses the results of diffraction patterns, The X'Pert High Score Plus software.

3. Results and Discussion

3.1 Flexural Strength

Figure 1 shows the flexural strength of fly ash geopolymer ceramic with different NaOH concentration. The flexural strength increased with the increasing of NaOH concentration and unsintered sample was expected to have a lowest flexural strength at 3.13 MPa compared to the sintered samples. As for sintered samples, the flexural strength at 8 M having the lowest strength among the other NaOH concentration which at 48 MPa due to lower alkalinity and insufficient of Na^+ and OH^- at for the dissolution of fly ash. Therefore, would cause poor geopolymerization due to low leaching rate of Al 3^+ and Si 4^+ from fly ash thus lowering the flexural strength of the samples [11]. The maximum of flexural strength of 61.60 MPa was achieved at 12 M due to sufficient alkalinity (Na concentration) for activating the fly ash geopolymer. However, the flexural strength started to decrease at 14 M with 49.73 MPa contributes the excess of Na^+ cation could weaken the structure [12].

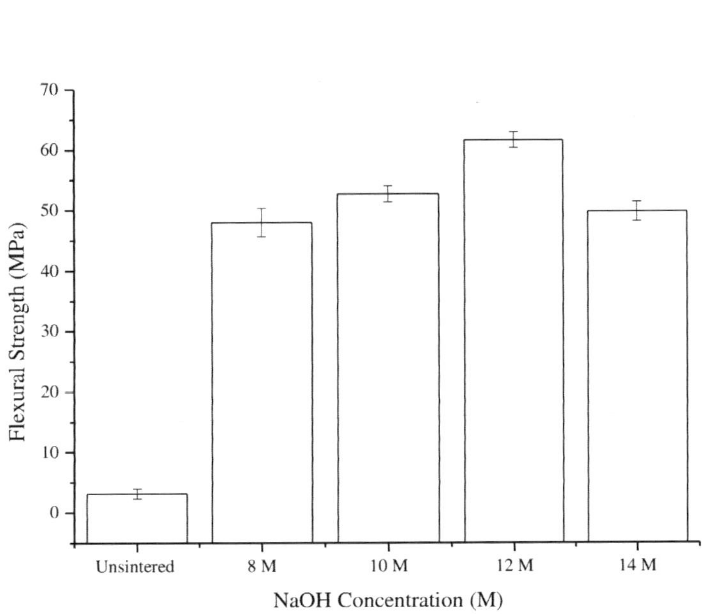

Figure 1. Flexural strength of fly ash geopolymer ceramic at different NaOH concentration.

3.2 Density

Figure 2 represents density of fly ash geopolymer ceramic as function of NaOH concentration. The densities vary in the range of 1.85 to 2.25 kg/m^3. Increasing of NaOH concentration from 8 M to 14 M increased the density of the fly ash geopolymer. As for sintered samples, the density at 8 M having the lowest density among the other NaOH concentration which at 2.05 g/cm^3 attributed to the less concentration of NaOH in the activating solution to produce a dense structure of fly ash geopolymer ceramic at low molarity. Concentration of NaOH at 14 M shows the highest density which is at 2.25 kg/m^3 while the lowest density was depicted for unsintered sample at 1.85 g/cm^3. High concentration of NaOH caused the greater dissolution process due to the leaching of alumina and silica which contribute to the increases in geopolymerization reaction and also produced denser structure [10].

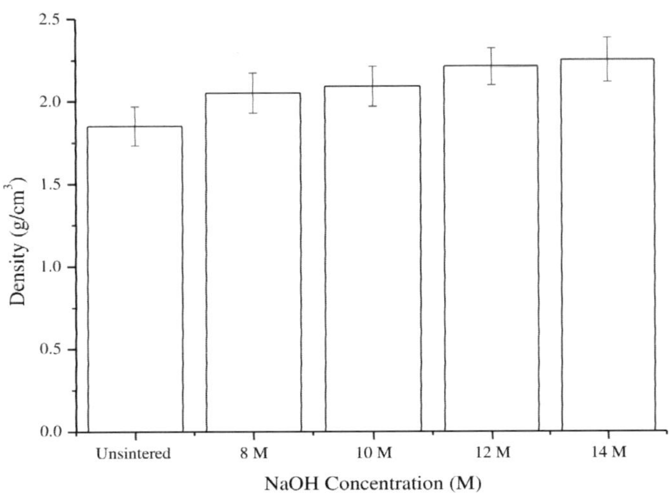

Figure 2. Density of fly ash geopolymer ceramic at different concentration of NaOH.

3.3 Morphology Analysis

Figure 3 shows the microstructure of the fly ash geopolymer and fly ash geopolymer ceramic at1000x and 2000x magnifications. According to illustrated micrograph, the microstructure of fly ash geopolymer are porous and heterogenous mixture of non or partly-reacted fly ash grains, consist of a dense continuous gel-like with microcracks and micropores which can be clearly observed on the surface. The micro-crack was cause by the drying shrinkage, a physical property, inherent of the gel. The gel shrinkage occurs because the removal of a water during polycondensation creates capillary tension within the gel matrices [13]. While the microstructure image of fly ash geopolymer ceramic at 12 M NaOH concentrations shows an appearance of pores. The samples appeared to have a smoother geopolymer matrix and denser as the concentration increases indicate that geopolymerization has occurred well contributes to high strength as proved by density and flexural strength results as well as the effect of the densification process during sintering.

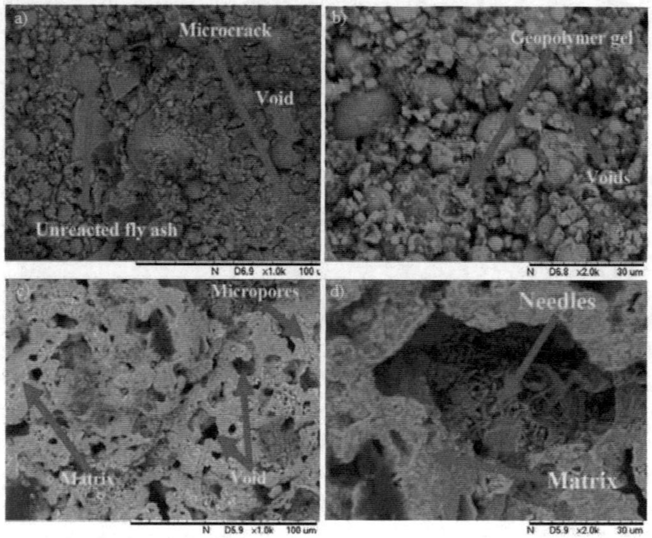

Figure 3. Microstructure images of fly ash geopolymer a) 1000x, b) 2000x and fly ash geopolymer ceramic at 12 M of NaOH concentration with different magnification c) 1000x, d) 2000x.

3.4 Phase Analysis

For the phase analysis, the sample with optimum flexural strength at 12 M of NaOH is shown in Figure 4. The broad reflection pattern of fly ash geopolymer begins to show a sharp peak in fly ash geopolymer ceramic indicates the transformation of amorphous to crystalline phase upon sintering. Fly ash consist of minerals such as quartz, mullite and hematite. The dilution effect cause the disappearance of quart, mullite and hematite peaks. During geopolymerization, new crystalline phase of formed. The crystalline phases (quartz, mullite) detected in the initial material remained apparently unaltered with activation. The use of NaOH as activators in geopolymers lead to the formation of quartz structure [14]. Geopolymers with low NaOH concentration shows low intensity of quartz formation. These indicate that the formation of geopolymer products were slow giving a low flexural strength. At higher NaOH concentrations, the quartz peak was detected to be high, gave a high level and better of geopolymerization has occurred [10]. A higher NaOH concentration have better result due to ability of dissolving the fly ash particles during geopolymerization process.

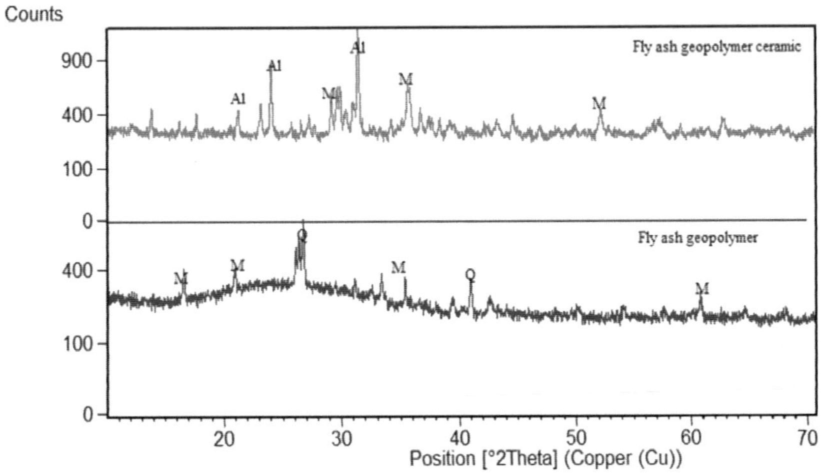

Figure 4. Phase analysis of fly ash geopolymer and fly ash geopolymer ceramic at 12 M NaOH concentration (Q= Quartz, M= Mullite, Al= Albite).

4. Conclusion

The aims of this paper is to synthesize and characterize fly ash geopolymer ceramic with the optimum sodium hydroxide based on flexural strength. The results concluded that the 12 M is the optimum NaOH concentration to produce highest flexural strength at 61.60 MPa for fly ash geopolymer ceramic. The broad, amorphous peak showed by phase analysis indicates that geopolymerization has occurred. Microstructure image shows smoother and having geopolymer matrix which gives denser structure as supported by excellent density.Thus, it can be conclude that the concentration of NaOH solution affects the dissolution ability of the fly ash as the higher of concentration provides a better dissolving ability of raw materials and accelerate the monomer condensation, hence increase the geopolymer bonding strength.

5. References

[1] Askeland, D. and Wright, W. (2016). The science and engineering of materials, 7th ed. Boston: Cengage Learning, pp. 532–564.

[2] Van Jaarsveld, J. G. S., Van Deventer, J. S. J., & Lukey, G. C. (2003). The characterisation of source materials in fly ash-based geopolymers. Materials Letters, 57(7), 1272–1280.

[3] Tabassum, R. K., Khadwal, A., & Ash, F. (2015). Effect of sodium hydroxide concentration on various properties of geopolymer concrete, 0869(10), 28–32.

[4] Bakri, A. M. M. Al, Kamarudin, H., Bnhussain, M., Nizar, I. K., Rafiza, A. R., & Zarina,Y. (2012). The processing, characterization, and properties of fly ash based geopolymer concrete, 90–97.

[5] Yahya, Z., Abdullah, M. M. A. B., Hussin, K., Ismail, K. N., Sandu, A. V., Vizureanu, P., & Razak, R. A. (2013). Chemical and physical characterization of boiler ash from palm oil industry waste for geopolymer composite. Revista de Chimie, 64(12), 1408–1412.

[6] Wazien, A.Z.W., Abdullah, M.M.A.B., Abd Razak, R., Rozainy, M.A.Z.M.R., Tahir, M.F.M. Strength and Density of Geopolymer Mortar Cured at Ambient Temperature for Use as Repair Material, (2016) IOP Conference Series: Materials Science and Engineering, 133 (1)

[7] Heah, C. Y., Kamarudin, H., Bakri, A. M. M. Al, Luqman, M., & Nizar, I. K. (2011). Potential Application of Kaolin Without Calcine as Greener Concrete : A Review, *Australian Journal of Basic and Applied Sciences,*5(7), 1026–1035.

[8] Liew, Y. M., Heah, C. Y., Li, L. yuan, Jaya, N. A., Abdullah, M. M. A. B., Tan, S. J., & Hussin, K. (2017). Formation of one-part-mixing geopolymers and geopolymer ceramics from geopolymer powder. Construction and Building Materials, 156, 9–18.

[9] Muhammad Faheem, M.T., Mustafa Al Bakri, A.M., Kamarudin, H., Ruzaidi, C.M., Binhussain, M., Izzat, A.M. The relationship of Na2SiO3/NaOH ratio, Kaolin/Alkaline activator ratio and sand/kaolin ratio to the strength of kaolin-based non load bearing Geopolymer Brick(2013) International Review of Mechanical Engineering, 7 (1), pp. 161-166.

[10] Sani, N. A. M., Man, Z., Shamsuddin, R. M., Azizli, K. A., & Shaari, K. Z. K. (2016). Determination of excess sodium hydroxide in geopolymer by volumetric analysis. Procedia Engineering, 148, 298– 301.

[11] Heikal, M., Nassar, M. Y., El-Sayed, G., & Ibrahim, S. M. (2014). Physico-chemical, mechanical,microstructure and durability characteristics of alkali activated Egyptian slag. *Construction and Building Materials*, *69*, 60–72.

[12] Alehyen, S., & Achouri, M. T. (2017). Characterization, microstructure and properties of fly ash based geopolymer. Journal of Materials and Environmental Sciences, 8(5), 1783–1796.

[13] Provis, J. L., Yong, C. Z., Duxson, P., & van Deventer, J. S. J. (2009). Correlating mechanical and thermal properties of sodium silicate-fly ash geopolymers. *Colloids and Surfaces A: Physicochemical and Engineering Aspects*, *336*(1–3), 57–63.

[14] Jaya, N. A., Al Bakri Abdullah, M. M., Ghazali, C. M. R., Binhussain, M., Hussin, K., & Ahmad, R. (2016). Effect of NaOH concentration on flexural strength, phase formation and microstructural development of kaolin geopolymer ceramic. *Materials Science Forum*, 857,405-411.

IOP Conf. Series: Materials Science and Engineering **743** (2020) 012015 doi:10.1088/1757-899X/743/1/012015

Properties of Blended Alkaline System Geopolymer- A Review

Fakhryna Hannanee Ahmad Zaidi[1], Romisuhani Ahmad[1*], Zarina Yahya[1], Muhammad Faheem Mohd Tahir[2], Wan Mastura Wan Ibrahim[1] and Ahmad Syauqi Saufi[1]

[1]Faculty of Engineering Technology, Universiti Malaysia Perlis (UniMAP), Perlis, Malaysia
[2]School of Material Engineering, Universiti Malaysia Perlis, Perlis, Malaysia

Email: * fakhryna0508@gmail.com

Abstract. Geopolymers are inorganic material that comprise of silicon(Si) and aluminium(Al) bonded by oxygen atom to form a polymer network. The binder material used for geopolymer such as fly ash and blast furnace are mostly the industrial waste or by-products containing high content of silica and aluminium which acted as precursor for geopolymerization. The raw material plays an important role in the formation of geopolymer for each material may result in different properties of geopolymer. To improve the performance of these binders, numerous studies have been focused on the production of mixes based on blends of reactive precursors. The blends usually involve a Ca-rich precursor such as granulated blast furnace slag (GGBS), and an aluminosilicate source such as metakaolin or low calcium fly ash, to promote the stable coexistence of calcium silicate hydrate (C–S–H) gels formed from the activation of the GGBS and the geopolymer gel (N–A–S–H) produced from the activation of the aluminosilicate. Thus, this paper is intended to review the properties of different type of mixes of blended alkaline system.

1. Introduction

Geopolymer was first introduced by Davidovits in 1970s to describe a group of minerals binders. The chemical structure of geopolymer comprise of silicon and aluminium bonded by oxygen atom to form a polymer network [1]. The raw material for geopolymer includes any pozzolanic compound or source of silica and alumina that is readily dissolve in the alkaline solution which acts as precursor that leads to geopolymerization [2]. Therefore, such material can also be called as alkali activated aluminosilicate binders or alkali activated cementitious material.

For each type of raw material, different chemical composition can be observed thus resulting to different properties of alkali activated binders [3]. As a function of chemical composition of raw materials, the alkaline binder is classified into two groups such that the binder synthesized from raw materials rich in calcium such as blast furnace slag will produce calcium silicate hydrate (C-S-H) gel. While binder synthesized from raw materials low in calcium and rich SiO_2 and Al_2O_3 such as metakaolin and class F fly ash will result in a formation of amorphous material (N-A-S-H) called geopolymer gel [4].

Geopolymers represents an attractive binding material known for high compressive strength, good durability in aggressive environments, low shrinkage and good thermal resistivity [5]. However, taking example of the most common raw material for geopolymer which is fly ash, has a limiting factor such as its low reactivity and consequent low strength gain when cured at room temperature [6]. To improve

Content from this work may be used under the terms of the Creative Commons Attribution 3.0 licence. Any further distribution of this work must maintain attribution to the author(s) and the title of the work, journal citation and DOI.
Published under licence by IOP Publishing Ltd

the performance of these binders, numerous studies have been focused on the production of mixes based on blends of reactive precursors. The blends usually combines the Ca-rich precursor and an aluminosilicate source to promote the stable coexistence of calcium silicate hydrate (C–S–H) gels and the geopolymer gel (N–A–S–H) [7]. The voids and pores within the geopolymeric binder become filled with the C-S-H gel and helps to bridge the gaps between the different hydrated phases and unreacted particles, thereby resulting the increase in mechanical strength for these binders [8]. This paper intends to review the properties of different type of blended alkaline system geopolymer.

2. Raw Material

2.1. Fly ash

Fly ash is a waste material generated abundantly during the burning of coal. Initially, the usage of fly ash is generally limited to 15-25% of the total cementitious materials for the purpose of reducing the heat of hydration [9]. Due to the high content of silica and aluminium, fly ash is nowadays utilized to produce geopolymer material [10]. The chemical composition of fly ash is tabulated in **Table 1**. Mustafa Al Bakri et al. (2011) stated that the fly ash consist mostly of glassy, hollow, spherical particles, which were cenospheres (thin-walled, hollow spheres); the appearance of the microstructures of the original fly ash agreed well with previous literatures [11].

Table 1. Chemical composition of fly ash .

Element	SiO_2	Fe_2O_3	Al_2O_3	CaO	K_2O	Na_2O_{eq}	SO_3	MgO	LOI
Fly ash [12]	55.23	10.17	25.95	1.32	1.59	1.59	0.18	0.31	5.25

2.2. Ground Granulated Blast Furnace Slag

Ground granulated blast furnace slag (GGBS) is a by-product of iron manufacturing. The main constituents of GGBS are CaO, SiO_2, Al_2O_3 and MgO such tabulated in **Table 2**. These are the minerals usually found in cementitious substances thus have the potential to improves the properties such as strength and durability when added into concrete.

Table 2. Chemical composition of GGBS.

Element	SiO_2	Fe_2O_3	Al_2O_3	CaO	K_2O	Na_2O_{eq}	SO_3	MgO	LOI
GGBS [13]	30.53	0.33	13.67	46.02	0.36	0.24	-	5.09	0.22

2.3. Metakaolin

Metakaolin is a supplementary cementitious material with high pozzolanic activity. To produce metakaolin, kaolin was heated up to 650–800 °C where the long-range order of silicon–aluminium layer structure deteriorate with structure deformation due to the loss of interlayer water and OH^- [3]. The chemical composition of kaolin and metakaolin is as tabulated **Table 3**.

Table 3. Chemical composition of Kaolin and Metakaolin [14].

Element	SiO_2	Fe_2O_3	Al_2O_3	TiO_2	K_2O	MnO_2	ZRO_2
Kaolin	54.5	4.32	32.4	1.33	5.58	0.09	0.08
Metakaolin	554.1	4.41	33.0	1.34	5.67	0.11	0.09

3. Properties of Blended alkaline system

Low-calcium fly ash (Class F) has been investigated as a suitable material for geopolymer because of its wide availability, high in silica and alumina composition and less water demand [15]. However, one challenge for wide application of fly ash based geopolymer is the requirement of curing at elevated temperature. Many researchers attempted to enhance the reactivity of fly ash by adding some calcium containing material [16]. The addition of calcium oxide (CaO) forms hydrated products such as calcium silicate hydrates (CSH) along with the aluminosilicate geopolymer network [17].

Typically, GGBS incorporated low calcium fly ash are intensely researched [18]–[20]. The main product of alkali activation of fly ash is alkaline amorphous aluminosilicate gel, N-A-S-H gel in which is limited due to its low reactivity and low strength when cured. While the main product of alkali activation slag is calcium silicate hydrate with Al in the structure, C-S-H gel also known to give high strength but poor workability [21].

For example, research by Nagaraj et al. (2018) uses fly ash and GGBS as the raw material and the different percentage addition of materials is investigated. As the percentage of fly ash increases, it also increases the workability which in return, reduces the strength. Oppositely, GGBS reduces the workability and increases the strength of the concrete. Therefore, according to the application of the concrete, the amount of each material have to be controlled [22].

Research by Marjanovic (2014) investigates the physical and mechanical properties of alkali activated fly ash- GGBS blends [21]. The optimal characteristics of investigated alkali-activated binder with exceptionally high compressive strength and acceptable setting time were achieved with the blend composition of 25 % FA and 75% GGBS.

Aside from incorporating GGBS with low calcium fly ash, metakaolin is also used providing similar role as low calcium fly ash as the aluminosilicate source. Research by Borges et al. (2016) investigates the performance of blended metakaolin and GGBS alkali activated mortar with 60/40 MK/GGBS and the main binder. The microstructure was assessed using Scanning Electron Microscopy (SEM) where results shows that the addition of GGBS significantly alters the microstructure of alkali-activated mortars thus promotes the reduction of porosity and capillary sorption [23].

Similarly, Huseien et al. (2018) prepares geopolymer by substituting metakaolin with range 0-15% into GGBS. The result shows that metakaolin with 5% substitution obtains the highest bonding strength in early 1 days and 28 days of testing [13]. Thus, it is noted that the blended alkaline system helps counterbalance the disadvantages and alter the properties of geopolymer according to the intended application.

4. Conclusion

This paper reviews on the properties of blended alkaline system geopolymer. The blends usually combine the Ca-rich precursor such as GGBS and an aluminosilicate source such as metakaolin and fly ash to promote the stable coexistence of calcium silicate hydrate (C–S–H) gels and the geopolymer gel (N–A–S–H). It is clear that blended alkaline system helps to improve the properties of geopolymer. With the further development, this understanding will give the ability to design geopolymers with specific applications in mind.

References

[1] N.A. Jaya, M. M.A.B Abdullah, C.M.R. Ghazali, M. Binhussain, K. Hussin,R. Ahmad, *Mater.Sci.Forum,* 857,405–411 (2016).
[2] N. A. Jaya, M. M. Al Bakri Abdullah, C. M. R. Ghazali, M. Binhussain, K. Hussin, R. Ahmad, *Key Eng. Mater.*, **700**, 3–11 (2016).
[3] Z. H. Zhang, H. J. Zhu, C. H. Zhou, and H. Wang, *Appl. Clay Sci.*, **119**, 31–41 (2016).
[4] D. Khale, R. Chaudhary, *J. Mater. Sci.*, **42**, 729–746 (2007).
[5] N. A. Jaya, M. M. Al Bakri Abdullah, C. M. R. Ghazali, M. Hussain, K. Hussin,R. Ahmad, *MATEC Web Conf.*, **78**, (2016).
[6] N. Marjanovic, M. Komljenovic, Z. Bascarevic, V. Nikolic, *Procedia Eng.*, **108**, 231–238 (2015).

[7] S.A.Bernal, R. Mejia de Gutierrez, J. L. Provis, *Constr. Build. Mater.*, **33**, 99–108 (2012).
[8] C. K. Yip, G. C. Lukey, and J. S. J. van Deventer, *Cem. Concr. Res.*, **35**, 1688–1697 (2005).
[9] M. Jalal, A. Pouladkhan, O. F. Harandi, D. Jafari, *Constr. Build. Mater.*, **94**, 90–104 (2015).
[10] M. M. Al Bakri Abdullah, *Rev Chim-Bucharest*, **64**, 382–387 (2013).
[11] O.A.K.A. Abdulkareem, H. Kamarudin, M. M. Al Bakri Abdullah, M. N. Norazian, A. R. Rafiza, C. M. R. Ghazali, *Appl. Mech. Mater.*, **110–116**, 734–739 (2011).
[12] A. Fauzi, M. Fadhil, A. B. Malkawi, M.M.A.B Abdullah, *Procedia Eng.*, **148**, 487–493, (2016).
[13] G. F. Huseien, J. Mirza, M. Ismail, S. K. Ghoshal, and M. A. M. Ariffin, *Ain Shams Eng. J.*, **9**, 1557–1566 (2018).
[14] Y. M. Liew *et al.*, *Constr. Build. Mater.*, **156**, 9–18 (2017).
[15] M.M.A. Abdullah, K. Hussin, M. Bnhussain, K.N. Ismail, W.M. . Ibrahim, *J. Asian Sci. Res.*, **1**, 1–12 (2011).
[16] V.S. Kumar, N. Ganesan, P. V. Indira, *IOP Conf. Ser. Earth Environ. Sci.*, **80**, 1 (2017)
[17] P. Nath, P. K. Sarker, *Constr. Build. Mater.*, **66**, 163–171 (2014).
[18] A. Karthik, K. Sudalaimani,nC.T.V. Kumar, *Constr. Build. Mater.*, **149**,338–349 (2017)
[19] T. Srinivas, N. V. R. Rao, *Int. J. Res. Chem. Metall. Civ. Eng.*, **4**, 60–63 (2017).
[20] X. Gao, Q. L. Yu, H. J. H. Brouwers, *Cem. Concr. Compos.*, **59**, 119–128 (2015).
[21] N. Marjanovic, M. Komljenovic, Z. Bascarevic, V. Nikolic, R. Petrovic, *Ceram. Int.*, **41**, 1421–1435 (2015).
[22] V.K. Nagaraj, D.L. Venkatesh Babu, *Asian J. Civ. Eng.*, **8**, 56 (2018).
[23] P. H. R. Borges, N. Banthia, H. A. Alcamand, W. L. Vasconcelos, and E. H. M. Nunes, *Cem. Concr. Compos.*, **71**, 42–52 (2016).

IOP Conf. Series: Materials Science and Engineering **743** (2020) 012016 doi:10.1088/1757-899X/743/1/012016

Spectrophotometric Analysis Assay for the Measurement of Arsenic Using Nanocomposite of Silica and Graphene

M. N. A. Uda [1,3,*], Uda Hashim[3], Subash C. B. Gopinath[1,3], M. N. Afnan Uda[2], N. A. Parmin[1,3], Hussain Adam[3] and Nur Hulwani Ibrahim[3]

[1] School of Bioprocess Engineering Group, Universiti Malaysia Perlis (UniMAP) 02600 Arau Perlis
[2]School of Microelectronic Engineering Group, Universiti Malaysia Perlis (UniMAP) Putra Campus, 02600 Pauh, Perlis Malaysia
[3] Institute of Nano Electronic Engineering (INEE), Universiti Malaysia Perlis (UniMAP), 01000 Kangar Perlis Malaysia

Email: * nuraiman@unimap.edu.my

Abstract. According to media reports in today's world, millions of people across the globe are suffering from arsenic (As) toxicity. Arsenic heavy metal occurs in different oxidative states in the environment and enters in the food chain through soil and water. In the agricultural field, irrigation with arsenic contaminated water has a higher level of arsenic contamination on the top soil, which may eventually affects the quality of crop production. The major crop like rice requires a considerable amount of water to complete its lifecycle. In Asian countries like Malaysia, Brunei, Indonesia, and Thailand, arsenic intake from rice diet is significantly higher, as rice plants have a special ability to take up arsenic from the soil and water used in irrigation. The transfer of arsenic from soil to plant systems is a serious issue that leads to considerable human exposure in a dangerous way. The applications of nanotechnology play a vital role in many research fields. And therefore, this research, employs the implementation of nanotechnology approach such as silica-grapheme nanocomposite that could be able to minimize the Arsenic heavy metal contamination issues.

1. Introduction

Rice is an essential food of vast people in the world. There are number of studies done by researchers concerning rice, however, rice issues still exist [1]–[7]. Nowadays, arsenic (As) contamination raises public awareness and concern globally because its affects human health as it is carcinogenic. This metalloid occurs naturally in paddy plantation area and major cause is coming from the excessive use of pesticides, fertilizers and, anthropogenic activities which led to increased pollution of soil [8]. Arsenic contamination is much more affected in rice plant compared to others crop due to anaerobic conditions. This condition will make faster movement for arsenic accumulation in paddy plant. And excess of this contamination will affect the growth of paddy significantly, such as lowering the titter, reducing the straw weight, reducing the height of plant and reducing crop yield [9-10].

Therefore, to overcome this problems, potential application of nanotechnology becomes vital field for today's innovation. Currently, nanotechnology covers all major sectors such as energy, medical healthcare, nano systems, environment, food science and agriculture [11]–[16]. By applying nanomaterials such as silica nanoparticles and graphene, this research can minimize the accumulation of arsenic from paddy plant. A part from that, both have capability in agriculture in terms of plant

Content from this work may be used under the terms of the Creative Commons Attribution 3.0 licence. Any further distribution of this work must maintain attribution to the author(s) and the title of the work, journal citation and DOI.
Published under licence by IOP Publishing Ltd

growth, plant disease resistance and high quality of yield due to the uniqueness of their size which suite nanoparticles size.

2. Material and method

2.1 Materials

Rice paddy straw was obtained from nearest rice mill, Perlis, Malaysia. Distilled water was used throughout the experiments. All chemicals were used as received without further purification.

2.2 Production of Silica Nanoparticles using paddy rice straw ash

The production for whole process as shown in figure 1. The paddy rice straw was washed and cleaned by distilled water to remove dirt and surfactant. It was dried naturally for 3 days under sun light. The washed rice straw obtained was designated as un leached rice straw. Then the cleaned rice straw was burned and ash was collected as main material for silica production. Roughly, 20g of ash was mixed together with 400 ml of 2.5 M of NaoH and continued with stirred at 100oC for 4 H. All mixed solution was filtered using Whatman filter paper. Then, the filtered solution was titrated slowly using 2 M of sulphuric acid and stopped until the solution reached at pH7. Subsequently, the stirring was continued for 18 h for adequate for equilibrium mixture. Afterward, it was stopped and waited until two separated layer gel was formed. This layer was separated using centrifugation process (6000 rpm, 5min) where pellet was washed three time with ethanol and two times with water. The pellet was dried at 80oC for 30 min.

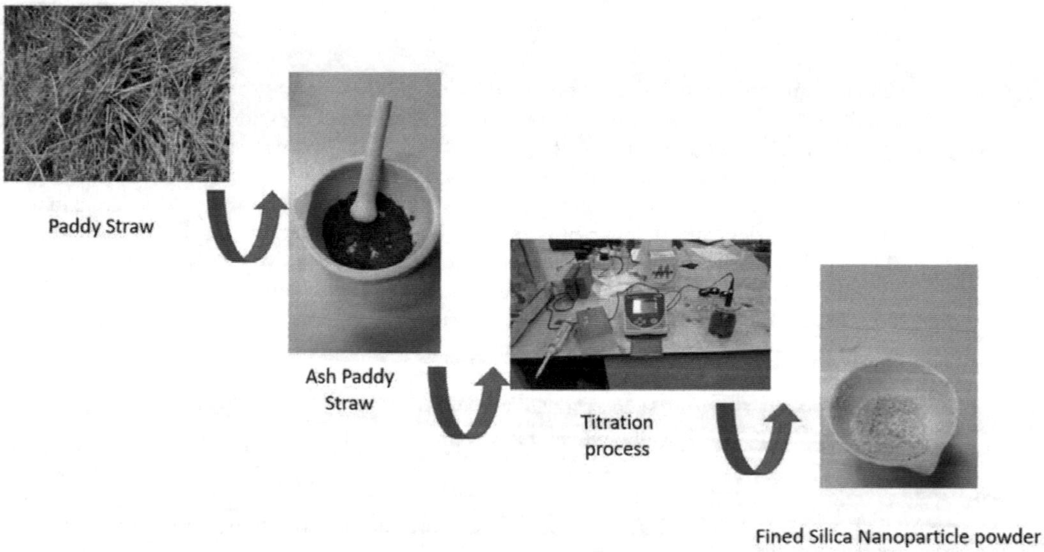

Figure 1. Experimental Setup for Production of Silica Nanoparticles.

2.3 Chemical synthesis process of Graphene from paddy straw

The chemical synthesis of graphene from ash paddy straw process is shown in Figure 2. Briefly, 3 gm of ash was mixed together with 12 gm of KOH using mortar and pestle. This mixture was compacted into white porcelain crucible. Next, all mixture undergoes annealing process where the top of porcelain crucible needs to close by using crucible cap. This sample was annealed at 700oC for 2 h in muffle furnace. At this stage, KOH act to active the treatment of mixture. This mixture was stirred continuously for 6 h. At last stage, the sample was washed using distilled water and filtered to remove any excess of KOH. The filtered sample was dried at 150oC for 24 h.

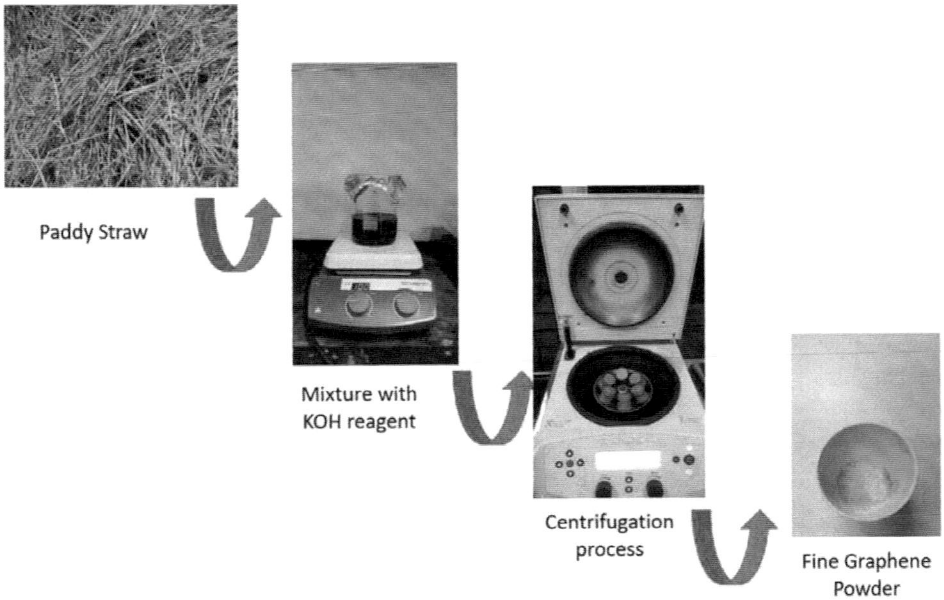

Figure 2. Experimental Setup for Chemical Synthesis of Graphene.

2.4 Preparation and Synthesis of Silica coated –graphene nanomaterial
Briefly, 500 mg of silica with 2% APTES was mixed together and incubated for 1 hr. Next the process continued with centrifugation process where it was washed two times with distilled water. At this stage, Pellet was taken and combines with 500 mg of graphene in 1 ml distilled water. Both mixtures were incubated for 1 hr. Subsequently, the solution was centrifuged and pellet was taken and dried at 70 oC for 2h.

2.5 Spectrophotometric study of the interaction between arsenic and adsorption material
In this analysis, the prepared readily fine silica, graphene and silica -graphene was distributed in eppendorf tube where each tube was filled with 1mg for each material. Next, arsenic was filled (1mg) and mixture together with each material. This mixture was incubated for 30 min to 1 hr for equilibrium mixture. Lastly, all mixture was ready for measurement. The absorbance spectrum was recorded

3. Result and Discussion
In this experiment, paddy straw has been selected for main resources for adsorption of arsenic analysis. This material is affordable and easy to find after harvested process in paddy field area because in Malaysia, rice plantation was covers about 600,000 ha of area, thus, utilization of this by product will lead to sustainability of environment. The by product is produced annually because in Malaysia most farmers plant paddy twice a year. Based on figure 3, the determination of range of arsenic by UV/Vis spectroscopy absorbance was recorded from 0-750 nm. The result shows that, 220 nm was suitable range for arsenic for the reactions where all the reaction with adsorption material will be conducted within this range. In this analysis, different range of concentrations of arsenic have been used, the graph shows a sequence pattern from high concentration (24m mg/ml) to lower concentration (0.75 mg/ml).

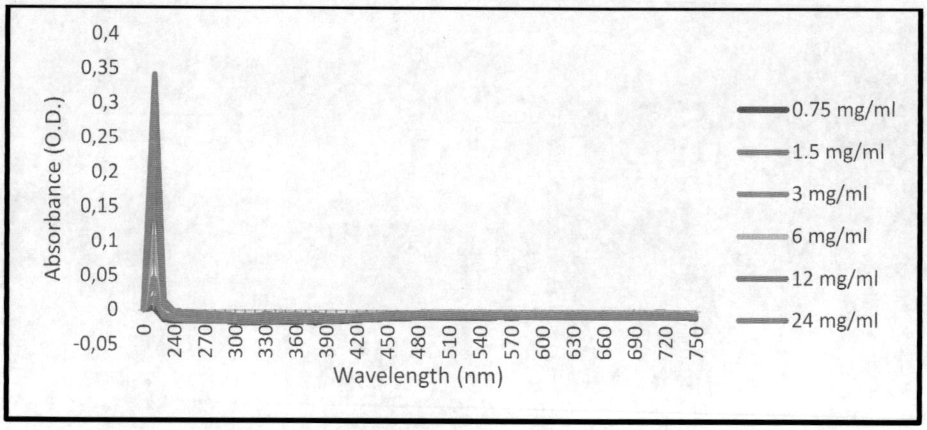

Figure 3. Scanning absorbance range for arsenic.

The spectrophotometric analysis assay continues with different material adsorption. Based on figure 4(a) silica nanoparticles has been used for this analysis. The particles size of this silica used in this analysis is measured 70nm where its consider as nanosized structure because nanomaterial only consider where the material is less than 100 nm [17]. The highest concentration give 0.35 O.D where the lowest concentration give 0.07 O.D. For figure 4(b), graphene has been used, the pattern shows similar as silica nanoparticles, however, the different only the result of absorbance where the highest absorbance located at 0.58 O.D and the lowest concentration located at 0.12 O.D. Lastly, the nanocomposite is consists of combination of silica nanoparticles and graphene. Based on this graph, its shows slightly high absorbance compared with the other graphs for high concentration 0.69 O.D and the lowest concentration is located at 0.1 O.D. This results clearly show that, nanocomposite play a tremendous properties because it has an offer for enhancement with their unique surface area and active sites for adsorption analysis [18].

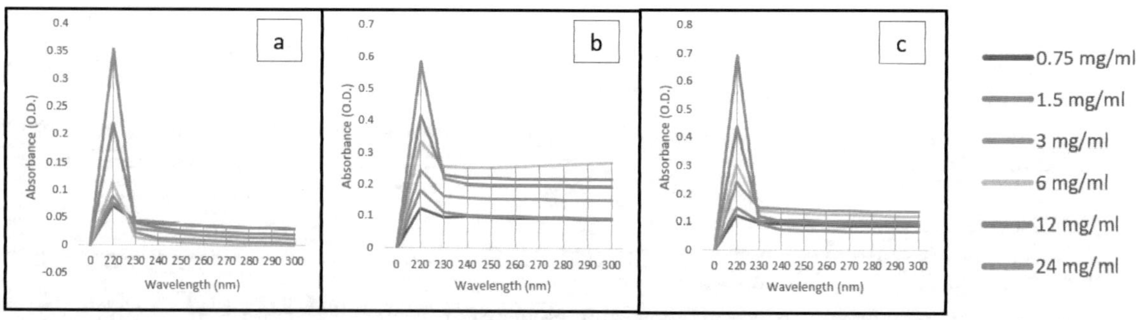

Figure 4. Spectrophotometric Analysis of Arsenic with (a) Silica Nanoparticles (b) Graphene (c) Silica Graphene Nanocomposite.

A part from that, based on figure 5, the graph shows that, the percentages of arsenic removal with different adsorption material. Three adsorptions have been compared to evaluate which of the materials has the most sorption capacity. This result confirmed that, nanocomposite consist of silica nanoparticles and graphene shows a greatest effect of adsorption of arsenic at 24 mg/ml, this nanocomposite is capable to reach 51% of removing the arsenic, whereas for graphene it only indicates for 41% removal of arsenic. For silica nanoparticles even though it shown a lowest capacity adsorption but still it has proven that silica nanoparticles can be used as material for arsenic removal adsorption. A part that, the sizing of silica nanoparticles plays vital impact for sorption capacity where the smallest the particles will have a big tendency of adsorption capacity. A study shown that if the material with less than 10nm being used for the reaction it will generates an electronic effect of the mechanism of the nanoparticles [19]

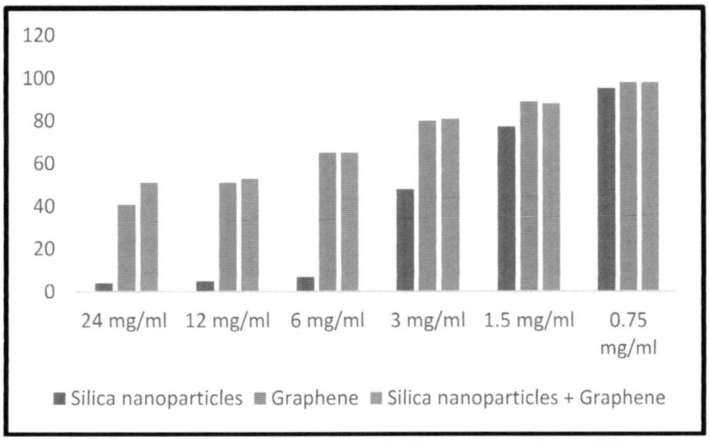

Figure 5: Percentages of Arsenic Removal after Interaction with Nanomaterial

4. Conclusion

Overall, silica nanoparticles, graphene, and nano composite silica graphene have facilitated a good adsorbent for arsenic removal using spectrophotometric analysis assay. Moreover, the synthesis and production of this nanomaterial is simple and be apple to penetrate and compete with other commercially adsorbent for arsenic.

Acknowledgement

The authors wish to thank Universiti Malaysia Perlis (UniMAP) for providing financial and technical staff who support to conduct this research.

References

[1] M. N. A. Uda *et al.*, "Immunosensor development for rice tungro bacilliform virus (RTBV) detection using antibody nano-gold conjugate," *AIP Conf. Proc.*, vol. 1808, no. October, 2017.

[2] M. N. A. Uda *et al.*, "Reviewed Immunosensor Format Using Nanomaterial for Tungro Virus Detection," vol. 832, no. November 2013, pp. 410–414, 2014.

[3] M. N. A. Uda *et al.*, "Antimicrobial Activity of Plant Extracts from Aloe Vera, Citrus Hystrix, Sabah Snake Grass and Zingiber Officinale against Pyricularia Oryzae that causes Rice Blast Disease in Paddy Plants," *IOP Conf. Ser. Mater. Sci. Eng.*, vol. 318, no. 1, 2018.

[4] M. N. . A. Uda *et al.*, "Comparative Study Between Elisa And Surface Plasmon Resonance (Spr) For Rice Tungro Disease Detection," vol. 9, no. 11, pp. 5568–5571, 2013.

[5] M. N. . Uda *et al.*, *A Disposable Biosensor Based on Antibody-Antigen Interaction for Tungro Disease Detection*. Elsevier Inc., 2019.

[6] M. N. A. Uda *et al.*, "Comparison Study of Two Different Isolation and Purification Method for Rice Tungro Bacilliform Virus (RTBV)," *Agric. Agric. Sci. Procedia*, vol. 2, no. August, pp. 107–112, 2014.

[7] M. N. A. Uda *et al.*, "Immunosensor development formatting for tungro disease detection using nano-gold antibody particles application," *AIP Conf. Proc.*, vol. 1808, no. 020057, 2017.

[8] L. J. Looi, Z. A. Ahmad, W. Y. Lim, and H. Hazzeman, "Bioconcentration and Translocation Efficiency of Metals in Paddy (Oryza sativa):," *Sains Malaysiana*, vol. 43, no. 4, pp. 521–528, 2014.

[9] M. T. Bhadha, J. H.; Vanweelden, "Arsenic Accumulation in Rice (Oryza sativa): An Overview," *Int. J. Environ. e Agric. Res.*, vol. 2, no. 4, pp. 72–75, 2016.

[10] A. Mitra, S. Chatterjee, R. Moogouei, and D. Gupta, "Arsenic Accumulation in Rice and

Probable Mitigation Approaches: A Review," *Agronomy*, vol. 7, no. 4, p. 67, 2017.
[11] S. C. B. Gopinath, S. Ramanathan, K. Hann Suk, M. Ee Foo, P. Anbu, and M. N. A. Uda, "Engineered nanostructures to carry the biological ligands," *MATEC Web Conf.*, vol. 150, p. 06002, 2018.
[12] M. A. M. Akhir *et al.*, "Ethylene Gas Sensing Properties of Tin Oxide Nanowires Synthesized via CVD Method," *IOP Conf. Ser. Mater. Sci. Eng.*, vol. 318, no. 1, 2018.
[13] N. A. Parmin *et al.*, "Voltammetric determination of human papillomavirus 16 DNA by using interdigitated electrodes modified with titanium dioxide nanoparticles," *Microchim. Acta*, vol. 186, no. 6, 2019.
[14] S. A. B. Ariffin, T. Adam, U. Hashim, S. Faridah Sfaridah, I. Zamri, and M. N. A. Uda, "Plant Diseases Detection Using Nanowire as Biosensor Transducer," *Adv. Mater. Res.*, vol. 832, pp. 113–117, 2013.
[15] T. Adam and U. Hashim, "Label Free Detection of Single Biomolecules," in *2014 Fifth International Conference on Intelligent Systems, Modelling and Simulation*, 2014, pp. 1–8.
[16] T. Adam and U. Hashim, "Fabrication of Micro-Mixer for Life Sciences Applications," 2014.
[17] M. E. Foo *et al.*, "Antimicrobial activity of functionalized single-walled carbon nanotube with herbal extract of Hempedu bumi," *Surf. Interface Anal.*, vol. 50, no. 3, pp. 354–361, 2018.
[18] M. Habuda-Stanić and M. Nujić, "Arsenic removal by nanoparticles: a review," *Environ. Sci. Pollut. Res.*, vol. 22, no. 11, pp. 8094–8123, 2015.
[19] M. Umadevi, T. Rani, T. Balakrishnan, and R. Ramanibai, "Antimicrobial Activity of Silver Nanoparticles Prepared Under an Ultrasonic Field," *Int. J. Pharm. Sci. Naotechnology*, vol. 4, no. 3, pp. 1491–1496, 2011.

IOP Conf. Series: Materials Science and Engineering **743** (2020) 012017 doi:10.1088/1757-899X/743/1/012017

Analysis on Gold Nanorod Interdigitated Elecrtrode Sensor Using Simple Photolithograpic Process

Hussaini Adam[1*], Subash C. B. Gopinath[1,2], M. N. A. Uda[1,2], Uda Hashim [1], Tijjani Adam[1] and M. N. Afnan Uda[3]

[1]Institute of Nano Electronic Engineering (INEE), Universiti Malaysia Perlis (UniMAP), 01000 Kangar Perlis Malaysia
[2] School of Bioprocess Engineering Group, Universiti Malaysia Perlis (UniMAP) 02600 Arau Perlis
[3] School of Microelectronic Engineering Group, Universiti Malaysia Perlis (UniMAP) Putra Campus, 02600 Pauh, Perlis Malaysia

Email: * hussainiadam@gmail.com

Abstract. With the higher demand of preventative healthcare in order to minimize costs and improve healthcare systems, the development and enhancement of sensor technology is vital. It is essential to develop a diagnostic device that can minimize time and lower task in testing, and can effectively reduce manufacturing and delivery costs because of portability of its designs. Here, we briefly describe the fabrication of aluminum interdigitated electrodes and deposition of gold nanorod on the fabricated microelectrode that can detect changes on the modified surface of the aluminum interdigitated electrode. Electrodes made from aluminum was employed for the fabrication because it is the most widely used electrode. Gold nanorod was deposited on the desired surface in order to enhance an enzymatic Response. The use of gold nanorod also enhances the sensitivity of detection due to the decrease of the thickness of probed zone. A simple and facile method for the deposition of gold nanorod colloid was described via a simple photolithographic technique on the interdigitated electrode (IDE). The gold nanoparticles pattern deposition on IDE was investigated by high power microscope (HPM), 3D Profilometer, and atomic force microscope (AFM).

1. Introduction

The discovery of modified electrodes is vital in creating a new device with high sensitivity and selectivity since the modifiers deliberate captivating properties that will lead to a specific recognition of analytes [1-6]. The study of modification of electrodes is important in achieving a better comprehension of the nature of charge transfer methods within the thin films [7-12]. The modification of conductive surfaces involved adsorption and covalent bond formation. Recently, gold nanonods have encouraged much interest in modifying the surface of the electrodes due to their intriguing physicochemical properties [13]. For example, gold nanorod possess exceptional electrical and catalytic properties, large surface area to volume ratio and large number of active sites which make them principally suitable for analytical purposes [14]. The properties of gold nanorod intensely depend on their sizes and shapes, therefore a procedure for their synthesis, regulating their growth and their morphologies are important and interesting [15-16]. Photolithographic technique is a simple way of fabrication and electrodeposition. This method can sometimes have some limitations on the nanomaterial dimensions and their morphologies, but it provides a lot of advantages, particularly related to the rapid synthesis

Content from this work may be used under the terms of the Creative Commons Attribution 3.0 licence. Any further distribution of this work must maintain attribution to the author(s) and the title of the work, journal citation and DOI.
Published under licence by IOP Publishing Ltd

IOP Conf. Series: Materials Science and Engineering **743** (2020) 012017 doi:10.1088/1757-899X/743/1/012017

time, the absence of chemical reductants [17]. Furthermore, when the modifier film is directly deposited on the electrode it allows a better adhesion on the surface of the IDE. The quality of biosensors is primarily recognized by the specificity of the biomolecule alongside the quality of the transducer [18]. Also the reliability in analyzing the interactions of biomolecules is essential so as to achieve the contemporary quality assurance criteria. For the detection of analytes, a biosensor must possess characteristics such as, linearity, sensitivity, selectivity, reproducibility, and stability [19]. Biosensors with a high sensitivity is capable of detecting diseases at their early states and significantly proving the chance of possibly life-saving detection and intervention [20-22].

2. Material and methods

2.1 Starting Material

The starting material used in this project is 4 inch silicon substrate. The process was started by cleaning the silicon wafer to wash away any impurity that might be present on the surface of the wafer which could eventually disrupt the process of wafer pattern transfer. The silicon wafer was washed using RCA1 and RCA2. Then, oxide layer of silicon was grown on the wafer surface. This insulating layer is to improve the final performances of the device [23].

2.2 Mask design

The purpose of designing the mask was to fabricate the microelectrode. The mask was designed using AutoCAD software. The design of the mask consists of comb like finger structure between the two electrodes to create a gap between them [24]. Figure1 shows the mask design using AutoCAD software.

Figure 1. Design of IDE using AutoCAD software.

2.3 Device fabrication

This part describes the fabrication of the aluminum interdigitated electrode. Before the fabrication process starts, oxidation of silicon was conducted using an evaporator machine. Simple lithography technique was used for the fabrication of aluminum interdigitated electrode [12]. AL IDE was generated by wet etching-assisted conventional lithography, with a gap between adjacent electrodes [13]. Photolithography technique is a technique that is used to transfer designed pattern from the mask to photoresist on the wafer surface [14]. After the aluminum was pasted on the silicon wafer, then followed by spin coating for 30s at 2500rpm was carried out. Afterward, the photo resists were exposed UV light by mask aligner. The time exposure was conducted for 12 sec to pattern transfer of the IDE mask on wafer surface. Then, the substrate was developed using RD6. Next, the photoresist pattern was etched in aluminum etchant and photoresist was stripped off using acetone. Finally, the desired shape and size of IDE were successfully exposed. The figure 2 bellow shows the Step-by-step fabrication process of the interdigitated electrode: (a) silicon substrate; (b) oxide layer on the silicon substrate; (c) placing the mask pattern on the photoresist layer; (d) transferring the mask pattern onto the photoresist layer; (e) depositing AL via electroplating; (f) removing the photoresist.

This part describes the fabrication of the aluminum interdigitated electrode. Before the fabrication process starts, oxidation of silicon was conducted using an evaporator machine. Simple lithography technique was used for the fabrication of aluminum interdigitated electrode [25]. AL IDE was generated by wet etching-assisted conventional lithography, with a gap between adjacent electrodes [26]. Photolithography technique is a technique that is used to transfer designed pattern from the mask to photoresist on the wafer surface [27]. After the aluminum was deposited on the silicon wafer, then followed by spin coating for 30s at 2500rpm was carried out. Afterward, the photo resists were exposed UV light by mask aligner. The time exposure was conducted for 12 sec to pattern transfer of the IDE mask on wafer surface. Then, the substrate was developed using RD6. Next, the photoresist pattern was etched in aluminum etchant and photoresist was stripped off using acetone. Finally, the desired shape and size of IDE were successfully achieved. The figure 2 bellow shows the Step-by-step fabrication process of the interdigitated electrode: (a) silicon substrate; (b) oxide layer on the silicon substrate; (c) placing the mask pattern on the photoresist layer; (d) transferring the mask pattern onto the photoresist layer; (e) depositing AL via electroplating; (f) removing the photoresist, as shown below :

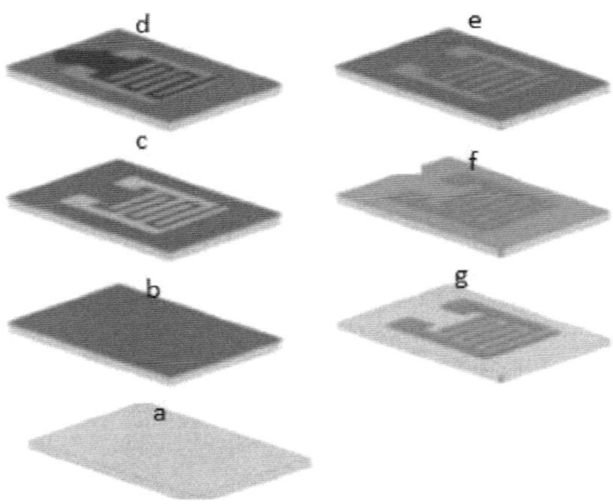

Figure 2. Step of fabrication of IDE.

2.4 Deposition of gold nanorod on the IDE

The deposition of gold nanoparticles on the surface of aluminum interdigitated electrodes is crucial [28], and the fundamental studies of such modified electrodes have been performed to achieve a better understanding of the nature of charge transfer inside the thin films [29]. Furthermore, Generation of hybrid nanostructures has been attested as a promising approach to develop high-performance sensing substrates [30]. Gold nanorod was deposited using conventional photolithographic process. After the device was fabricated, the surface of the IDE was covered with gold nanorod and spin coated for 20sec, after the spin coating process was done, the device was placed on a hotplate for 30min for the full deposition on the IDE to take place and the device was finally removed on the hotplate instrument after 30min. As shown in figure 3.

Figure 3. IDE deposited with gold nanorod.

2.5 Device charaterization

After AlIDE fabrication by a standard UV-photolithography process, HPM, SEM, AFM and 3D profilor were used to image the AL IDE. 3D and AFM were used to image the IDE before and after coated with gold nanorod. 3D-nanoprofilometer was used for three dimensional analysis of grains between AlIDE finger electrodes as shown in figure 4 below:

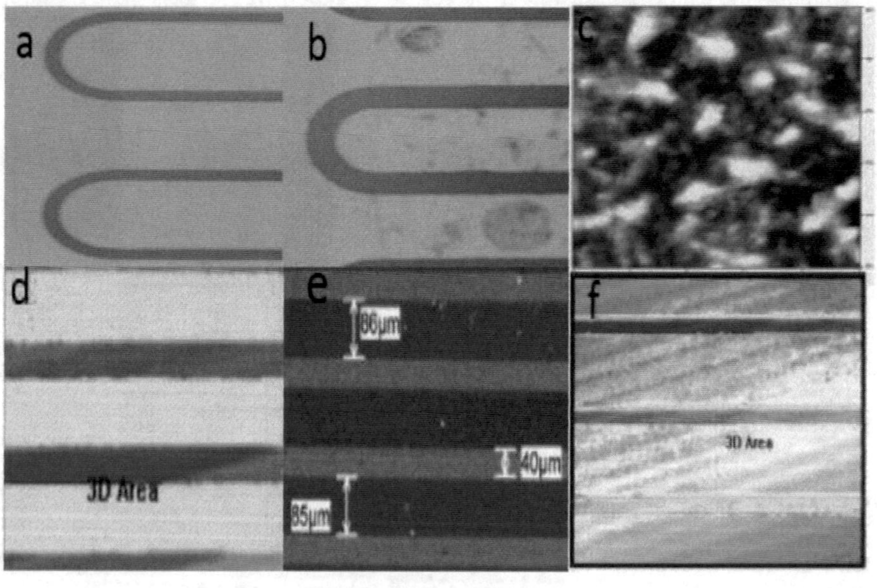

Figure 4. Characterization of IDE.

Table 1. Measurement of AL IDE

Parameter	Type of electrode	Description
With	250 µm	Finger electrode width
Gap	20 µm	Gap between finger electrodes

3. Result and Discussion

Aluminum IDE was successfully fabricated and coated with gold nanorod. After the fabrication was carried out, the device was inspected by using high power microscope, scanning electron microscope, 3D Profilometer, and atomic force microscope. The surface characterization is important in examining the IDE surface[31]. The HPM and SEM images show that the aluminum electrode has been successfully fabricated without any distortion and gold nanorod was successfully deposited (figure 3). The material used for the fabrication of IDE electrode is aluminum. The fabrication was done using photolithography process for resist development. Figure 4a indicates the bare IDE image taken using high power microscope. As it can be clearly seen that the characterization of the bare IDE image using high power microscope was successful since no any distortion was detected. The images of an IDE sensor taken using high power microscope after being deposited with gold nanorod is shown in figure 4b, the gold nanorods were well covered on the desired surface of the IDE as it can be observed as dusky spots on the IDE surface.

Figure 4c shows the image taken using 3D Profilometer of with gold nanorod deposited on the surface of the IDE, the surface roughness of the IDE can be seen clearly with the gold nanorod evenly dispersed on the IDE. Also figure 4d shows the image of IDE taken using 3D Profilometer, as it can be seen clearly that the IDE was patterned properly. Figure 4e indicates the image taken using SEM with a gap size and finger electrode lengths. The gap of the IDE image was revealed by SEM to be 40 µm and the finger electrode lengths were 86 µm and 85 µm respectively. Figure 4f indicates the 3D image of IED after gold nanoparticles were deposited on the surface of the IDE. AFM image uncovered gold surface. The coating of aluminum interdigitated electrode with gold nanorod was also controlled by atomic force microscope. The obtained layer is continuous, remarkably smooth, and densely uniform compared with bare IDE surface, without any error.

It is clearly shown that with the presence of gold nanorod, the surface becomes rough as in the above. The surface roughness improves the sensitivity of biomolecule detection as greater number of surface area was obtained [32]. Higher surface area to volume ratio allows higher sensitivity detection of biomolecules. It can be observed that the aluminum possesses very low surface roughness. Unlike gold nanorod, harsh surface with high porosity surface pattern formed. Basically, the rough surface provides higher surface to volume ratio that allows better sensitivity towards the biosensor devices [14]. The SEM imaging was used to analyze the morphological features of gold nanorod. Gold nanorod distinctly formed on aluminum interdigitated electrode due to the unique morphology of aluminum interdigitated electrode. The smooth surface of silicon wafer does not incite the formation rock-like structures. The aluminum interdigitated electrode grown via photolithography shows a unique irregularly spaced structure with rippled surfaces that cause the gold nanorod to self-assemble into rod-like structure. Table 1 shows the measurement IDE gap and finger electrode lengths. As it can be clearly seen that the gap is 40 µm and the finger electrode lengths of 86 and 85 µm. Based on the morphological characterizations conducted, we can conclude that the fabrication and deposition of gold nanorod were successful.

4. Conclusion

The advantages of using gold nanorod for surface modification is that, they have electrochemically active surface areas and the better accessibility of the analyte to the surface of the electrode. Fabrication of IDE and coating of gold nanorod were described based on photolithograpic method. The method serves as a coating for a conductive surface to develop an electrochemical sensor with an actively modified sensing surface. This method is simpler and cheaper than other modification techniques. The bond of the coating is better and gold nanorod are already adhered to the IDE surface without the application of any supplementary chemical that might inhibit subsequent sensing materials. Nevertheless, this method produces a low amount of material on the electrode surface, making the characterization more complex. Hence, further optimization is needed for the fabrication of submicron of IDEs in order to improve the sensitivity.

Acknowledgement

The author would like to thank all staff members Institute of Nano Electronic Engineering (INEE) and School of Bioprocess Engineering, Universiti Malaysia Perlis for their technical assistance and contribution.

References

[1] D. Tonelli, E. Scavetta, and I. Gualandi, *Sensors (Switzerland)*, **19**, 5 (2019).
[2] T. Adam and U. Hashim, "Fabrication of Micro-Mixer for Life Sciences Applications," 2014.

IOP Conf. Series: Materials Science and Engineering **743** (2020) 012017 doi:10.1088/1757-899X/743/1/012017

[3] T. Adam and U. Hashim, "Label Free Detection of Single Biomolecules," in *2014 Fifth International Conference on Intelligent Systems, Modelling and Simulation*, 2014, pp. 1–8.
[4] S. C. B. Gopinath, U. Hashim, M. N. A. Uda, *AIP Conf. Proc.*, **1808**, 1–5 (2017).
[5] S. A. B. Ariffin, T. Adam, U. Hashim, S. Faridah, I. Zamri, 2018 (2013).
[6] T. Adam, U. Hashim, Thikra S Dhahi, M.N.A.Uda, in *2014 Fifth International Conference on Intelligent Systems, Modelling and Simulation*, 722–725 (2014).
[7] S. N. Bunawan *et al.*, *Procedia Chem.*, **20**, 121–123 (2016).
[8] V. V Malev, V. V Kondratiev, *Russ. Chem. Rev.*, **75**, 147–160 (2006).
[9] M. N. A. Uda *et al.*, **020058**, 020058 (2017).
[10] M. N. A. Uda *et al.*, *AIP Conf. Proc.*, **1808**, 020057 (2017).
[11] M. . N. A. Uda *et al.*, *Adv. Mater. Res.*, **832**, 410–414 (2014).
[12] M. N. A. Uda *et al.*, *AIP Conf. Proc.*,**1808** (2017).
[13] S. Moraes Silva, R. Tavallaie, L. Sandiford, R. D. Tilley, and J. J. Gooding, *Chem. Commun.*, **52**, 7528–7540 (2016).
[14] L. Qin *et al.*, *Sci. Total Environ.*, **652**, 93–116 (2019).
[15] J. Jeevanandam, A. Barhoum, Y. S. Chan, A. Dufresne, M. K. Danquah, *Beilstein J. Nanotechnol.*, **9**, 1050–1074 (2018).
[16] S. C. B. Gopinath, S. Ramanathan, K. Hann Suk, M. Ee Foo, P. Anbu, M. N. A. Uda, *MATEC Web Conf.*, **150**, 06002 (2018).
[17] S. Mourdikoudis, R. M. Pallares, N. T. K. Thanh, *Nanoscale*, **10**, 12871–12934 (2018).
[18] N. Bhalla, P. Jolly, N. Formisano, P. Estrela, **1**, 8 (2016).
[19] W. R. Adams, *PLoS One*, **12**, 1–20 (2017).
[20] M. Drahansky *et al.*, *Intech*, **1**,13 (2016).
[21] N. A. Parmin *et al.*, *Int. J. Biol. Macromol.*, **126**, 877–890 (2019).
[22] N. A. Parmin *et al.*, *Microchim. Acta*, **186**, 6 (2019).
[23] V. Von, L. Und, R. Des, O. Der, A. G. D. E. R. Wissenschaftlichen, **207**, 651–656 (2010) .
[24] J. Oberländer *et al.*, **3**. 26115–26127, (2015).
[25] M. A. Farehanim *et al.*, *Adv. Mater. Res.*, **1109**, 118–122 (2015).
[26] R. Haarindraprasad, U. Hashim, S. C. B. Gopinath, V. Perumal, W. W. Liu, and S. R. Balakrishnan, *Anal. Chim. Acta*, **925,** 70–81 (2016).
[27] A. K. M. Muaz, U. H. Member, W. Liu, F. I. Member, K. L. Thong, M. S. M. Member, 8–10, (2014).
[28] V. Perumal *et al.*, *Biosens. Bioelectron.*, **116,** 116–122 (2018).
[29] M. Citartan, S. C. B. Gopinath, J. Tominaga, T. H. Tang, *Analyst*, **138**, 3576–3592 (2013).
[30] V. Perumal *et al.*, *Nanoscale Res. Lett.*, **30**, 49 (2016).
[31] N. Z. Natasha, R. D. A. A. Rajapaksha, M. N. A. Uda, U. Hashim, **020235**, (2017).
[32] Z. Li, C. Leung, F. Gao, Z. Gu, 22473–22489, (2015).

IOP Conf. Series: Materials Science and Engineering **743** (2020) 012018 doi:10.1088/1757-899X/743/1/012018

New Development Quantification Methods for Salt Iodine and Urinary Iodine Using Microfluidics Based Nanotechnology

Nur Hulwani Ibrahim[1] *, M. N. Salimi[3], M. N. A. Uda[1,2], N. A. Parmin[1], Uda Hashim[1] and M. N. Afnan Uda[3]

[1]Institute of Nano Electronic Engineering (INEE), Universiti Malaysia Perlis (UniMAP), 01000 Kangar Perlis Malaysia.
[2]School of Bioprocess Engineering Group, Universiti Malaysia Perlis (UniMAP) 02600 Arau Perlis
[3]School of Microelectronic Engineering Group, Universiti Malaysia Perlis (UniMAP) Putra Campus, 02600 Pauh, Perlis Malaysia

Email:* hulwaniibrahim@gmail.com

Abstract. In Malaysia, the first Iodine Deficiency Disorders (IDD) survey was conducted in 1996 and it was discovered that Peninsular Malaysia did not have IDD problem until latter studies showed goitre occurrence of 34.7% in Hulu Langat district and urinary iodine lower than the adequate level of (100-199 ug/L) in Perak and Pahang states (Selamat et al., 2010). Baseline and periodical sampling of children and pregnant woman urine and imported salt commodities for the consumption of the population is mandatory for iodine measurement. Thus, development of quantitative methods of measurement of salt and food iodine is crucial for implementation of the USI program nationwide. In this study, interdigitated electrode (IDE) biosensor, a rapid, sensitive and selective method has been developed to determine the iodine content in both urine and salt. This method includes functionalization and silanization step using 3-aminopropyl triethoxysilane (APTES). The I-V characterization of IDE biosensor was performed using (Keithley 2450), Kickstart software and Probestation. It measures the amount of current flow through IDE which is directly proportional to the concentration of iodine in both urine and salt. Hence, IDE biosensor is proven to be a rapid, selective, sensitive method and can be developed as a new nanotechnology for the elimination of Iodine Deficiency Disorders (IDD) among children and pregnant woman.

1. Introduction

Iodine is an essential nutrient involved in the production of thyroid hormones and is required for the development of fetal nervous system during pregnancy [1]. Pregnant women are mostly susceptible to iodine deficiency because of its higher demand during pregnancy. Iodine deficiency can lead to major health problems, including endemic goiter, cretinism, and growth retardation. Besides that, it can also cause perinatal problem such as congenital anomalies, miscarriage and stillbirth [2]. Iodine deficiency is a serious and preventable health problem affecting more than 2 billion people worldwide [3].

In order to reduce the impact of iodine deficiency, it is essential to determine the population iodine nutrition status and promote adequate iodine intake during pregnancy. Children with adequate iodine nutrition were also shown to have higher intelligent quotient (IQ) compared to those with low iodine nutrition [4].

Content from this work may be used under the terms of the Creative Commons Attribution 3.0 licence. Any further distribution of this work must maintain attribution to the author(s) and the title of the work, journal citation and DOI.
Published under licence by IOP Publishing Ltd

Measurement of urine iodine need to be carried out since low level of it has been reported to be correlated with the occurrence of Iodine Deficiency Disorders (IDD). The population median urinary iodine values have been classified accordingly by the World Health Organization (WHO) and urinary iodine concentration (UIC) in 2007 (Table 1) to severe, moderate and mild iodine deficiency, adequate iodine nutrition, slight and excessive risk for iodine-induced hyperthyroidism and autoimmune thyroid diseases.

Table 1. The classifications of median urinary iodine values related to iodine nutrition status (Source: WHO, 2007).

Median urinary iodine (μg/L)	Iodine intake	Iodine nutrition status
School-age children (6 years or older)		
<20	Insufficient	Severe iodine deficiency
20–49	Insufficient	Moderate iodine deficiency
50–99	Insufficient	Mild iodine deficiency
100–199	Adequate	Adequate iodine nutrition
200-299	Above requirement	Risk of iodine-induced hyperthyroidism in susceptible groups
≥300	Excessive	Risk of adverse health consequences (iodine-induced hyperthyroidism, autoimmune thyroid disease)
Pregnant women		
<150	Insufficient	
150–249	Adequate	
250–499	Above requirements	
≥500	Excessive	
Lactating women and children aged less than 2 years		
<100	Insufficient	
≥100	Adequate	

* The term "excessive" means in excess of the amount required to prevent and control iodine deficiency

One of the most implemented interventions in combating the IDD is through Universal Salt Iodization (USI).Thus, the quantitation of the iodine in salt consignments is crucial upon enactment of USI (Hetzel, 1983). Quantitation of food iodine is also important for food science research including for compilation of food composition database for public use. The International Council for control of iodine Deficiency Disorders (ICCIDD) had proposed the standard method of titration for determining iodine content in salt. Commercially, iodine concentration is exclusively measured calorimetrically by the Sandell-Kolthoff method in which yellow cerium (IV) is reduced to cerium (III) in the presence of iodide ions. This method requires complicated sample preparation, involves a slow response time, and utilizes a toxic reagent that must be properly stored and disposed [5]. Other methods including the Inductively-Coupled Plasma-Mass Spectrometry (ICP-MS) is more sensitive and precise but is very expensive and may not be affordable by small/medium-scaled laboratories. However, all these methods are laboratory-based methods and salt sample have to be sampled and taken to the laboratory for analysis.

Recently, researchers use an integrated approach by combining nanoscience technology, electronics, computers and biology to create portable biosensors with extraordinary sensing capabilities

[6-11]. Thus, in this study nanotechnology biosensor will be develop for a rapid determination of iodine in both urine and salt. This kind of nanotechnology has been widely used in other medical field [12-14], agriculture [14-18] food industry [15-16] ,and Environment [22].

2. Material and method

2.1 Materials

Ethanol, 3-aminopropyl triethoxysilane (APTES), sodium chloride, commercial salt: rock salt and iodised salt. All analytical grade reagents were purchased from Sigma Aldrich.

2.2 Preparation of standard solution of iodine

A series of eight standard solution of iodine solutions were prepared by diluting 1000 μg/L of their stock solution with distilled water until calibration mark. The standard solutions were prepared at eight different concentrations depending on the sensitivity of the interdigitated electrode (IDE). The lowest concentration of iodine at 0.001 μg/L was prepared until 1000 μg/L.

2.3 I-V Characterization for detection of iodine concentration using interdigitated electrode (IDE)

Firstly, the IDE electrode was cleaned with ethanol and distilled water and left to dry for 5 minutes. After that, 2μL of APTES solution was pipetted on the active area of IDE electrode and left to dry in the dry cabinet at room temperature of 27 °Cfor 30 minutes. It builds an active layer on the IDE. Then, it was washed with distilled water to remove the unbounded APTES. APTES is an aminosilane frequently used in process of salinization and functionalization of the surfaces with alkoxysilane molecules. Lastly, 2μL of standard iodine and was pipetted on the active area of IDE electrode and kept inside the dry cabinet for 30 minutes. This step was repeated with 2μL salt solution. The I-V characterization then was completed using (Keithley 2450), Kickstart software and Probestation.

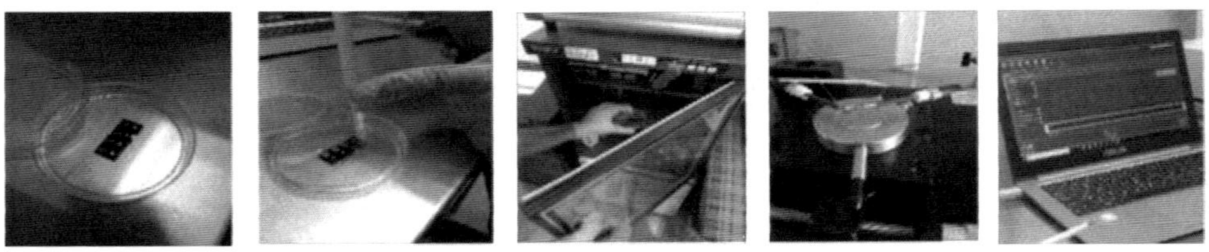

(1) (2) (3) (4) (5)

Figure 1. I-V characterization of iodine and salt solution.

3. Result and Discussion

The I-V characterization was performed using (Keithley 2450), Kickstart software and Probestation with the IDE biosensor to get the I-V characteristic as shown in Figure 2. The I-V characterization measures the amount of current flow through IDE. During the I-V characterization, picoammeter voltage was supplied from 0 to 1 V only, if the voltage applied more than that suitable range, it will damage the IDE biosensor [20]. Figure 2 shows the I-V characteristic for bare IDE, APTES and different concentration of standard iodine ranged from 0 ppb to 1000 ppb.

From the I-V measurement (Figure 2), the bare IDE gives the current value 2.27×10^{-11} A whereas for the functionalize IDE with APTES gives the current value 1.19×10^{-9}. Then, the current value recorded for 0 ppb, 0.001 ppb, 0.01 ppb, 0.1 ppb, 1 ppb, 10 ppb, 100 ppb and 1000 ppb were 1.01×10^{-9} A, 1.51×10^{-9} A, 2.94×10^{-9} A, 3.34×10^{-9} A, 6.92×10^{-9} A, 9.01×10^{-9} A, 1.01×10^{-8} A and 2.39×10^{-8} A respectively. It can be seen that, the current values increase as the concentration of standard iodine increase. Hence, IDE biosensor has high sensitivity that can measure until the lowest concentration of iodine. The usage of the IDE sensors in the biosensor field is tremendous in these days because of the large number of comb structured finger electrodes that gain high sensitivity [23].

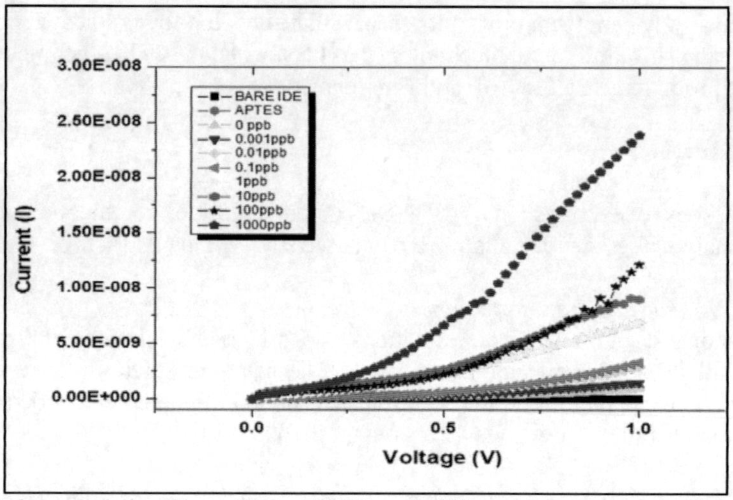

Figure 2. I-V characteristic of bare IDE, APTES and different concentration of standard iodine.

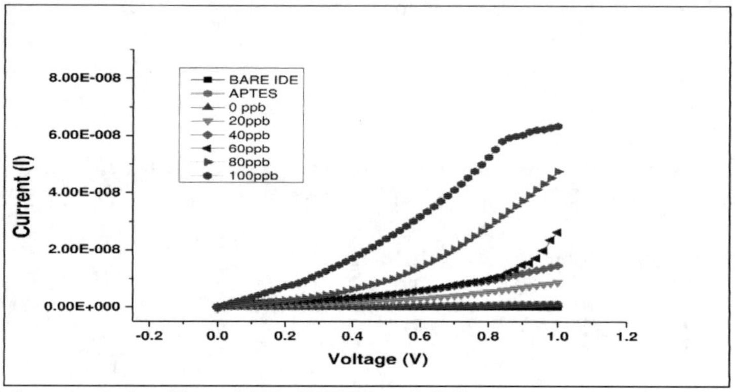

Figure 3. I-V characteristic of bare IDE, APTES and different concentration of standard iodine.

The I-V measurement of standard iodine then was repeated using 0 ppb, 20 ppb, 40 ppb, 60 ppb, 80 ppb and 100 ppb. The current values were 1.01×10^{-9}, 8.83×10^{-9}, 1.47×10^{-8}, 2.62×10^{-8}, 2.63×10^{-8}, 4.76×10^{-8} and 6.34×10^{-8} respectively. It shows that in Figure 3, the concentration of iodine is directly proportional to the current value, the current will continue to increase as the concentration of iodine increase. Thus, IDE biosensor capable for detection starting from low to high concentration of iodine.

Figure 4 shows the I-V characterization for different brands of salt where sodium chloride, NaCl act as a control. The current value for NaCl, sea salt and rock salt were 2.37×10^{-8}, 4.76×10^{-8} and 2.42×10^{-8} respectively. From the graph above, it clearly shown that, sea salt gives the highest current value which is 4.76×10^{-8} compare to others. Sea salt is iodized salt which contain the highest amount of iodine among the other salt. IDE biosensor is proven to be a very sensitive and rapid method to detect the present of iodine in the samples. This advance nanotechnology biosensor can be applied in biomedical field for monitoring and intervention programme of Universal Salt Iodization (USI) for elimination of Iodine Deficiency Disorders (IDD) in pregnant woman and schoolchildren.

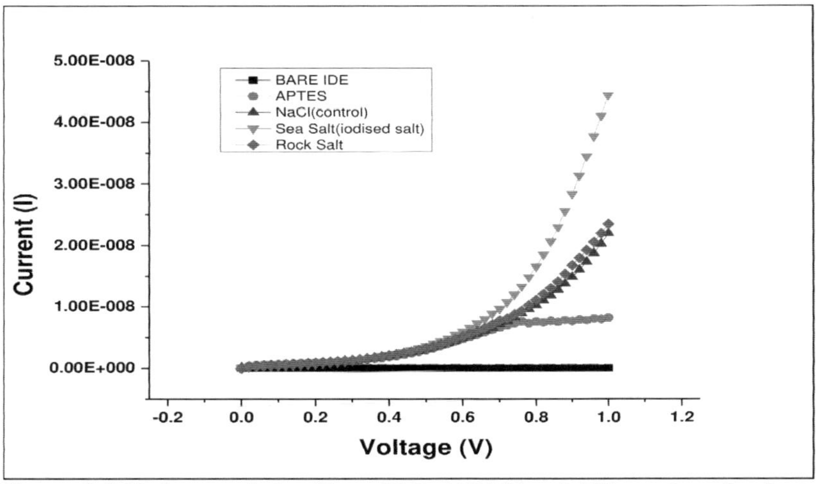

Figure 4. I-V characteristic of bare IDE, APTES and different brands of salt.

4. Conclusion

In this research, iodine was successfully detected using IDE biosensor. This nanotechnology biosensor can be developed and applied in biomedical field due to the simple, rapid, selective and sensitive method compared to the conventional method that requires complex laboratory handling methods and instrumentation.

Acknowledgement

The author would like to thank all staff members Institute of Nano Electronic Engineering (INEE) and School of Bioprocess Engineering, Universiti Malaysia Perlis for their technical assistance and contribution.

References

[1] M. B. Zimmermann, "Iodine requirements and the risks and benefits of correcting iodine deficiency in populations," *J. Trace Elem. Med. Biol.*, vol. 22, no. 2, pp. 81–92, 2008.

[2] A. A. Shamim *et al.*, "Iodine status in pregnancy and household salt iodine content in rural Bangladesh," *Matern. Child Nutr.*, vol. 8, no. 2, pp. 162–173, 2012.

[3] M. B. Zimmermann, "The role of iodine in human growth and development," *Semin. Cell Dev. Biol.*, vol. 22, no. 6, pp. 645–652, 2011.

[4] M. Qian *et al.*, "The effects of iodine on intelligence in children: A meta-analysis of studies conducted in China," *Asia Pac. J. Clin. Nutr.*, vol. 14, no. 1, pp. 32–42, 2005.

[5] E. Cho, A. Perebikovsky, O. Benice, S. Holmberg, M. Madou, and M. Ghazinejad, "Rapid iodine sensing on mechanically treated carbon nanofibers," *Sensors (Switzerland)*, vol. 18, no. 5, pp. 1–9, 2018.

[6] S. A. B. Ariffin, T. Adam, U. Hashim, S. Faridah Sfaridah, I. Zamri, and M. N. A. Uda, "Plant Diseases Detection Using Nanowire as Biosensor Transducer," *Adv. Mater. Res.*, vol. 832, pp. 113–117, 2013.

[7] U. Hashim, M. F. M. Fathil, M. K. M. Arshad, S. C. B. Gopinath, and M. N. A. Uda, "Characterization of zinc oxide thin film for pH detector," *AIP Conf. Proc.*, vol. 1808, 2017.

[8] M. N. A. Uda *et al.*, "Determination of set potential voltages for cucumber mosaic virus detection using screen printed carbon electrode," *AIP Conf. Proc.*, vol. 1808, no. March, 2017.

[9] T. Adam and U. Hashim, "Fabrication of Micro-Mixer for Life Sciences Applications," 2014.

[10] T. Adam, U. Hashim, Thikra S Dhahi, and M.N.A.Uda, "Fabrication of Semi-Conductor Silicon Nanowire by BOE Based Micro-Nano Indentation," in *2014 Fifth International Conference on Intelligent Systems, Modelling and Simulation*, 2014, pp. 722–725.

[11] S. A. B. Ariffin, T. Adam, U. Hashim, S. Faridah, and I. Zamri, "Fabrication of Nanostructure Based Biosensor for Biomolecules Detection Fabrication of Nanostructure Based Biosensor for Biomolecules Detection," no. March 2018, 2013.

[12] N. A. Parmin *et al.*, "Human Papillomavirus E6 biosensing: Current progression on early detection strategies for cervical Cancer," *Int. J. Biol. Macromol.*, vol. 126, pp. 877–890, 2019.

[13] S. C. B. Gopinath, S. Ramanathan, K. Hann Suk, M. Ee Foo, P. Anbu, and M. N. A. Uda, "Engineered nanostructures to carry the biological ligands," *MATEC Web Conf.*, vol. 150, p. 06002, 2018.

[14] N. A. Parmin, U. Hashim, S. C. B. Gopinath, and M. N. A. Uda, *Biosensor Recognizes the Receptor Molecules*. Elsevier Inc., 2019.

[15] M. N. A. Uda *et al.*, "Immunosensor development for rice tungro bacilliform virus (RTBV) detection using antibody nano-gold conjugate," *AIP Conf. Proc.*, vol. 1808, no. October, 2017.

[16] M. N. A. Uda *et al.*, "Reviewed Immunosensor Format Using Nanomaterial for Tungro Virus Detection," vol. 832, no. November 2013, pp. 410–414, 2014.

[17] T. Adam and U. Hashim, "Label Free Detection of Single Biomolecules," in *2014 Fifth International Conference on Intelligent Systems, Modelling and Simulation*, 2014, pp. 1–8.

[18] M. N. . Uda *et al.*, *A Disposable Biosensor Based on Antibody-Antigen Interaction for Tungro Disease Detection*. Elsevier Inc., 2019.

[19] M. N. A. Uda *et al.*, "Immunosensor development formatting for tungro disease detection using nano-gold antibody particles application," *AIP Conf. Proc.*, vol. 1808, no. 020057, 2017.

[20] M. N. A. Uda, R. D. A. A. Rajapaksha, M. N. A. Uda, U. Hashim, and A. B. Jambek, "Selective detection of E.coli O157:H7 bacteria DNA using electrical based aluminium interdigitated electrode biosensor," *AIP Conf. Proc.*, vol. 2045, no. December, 2018.

[21] N. Z. Natasha, R. D. A. A. Rajapaksha, M. N. A. Uda, and U. Hashim, "Electrical DNA Biosensor using Aluminium Interdigitated Electrode for E . Coli O157 : H7 detection," vol. 020235, 2017.

[22] M. A. M. Akhir *et al.*, "Ethylene Gas Sensing Properties of Tin Oxide Nanowires Synthesized via CVD Method," *IOP Conf. Ser. Mater. Sci. Eng.*, vol. 318, no. 1, 2018.

[23] R. D. A. A. Rajapaksha, M. N. A. Uda, U. Hashim, S. C. B. Gopinath, and C. A. N. Fernando, "Impedance based Aluminium Interdigitated Electrode (Al-IDE) biosensor on silicon substrate for salmonella detection," *IEEE Int. Conf. Semicond. Electron. Proceedings, ICSE*, vol. 2018-Augus, pp. 93–96, 2018.

Development of Voltage Amplifier Electronic Reader for Multiplex Detection of Two Electrode Electrical Biosensors

M. N. Afnan Uda[1]*, Asral Bahari Jambek[1], U. Hashim[1, 2] and M. N. A. UDA[3]

[1]School of Microelectronic Engineering Group, Universiti Malaysia Perlis (UniMAP) Putra Campus, 02600 Pauh, Perlis Malaysia
[2]Institute of Nano Electronic Engineering (INEE), Universiti Malaysia Perlis (UniMAP), 01000 Kangar Perlis Malaysia
[3]School of Bioprocess Engineering Group, Universiti Malaysia Perlis (UniMAP) 02600 Arau Perlis

Email: * nurafnan92@yahoo.com

Abstract. Amperometric electrical biosensors have small current variations at nano to micro range. There are limited number of electrical based two electrode electronic readers for biosensors are commercial available because of the amplification and noise issues at nano to micro ampere current range. The electronic reader focused on design a voltage amplifier circuit to capture and amplify three different range of current as nano, micro and mili ampere and convert it to detectable voltage range as an output voltage signal. Current input as nano, micro and mili current were flown through 10 KΩ, 10 Ωand 10 mΩresistors, respectively to convert different current inputs to the similar range in micro voltage. Then, MAX 4238 op-amp IC was used to amplify micro voltage to mili voltage. Arduino Uno circuit was act as the process and control circuit to read the output voltage from the amplifier circuit. Arduino Uno circuit will convert analog signal to digital signal and then the output voltage value is display in the LCD screen. The Proteus 8 Pro software was used to design, simulate and calibrate the amplifier circuit and Arduino Uno circuit. While, Arduino software was used to create a programming code and to upload in Arduino Uno circuit.Start your abstract here.

1. Introduction

The experiments of amperometric electrical based biosensor are expensive and time consuming [1], [2]. Because of the detectable range of biomarkers usually in pico molar (pM) to micro molar (μM) concentration and limit of detection (LOD) even goes down to femto molar concentrations for some early detection cases [3]–[5]. Thus, current variation for these concentrations, usually in nano ampere to micro ampere range and some time it goes down to pico ampere range. Therefore high precise equipment is necessary for biosensor research. Commercialization of the biosensors are critical because of the complexity and the large size of the detector systems [6]–[9]. Moreover, most of the electronic reader design only for 3 electrode type instead 2 electrode type for electrical biosensors. Usually most of the electronic readers used 3 electrode type were blood glucose meter which integrated with micro range amplifiers [3], [10]–[14] Nowadays, very few electronic readers used 2 electrode type for electrical biosensors and capability to detect low current until nano range in commercial. In addition, the use of biosensors included hand held electronic reader provide more potential for commercialization [15]–[19].

Conventional methods of biomarker detection have many draw backs as time consuming, low detectable range, high expensive, depend on lab base experiment and required knowledgeable people

Content from this work may be used under the terms of the Creative Commons Attribution 3.0 licence. Any further distribution of this work must maintain attribution to the author(s) and the title of the work, journal citation and DOI.
Published under licence by IOP Publishing Ltd

for testing. Usually this type of methods can identify single disease per time. By using electrical biosensor, the device able go to multiplex analyse [20], [21]. Apart from that, with an electronic reader, lab result can be retrieved much more rapidly, thus saving time and money and in case of emergency, these readers can provide critical information to emergency care provider with the accurate result [22]. Simple devices and techniques such as using proper grounding method, shielded and twisted wire, signal averaging methods, filter, and differential input voltage amplifiers can control the noise in most measurements. Some techniques prevent noise from entering the system, while other removes extraneous noise from the signal [23]–[25].

Almost every industry is now computerized and digitized for rapid data retrieval and trend analysis. In recent years, Arduino microcontroller based open-source hardware projects now popular for control the electrical systems. Moreover, this Arduino microcontroller is not a separate piece but a single board, used to code and upload computer code to the physical board. Other than that, in order to load new code onto Arduino microcontroller can simply use a USB cable. The most important, the Arduino microcontroller hardware is reasonable priced and development software is open-source [26]–[29]. Thus, it can be easily used.

In this paper, we have designed and developed hand held electronic reader which included voltage amplifier and Arduino Uno microcontroller unit (MCU). The reader is specific to two electrode amperometric electrical biosensors and capability to detect until nano, micro and mili ranges. Apart from that, the electronic reader can be used for multiplex analyse. The voltage amplifier is amplified extra small current through sensors from nano-micro-mili ampere range to mili voltage. The MAX 4238 op-amp is used as the amplifier. The mili output voltage through to analog input port of Arduino Uno board. Arduino Uno convert analog signal to digital signal and display the result to LCD display.

2. Material and methods

2.1. *Apparatus, equipment and software*

The MAX 4238 op-amp was used as operational amplifier to amplify micro voltage to mili voltage. The voltage converter using shunt resistor was used to convert input current into input voltage to the amplifier. Then, LM 358 operational amplifier was used as to supply virtual ground for MAX 4238 op-amp and TPS 3809 was used as battery testing. The external voltage source 9 V DC battery was used to supply power for Arduino Uno circuit. The Arduino Uno Board was act as control circuit and isolate internal voltage source to others circuit. Arduino Uno internal voltage source was used to supply 3 V to amplifier circuit, 1 V to sensor circuit and 5 V to LCD display. Arduino Uno control circuit was used to read the amplify output voltage from MAX 4238 op-amp. LCD display was used to display the amplify output voltage and analog sensor value. Aluminium clip was used as the two electrode electrical biosensor connector. The voltage source was used to supply voltage through variable resister that acts as a replacement of biosensors to input current to the sensor ports. The Arduino Uno software was used to write program code and uploaded it into microcontroller Arduino Uno board. The Proteus 8 Pro software was used to design and simulate the circuit.

2.2. *The Circuit design*

Figure 1 shows the detail block diagram for a voltage amplifier for three different range as nano, micro and mili for electrical biosensors. There are six major parts of electronic reader device consists of aluminium clip sensor connection, switches sensors, low noise amplifier, Arduino Uno circuit, DC power supply, and LCD display. The amplifier circuit as voltage converter and voltage amplifier. Sensor circuit input system has included sensor input connectors and three dual select switches to select appropriate sensor to resistors. Arduino Uno has included in isolate power supply and main control circuit. As shown in the figure, amplified voltage output can be detected using microcontroller Arduino Uno and LCD display, acts as a replacement detected using multimeter.

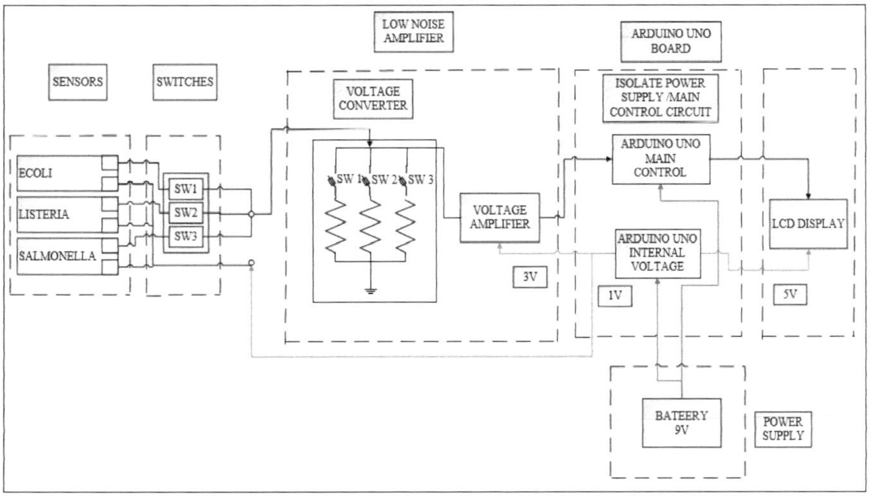

Figure 1. Detail block diagram for electronic reader.

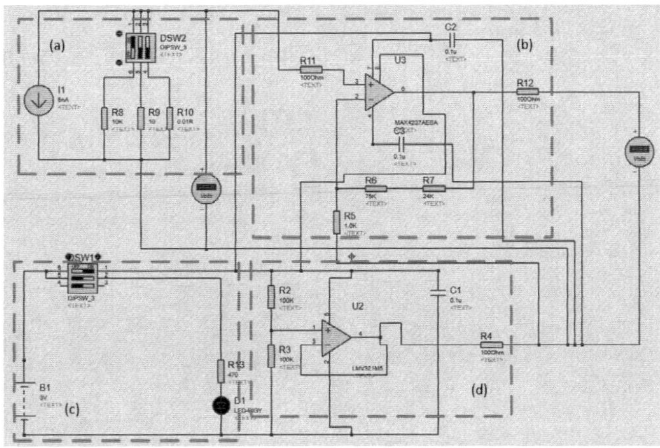

Figure 2. Voltage Amplifier circuit design using Proteus 8 Pro software.

Voltage amplifier circuit was designed using Proteus 8 Pro software as shown in Figure 2. There are four major voltage amplifier circuit parts, voltage converter, MAX 4238 op-amp, power supply circuit, and virtual ground suppling circuit. Figure 2 (a) shows the sensor selection circuit with current to voltage converter by using shunt resistor. There are three switches to select the respective resistor for selected sensors in three different current as nano, micro and mili ampere. Figure 2 (b) shows the MAX 4238 op-amp circuit and Figure 2 (c) shows the power supply circuit which was included with TPS 3809 as battery testing circuit. Figure 2 (d) shows the virtual ground suppling circuit to MAX 4238 op-amp IC. The voltage amplifier circuit was simulated using Proteus 8 Pro software.

Arduino circuit design was act as replacement for multimeter to read the mili output voltage amplifier. As shown in Figure 3, there are four major Arduino Uno circuit parts, included Arduino Uno main control circuit, Arduino Uno internal voltage, LCD display and three selection switches. Figure 3 (a) shows the "ELS ELECTRONIC READER" on LCD display after the Arduino Uno circuit switch ON. Figure 3 (b) shows the "ECOLI" pathogen type with analog value and mili output voltage amplifier on LCD display after the first selection switches been selected. While, Figure 3 (c) and Figure 3 (d) shows the "LISTERIA" and "SALMONELLA" pathogen type with analog value and mili output voltage amplifier on LCD display after the second and third selection switches been selected.

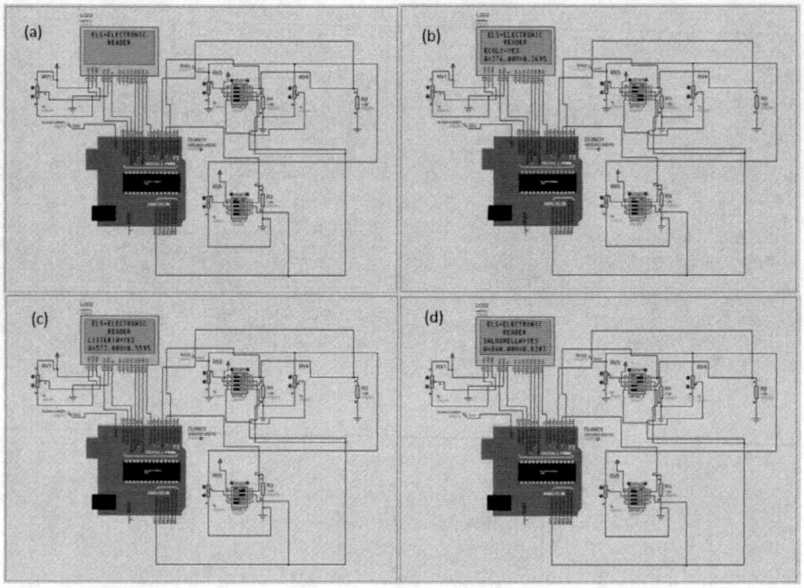

Figure 3. Arduino circuit design simulation using Proteus 8 Pro software.

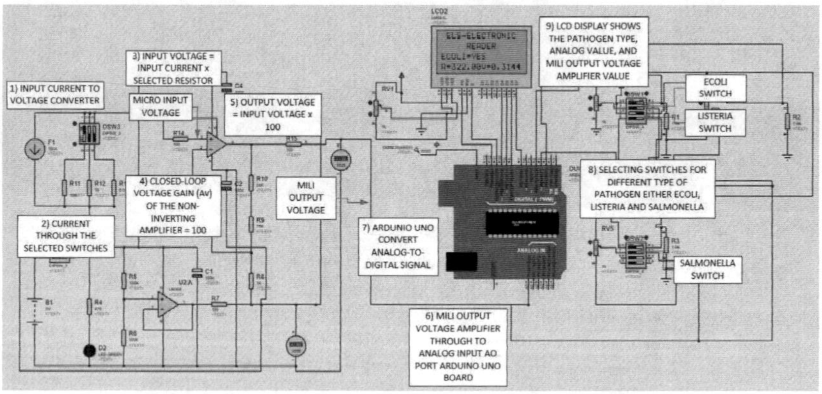

Figure 4. Complete operation of electronic reader.

Figure 4 shows the complete operation of electronic reader circuit using Proteus 8 Pro software. Electronic reader consists of combination voltage amplifier circuit and Arduino circuit. The electronic reader design was simulated using simulator on Proteus 8 Pro software. The theoretical and experimental result were recorded and compared to identify the accuracy of the electronic reader circuit [29], [30]. These formula was used to calculate the theoretical output voltage through the voltage amplifier circuit and Arduino Uno microcontroller. Based on the equation, sensor input current Ii (A), shunt resistance for appropriate range R (Ω), and input voltage Vi (V) as defined. As shown in Equation (1), the calculation to find the input current to sensor switches port using voltage source and variable resistor. Equation (2) shows the calculation to find the input voltage.

$$Ii = \frac{VR}{R} \tag{1}$$

$$Vi = Ii \times R \tag{2}$$

After converting input current to input voltage, the input voltage will apply directly to the non-inverting (+) input terminal as shown in Figure 5. It means the output signal is in-phase with the input signal. The signal at the output is not inverted when compared to the input. Moreover, the feedback control of the non-inverting operational amplifier is done by applying a small part of the output voltage signal back to the inverting input terminal via an $R_f - R_2$ voltage divider network [31]. This closed-loop configuration produces a non-inverting amplifier circuit with very good stability, a very high input impedance, as no current flows into the positive input terminal and a low output impedance [16]. Equation (3) and Equation (4) shows the equation of the gain and the output voltage (Vo). Based on the calculation, the gain of MAX 4238 op-amp IC is 100. The voltage drop across shunt resistor in micro voltage, the analog to digital in microcontroller Arduino read only in mili voltage range. Then, the amplifier need to amplify the signal with gain 100 [3].

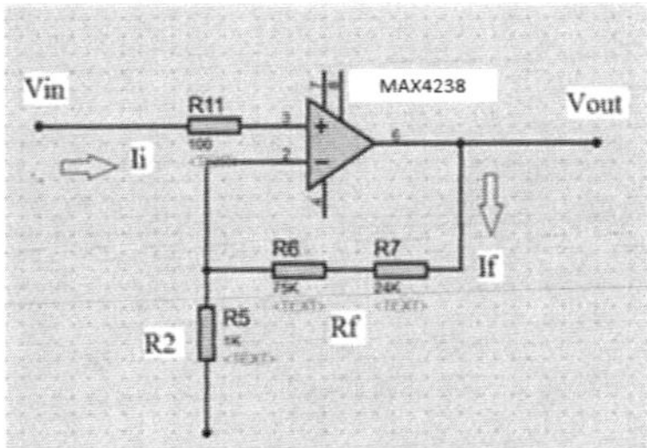

Figure 5. Non-inverting Operational Amplifier.

$$\text{Gain, Av} = \frac{\text{Vout}}{\text{Vin}} = \frac{\text{R2} + \text{Rf}}{\text{R2}} = 1 + \frac{\text{Rf}}{\text{R2}} \tag{3}$$

$$\text{Vo} = \text{Vi} \times \text{G} \tag{4}$$

After amplify the micro input voltage to mili output voltage, the voltage value will through to Arduino Uno microcontroller on analog input "AO" port. Arduino Uno converted the analog voltage to discrete value. The reference voltage (Vref) used in the Arduino equal to 1 V. Equation (5) shows the calculation for microcontroller Arduino Uno read the output voltage amplifier and display to LCD display.

$$\text{Sensor Value} = \text{Analog Read}$$
$$\text{Output Voltage} = (\text{Sensor Value} \ / \ 1024) \ * \text{Vref} \tag{5}$$

Figure 6 shows the electronic reader hardware setup consists of low noise amplifier, microcontroller Arduino Uno, LCD display and aluminium clips included sensor connectors. Figure 6 (a) shows the separate circuits of the electronic reader. After testing of each separate circuit, the separate circuit were assembled to make one complete device as shown in Figure 6 (b). After designing the software and hardware circuit, Proteus 8 Pro software was used to convert the electronic reader schematic design to PCB design layout.

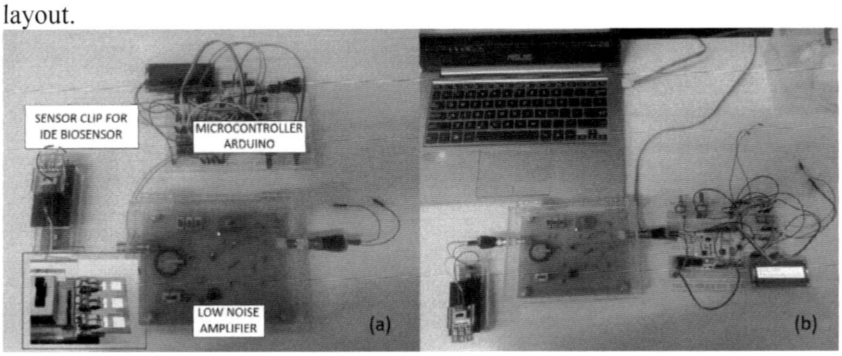

IOP Conf. Series: Materials Science and Engineering **743** (2020) 012019 doi:10.1088/1757-899X/743/1/012019

Figure 6. Electronic reader hardware setup, (a) separate circuit, (b) integration of separate circuit.

2.3 Fabrication and assembly of PCB
Proteus 8 Pro software was used to convert the electronic reader schematic design to PCB layout as shown in Figure 7 (a). Then, the PCB layout convert to 3D visualizer layout as shown in Figure 7 (b). The circuit design layout was transferred onto copper clad board to make PCB, after removed first and second layer as shown in Figure 7 (c) and Figure 7 (d). After that, assembled all the components to make the portable amplifier device as shown in Figure 7 (e) and Figure 7 (f).

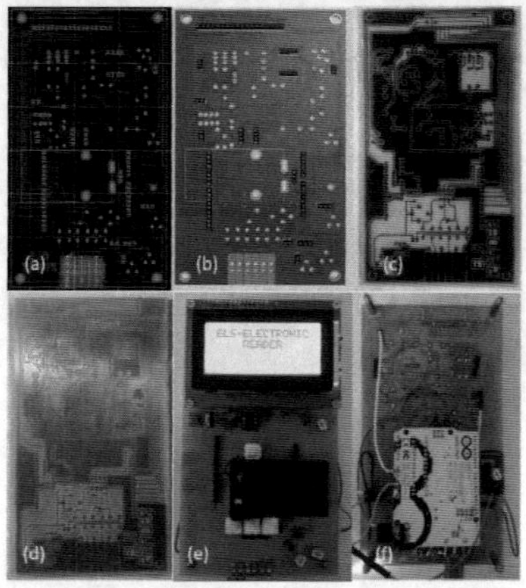

Figure 7. Fabrication process of the electronic reader PCB board. (a) PCB layout of the electronic reader circuit, (b) PCB layout convert to 3D visualizer layout view, (c) copper clad board without first layer, (d) copper clad board without second layer, (e) top view of the electronic reader and (f) bottom view of the electronic reader.

3. Result and Discussion
Figure 8 shows the measurement setup for testing the performance of the electronic reader circuit. This project was focus in three different current ranges consists of nano, micro and mili current. Voltage source and variable resistor acts as a replacement of biosensor input current. The electronic reader was validated by comparing the output voltage on LCD display with theoretical output.

As shown in Figure 8, a different current input was supplied using the voltage source that supply through the different range of variable resistor 20 MΩ for nano ampere range and 10 KΩ for micro ampere range. While, 100 Ω and 10 Ω used for mili ampere range. Therefore, the input current shows in range 0 to 1000 μA for nano and micro. While, the input current for mili range shows in range 0 to 300 μA. The current flow through voltage converter by using shunt resistor to produce input voltage in micro voltage value. Apart from that, there are three different switches on the voltage converter to select as the input current range as nano, micro and mili ampere with different value of resisters as 10 KΩ, 10 Ω and 10 mΩ respectively to convert three different range values to same range in micro input voltage.

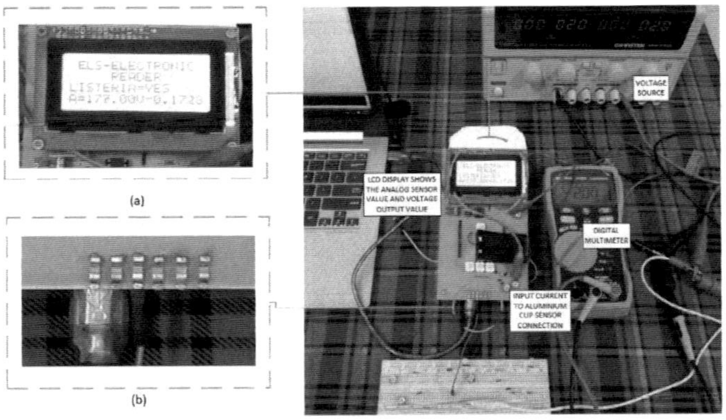

Figure 8. Measurement setup for testing the performance of the electronic reader circuit.

Table 1. Theoretical and experimental result for nano ampere range.

		Theoretical Result			**Experimental Result**		**Difference**	
Voltage Regulator (V)	Resistor (Ω)	Input Current (nA)	Input Voltage (μV)	Output Voltage (mV)	Analog Value	Output Voltage (mV)	Output Voltage (mV)	Percent Error (%)
0	20M	0	0	0	0	0	0	0
2	20M	100	1000	100	77	97.43	2.57	2.57
4	20M	200	2000	200	179	196.20	3.80	1.90
6	20M	300	3000	300	281	294.60	5.40	1.80
8	20M	400	4000	400	383	393.50	6.50	1.63
10	20M	500	5000	500	484	492.30	7.70	1.54
12	20M	600	6000	600	592	591.80	8.20	1.36
14	20M	700	7000	700	692	690.70	9.30	1.33
16	20M	800	8000	800	794	789.40	10.60	1.32
18	20M	900	9000	900	895	887.80	12.20	1.35
20	20M	1000	10000	1000	995	971.80	28.20	2.82
							Average Percent Error (%)	1.76

Table 1 shows the theoretical and experimental result of output voltage for input current in nano ampere range. The different voltages through 20 MΩresistor were supplied the input current. The range of input current was between 0 to 1000 nA. The result shows a small different in mili voltage with 1.76 % average percent error. While, for micro and mili ampere input current were followed same calculation procedure. The output voltage for theoretical and experimental result for micro and mili current input were shown average percent error was 1.8 % and 3.15%. The factors for overall voltage amplifier electronic reader accuracy depend on PCB design, components grounding, thermoelectric EMFs, and electrostatic interference [20], [21]. Hence, the results were obviously shown that the voltage amplifier electronic reader capable to detect biomolecules in biosensor application. Figure 9 shows the graph comparison between theoretical and experimental result for nano, micro and mili ampere range.

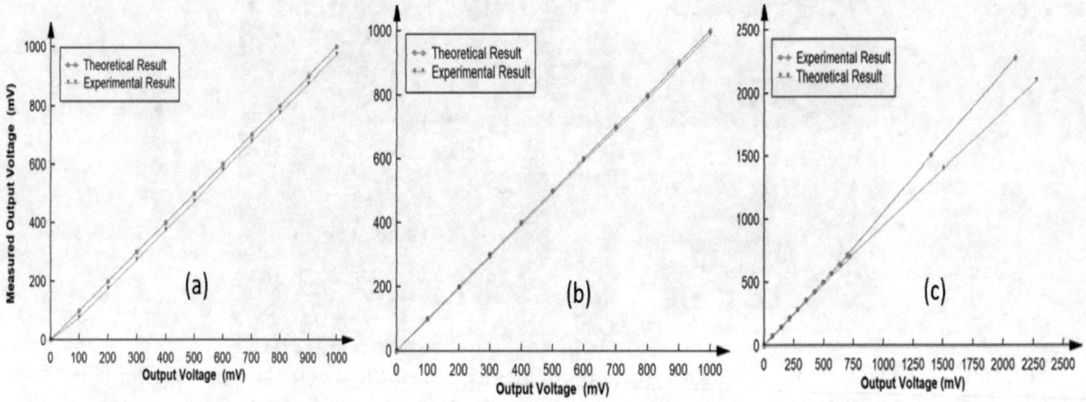

Figure 9. Graph comparison between theoretical and experimental result, (a) nano ampere range, (b) micro ampere range and (c) mili ampere range.

4. Conclusion

An electronic reader for detection of different detectable range as nano, micro and mili ampere current was designed and fabricated successfully. The design and programming was done using Proteus 8 Pro software and Arduino software, respectively. Electronic reader was tested for different current inputs with different ranges via voltage source coupled with variable resistors. Apart from that, the theoretical and experimental results of electronic reader were recorded and compared. The average percent error for nano, micro and mili range were 1.76 %, 1.8 % and 3.15 %, respectively. Appropriate precautions should be taken to reduce the noises and error. The experimental results were approximately similar with theoretical result. The processing and control part was successfully done using Arduino software and results were displayed in LCD screen successfully. According to the results, the electronic reader can successfully function to detect and produce output voltage in mV for two electrode based biosensor applications.

5. Acknowledgement

The author would like to thank all staff members of the Institute of Nanoelectronic Engineering in Universiti Malaysia Perlis (UniMAP) for their technical advice and contributions, directly and indirectly. Collaborative Research in Science and Technology Center (CREST) is acknowledged for providing grant for this research program.

References

[1] M. N. A. Uda, R. D. A. A. Rajapaksha, M. N. A. Uda, U. Hashim, A. B. Jambek, *AIP Conf. Proc.*, **2045**, 2018.

[2] R. D. A. A. Rajapaksha, M. N. A. Uda, U. Hashim, S. C. B. Gopinath, C. A. N. Fernando, *IEEE Int. Conf. Semicond. Electron. Proceedings, ICSE*, **2018**, 93–96, (2018).

[3] N. A. N. Yahaya, R. D. A. A. Rajapaksha, M. N. A. Uda, and U. Hashim, *AIP Conf. Proc.*, **1885**, (2017).

[4] M. N. A. Uda *et al.*, *AIP Conf. Proc.*, **1808**, 10–14, (2017).

[5] R. D. A. A. Rajapaksha, U. Hashim, M. N. Afnan Uda, C. A. N. Fernando, S. N. T. De Silva, *Microsyst. Technol.*, **6**, 70 (2017).

[6] U. Hashim, M. F. M. Fathil, M. K. M. Arshad, S. C. B. Gopinath, M. N. A. Uda, *AIP Conf. Proc.*, **1808**, (2017).

[7] S. C. B. Gopinath, S. Ramanathan, K. Hann Suk, M. Ee Foo, P. Anbu, M. N. A. Uda, *MATEC Web Conf.*, **150**, 06002, (2018).

[8] R. D. A. A. Rajapaksha, U. Hashim, N. Z. Natasha, M. N. A. Uda, V. Thivina, C. A. N. Fernando,

Proc. 2017 IEEE Reg. Symp. Micro Nanoelectron. RSM 2017, **5**, 191–194 (2017).
[9] M. N. A. Uda *et al.*, *AIP Conf. Proc.*,**1808**, 2017.
[10] W. Bai and Z. Zhu, **51**, 30–37, (2016).
[11] D. Senthilkumar, U. Pandit, and P. S. Jagtap, **3(1)**, 403–408 (2013).
[12] T. Nguyen, C. Kim, G. Ihm, M. Yang, S. Lee, **52(5)**, 1433–1442 (2004).
[13] M. N. . Uda *et al.*, Elsevier Inc., 2019.
[14] M. N. A. Uda *et al.*, *AIP Conf. Proc.*, **1808**, (2017).
[15] M. G. Yanez, *Free. Semicondutor*, 1–34, (2013).
[16] U. Hashim *et al.*, *Proc. - 2015 2nd Int. Conf. Biomed. Eng. ICoBE 2015*, 30–31, (2015).
[17] R. D. A. A. Rajapaksha, N. A. N. Yahaya, M. N. A. Uda, U. Hashim, in *BOND 21*, 7 (2018).
[18] M. Zaki *et al.*, *ICOBE*, 2015, 4 (2015).
[19] P. Intra, N. Tippayawong, *C. J. Nat. Sci*, **6(2)**, 313–320, (2007).
[20] R. A. Buda, M. M. A. I. Member, 8–10, (2014).
[21] S. Kumar, T. Van Oordt, E. M. Schneider, R. Zengerle, F. Von Stetten, J. H. T. Luong, *Biosens. Bioelectron.*, **67**, 248–255 (2015).
[22] R. D. A. A. Rajapaksha, U. Hashim, M. N. A. Uda, C. A. N. Fernando, **10(1)**, 61–64, (2017).
[23] R. D. A. A. Rajapaksha, N. A. N. Azman, M. N. A. Uda, U. Hashim, S. C. B. Gopinath, C. A. N. Fernando, *AIP Conf. Proc.*, **2045**, (2018).
[24] M. . N. A. Uda *et al.*, *Adv. Mater. Res.*, **832**, 410–414, (2014).
[25] S. C. B. Gopinath, U. Hashim, M. N. A. Uda, *AIP Conf. Proc.*, **1808**, 1–6, (2017).
[26] B. Mallick and A. K. Patro, *International Journal of Science, Engineering and Technology Research (IJSETR),* **5(1)**, 84–89 (2016).
[27] L. Acho, *J. Franklin Inst.*, **352(8)**, 3113–3121, (2015).
[28] A. Sheinin, A. Lavi, I. Michaelevski, *J. Neurosci. Methods*, **243**, 8–17 (2015).
[29] O. A. Olumodeji and M. Gottardi, *Integr. VLSI J.*, **58**, 438–445 (2017).
[30] F. Al-Obaidy, F. Yazdani, F. A. Mohammadi, *Comput. Electr. Eng.*, **58**, 88–100 (2017).
[31] R. D. A. A. Rajapaksha, N. A. N. Yahaya, M. N. A. Uda, U. Hashim, *AIP Conf. Proc.*, **2045**, 1–7 (2018).

Development of Internet of Things (IOT) Based Electronic Reader for Medical Diagnostic System

M. N. Afnan Uda[1,*], Asral Bahari Jambek[1], U Hashim[1, 2], M. N. A. UDA[3] and M. A. F. Bahrin[1]

[1]School of Microelectronic Engineering Group, Universiti Malaysia Perlis (UniMAP) Putra Campus, 02600 Pauh, Perlis Malaysia
[2]Institute of Nano Electronic Engineering (INEE), Universiti Malaysia Perlis (UniMAP), 01000 Kangar Perlis Malaysia
[3]School of Bioprocess Engineering Group, Universiti Malaysia Perlis (UniMAP) 02600 Arau Perlis

Email: nurafnan92@yahoo.com

Abstract. This paper is about an experiment for performing foodborne pathogens electronic reader using wireless sensing Internet of Thing (IoT). There are limited number of electronic readers for biosensors application with wireless internet connection. This research is to overcome the problem of commercial available electronic reader based on biosensor application method that only can be perform in offline or standalone device. This paper shows a complete system on how the data from electronic reader can be collected, easily understand by user and transfer data through the wireless internet connection via platform of IoT. There are three stages that is coding modification, android application development and transmit data to cloud storage. The NodeMCU microcontroller was used as a transfer medium for transfer data to internet. The Android Studio software was used for mobile application development. While, Arduino software was used to create a programming code to upload in NodeMCU microcontroller.

1. Introduction

Internet of Things (IoT) is an open and comprehensive network of smart objects capable of self-organizing, sharing information, data and resources, reacting and acting in the face of environmental situations and changes. The IoT in the IT arena is a novel paradigm shift. The phrase "Internet of Things" is coined from the two words, the first word is "Internet" and the second word is "Things". The Internet is a global computer network system using the standard Internet protocol suite to serve billions of users around the world. It is a global network system comprising millions of private, public, academic, business and government networks, ranging from local to global, connected by a wide range of electronic, wireless and optical networking technologies [1-2]. IoT is a technology that uses physical objects or 'things' embedded with communication capabilities, and facilitates intelligent data exchange among such devices, or between these devices and other internet-enabled systems. This is a new era of advanced technology, where each and everything is connected with each other and information can be accessed with existing of internet connection [3-5].

Currently, many of commercial electronic reader based biosensor application are standalone devices and do not have the capability to transmit information with internet connectivity [6-14]. Hence, the problem is the process of transmitting data from the electronic reader to the relevant party becomes slow

Content from this work may be used under the terms of the Creative Commons Attribution 3.0 licence. Any further distribution of this work must maintain attribution to the author(s) and the title of the work, journal citation and DOI.
Published under licence by IOP Publishing Ltd

[15-16]. Thus, no immediate action cannot be taken. In order to overcome this problem, the existing electronic reader-based biosensor need to be designed and improved the systems by embedded with the IoT system for improvement of connectivity capability of device. The use of wireless input and output terminal capable of accessing the internet to transmit and receive data from the electronic reader when sensors are attached to the electronic reader [17]–[19].

This research is proposed to design a smart electronic reader-based biosensor by embedding the IoT feature into the system. This project will focus on hardware and software according to electronic reader circuitry with ESP 8266 Wi-Fi module were used as microcontroller. Then, to design the algorithm to build a smart device application for user interface by using android apps maker such as "Android Studio". The expected outcome of this project is to develop and introduce the electronic reader for biosensor application with IoT system. With this capability now the reading result of electronic reader can be access from the smart phone through mobile app. In addition, the result will be transmit to computer for display and record data. This system can develop a rapid news for taking an action in order to ensure the cleanliness and health awareness of pathogenic bacteria that caused food poisoning.

2. Material and methods

In the era of technology innovation, everything requires monitoring and controlling. Therefore, this research is about to offers an IoT framework for assisting to the electronic reader based on foodborne pathogens detection for food safety monitoring system in order to food seller serve the customers with quality and safe food. The system is focus on the rapid detection, the data can be transfer to the medical officer for the record and analysis the data for fast action. Hence, making the medical officer to manage the data in real time using personal computer or mobile phone. Figure 1.0 shows the complete system via Internet of Things (IoT).

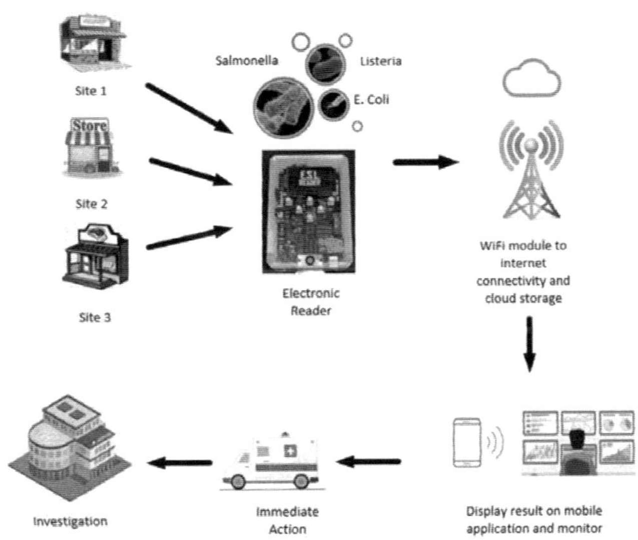

Figure 1. Complete system via Internet of Things (IoT).

2.1. Coding Modification

The first step is to modify the current coding. It needs to enable Wi-Fi connection to push the data at cloud server that need internet connection. The NodeMCU ESP8266 can communicate in two different ways via Wi-Fi. It can be linked to an existing wireless hot spot or access point which same method the way people connected their phone or computer to the Internet. This is stated to as "station" mode. The NodeMCU can reach the internet in station mode and can also connect with other devices along with other NodeMCU that have been connected to the same system. Figure 2.0 shows the example coding that the user needs to replace the SSID and password variables to wireless network ID and password.

```
#include <ESP8266WiFi.h>
#include <PubSubClient.h>

// Update these with values suitable for your network.

const char* ssid = "........";
const char* password = "........";
```

Figure 2. Coding Architecture Connect the NodeMCU ESP8266 Unit to an Existing Wi-Fi Access Point.

The existing coding of the electronic reader need to be modified based on the IoT architecture. The objective of this method is to achieve how to send JSON messages over MQTT using the ESP8266. Thus, the user will be using two libraries that handle the low-level details and expose us both the JSON encoding and the MQTT publishing functionalities in easy to use interfaces. These libraries are the PubSubClient, for the MQTT related functionality, and the ArduinoJson, for the JSON related functionality [19-21]. Figure 3.0 shows the coding modification to push the data transfer on cloud server.

```
if (selectCurrent_number == 1) {
    String type = "";
    if(selectSensor_number == 0) type = "0";
    if(selectSensor_number == 1) type = "1";
    if(selectSensor_number == 2) type = "2";

    if (output_value > 40) {
      if(wifi_state == 1 && sended == false){
        String sendstring = "{\"identifier\"=\"ESL READER 1\",\"detected\" = \"true\",\"type\"=\"" +type+"\"}";
        char copy[100];
        sendstring.toCharArray(copy,100);
        client.publish("foodDetector", copy);
        sended = true;
      }
      display.setCursor(0, 23);
      display.print("DETECTION: YES <<");
    }
```

Figure 3. After Modification of IoT System Architecture.

2.2 Mobile Apps development

Android Studio is the official IDE development environment for create android application. It's designed for android to speed up development and help a user build the highest quality apps for each android device it provides android developer specific tools, including rich code editing, debugging, testing, and profiling tools. Firstly, create an Android project by installing the latest version of Android Studio. Then, create a new Android project for your first app. This app displays the string the electronic reader reading result on the screen of the Android virtual or physical device as shown in Figure 4.0.

Usually, each screen in Android app is associated with one Java class known as an main activity. The single screen with "Hello World" displayed is created by one activity, called MainActivity.java. This was generated for user when created the new project. Each visible activity in an Android app has a layout that defines the user interface for the activity. Android Studio has a layout editor where it can create and define layouts.

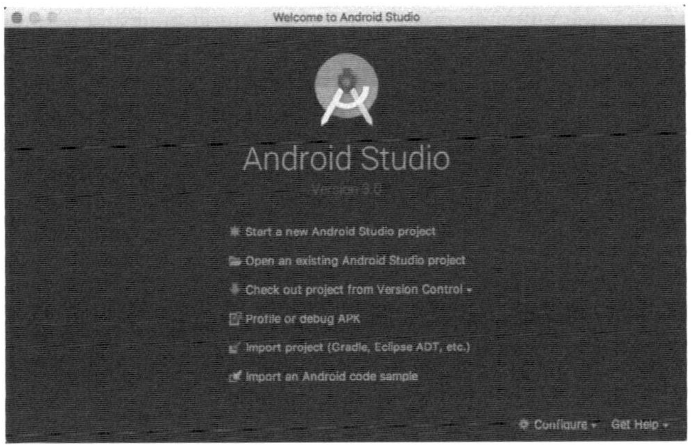

Figure 4. Android studios create new project windows.

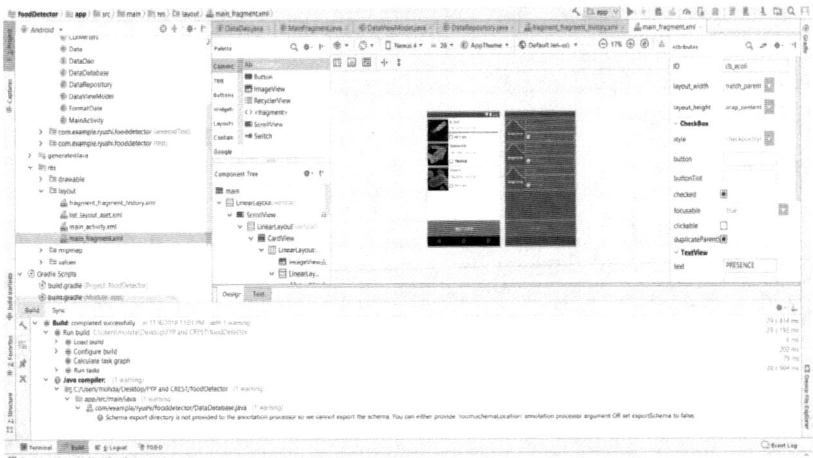

Figure 5. Developing layout of ESL READER application.

Figure 5 shows the developing layout of ESL READER application. The layouts are defined in XML. A part from that, the layout editor lets the user define and modify the layout either by coding XML or by using the interactive visual editor. Every element in a layout is a view. In this project, it needs an android device such as a android phone to run the apps. A data cable is need to connect the android phone with the computer via the USB port. After that, turn on USB debugging on the android phone to allow the android studio to communicate with android phone as shown in Figure 6.0. Then, install the app on android phone and run it.

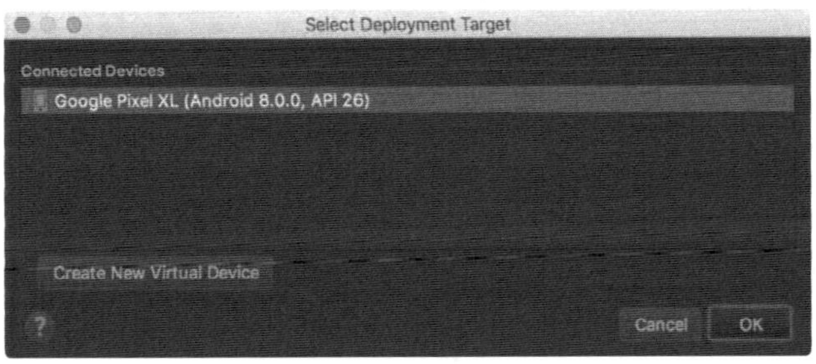

Figure 6. Select deployment target panel.

2.3 MQTT Cloud Setup

MQTT is the machine-to-machine protocol of the future. It is ideal for the IoT world of connected devices. Moreover, MQTT protocol is a light weight and an extremely simple protocol. MQTT has a few advantages including providing the relevant data to any intelligent and decision-making resource that can use it [18]. In additions, it allows management and deployment to IoT solutions to be extremely scalable. To create the MQTT cloud server, login to the CloudMQTT website and create an account as shown in Figure 7.

Figure 7. Main CloudMQTT Panel.

The instance is provided instantly after sign up and can view the instance details, such as connection information on the details page. When connecting via client libraries, the user is often forced to format a connection URL as shown in Figure 8.

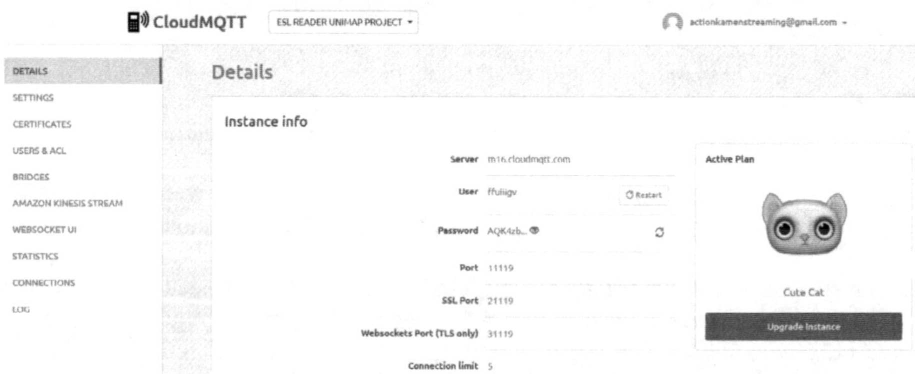

Figure 8. Information details panel.

Figure 9 shows how the data will be publish from electronic reader and the data will be retrieve by subscriber which is the smart phone. Therefore, the electronic reader need to connect with the smart phone Wi-Fi for internet connection. After that, the electronic reader automatically linked with cloudMQTT. The MQTT will hold the data topic in data base. Then, the data topic will be display in the smart phone through mobile application. MQTT is based on clients and a server. Similarly, the server is responsible for handling the client's requests of receiving or sending data between each other. MQTT server is called a broker and the clients are simply the connected devices. For example, if a device (a client) wants to send data to the broker, its call this operation a "publish". While, if a device (a client) wants to receive data from the broker, its call this operation a "subscribe".

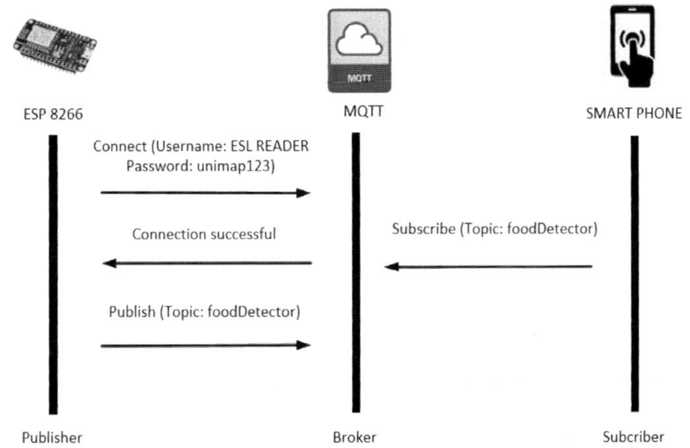

Figure 9. Schematic diagrams of publish and subscribe data flow.

3. Result and Discussion

After turn on the Wi-Fi by pressing Wi-Fi button on the electronic reader, it takes around 5 seconds to connect the electronic reader to the Wi-Fi. If the electronic reader failed to connect to the Wi-Fi, the electronic reader will reconnect again unless the user pushes off the Wi-Fi button. Figure 10.0 shows the serial monitor that shows the electronic reader has successfully connected to the Wi-Fi hotspot of the smartphone device. Then, the electronic reader has the capability to send messages to the MQTT server.

```
waiting to connect
WiFi connected
IP address:
(IP unset)
In reconnect...
Attempting MQTT connection...failed, rc=-2 it will try again
35||4.06        inf||-0.13       4.19||12489     34||0.00
0
In reconnect...
Attempting MQTT connection...connected
```

Figure 10. Serial monitor shows the reader successfully connected to hotspot Wi-Fi.

After completed the inspection using the electronic reader to detect the presence of foodborne pathogen using Al IDE biosensor, the result gained will be sent to the cloud server according to the condition specified in the coding. There are three sensors used to detect food pathogen which is E. coli, Salmonella and Listeria. The cloud server monitor shows the type of sensors and the presence of the foodborne pathogen from the sample. Figure 11 shows the data that sent by electronic reader has been successfully received. Based on the JSON format that was sent to the cloud server it shows some conditions in the following table below.

Table 1. The specified condition by using JSON format to send the data to cloud server.

Type	**E. coli**		**Salmonella**		**Listeria**	
Sensor Number	0		1		2	
Condition Value (nA)	< 3.5	>3.5	< 3.5	>3.5	< 3.5	>3.5
Reading Result	False	True	False	True	False	True

Table 1 shows the specified condition format data to detect the food pathogens which are E. coli, Salmonella and Listeria. The value of output current condition has been set in the Arduino programming, based on the average value from the previous I-V characterization target ssDNA E. coli, Salmonella and Listeria biosensor reading [22-24]. All the detection reading is in nano ampere range.

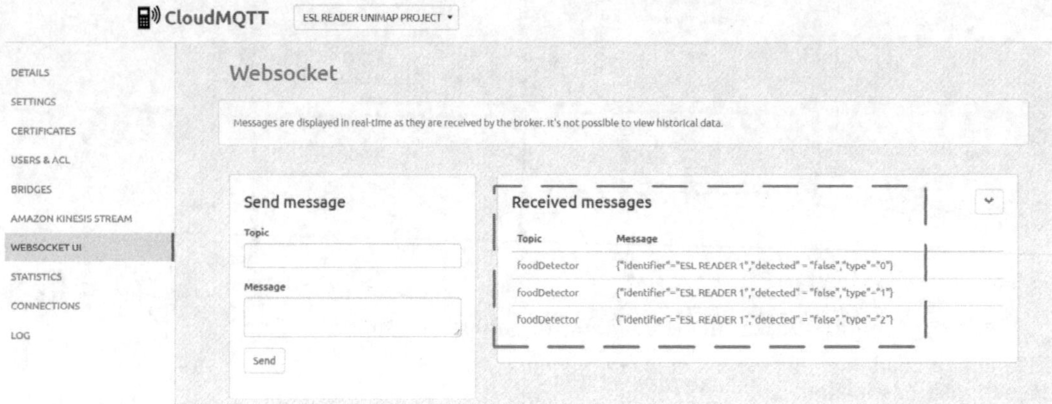

Figure 11. Received data from electronic reader.

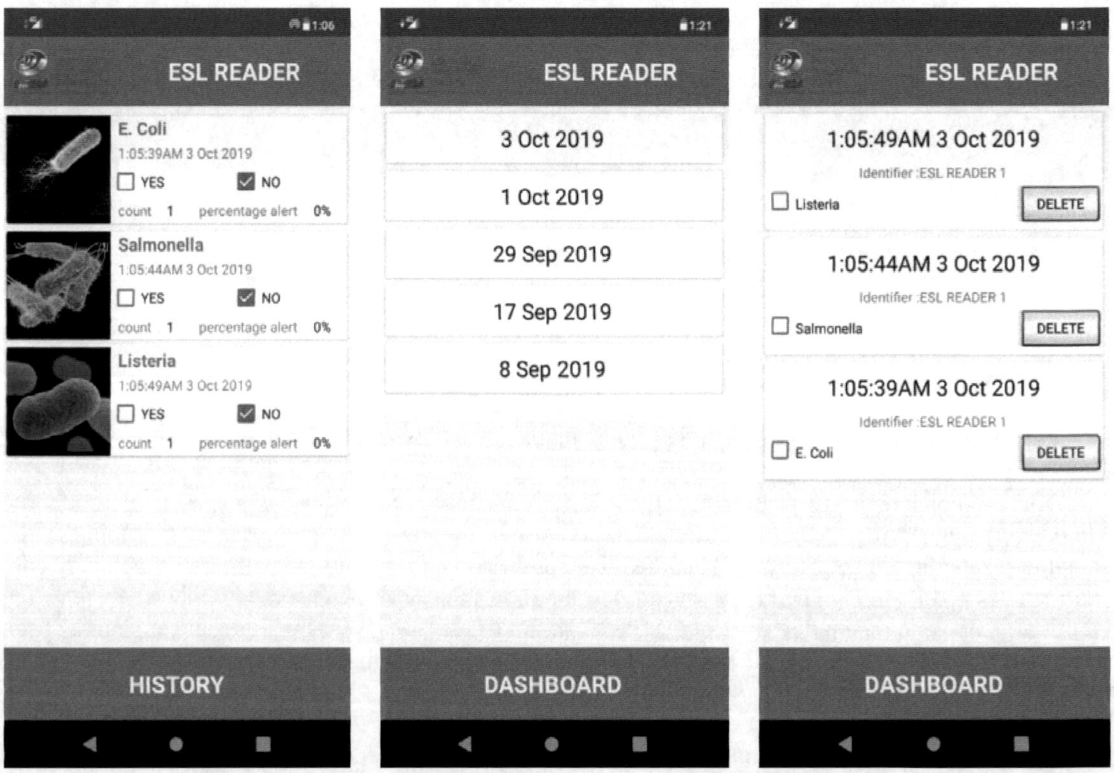

(a) (b) (c)

Figure 12. Mobile application (a) Home Page, (b) organize folder page and (c) record page.

Figure12 shown the layout editor for ESL READER mobile application. The ESL READER mobile application system was designed and tested successfully. The mobile application system focusing for Android supportive mobile devices in Android Studio software using the Java programming language.

The ESL READER mobile application comprises user interface layout for observing the reading result from the electronic reader. Each layout serving various purpose of information. The main layout interface consists of electronic reader reading result such as the type of food pathogen, the presence state indicator (YES/NO), the count sample and the percentage alert.

4. Conclusion

The Internet of Things (IoT) system for electronic reader has been designed and established. It was able to enhance the communication between electronic reader and internet connectivity through mobile application for better efficiency. The IoT system were assisting to the electronic reader to transmit the reading result from the electronic reader to smart phone through mobile application for record, control and monitoring. It is providing the real time monitoring system. Last but not least, the data can be transfer faster to medical officer about breeding bacteria in food that caused illnesses for taking immediate action.

Acknowledgement

The author would like to thank all staff members of the Institute of Nanoelectronic Engineering in Universiti Malaysia Perlis (UniMAP) for their technical advice and contributions, directly and indirectly. Collaborative Research in Science and Technology Center (CREST) is acknowledged for providing grant for this research program.

References

[1] Y. Zhu and P. Liang, *IOP Proceedings*, **908**, 78 (2017).
[2] G. Rateni, P. Dario, and F. Cavallo, "Smartphone-Based Food Diagnostic Technologies: A Review," 2017.
[3] J. Gubbi, R. Buyya, S. Marusic, and M. Palaniswami, *Futur. Gener. Comput. Syst.*, **29**, 1645–1660, (2013).
[4] A. Whitmore, A. Agarwal, and L. Da Xu, **60**, 261–274 (2015).
[5] B. Liu, "Research on Application of Internet of Things in Electronic commerce," pp. 13–16, 2010.
[6] R. D. A. A. Rajapaksha, N. A. N. Yahaya, M. N. A. Uda, U. Hashim, *AIP Conf. Proc.*, **2045**, 1–7 (2018).
[7] R. D. A. A. Rajapaksha, M. N. A. Uda, U. Hashim, S. C. B. Gopinath, and C. A. N. Fernando, *IEEE Int. Conf. Semicond. Electron. Proceedings, ICSE*, **2018**, 93–96 (2018) .
[8] M. . N. A. Uda *et al.*, *Adv. Mater. Res.*, **832**, 410–414, (2014) .
[9] M. N. A. Uda, R. D. A. A. Rajapaksha, M. N. A. Uda, U. Hashim, A. B. Jambek, *AIP Conf. Proc.*, **2045**, 2018.
[10] R. D. A. A. Rajapaksha, U. Hashim, N. Z. Natasha, M. N. A. Uda, V. Thivina, C. A. N. Fernando, *Proc. 2017 IEEE Reg. Symp. Micro Nanoelectron. RSM 2017*, **5**, 191–194 (2017).
[11] S. A. B. Ariffin, T. Adam, U. Hashim, S. Faridah Sfaridah, I. Zamri, and M. N. A. Uda, *Adv. Mater. Res.*, **832**, 113–117 (2013).
[12] M. N. . Uda *et al.*, Elsevier Inc., 2019.
[13] M. N. A. Uda *et al.*, *AIP Conf. Proc.*, **1808**, (2017).
[14] M. N. A. Uda *et al.*, *AIP Conf. Proc.*, **1808**, 10–14, (2017).
[15] N. A. N. Yahaya, R. D. A. A. Rajapaksha, M. N. A. Uda,U. Hashim, *AIP Conf. Proc.*, **1885**, (2017).
[16] R. D. A. A. Rajapaksha, U. Hashim, M. N. A. Uda, and C. A. N. Fernando, *Mater. Sci. Eng. B*, **10(1)**, 61–64 (2017) .
[18] R. K. Kodali and S. Soratkal, *Automation System*, **23**, 35 (2015).
[19] R. D. A. A. Rajapaksha, N. A. N. Azman, M. N. A. Uda, U. Hashim, S. C. B. Gopinath, C. A. N. Fernando, *AIP Conf. Proc.*, **2045**, (2018).
[20] S. Grassini, S. Corbellini, M. Parvis, E. Angelini, and F. Zucchi, *Meas. J. Int. Meas. Confed.*,

114, 508–514 (2018).
[21] I. Luiz, B. De Moura, L. Carlos, D. S. Monteiro, F. A. Soares, *Int. J. Electr. Robot. Electron. Commun. Eng.*, **8**, 310–314 (2014).
[22] O. A. Olumodeji and M. Gottardi, *Integr. VLSI J.*, **58**, 438–445, (2017).
[23] R. D. A. A. Rajapaksha, U. Hashim, M. N. Afnan Uda, C. A. N. Fernando, S. N. T. De Silva, *Microsyst. Technol.*, **6**, 70 (2017).
[24] R. D. A. A. Rajapaksha, U. Hashim, M. N. Afnan Uda, C. A. N. Fernando, S. N. T. De Silva, *Microsyst. Technol.*, **6**, 70 (2017).
[25] N. Z. Natasha, R. D. A. A. Rajapaksha, M. N. A. Uda, and U. Hashim, **020235**, (2017).

Nano-micro-mili Current to Mili Voltage Amplifier for Amperometric Electrical Biosensors

M. N. Afnan Uda[1,*], N. A. Parmin[2], Asral Bahari Jambek[1], U. Hashim[1,2], M. N. A. Uda[3], S. N. A. Shaharuddin[2] and Hussaini Adam[2]

[1]School of Microelectronic Engineering Group, Universiti Malaysia Perlis (UniMAP) Putra Campus, 02600 Pauh, Perlis Malaysia
[2]Institute of Nano Electronic Engineering (INEE), Universiti Malaysia Perlis (UniMAP), 01000 Kangar Perlis Malaysia
[3]School of Bioprocess Engineering Group, Universiti Malaysia Perlis (UniMAP) 02600 Arau Perlis

Email: * nurafnan92@yahoo.com

Abstract. Amplification of nano and mircoampere electrical signal to the detectable range is essential in the biosensor field. This research is mainly focused on design an amplifier circuit to capture and amplify three different range of current as nano, micro and mili ampere and convert it to detectable voltage range as an output signal to the processing circuit. The Proteus 8 Pro software was used to design, simulate and calibrate the amplifier circuit. Firstly, current input as mili, micro and nano current were flown through 0.1 mΩ, 10 Ω and 10 KΩ resistors, respectively to convert different current inputs to the similar range in micro voltage. The MAX 4238 opamp IC was used to amplify micro voltage to mili voltage. LM 358 dual operational amplifier was used to supply virtual ground to MAX 4238 amplifier. The amplified output voltage of three different current inputs as nano, micro and mili were nearly equal to theoretical outputs.

1. Introduction

The development of simple and portable electronic readers have elevated potential for automation of sensor-based applications. Normally biosensors give extra small electrical output and most of them are amperometric, impedometric and potentiometric [1], [2]. To measure this small electrical outputs, bench type source measurement units and impedance analyzers were widely used [3], [4]. Requirement of portable electronic readers enhance the convenience to retrieve results much more easily and rapidly. Thus, using electronic readers can be saved time, handle out field and save money [5], [6]. Direct current (DC) voltage, DC current and resistance were measured most frequently using present electronic reader and other equipment [7]. Moreover, most industries are still based on the desktop computer and user need to complete the appropriate actions in the laboratory [8][9]. Furthermore, the equipment that used to measure low current was so expensive and inefficient because of the limitation of the detectable range [10]–[13]. In low current amplification there should be considered many factors in order to produce the accurate result such as leakage current or burden voltage. The volume of leakage current or burden voltage will affect the voltage amplifier circuit measured that will be corrupting the measurement contrary with theoretical measurement. Hence, it should be as low as possible the value or leakage current or burden voltage. Moreover, incorrect human handling of PCB also were intentionally contaminated before testing, will affected the measurement. This was similarly a potential source of

Content from this work may be used under the terms of the Creative Commons Attribution 3.0 licence. Any further distribution of this work must maintain attribution to the author(s) and the title of the work, journal citation and DOI.
Published under licence by IOP Publishing Ltd

IOP Conf. Series: Materials Science and Engineering **743** (2020) 012021 doi:10.1088/1757-899X/743/1/012021

contamination resulting from fingerprint, skin oil, and saliva [14]–[16].In this paper, we designed and developed an amplifier to amplify extra small current through sensors from nano ampere range to mili ampere range. The different range input current has converted to micro voltage and supply to voltage amplifier as input. The MAX 4238 is used to amplify the micro voltage to detectable mili voltage output which can be detect from the multimeter or microcontroller-based processing circuits.

2. Material and methods

2.1 Apparatus, equipment and software

The MAX 4238 was used as operational amplifier to amplify micro voltage to mili voltage. The LM 358 operational amplifier was used as to supply virtual ground for MAX 4238 opamp IC. The TPS 3809 was used as battery testing circuit. The 3 V coin battery was used as the power source to supply power for amplifier circuit and sensors. All ICs and resistors were used in this prototype are from Texas instruments. The Keithley Source Measurement Unit (SMU) was used to compare the resultant current with prototype, in three different current ranges as nano, micro and mili ampere ranges. The Proteus 8 Pro software was used to design and simulate the circuit.

2.2 The Circuit design

Figure 1 shows the detail block diagram for a voltage amplifier for three different range as mili, micro, and nano for electrical biosensors. As shown in the figure there are three major parts included in the amplifier circuit as DC power supply, voltage converter and voltage amplifier. Sensor input system has included sensor input connectors and three dual select switches to select appropriate sensor to resistors. The amplified voltage output can be detected using multimeters or supplied as a input signal for a processing circuit.

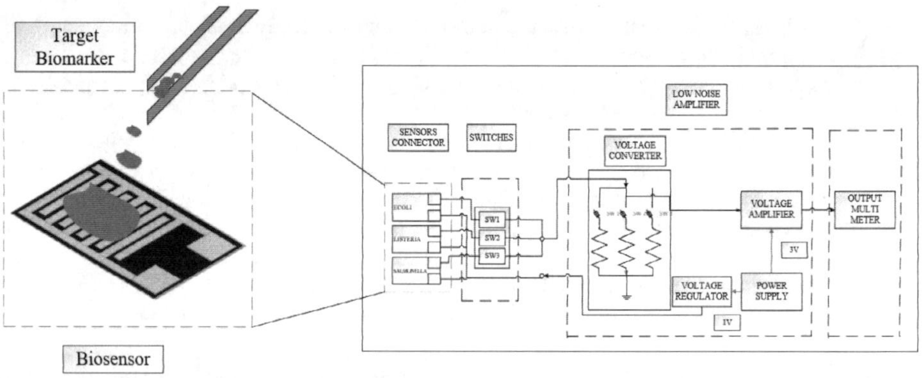

Figure 1. Detail block diagram for voltage amplifier.

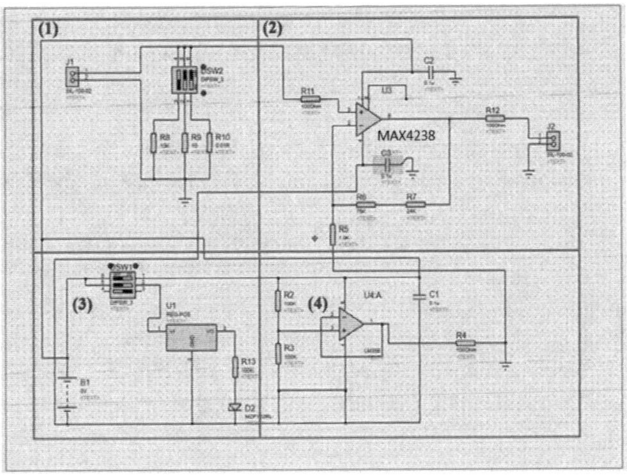

Figure 2. Voltage Amplifier circuit design using Proteus 8 Pro software.

Figure 2 shows the whole schematic design for voltage amplifier circuit using Proteus 8 Pro software. Figure 2 (1) shows the sensor selection circuit with current to voltage converter. Three switches have designed to select the respective resistor for selected sensors in three different current range as nano, micro and mili ampere. Figure 2 (2) shows the MAX 4238 opamp IC circuit. Figure 2 (3) shows the power supply circuit which was included with TPS 3809 as battery testing circuit and Figure 2 (4) shows the virtual ground suppling circuit to the MAX 4238 amplifier IC. After design the circuit, it was simulated using in-built simulator on Proteus 8 Pro software.

2.3 Fabrication and assembly of PCB

The in-built ARES feature in Proteus 8 Pro software was used to convert the circuit diagram to PCB layout as shown in Figure 3 (1). The connection of the component in printed circuit board (PCB) layout can be adjusted the component position and after finish the process, the circuit was transferred to copper clad board to make PCB as shown in Figure 3 (2). Figure 3 (3 and 4) shows the bottom and top view of the PCB after assembled the components to make the portable amplifier device.

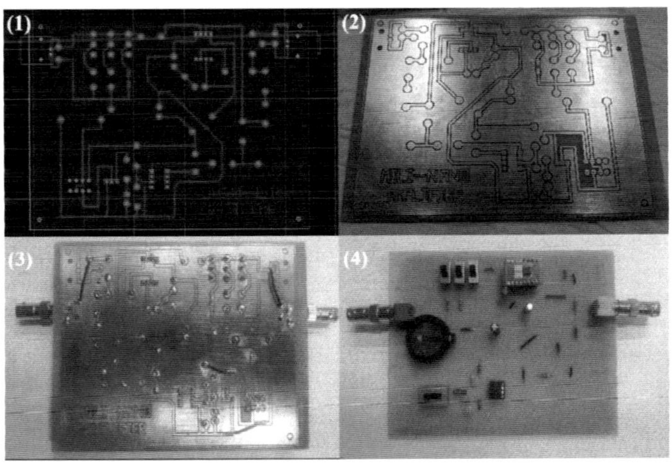

Figure 3. Fabrication process of the circuit board, (1) PCB layout of the amplifier circuit, (2) developed PCB on a copper clad, (3) bottom view of the circuit and (4) top view of the circuit.

3. Result and Discussion

Figure 4 shows the measurement setup for testing the performance of the amplifier circuit. Three different current ranges were supplied separately through the sensor connectors while changing the appropriate selection switches. The circuit was validated by comparing the output voltage through the multimeter value with theoretical output.

As shown in the figure 4, current inputs in different ranges were supplied using the external voltage source through the different range of variable resistors (1 GΩ for nano ampere range, 1 MΩ for micro ampere range and 1 KΩ for mili ampere range). The voltage source and variable resistor acts as a replacement of biosensor to input current to the sensor ports. In the voltage converter, the input current will flow through shunt resistor to produce input voltage in micro voltage value. There are three different switches to select the input current range as mili ampere, micro ampere and nano ampere by coupling three resistors as 0.1 Ω, 10 Ω and 10 KΩ, respectively to convert three different range values to same range in micro voltage. The low noise amplifier get this value as input voltage and amplifies the input voltage from micro voltage value to output voltage in mili voltage value. The gain of MAX 4238 opamp IC is 100, the input voltage times with 100 gives the value of output voltage in mili voltage value.

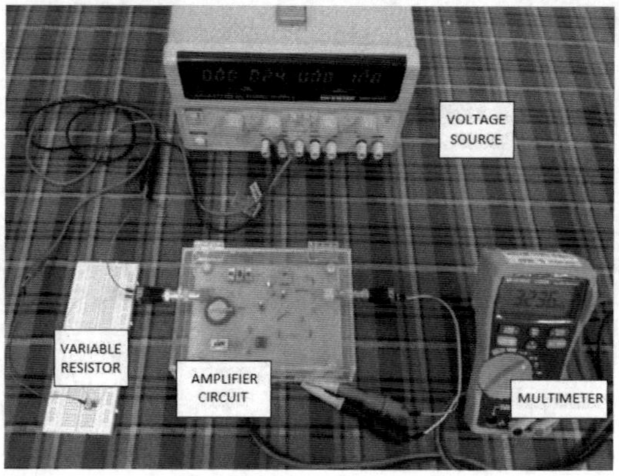

Figure 4. Measurement setup for testing the performance of the amplifier circuit.

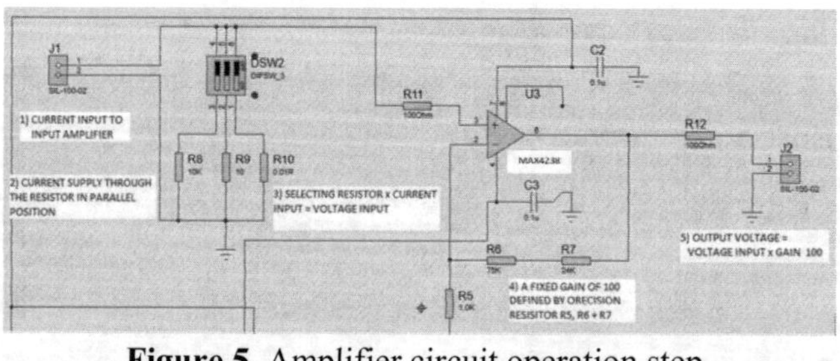

Figure 5. Amplifier circuit operation step.

Figure 5 shows the operation steps of the amplifier circuit. The theoretical result and experimental were recorded and compared to identify the accuracy of the amplifier circuit. Following formula was used to calculate the theoretical output voltage through the amplifier circuit. The Ohm's law was used to calculate the input voltage to the amplifier [17]. According to the equation, sensor input current Ii (A), shunt resistance for appropriate range is R (Ω) and input voltage for sensor Vi (V) as defined. Equation 1 shows the calculation to find the input current to sensor port using power generator and variable resistor. Equation 2 shows the calculation to find the input voltage (V) to voltage amplifier from voltage converter.

$$Ii = \frac{VR}{R} \tag{1}$$

$$Vi = Ii \times R \tag{2}$$

Figure 6 shows the opamp circuit in non-inverting mode. After converted current to voltage, it will apply directly to the non-inverting (+) input terminal which means output signal is in-phase with the input signal. Feedback control of the non-inverting operational amplifier is achieved by applying a small part of the output voltage signal back to the inverting input terminal via an R_f – R_1 voltage divider network. In addition, this closed-loop configuration produces a non-inverting amplifier circuit with very good stability [18]. Equation 3 shows the equation of the gain (G). Equation 4 shows the equation of output voltage (V).

Before integrate the amplifier circuit, burden voltage or leakage current have to be considered [19]–[21]. It will affect the performance of amplifier circuit and contradict with theoretical measurement. Apart from that, poor human handling as a fingerprint, saliva, dust, chemicals, temperature surrounding, can affect the performance of amplifier circuit [22]–[24]. Those contaminants could be reduced by applying PCB protection coating or making a proper casing for the circuit [25]. Equation 5 shows the calculation for difference value for theoretical result and experimental result and Equation 6 shows the percent error for theoretical result and experimental result.

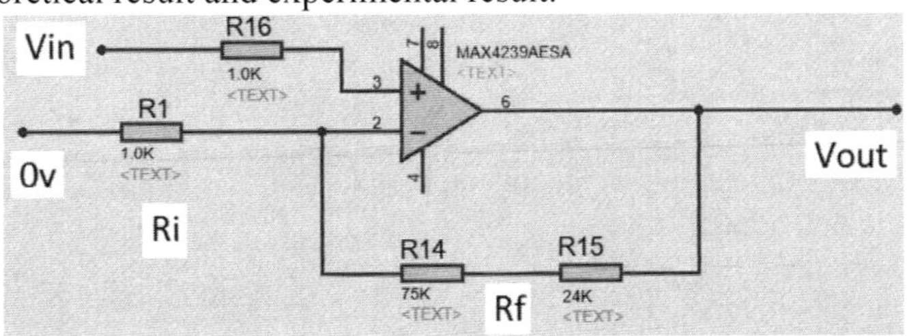

Figure. 6 Non-inverting operational amplifier circuit

$$Gain, G = \frac{Vout}{Vin} = \frac{R1 + Rf}{R1} \tag{3}$$

$$Vo = Vi \times G \tag{4}$$

$$\text{Difference} = \text{Theoretical Result} - \text{Experimental Result} \tag{5}$$

$$\text{Percent error} = \left|\frac{Theoretical\ Result - Experimental\ Result}{Theoretical\ Result}\right| \text{x } 100 \tag{6}$$

3.1 Comparison of theoretical and experimental values

Table 1. Theoretical result and experimental result for nano ampere range.

		Theoretical Result			**Experimental Result**	Difference	
Voltage Regulator (V)	Resistor (Ω)	Input Current (nA)	Input Voltage (μV)	Output Voltage (mV)	Output Voltage (mV)	Output Voltage (mV)	Percent Error (%)
0	1G	0	0	0	0	0	0

0.4	1G	0.4	4	0.4	0.42	-0.02	5
0.8	1G	0.8	8	0.8	0.85	-0.05	6.25
1.2	1G	1.2	12	1.2	1.28	-0.08	6.67
1.6	1G	1.6	16	1.6	1.68	-0.08	5.00
2.0	1G	2.0	20	2.0	2.11	-0.11	5.50
2.4	1G	2.4	24	2.4	2.48	-0.08	3.33
2.8	1G	2.8	28	2.8	2.89	-0.09	3.21
3.2	1G	3.2	32	3.2	3.29	-0.09	2.81
3.6	1G	3.6	36	3.6	3.75	-0.15	4.16
4.0	1G	4.0	40	4.0	4.15	-0.15	3.75
4.4	1G	4.4	44	4.4	4.51	-0.11	2.50
4.8	1G	4.8	48	4.8	4.87	-0.07	1.45
5.2	1G	5.2	52	5.2	5.28	-0.08	1.53
5.6	1G	5.6	56	5.6	5.73	-0.13	2.32
6.0	1G	6.0	60	6.0	6.15	-0.15	2.50
6.4	1G	6.4	64	6.4	6.48	-0.08	1.25
6.8	1G	6.8	68	6.8	6.93	-0.13	1.91
7.2	1G	7.2	72	7.2	7.36	-0.16	2.22
7.6	1G	7.6	76	7.6	7.81	-0.21	2.76

Table 1 shows the theoretical and experimental result of output voltage for input current in nano ampere range. The different voltages through 1 GΩresistor were supplied the input current. The range of input current for theoretical and experimental result was between 0 to 7.6 nA. While, the output voltage for theoretical and experimental result were shown a small different in mili voltage with 6.67 % average percent error. The theoretical and experimental result of output voltage for micro and mili ampere input current were followed same calculation procedure. The average percent error for micro and mili ampere range was 2.26% and 1.04%. The factors for overall accuracy depend on leakage resistance, input loading, thermoelectric EMFs, and electrostatic interference [25]. Therefore, the results were clearly shown that the system can be used to identify biomolecules in biosensor application. Figure 7 shows the graph comparison between theoretical and experimental result for nano, micro and mili ampere range.

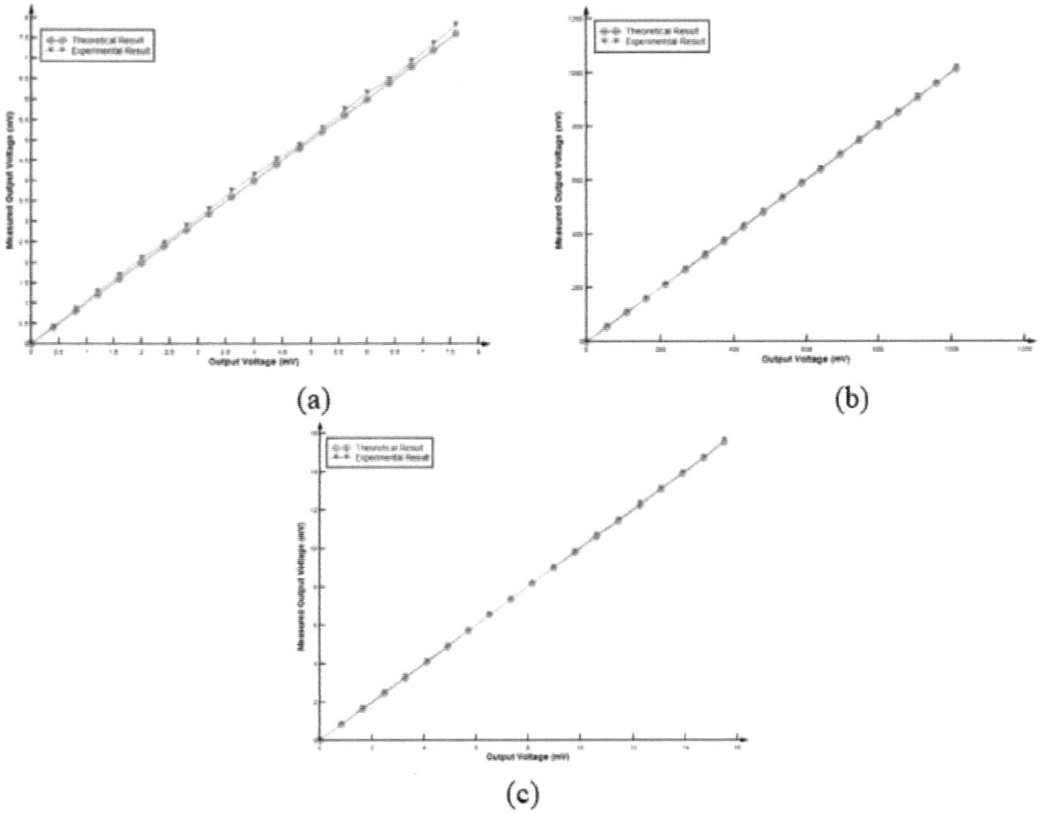

Figure 7. Graph comparison between theoretical and experimental result, (a) nano ampere range, (b) micro ampere range and (c) mili ampere range.

4. Conclusion

Nano, micro and mili ampere current to mili voltage amplifier for electrical biosensors has been designed and simulated using Proteus 8 Pro software. The amplifier is high precision in the detectable range. Actual device was tested with dummy current source via voltage source coupled with resistors. Appropriate precautions should be taken to reduce the noises and error in nA range. The theoretical result and experimental result of amplified voltages have been recorded and compared. The results were indicated a small different in each ranges were in the linear pattern. The changes may happen due to the interferences of the connections, components and humidity. Finally, in this experiment shows that the system can successfully detect and produce output voltage in mV for two electrode based amperometric biosensor applications.

Acknowledgement

The author would like to thank all staff members of the Institute of Nanoelectronic Engineering in Universiti Malaysia Perlis (UniMAP) for their technical advice and contributions, directly and indirectly. Collaborative Research in Science and Technology Center (CREST) is acknowledged for providing grant for this research program.

References

[1] R. D. A. A. Rajapaksha, N. A. N. Yahaya, M. N. A. Uda, U. Hashim, *AIP Conf. Proc.*, **2045**, 1–7 (2018).
[2] N. A. N. Yahaya, R. D. A. A. Rajapaksha, M. N. A. Uda,U. Hashim, *AIP Conf. Proc.*, 1885,

2017..

[3] R. D. A. A. Rajapaksha, U. Hashim, N. Z. Natasha, M. N. A. Uda, V. Thivina, and C. A. N. Fernando, *Proc. 2017 IEEE Reg. Symp. Micro Nanoelectron. RSM 2017*, 191–194 (2017).

[4] R. D. A. A. Rajapaksha, M. N. A. Uda, U. Hashim, S. C. B. Gopinath, and C. A. N. Fernando, *IEEE Int. Conf. Semicond. Electron. Proceedings, ICSE*, **2018**, 93–96 (2018).

[5] J. Shim and J. Jeong, *Microelectronics J.*, **65**, 78–83 (2017).

[6] B. Linares-Barranco, T. Serrano-Gotarredona, R. Serrano-Gotarredona, C. Serrano-Gotarredona, *Analog Integr. Circuits Signal Process.*, **38**, 103–119 (2004).

[7] M. N. A. Uda, R. D. A. A. Rajapaksha, M. N. A. Uda, U. Hashim, A. B. Jambek, *AIP Conf. Proc.*, **2045** (2018).

[8] R. D. A. A. Rajapaksha, U. Hashim, M. N. A. Uda, and C. A. N. Fernando, *Mater. Sci. Eng. B*, **10(1)**, 61–64 (2017).

[9] R. D. A. A. Rajapaksha, N. A. N. Azman, M. N. A. Uda, U. Hashim, S. C. B. Gopinath, C. A. N. Fernando, *AIP Conf. Proc.*, **2045**, (2018).

[10] U. Hashim, M. F. M. Fathil, M. K. M. Arshad, S. C. B. Gopinath,M. N. A. Uda, *AIP Conf. Proc.*, **1808** (2017).

[11] M. N. A. Uda *et al.*, *AIP Conf. Proc.*, **1808**, (2017).

[12] S. C. B. Gopinath, U. Hashim, M. N. A. Uda, *AIP Conf. Proc.*, **1808**, 1–5 (2017).

[13] N. Z. Natasha, R. D. A. A. Rajapaksha, M. N. A. Uda, U. Hashim, *AIP Conf. Proc.*, **1885**, (2017).

[14] U. Hashim *et al.*, *Proc. - 2015 2nd Int. Conf. Biomed. Eng. ICoBE 2015*, 30–31, (2015).

[15] S. Carrara *et al.*, *Sensors (Switzerland)*, 12(8), 11013–11060, (2012).

[16] I. Zaccari, A. G. Davies, C. Walti, and S. X. Laurenson, *Proc. 7th Cairo Int. Biomed. Eng. Conf. CIBEC 2014*, (2015).

[17] T. Nguyen, C. Kim, G. Ihm, M. Yang, S. Lee, **52(5)**, 1433–1442, (2004).

[18] D. Senthilkumar, U. Pandit, P. S. Jagtap, **3(1)**, 403–408, (2013).

[19] P. Mohseni, K. Najafi, *IEEE Trans. Biomed. Eng.*, **51(5)**, 832–837 (2004).

[20] J. C. Stanton, *IEEE Trans. Nucl. Sci.*, vol. 36, no. 1 pt 1, pp. 522–527, 1989.

[21] M. Zhao *et al.*, *Proc. - IEEE Int. Symp. Circuits Syst.*, **2015**, 337–340 (2015).

[22] B. Goldstein *et al.*, *2011 Ieee Iscas*, **6(2)**, 1017–1020, (2011).

[23] B. D. DeBusschere and G. T. A. Kovacs, *Biosens. Bioelectron.*, **16**, 543–556, (2001).

[24] T. Regan, J. Munson, G. Zimmer, M. Stokowski, *Analog Circuit Des.* **2**, 1111–1215, (2013).

[25] P. Intra and N. Tippayawong, "An ultra-low current meter for aerosol detection," *C. J. Nat. Sci*, **6(2)**, 313–320 (2007).

IOP Conf. Series: Materials Science and Engineering **743** (2020) 012022 doi:10.1088/1757-899X/743/1/012022

Electrical DNA Biosensor Using Aluminium Interdigitated Electrode for Salmonella Detection

M. N. Afnan Uda[1,*], Asral Bahari Jambek[1], U. Hashim[1,2] and M. N. A. UDA[3]

[1]School of Microelectronic Engineering Group, Universiti Malaysia Perlis (UniMAP) Putra Campus, 02600 Pauh, Perlis Malaysia
[2]Institute of Nano Electronic Engineering (INEE), Universiti Malaysia Perlis (UniMAP), 01000 Kangar Perlis Malaysia.
[3]School of Bioprocess Engineering Group, Universiti Malaysia Perlis (UniMAP) 02600 Arau Perlis

Email: * nurafnan92@yahoo.com

Abstract. Nowadays there are many alternative methods that have been discovered and developed for the rapid detection of foodborne pathogens that can cause food poisoning. Unfortunately, majority of them still requires improvement in sensitivity and selectivity issues to be of any practical use daily. In this research, biosensors was prepared from 5 μm gap Aluminium interdigitated electrode (Al IDE) to detect Salmonella enterica typhi (S. typhi). The IDE sensors in the biosensor field is extremely interest in these days due to the high number of finger electrodes as comb structure which can gain high sensitivity through electrical measurements. S. typhi is a serious food borne pathogen, makes typhoid disease which causes many deaths annually in worldwide. Functionalization steps of the Al IDE to create biosensor was based on silanization by APTES, immobilization with carboxylic functionalized S. typhi ssDNA probes and blocking agent with tween-20 were the major functionalization steps. The functionalized steps were electrically characterized using current voltage measurements. The selectivity measurement was performed with specific target was identified electrically using complementary, non-complementary and single base mismatch ssDNA target.

1. Introduction

Foodborne outbreaks have been reported with significant morbidity worldwide and pose risk towards human population [1-2]. Most commonly, the outbreaks take place due to the ingestion of pathogenic bacteria like Salmonella [3-4]. Salmonella enterica typhi (S. typhi) is one of the most commonly reported serotypes which causes typhoid fever, affects roughly 17 million people annually and causing nearly 600,000 deaths. These serotypes which can be transferred from animal-to-human or from human-to-human [5]. Therefore, generation of the right sensing approach for Salmonella will pave the way for easier detection and curing.

Most of the conventional method for Salmonella detection still lack of sensitivity [6]. Apart from that, it can only be detected after the bacteria separated all over the body. Due to that, electrical based biosensor technology with nano amperes range will help healthcare and food industry to get early and accurate detection [6-8]. Commercial application of biosensors are given significant impact on the medical and food analysis to overcome the drawbacks in conventional methods [9-11]. Nowadays, biosensor researches are widely concentrated, numerous research and excellent reviews have been

Content from this work may be used under the terms of the Creative Commons Attribution 3.0 licence. Any further distribution of this work must maintain attribution to the author(s) and the title of the work, journal citation and DOI.
Published under licence by IOP Publishing Ltd

published [12-21]. Moreover, the high sensitivity, selectivity, repeatability and real-time in field detection are the main advantages of biosensors [22-27].

In this research, electrical based Al IDE on silicon substrate is fabricated. The fabricated Al IDE is physically characterized using a low power microscope (LPM) and high power microscope (HPM). The functionalization steps as biosensor are, silanization using APTES, immobilization using carboxylic ssDNA probes and blocking the non-immobilized area using tween-20. I-V characteristic were performed for selectivity detection by hybridizing with synthetic ssDNA samples as complementary, non-complementary and single base mismatch targets by using Keithley 2450, Kickstart software and probe station.

2. Material and methods

2.1 Chemicals, Reagents and Instruments

Ethanol (C2H5OH), Sodium Hydroxide (NaOH), (3-Aminopropyl) Triethoxysilane (APTES) (C9H23NO3Si), Deionized Water (DI) was obtained from Sigma Aldrich, USA. All of the other chemicals were analytical reagent grade and purchased commercially. Deionized Distilled Water (DDI – water) was used throughout this experiment. The 30-base synthetic oligonucleotides were purchased from AIT Biotech Pte. Ltd (Singapore). IDE mask was designed using the aid of AutoCAD Software for the detection of ssDNA which an extremely small scale in size before being transferred to commercial chrome mask. The design and mask then sent to Silterra (M) Sdn Bhd to fabricate the Al IDE. Al IDE surface was checked using Low Power Microscopy (LPM) and High Power Microscopy (HPM) for surface characterization. The measurements for I-V characteristics were carried out by using Keithley 2450, Kickstart software and probe station.

2.2 Al IDE surface functionalization

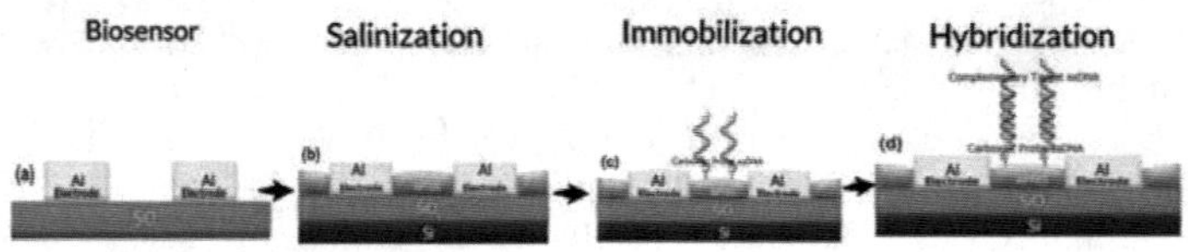

Figure 1. Functionalization steps of biosensor (a) Bare Al IDE, (b) Salinization Process using APTES, (c) Immobilization with carboxylic Probe Salmonella ssDNA and (d) Hybridization with ssDNA target of Salmonella.

Figure 1 shows the schematic diagram of the functionalization process of biosensor. First, the active area of the bare Al IDE surface was washed with NaOH to enhance hydroxyl on the SiO2 layer. Then, the bare Al IDE surface was functionalized with 0.1 M, 2µl of APTES. Salinization is the covering of a surface with organofunctional alkoxysilane molecules. Mineral components of Silicon dioxide surfaces can all be salinized because they contain hydroxyl groups which attack and displace the alkoxy groups on the saline thus forming a covalent –Si-O-Si- bond. The silanized sensor was incubate 60 min and DI water was used to wash the unattached APTES molecules. After that, 10 µM, 2 µl of carboxylic probe ssDNA was used for immobilization process. The immobilized sensor was incubate for 60 min and DI water was used to wash the unattached ssDNA probes from the APTES layer on the active area. The stock solutions of all oligonucleotides were prepared in autoclaved ultrapure water (> 18MΩ) to obtain a 10µM solution and kept frozen (-20°C). Then, tween-20 was used to block the non-immobilized area to avoid the attachment of non-specific targets binding. The sensor was incubate for 30 minute and DI water was used to wash the unattached tween-20 molecules. The hybridization reaction was tested with the complementary, non-complementary and mismatch target ssDNA by droping 10 µM, 2µl to the active surface area in the biosensor surface. The hybridized sensor was incubate 30 min. After that, the surface was washed with DI water to remove non-hybridized ssDNA target from the biosensor surface.

Then electrically characterize by Keithley I-V measurement to determine and investigate the current changes after each process.

3. Result and Discussion

3.1 Surface Characterization

The fabricated Al IDE was morphologically characterized using LPM and HPM for 20x resolution power. It is significant to make sure the fabrication process is done with circumspect to ensure the measurement is not affected during the characterization process [4]. Figure 2(a) shows the LPM image surface of the well fabricated IDE which has two electrodes that connected with finger comb structured electrode surface. Large number of fingers was fabricated properly. Figure 2(b) shows the HPM image surface topography shows uniform structure without shortage and any contaminants on the surface.

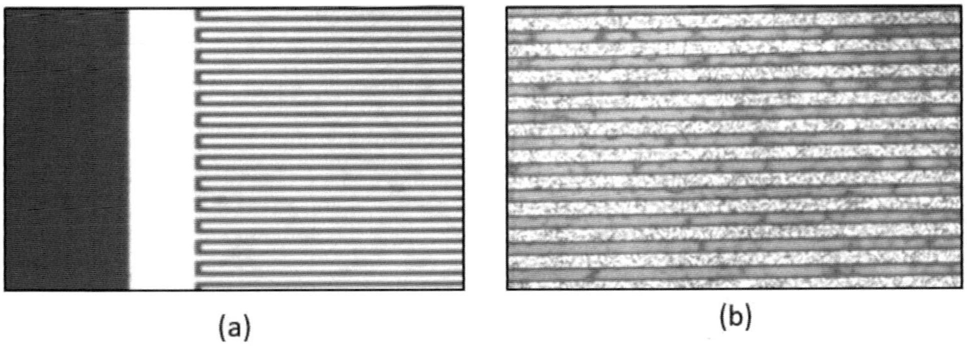

Figure. 2 Al IDE surface characterization using (a) Low power Microscopic (LPM) and (b) High power Microscopic (HPM) images.

3.2 Electrical Characterization

Electrical characterization was done using picoameter voltage source Keithley 2450, Kickstart software and probe station to get I -V characteristics as shown in Figure 3. The voltage between two electrodes was being prescribed from 0V to 1V. If voltage applied is higher than operating voltage, then sensor maybe damage [14],[28].

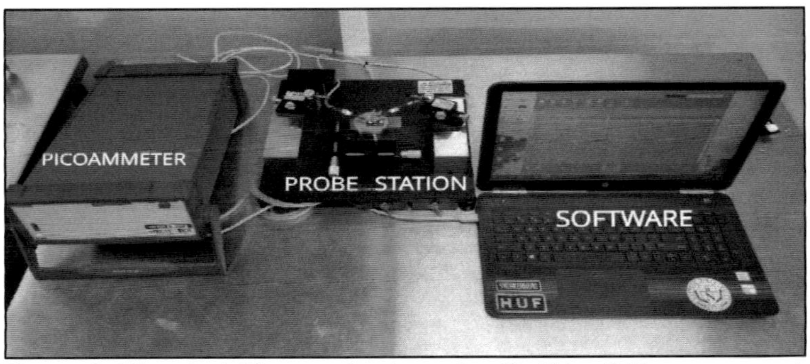

Figure 3. Electrical Characterization using Keithley I-V measurement.

In this work, electrical measurement of Al IDE's structured is proposed and the result can be determined by electrical characterization by current voltage I-V characterization for more accurate measurement. Figure 4 shows the graph for two Al IDEs has been electrically characterized. Both Al IDEs shows the slight voltage variation at 1V. The maximum current for first and second IDEs are 2.655×10^{-11} A and 2.747×10^{-11} A. Due to that current range an IDE are well fabricated and processed without any shortage. In additions, these results were confirmed that Al IDE was prepared with nearly similar dimensions and parameters. The probability of the IDEs to be shorted is high during the production process due to small finger comb dimension. So, its need to be measured repeatedly to ensure no shortage happens before proceeds with other process.

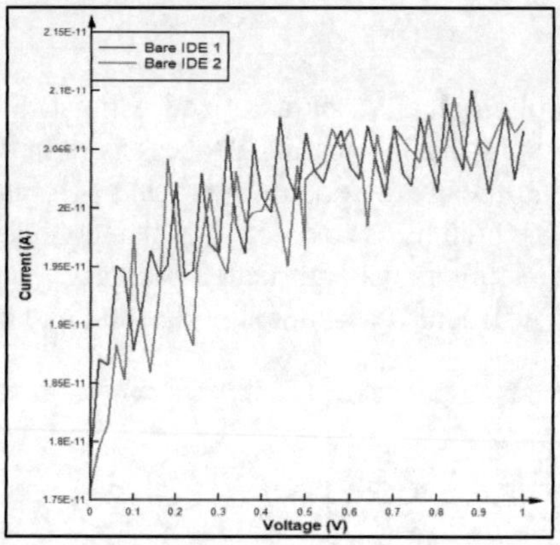

Figure 4. Graph of Current Vs Voltage for Bare Aluminum Interdigitated Electrode (Al IDE).

Figure 5 shows the I-V characteristics of biosensor preparation process of different steps, after salination, after immobilization and test with complementary ssDNA which is hybridization process. After silanization with APTES the current increase to nA range. APTES to provide contact between the organic and inorganic surfaces of a single-stranded DNA probe. APTES has an amine group that rich with positively charge ion. As the result, the current keep increases when drop the APTES on the bare of Al IDEs surface. The current captured at 1V shows the value which was 1.1×10^{-9} A. When APTES was dropped onto the IDE surface the length and area between comb fingers of an IDE is fixed then the density of an APTES become increase due to the increasing the number of hydroxide ions (OH^-). The current value then keeps increase after ssDNA probe was immobilized on the top of APTES layer. The current captured at 1V shows the value which was 2.5×10^{-9} A. Thus, when applied potential difference between two electrodes then the voltage gradually increases, accumulated positive charge start to move to negative side. Then, it causes an electric field increase. The current increases due to the increasing of the electric field. The current generation mechanism can be explained using following equations.

$$Conductivity\,,\ \sigma = \frac{l}{p} \tag{1}$$

$$Resistivity,\ R = \frac{pl}{A} \tag{2}$$

$$Electric\ field,\ E = \frac{k.Q}{d^2} \tag{3}$$

Finally, the 2 µl of 10 µM concentrated Salmonella ssDNA target was hybridized on the probe Salmonella. The current captured of the target at 1V is 4.3×10^{-9} A. The physical characteristics of ssDNA could offer substantial advantages as nucleic acid capture moieties in solid support based hybridization systems. DNA is negatively charged and after hybridization with complementary target ssDNA, free total positive charge carrier in APTES layer is increased because of the increment of the total negative charge of DNA probe and target. This sensing component changes the resistivity in the APTES layer according to the concentration of target ssDNA which are captured by ssDNA probes. Hybridization of synthetic target ssDNA can cause electric field increase. So current will decrease when electric field is increase.

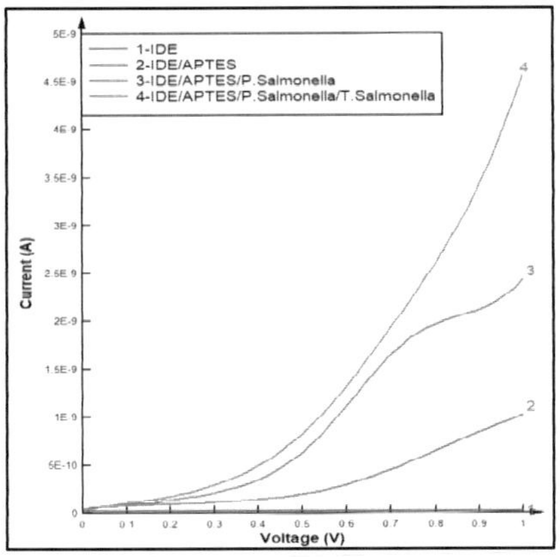

Figure 5. I-V characteristics for functionalization steps: (1) bare IDE, (2) silanization with APTES, (3) ssDNA probe salmonella and (4) shows I-V characteristics after hybridized with complementary Salmonella target ssDNA.

Figure 6 shows the selectivity measurement that was tested for complementary, non-complementary and single base mismatches target of salmonella separately. 10µM concentrated targets and 2µl volume was used for selective measurements. Curve 3 until 6 shows the Probe Salmonella, complement, non-complementary and mismatch measurement captured at 1V are 2.5×10^{-9} A, 4.3×10^{-9} A, 1.7×10^{-9} A and 2.2×10^{-9} A respectively. As recovery, the accessibility of immobilized probes to complementary target sequences can be enhanced by treating the surface with a small molecule blocking agents. The carboxylic group rapidly displaces the weaker adsorptive contact between oligonucleotides leaving the probes was moored probe through carboxylic end group. For probe and non-complement current curve shows nearly same value. It is because non-complementary ssDNA cannot be bind with immobilized ssDNA probes. As well as they cannot bind with APTES layer because of the blocking agents cover the area which was unbounded by immobilized ssDNA. Hence, non-complementary target ssDNA thoroughly removed after washing with deionized water.

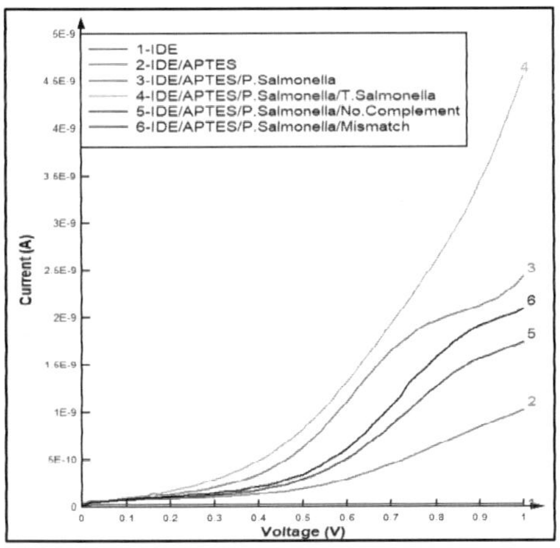

Figure. 6 Selectivity measurements: (1) Bare Al IDE, (2) Silanization with APTES, (3) Immobilized ssDNA probe salmonella, (4) hybridization with complementary salmonella target ssDNA, (5) Non-complementary target ssDNA and (6) single base mismatch target ssDNA.

4. Conclusion

As conclusion, this research was totally described an approach for Salmonella target ssDNA via Al IDE based biosensor. I-V characteristic were performed for different functionalization process of the biosensor such as bare Al IDE, silanization with APTES and immobilization with ssDNA salmonella probe. Blocking step with tween-20 was important to detect target specifically. Different current variation shows for different concentration, same concentration, complementary, non-complementary and mismatch ssDNA target. Due to the point above, this method of patient diagnosis provides rapid diagnosis which allows for faster and more efficacious therapeutic intervention, thereby preventing full-blown infection and also decreasing the spread of disease.

Acknowledgement

The author would like to thank all staff members of the Institute of Nanoelectronic Engineering in Universiti Malaysia Perlis (UniMAP) for their technical advice and contributions, directly and indirectly. Collaborative Research in Science and Technology Center (CREST) is acknowledged for providing grant for this research program.

References

[1] R. D. A. A. Rajapaksha, U. Hashim, N. Z. Natasha, M. N. A. Uda, V. Thivina, C. A. N. Fernando, *Proc. 2017 IEEE Reg. Symp. Micro Nanoelectron. RSM 2017*, 191–194, (2017).
[2] N. Z. Natasha, R. D. A. A. Rajapaksha, M. N. A. Uda, U. Hashim, *AIP Conf. Proc.*, **1885**, (2017).
[3] R. D. A. A. Rajapaksha, M. N. A. Uda, U. Hashim, S. C. B. Gopinath, C. A. N. Fernando, *IEEE Int. Conf. Semicond. Electron. Proceedings, ICSE*, **2018**, 93–96 (2018).
[4] M. N. A. Uda, R. D. A. A. Rajapaksha, M. N. A. Uda, U. Hashim, A. B. Jambek, *AIP Conf. Proc.*, **2045** (2018).
[5] R. D. A. A. Rajapaksha, U. Hashim, M. N. Afnan Uda, C. A. N. Fernando, S. N. T. De Silva, *Microsyst. Technol.*, **6**, 70 (2017).
[6] R. D. A. A. Rajapaksha, U. Hashim, M. N. A. Uda, and C. A. N. Fernando, *Mater. Sci. Eng. B*, **10(1)**, 61–64 (2017) .
[7] R. D. A. A. Rajapaksha, N. A. N. Azman, M. N. A. Uda, U. Hashim, S. C. B. Gopinath, C. A. N. Fernando, *AIP Conf. Proc.*, **2045**, (2018) .
[8] N. A. N. Yahaya, R. D. A. A. Rajapaksha, M. N. A. Uda,U. Hashim, *AIP Conf. Proc.*, **1885**, (2017).
[9] S. C. B. Gopinath, S. Ramanathan, K. Hann Suk, M. Ee Foo, P. Anbu, M. N. A. Uda, *MATEC Web Conf.*, **150**, 06002, (2018).
[10] M. . N. A. Uda *et al.*, *Adv. Mater. Res.*, **832**, 410–414 (2014).
[11] S. C. B. Gopinath, U. Hashim, M. N. A. Uda, *AIP Conf. Proc.*, **1808**, 1–5 (2017).
[12] R. D. A. A. Rajapaksha, N. A. N. Yahaya, M. N. A. Uda, U. Hashim, *AIP Conf. Proc.*, **2045**, 1–7 (2018) .
[13] S. Nadzirah, N. Azizah, U. Hashim, S. C. B. Gopinath, M. Kashif, *PLoS One*, **10(10)**, 1–15 (2015).
[14] M. Xu, R. Wang, Y. Li, "*Talanta*, **162**, 511–522, (2017).
[15] R. D. A. A. Rajapaksha, U. Hashim, M. N. Afnan Uda, C. A. N. Fernando, S. N. T. De Silva, *Microsyst. Technol.*, **6**, 70 (2017).
[16] P. Poltronieri, V. Mezzolla, E. Primiceri, and G. Maruccio, *Foods*, **3**, 511–526 (2014).
[17] M. N. A. Uda *et al.*, *AIP Conf. Proc.*, **1808**, (2017).
[18] M. N. A. Uda *et al.*, *AIP Conf. Proc.*, **1808**, 10–14, (2017).
[19] T. Adam and U. Hashim, *2014 Fifth International Conference on Intelligent Systems, Modelling and Simulation*, 1–8 (2014).
[20] M. N. . Uda *et al.*, Elsevier Inc., (2019).
[21] S. A. B. Ariffin, T. Adam, U. Hashim, S. Faridah Sfaridah, I. Zamri, and M. N. A. Uda, *Adv. Mater. Res.*, **832**, 113–117 (2013).

IOP Conf. Series: Materials Science and Engineering **743** (2020) 012022 doi:10.1088/1757-899X/743/1/012022

[22] M. G. Yanez, *Free. Semicondutor*, 1–34, (2013).
[23] U. Hashim *et al.*, *Proc. - 2015 2nd Int. Conf. Biomed. Eng. ICoBE 2015*, 30–31, (2015) .
[24] N. A. Abdul-Mutalib, A. N. Syafinaz, K. Sakai, and Y. Shirai, *Int. Food Res. J.*, **22(3)**, 896–901, (2015).
[25] W. P. Sharifa Ezat, D. Netty, G. Sangaran, *Malaysian J. Public Heal. Med.*, **13**, 1–7, (2013).
[26] B. S. Rao and U. Hashim, *AIP Conf. Proc,* **795**, 388–392, (2013).
[27] N. A. Parmin *et al.*, *Microchim. Acta*, **186(6)**, (2019).
[28] N. A. Parmin *et al.*, *Int. J. Biol. Macromol.*, **126**, 877–890, (2019).
[29] P. Intra and N. Tippayawong, “An ultra-low current meter for aerosol detection,” *C. J. Nat. Sci*, **6(2)**, 313–320 (2007).

IOP Conf. Series: Materials Science and Engineering **743** (2020) 012023 doi:10.1088/1757-899X/743/1/012023

Development of Geopolymer Ceramic as a Potential Reinforcing Material in Solder Alloy: Short review

Nadiah 'Izzati Zulkifli[1], Mohd Mustafa Al Bakri Abdullah[1], Mohd Arif Anuar Mohd Salleh[1], Romisuhani Ahmad, Andrei Victor Sandu[3] and Nurul Aida Mohd Mortar[1]

[1]Center of Excellence Geopolymer and Green Technology, School of Materials Engineering, Universiti Malaysia Perlis (UniMAP), Perlis, Malaysia
[2]Faculty of Engineering Technology, Universiti Malaysia Perlis (UniMAP), Perlis, Malaysia
[3]Gheorghe Asachi Technical University of lasi, Faculty of Materials Science and Engineering, lasi, Romania

Email: nadiahizzati.zulkifli@gmail.com

Abstract. Nowadays, the consumption of lead-free solder has been widely used around the world since the utilization of SnPb solder has been banned and restricted by European Union. Variety of studies have been conducted by the researchers to find an alternative to replace the usage of SnPb such as SnCu, SAC, SnAg and etc. However, the development of plain lead-free solder was declared to provide low mechanical, thermal, and electrical properties in terms of interfacial intermetallic compound and wettability towards its solder joint compare to the traditionally monolithic SnPb solder alloy. Mostly, previous studies stated that addition of some additives such as ceramic particles (Si_3N_4, TiO_2, SiC, NiO and etc) may improves the solder joint reliability. At the same time, no major studies were done using geopolymer ceramic as reinforcing agent in plain matrix alloy. Therefore, this paper reviews the fabrication process of multiple geopolymer-based ceramic such as fly ash, kaolin, and slag as reinforcement in solder alloy. The development process includes the processing method of geopolymer ceramic and the characterization of geopolymer ceramic as reinforcing material consist of; i) chemical composition, and ii) phase identification.

1. Introduction

Soldering process becomes necessary in electronic packaging industries as one part of mechanically support and joint the components on the circuit board to assembly together and thus, providing electrical support by transmitting electrical connection within the solder joint between solder and circuit [1]. As solder will remain as primary assembly and interconnection technology for several period of time, the soldering process has been a fundamental aspect in realisation of all electronic products for a functioning lifespan [2].

Formerly, the soldering industries used Tin lead (SnPb) eutectic alloy solder material in the production of solder material for the past electronic devices. In fact, the used of SnPb eutectic alloy in soldering industries has successfully produced exclusive combination of chemical, physical, mechanical and electrical properties in producing good reliable solder alloy with reasonable price for many years [2], [3]. Unfortunately, the used of SnPb was reported to release toxicity and might be harmful to human's health and environment if it is used for long-term usage [4]. As a result, the

Content from this work may be used under the terms of the Creative Commons Attribution 3.0 licence. Any further distribution of this work must maintain attribution to the author(s) and the title of the work, journal citation and DOI.
Published under licence by IOP Publishing Ltd

legislation on restriction of the toxic SnPb by European Union (EU) Waste of Electrical, Electronic Equipment (WEEE) and RoHS (Restriction of Hazardous Substances) have strictly banned the usage due to health and environmentally concern since it might give negative impact to the human's health and environment surrounding [3], [5].

Therefore, researchers find other alternatives in order to replace traditionally SnPb eutectic alloy with lead-free solder alloy. In soldering process, it is difficult to control the robustness of solder joint in electronic device. Moreover, the rapid modernity in electronic application is a great challenge faced by the electronic packaging industries as they need to keep in track with the emerging advances while struggling to stay within manufacturing cost limit [6]. addition of other reinforcement such as from ceramic based are widely acknowledge by the researchers as it can increase the mechanical and thermal properties of the solder joint. Recently, many investigation about Sn-0.7/Cu/Si_3N_4, Sn-0.7Cu-0.05Ni/TiO_2, Sn-3.0Ag-0.5Cu/SiC, Sn-3.0Ag-0.5Cu/ NiO and other related to this study has been observe to have various properties for the matrix alloy solder [7], [8] Ceramic materials demonstrate enhanced properties in terms of microstructure, thermal properties, tensile strength, liquidous temperature and others under demanding conditions became one of the factors in choosing ceramic as reinforcement as it is reported to decompose β-Sn microstructure [9].

2. Processing of geopolymer for solder reinforcement

The history of alkaline activator was implemented since 1940s. In 1950s, Krivenko and Glukhovsky had invented the used of alkaline activator in cement system that comprised of alumino-silicate phases and calcium silicate hydrated (CSH) [10], [11]. Later on, geopolymer was introduced by J. Davidots in 1978 after 30 years of studies where the process occurs the combination of inorganic material such as silica and alumina with an alkaline activator [12]. The terms geopolymer came from an analogy of inorganic materials and organic polymers consist a group of mineral binders with some chemical composition similar to zeolite but in an amorphous structure. The mechanism of geopolymerization consist of dissolution, transportation or orientation, polycondensation and going through an exorthemic process [13], [14] The selection for solid raw material geopolymer-based could be natural minerals rich in silicone (Si) and aluminium (Al) such as clays, kaolinite, or also called waste materials like red mud, slags, fly ash etc. While the liquid for the alkaline activator usually consist of NaOH or KOH with a soluble alkali material such as Na_2SiO_3 [15]. Different fabrication methods in producing geopolymer ceramic gives different mechanical properties and relative densities. In processing geopolymer as reinforcement, this paper will focus more on mix design that provided optimum properties toward the geopolymer performance previously.

Nmiri et. al found that a maximum strength of metakaolin-based geopolymer produced when using NaOH solution as alkaline activator compare to KOH solution during thermal treatment at 700°C [12]. For fly ash geopolymer, a few researchers agreed that 12 M of NaOH solution as an alkaline activator gave the best performance in terms of mechanical properties [16]–[18] Other than that, a concentration that is high than 12 M is difficult to be handle since it might be harmful when in contact to direct skin and a little bit costly due to the high usage of NaOH pallets [19]. In term of solid to liquid (S/L) and $NaSiO_3$/NaOH ratio, mostly researchers also approved that S/L ratio of 2.0 and 2.5 was the best mixture workability in providing high flexural strength [20], [21] Basically, fly ash geopolymer and slag geopolymer can be cured at an ambient temperature. But due to the slow reactivity of kaolin geopolymer, a high cured temperature at 60°C was proposed by Heah et. al since it helps the geopolymerization to speed up [22], [23].

Commonly, kaolin refers to a family of kaolinitic clays that consist of 1:1-type clay mineral with one tetrahedral sheet and one octahedral sheet [11], [24]. The most important parameters that need to be emphasize in processing kaolin geopolymer is their low workability. Several researchers found out that S/L ratio of 1.0 with NaOH molarity of 8M gives the best result where it does not crack and were stable when exposed to the water for several days [22], [25]–[29]. A higher S/L ratio could lead to low workability and make it difficult during mixing process. Heah et. al concluded that a ratio of 0.32 Na_2SiO_3/NaOH could give an increment in strength [10], [14].

Slag is a non-toxic material which can be insert as a raw material in geopolymer production for making high performance of geopolymer product due to their high content of CaO as shown in table 1 below [30]. Processing slag geopolymer might be tough since the CaO content is normally high causing low workability and fast setting during processing. Previously, slag was used as additive to increase the reactivity or speed up the geopolymerization rate of base material such as fly ash class f and kaolin. Based on Aziz et. al, 10 M concentration of NaOH with Na_2SiO_3/NaOH ratio of 2.5 provided maximum compressive strength than low concentration of NaOH and alkaline activator. S/L ratio was the main contributor in order to produce good geopolymer properties. According to their studies, S/L ratio of 3.0 promotes higher compressive strength compare to S/L = 1.0 [31].

Table 1. Chemical compound base on XRF result of raw material for geopolymer-based [32]–[34].

Compound	Mass (%)		
	FAC	Kaolin	Slag
SiO2	21.23	54.5	33.56
Al2O3	25.16	32.4	8.44
Fe2O3	4.86	4.32	-
CaO	32.86	-	47.49
MgO	4.95	-	4.74
SO3	9.28	-	4.14
Na2O	0.43	-	0.37
K2O	-	5.58	-
MnO3	-	0.09	0.37
TiO2	0.95	1.33	0.9
ZrO2	-	0.08	-

3. Geopolymer Ceramic Powder

Geopolymer ceramic produced when the geopolymer was introduced to high temperature as high as 500°C and above depends on its material's melting point. The removal of carbon in fly ash class f are also known as carbon burn-out (CBO) system. At temperature of 500°C, active carbon in fly ash class f was removed with a little change in inorganic portion. All carbon content was said to remove at thermal treatment range of 700°C. At this state, the crystalline structure already slightly appeared proved that ceramic-based already achieved [35]–[37] Nevertheless, only a few crystalline background on the XRD pattern produced at this state [38]. According to K. Zulkifli et. al and Heah et. al, the crystalline phase started to increase when fly ash geopolymer and metakaolin geopolymer was subjected to 800°C. However, at temperature 1000°C, the amorphous phase dropped from 66% to 43% due to the crystallinity appearance of nepheline ($NaAlSiO_4$) and anortoclase ($NaAlSi_3O_8$) [23], [39]. John Fox elucidated that fly ash class f started to crystalline starting at 700°C to 800°C while fly ash class c shows its appearance of crystalline phase after 800°C and becomes almost fully crystalline at 1000°. Due to the high percentage of amorphous phase in class f fly ash, John Fox concluded that higher amount of CaO content in class c fly ash contribute to high crystallization degree after being heated to 1000°C and above [38].

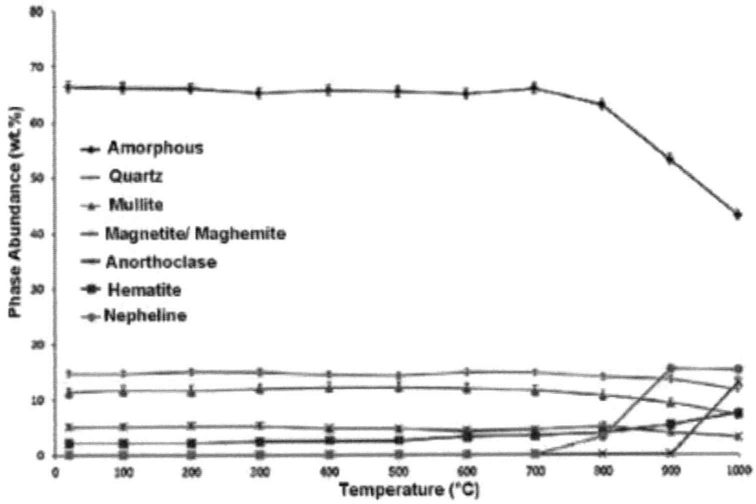

Figure 1. Fly ash geopolymer (16.4% iron oxide) fired under various temperature [39].

A study by Y. Liew found that the water absorption of kaolin geopolymer ceramic (KGC) reduced, produced maximum flexural strength when subjected to 1200°C temperature due to transformation of crystalline phases that help to decline water absorption [40], [41]. Sintering to a high temperature up to 1200°C resulted the presence of high nepheline phase which help the enhancement of mechanical strength due to the high amount of silicon-rich and polymerized species [42]. Previous study of an alkali activated shale-slag binder indicated that the amorphous content was gradually disappearing and transform into crystalline state starting from 800°C and above [43]. By sintering at high temperature, densification of structure will occur without stimulating the grain growth and therefore allow the refinement of microstructure into a dense ceramics [44].

4. Conclusion

This review paper summarized that it is possible to process numerous geopolymer powder as reinforcement in solder alloy. Although there aren't many studies focused on this, it is possible to fabricate geopolymer reinforced solder alloy to characterize the difference in chemical composition either it might alter the solder joint performance or neither. Besides, the different geopolymer-based material also led to different phase diagram and chemical composition when subjected to a certain temperature. Depending on the raw material selection, geopolymer exhibit a wide variety of properties such in terms of compressive strength, low thermal conductivity, fire resistance, acid resistance and many other more. Geopolymer are considered as fine engineering materials with a broad application. Therefore, different characteristics and properties of geopolymer towards solder alloy can be explores further.

Acknowledgement

This work was supported by Centre of Excellence Geopolymer & Green Technology (CeGeogTech), Universiti Malaysia Perlis, Nihon Superior Co. Ltd and Institutional Links grant, ID 332397914, under the Newton-Ungku Omar Fund partnership. The grant is funded by the UK Department of Business, Energy and Industrial Strategy (BEIS) and Malaysia and delivered by the British Council.

References

[1] N. Nadiah 'Izzati Binti Zulkifli, M. Mustafa Al Bakri Abdullah, and M. Arif Anuar Mohd Salleh, "Fabrication of Novel Geopolymer Reinforced Tin Copper Solder in Suppressing Intermetallic Layer Growth," *IOP Conf. Ser. Mater. Sci. Eng.*, vol. 551, p. 012091, 2019.

[2] H. R. Kotadia, P. D. Howes, and S. H. Mannan, "A review: On the development of low melting temperature Pb-free solders," *Microelectron. Reliab.*, vol. 54, no. 6–7, pp. 1253–1273, 2014.

[3] M. Aamir, R. Muhammad, N. Ahmed, M. Sadiq, and M. Waqas, "Mechanical Properties of Lead Free Solder Alloy for Green Electronics Under High Strain Rate and Thermal Aging," vol. 36, no. 1, pp. 1–9, 2017.

[4] M. I. I. Ramli, N. Saud, M. A. A. M. Salleh, M. N. Derman, and R. M. Said, "Effect of TiO2 additions on Sn-0.7Cu-0.05Ni lead-free composite solder," *Microelectron. Reliab.*, vol. 65, pp. 255–264, 2016.

[5] S. Cheng, C. M. Huang, and M. Pecht, "A review of lead-free solders for electronics applications," *Microelectron. Reliab.*, vol. 75, pp. 77–95, 2017.

[6] M. A. A. Mohd Salleh, S. McDonald, and K. Nogita, "Non-Metal Reinforced Lead-Free Composite Solder Fabrication Methods and its Reinforcing Effects to the Suppression of Intermetallic Formation: Short Review," *Appl. Mech. Mater.*, vol. 421, no. September, pp. 260–266, 2013.

[7] M. A. A. Mohd Salleh, S. D. McDonald, H. Yasuda, A. Sugiyama, and K. Nogita, "Rapid Cu6Sn5growth at liquid Sn/solid Cu interfaces," *Scr. Mater.*, vol. 100, pp. 17–20, 2015.

[8] S. Chellvarajoo, M. Z. Abdullah, and C. Y. Khor, "Effects of diamond nanoparticles reinforcement into lead-free Sn-3.0Ag-0.5Cu solder pastes on microstructure and mechanical properties after reflow soldering process," *Mater. Des.*, vol. 82, pp. 206–215, 2015.

[9] E. A. Eid, A. N. Fouda, and E. S. M. Duraia, "Effect of adding 0.5wt% ZnO nanoparticles, temperature and strain rate on tensile properties of Sn-5.0wt% Sb-0.5wt% Cu (SSC505) lead free solder alloy," *Mater. Sci. Eng. A*, vol. 657, pp. 104–114, 2016.

[10] Y. M. Liew *et al.*, "Optimization of solids-to-liquid and alkali activator ratios of calcined kaolin geopolymeric powder," *Constr. Build. Mater.*, vol. 37, pp. 440–451, 2012.

[11] C. Y. Heah, H. Kamarudin, A. M. Mustafa Al Bakri, M. Luqman, I. Khairul Nizar, and Y. M. Liew, "Potential application of kaolin without calcine as greener concrete: A review," *Aust. J. Basic Appl. Sci.*, vol. 5, no. 7, pp. 1026–1035, 2011.

[12] A. Nmiri, N. Hamdi, O. Yazoghli-Marzouk, M. Duc, and E. Srasra, "Synthesis and characterization of kaolinite-based geopolymer: Alkaline activation effect on calcined kaolinitic clay at different temperatures," *J. Mater. Environ. Sci.*, vol. 8, no. 2, pp. 276–290, 2017.

[13] N. P. Rajamane, M. C. Nataraja, and N. Lakshmanan, "An introduction to geopolymer concrete," *Indian Concr. J.*, vol. 85, no. 11, pp. 25–28, 2011.

[14] A. Romisuhani, A. M. M. Al Bakri, and J. Nur Ain, "Review on development of clay based geopolymer ceramic composites," *Mater. Sci. Forum*, vol. 803, no. June 2015, pp. 37–43, 2015.

[15] D. D. Burduhos Nergis, M. M. A. B. Abdullah, P. Vizureanu, and M. F. Mohd Tahir, "Geopolymers and Their Uses: Review," *IOP Conf. Ser. Mater. Sci. Eng.*, vol. 374, no. 1, 2018.

[16] L. A. Sofri, M. M. A. B. Abdullah, M. R. M. Hasan, and Y. Huang, "Effect of NaOH concentration and fly ash/alkaline activator ratio on the compressive strength of road base material," *AIP Conf. Proc.*, vol. 2030, no. November, 2018.

[17] A. M. M. Al Bakri, H. Kamarudin, O. A. K. A. Abdulkareem, C. M. Ruzaidi, A. R. Rafiza, and M. N. Norazian, "Optimization of Alkaline Activator/Fly ASH Ratio on the Compressive Strength of Manufacturing Fly ASH-BASED Geopolymer," *Appl. Mech. Mater.*, vol. 110–116, no. October 2011, pp. 734–739, 2011.

[18] G. F. Huseien, J. Mirza, M. Ismail, M. W. Hussin, M. A. M. Arrifin, and A. A. Hussein, "The Effect of Sodium Hydroxide Molarity and Other Parameters on Water Absorption of Geopolymer Mortars," *Indian J. Sci. Technol.*, vol. 9, no. 48, 2016.

[19] S. V. Joshi and M. S. Kadu, "Role of Alkaline Activator in Development of Eco-friendly Fly Ash Based Geo Polymer Concrete," *Int. J. Environ. Sci. Dev.*, vol. 3, no. 5, pp. 417–421, 2012.

[20] Muhammad Faheem, M.T., Mustafa Al Bakri, A.M., Kamarudin, H., Ruzaidi, C.M., Binhussain, M., Izzat, A.M. The relationship of Na2SiO3/NaOH ratio, Kaolin/Alkaline activator ratio and sand/kaolin ratio to the strength of kaolin-based non load bearing Geopolymer Brick

International Review of Mechanical Engineering, 7 (1), pp. 161-166, 2013.

[21] M. M. Al Bakri Abdullah *et al.*, "Design, Processing and Chatacterization of Fly Ash Based Lightweight Geopolymer Concrete Using Foaming Agent Technology," *Rev Chim-Bucharest*, vol. 64, no. 4, pp. 382–387, 2013.

[22] C. Y. Heah *et al.*, "Effect of curing profile on kaolin-based geopolymers," *Phys. Procedia*, vol. 22, pp. 305–311, 2011.

[23] H. Cheng-Yong, L. Yun-Ming, M. M. A. B. Abdullah, and K. Hussin, "Thermal Resistance Variations of Fly Ash Geopolymers: Foaming Responses," *Sci. Rep.*, vol. 7, no. November 2016, pp. 1–11, 2017.

[24] Z. H. Zhang, H. J. Zhu, C. H. Zhou, and H. Wang, "Geopolymer from kaolin in China: An overview," *Appl. Clay Sci.*, vol. 119, pp. 31–41, 2016.

[25] C. Y. Heah *et al.*, "Study on solids-to-liquid and alkaline activator ratios on kaolin-based geopolymers," *Constr. Build. Mater.*, vol. 35, pp. 912–922, 2012.

[26] Wazien, A.Z.W., Abdullah, M.M.A.B., Abd Razak, R., Rozainy, M.A.Z.M.R., Tahir, M.F.M. Strength and Density of Geopolymer Mortar Cured at Ambient Temperature for Use as Repair Material, IOP Conference Series: Materials Science and Engineering, 133 (1), 2016.

[27] A. D. Hounsi, G. L. Lecomte-Nana, G. Djétéli, and P. Blanchart, "Kaolin-based geopolymers: Effect of mechanical activation and curing process," *Constr. Build. Mater.*, vol. 42, pp. 105–113, 2013.

[28] R. M. Said, M. A. A. M. Salleh, N. A. A. Amran, and M. I. I. Ramli, "Microstructure Evolution of Sn-Cu Based Solder Paste with Different Cu Concentration Subjected to Multiple Reflows," *Solid State Phenom.*, vol. 280, pp. 206–211, 2018.

[29] C. Substrates, C. M. Ruzaidi, A. M. M. Al Bakri, M. Binhusain, and M. S. S. Salwa, "Study on Properties and Morphology of Kaolin Based Geopolymer Coating on Study on Properties and Morphology of Kaolin Based Geopolymer Coating on Clay Substrates," no. May, 2014.

[30] T. W. Cheng and J. P. Chiu, "Fire-resistant geopolymer produce by granulated blast furnace slag," *Miner. Eng.*, vol. 16, no. 3, pp. 205–210, 2003.

[31] I. H. Aziz, M. M. A. B. Abdullah, H. C. Yong, L. Y. Ming, D. Panias, and K. Sakkas, "Correlation of the Processing Parameters in the Formation of Granulated Ground Blast Furnace Slag Geopolymer," *IOP Conf. Ser. Mater. Sci. Eng.*, vol. 209, no. 1, 2017.

[32] N. A. Jaya *et al.*, "Kaolin Geopolymer as Precursor to Ceramic Formation," vol. 01061, 2016.

[33] S. Parvathy S, A. K. Sharma, and K. B. Anand, "Comparative study on synthesis and properties of geopolymer fine aggregate from fly ashes," *Constr. Build. Mater.*, vol. 198, pp. 359–367, 2019.

[34] M. A. Yazdi, M. Liebscher, S. Hempel, J. Yang, and V. Mechtcherine, "Correlation of microstructural and mechanical properties of geopolymers produced from fly ash and slag at room temperature," *Constr. Build. Mater.*, vol. 191, pp. 330–341, 2018.

[35] J. Davidovits and M. Davidovics, "Geopolymer. Ultra-high temperature tooling material for the manufacture of advanced composites," *Int. SAMPE Symp. Exhib.*, vol. 36, no. pt 2, pp. 1939–1949, 1991.

[36] R. Ctvrtlik, V. Kulikovsky, V. Vorlicek, P. Bohac, and M. Stranyánek, "Effect of structure on mechanical properties of covalent ceramics," *Nanocon 2010 Conf.*, 2010.

[37] R. Ahmad, M. M. Al Bakri Abdullah, K. Hussin, A. V. Sandu, M. Binhussain, and N. A. Jaya, "Processing and properties of polymer filled geopolymer ceramics fabricated via powder metallurgy method: A review," *Rev. Adv. Mater. Sci.*, vol. 44, no. 1, pp. 26–32, 2016.

[38] J. M. Fox, "Changes in Fly Ash With Thermal Treatment," 2005.

[39] K. Zulkifly, H. C. Yong, M. M. A. B. Abdullah, L. Y. Ming, D. Panias, and K. Sakkas, "Review of Geopolymer Behaviour in Thermal Environment," *IOP Conf. Ser. Mater. Sci. Eng.*, vol. 209, no. 1, 2017.

[40] Y. M. Liew *et al.*, "Formation of one-part-mixing geopolymers and geopolymer ceramics from geopolymer powder," *Constr. Build. Mater.*, vol. 156, pp. 9–18, 2017.

[41] N. A. Jaya, M. M. Al Bakri Abdullah, and R. Ahmad, “Reviews on clay geopolymer ceramic using powder metallurgy method,” *Mater. Sci. Forum*, vol. 803, no. August 2015, pp. 81–87, 2015.
[42] A. Sabbatini, L. Vidal, C. Pettinari, I. Sobrados, and S. Rossignol, “Control of shaping and thermal resistance of metakaolin-based geopolymers,” *Mater. Des.*, vol. 116, pp. 374–385, 2017.
[43] D. Rieger *et al.*, “Effect of thermal treatment on reactivity and mechanical properties of alkali activated shale-slag binder,” *Constr. Build. Mater.*, vol. 83, pp. 26–33, 2015.
[44] A. Romisuhani, M. M. Albakri, H. Kamarudin, and S. V. Andrei, “The influence of sintering method on kaolin-based geopolymer ceramics with addition of ultra high molecular weight polyethylene as binder,” *IOP Conf. Ser. Mater. Sci. Eng.*, vol. 267, no. 1, 2017.

A Review of Morphology Analysis on Dolomite as an Additive Material in Geopolymer

Ahmad Syauqi Sauffi[1,*], Wan Mastura Wan Ibrahim[2], Romisuhani Ahmad[2], Nurul Aida Mohd Mortar[2], Fakhryna Ahmad Zaidi[1] and Warid Wazien Ahmad Zailani[3]

[1]Faculty of Engineering Technology, Universiti Malaysia Perlis (UniMAP), P. O. Box 77, d/a Pejabat, Pos Besar, 01007 Kangar, Perlis, Malaysia
[2]Centre of Excellence Geopolymer and Green Technology, School of Material Engineering, Universiti Malaysia Perlis (UniMAP), P. O. Box 77, d/a Pejabat, Pos Besar, 01007 Kangar, Perlis, Malaysia
[3]Faculty of Civil Engineering, Universiti Teknologi MARA, 40450 Shah Alam, Selangor

Email: * syauqichi@gmail.com

Abstract. Dolomite is a carbonate mineral in nature. Dolomite ($CaMg(CO_3)_2$) is an anhydrous carbonate mineral composed of calcium, magnesium, and carbonate. The word dolomite is also used to describe the sedimentary carbonate rock, which is composed predominantly of the mineral dolomite (also known as dolostone). Dolomite had been one of the frequent used materials in most of researches due to its accessibility to acquire and the properties. The composition of Ca and Mg became one of the main attractions for the usage of this natural mineral. One of the main properties needed to be studied when using any materials is morphology of the martial. Morphology analysis can give much valuable information about the surface topography and composition of the sample. Other than that, detailed three-dimensional and topographical imaging can also be obtained. This can helps the researches to finds the core value or the novelty in their study. Thus, reviews for the morphology analysis of dolomite were done to investigate how the structures of dolomite develop for certain study and usage. Besides that, dolomite as an addition into certain materials were reviewed and compared to observe the interactions and changes happened.

1. Introduction

Every year, more than 13 billion tons of carbon dioxide (CO_2) was produced by the production of Portland cement (OPC). From the total CO_2 emission of CO_2 in this world, 7% of them were from the CO_2 produced by OPC. This high CO_2 emission may lead to environmental problems such as global warming and greenhouse effect [1][2]. To overcome this problem, geopolymer in introduced. Geopolymer is an inorganic materials that containing aluminosilicate materials which are Si and Al that undergo polymerization of the materials in strong alkali environment [3]. Besides that, geopolymer have attracted increasing attention in recent years for their potential uses as construction materials that would partly replace ordinary Portland cement (OPC) usage these days [4].

Geopolymers are amorphous to semi-crystalline three-dimensional aluminosilicate materials synthesized at low temperature in a short time [5]. Geopolymerization occurs in highly alkaline solution with aluminosilicate oxides and silicates as the reactants. It is carried out by putting the

Content from this work may be used under the terms of the Creative Commons Attribution 3.0 licence. Any further distribution of this work must maintain attribution to the author(s) and the title of the work, journal citation and DOI.
Published under licence by IOP Publishing Ltd

aluminosilicate materials in contact with the alkaline activator solution which produce the polymeric chain of Si-O-Al-O bonds [6][7][8]. The most used alkaline solution for geopolymer is sodium hydroxide or potassium hydroxide and sodium silicate or potassium silicate solution. Based on latest study, alkaline activator that would give the most effective reaction towards geopolymerization would be the usage of sodium hydroxide solution with sodium silicate solution [9]. During geopolymerization, removal of water from the system occurs which makes that reaction exothermic in nature due to polycondensation [10]. Geopolymer synthesis involves dissolution, hydrolysis and condensation of aluminate and silicate species to form geopolymeric structure which occurs at room temperature [11]. The types of geopolymer was shown in Figure 1 which includes poly(sialate), poly(sialatesiloxo) and poly(sialate-dixoloxo) [12].

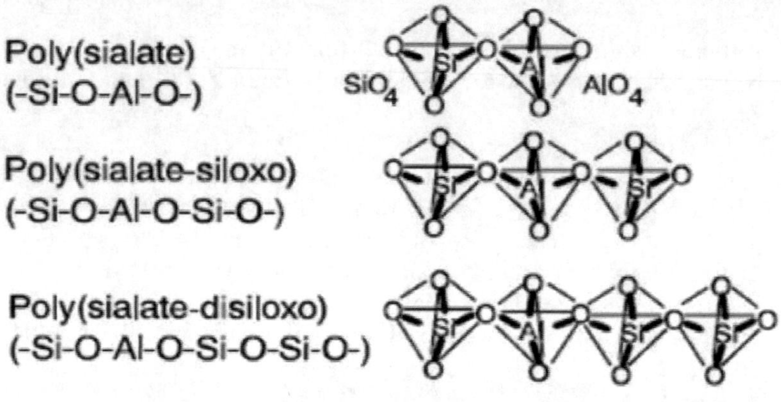

Figure 1: The types of geopolymer [13].

Geopolymers have a wide range of properties and characteristics that depends on the raw materials used and the processing conditions. The characteristics of geopolymer includes high compressive strength, low shrinkage, fast or low setting, acid resistance, fire resistance, and low thermal conductivity[14]. With these advantages, geopolymers are considered to be suitable for a broad range of engineering materials applications [15][16].

2. Dolomite as a raw materials

Dolomite is one of the most common carbonate minerals in the geologic record. It is an anhydrous carbonate mineral composed of calcium, magnesium, and carbonate, ideally $CaMg(CO_3)_2$ [14][15]. Dolomite is widespread sedimentary rocks that are abundant and generally inexpensive natural minerals [19]. There are two main mechanisms by which dolomite can form in the natural environment: (1) “directly from solution” or primary dolomite, and (2) dolomites that formed during digenesis or burial, most often through secondary replacement (dolomitization) of previously precipitated calcite or aragonite [20]. Dolomite is commonly used in road and building construction for foundation bases and as an aggregate for cement and asphalt concrete. Table 1 shows the chemical composition of dolomite.

Table 1. Chemical composition of dolomite [21].

CaO	SiO_2	Al_2O_3	Fe_2O_3	MgO	K_2O	Na_2O
33.4	2.5	0.7	0.3	17.1	0.1	0.1

Based on the chemical compositions, the highest composition content in dolomite is CaO with 33.4% and then followed by MgO with 17.1%. Besides that, dolomite also maintains SiO_2 and Al_2O_3 which is the backbone for the geopolymerization to occur. Due to the low Si and Al in the dolomite, the usage in geopolymer is quiet rare. Therefore, the study of dolomite in geopolymer would be a great interest. The study for the compressive strength, morphology, phases, and chemical bindings can be discovering as a novelty.

3. Morphology analysis

3.1. Dolomite morphology

Figure 22 shows the morphology of dolomite war material from different study. Figure 22a) show the finding from Ehud Cohen et al., [22]. The surfaces of dolomite on this study showed that it has a very rough surface with some of it have pointy shapes at the end. This dolomite have irregular shape and different of size.

Meanwhile in Figure 22 b), finding from Ana I. Casado et al., [23] shows that the dolomite had a smoother surface compared dolomite before. Although they have irregular shape, but here are no pointy ends that can be observed on the surface of the dolomite form Figure 22 b). Both of dolomite observed, have different surface. This will affect the reaction occurred when be used as a geopolymer.

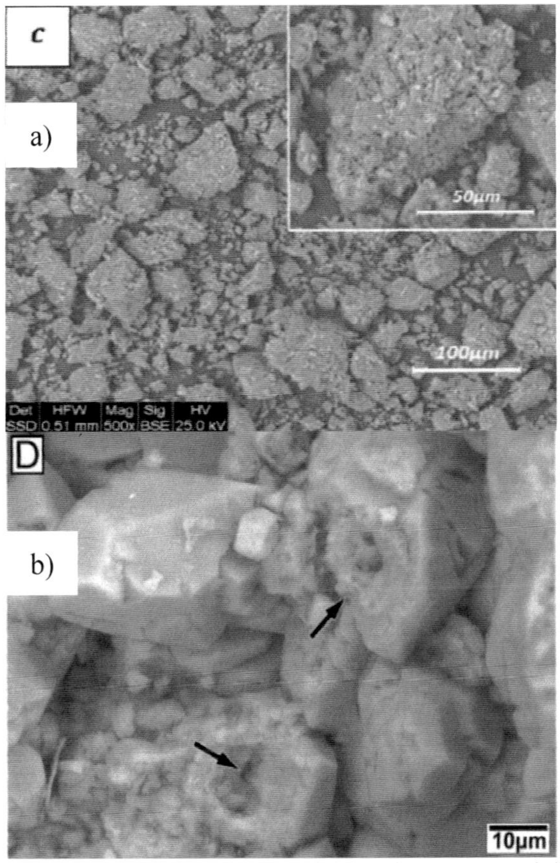

Figure 2. The morphology of dolomite raw material.

3.2. Fly Ash/Dolomite Geopolymer Morphology

Figure 33 shows the morphology for dolomite/fly ash geopolymer for solid to liquid ratio of 1.0, 2.5, and 3.0 obtained from E. A. Aizat et al. It shows that 1.0 S/L ratio non-homogeneous reaction because most of the dolomite and fly ash particle were not fully surrounded by the geopolymer matrix. The 1.0 S/L also shows a large void is present in the surface of the geopolymer. Another S/L ratio is 2.5. This S/L shows most of the particles in the geopolymer were surrounded by the geopolymer gel. This proved that the reaction for this S/L ratio is better than 1.0 S/L. Besides that, the voids in 2.5 S/L became smaller compared to 1.0 S/L. The morphology analysis for 3.0 S/L observed had different interaction. Some agglomeration occurred that can be observed and the material and geopolymer gel seems to be separated away from each other [24].

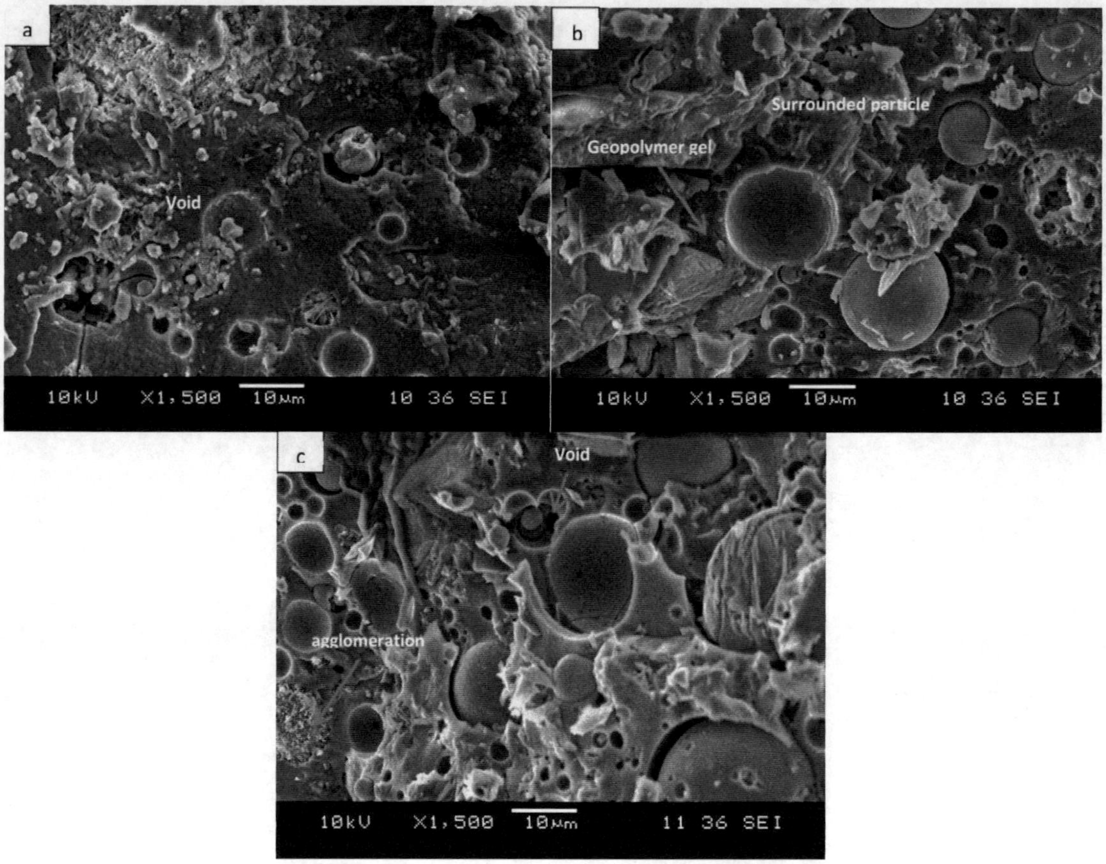

Figure 3. Morphology for dolomite/fly ash geopolymer for solid to liquid ratio of a) 1.0, b) 2.5, and c) 3.0.

3.3. Metakaolin/Dolomite Geopolymer

Figure 4 shows the morphology of metakaolin/dolomite geopolymer. The results were based on the optimum compressive strength obtained by the addition of dolomite into metakaolin .There was a few these that can be observed. Firstly, A is the geopolymeric gel. This shows that the geopolymerization had been occurred there Most of the surfaces were covered by the geopolymer gel. Label B show the unreacted of silica and alumina from the raw materials. There are only a small amount of unreacted silica and alumina. The unreacted is not due to the low reactivity, it was due to geopolymerization haven't fully completed.

Besides that, label D scattering of sodium aluminosilicate material with a high proportion of calcium. The pores can't be seen in this morphology due to the best mix ratio and dolomite addition into the systems. The morphology of fly ash/dolomite and metakaolin dolomite can't be compared because of different surface appearance. However, it still can be seen that both of the surface become cloud shape at certain places.

Figure 4. The morphology of metakaolin/dolomite geopolymer [19].

4. Conclusion

As for conclusion, dolomite can be used as a geopolymer material and it gave different effect towards different addition usage. Besides that, dolomite addition will produce cloud shape to the geopolymer surface. Besides that, there are less study on the dolomite that the morphology can be fully compared and studied. Thus, more study should be done with dolomite as a geopolymer material which can give novelties to the researchers.

References

[1] M. M.A.B. Abdullah ., *Rev Chim-Bucharest*, **64**, 382–387 (2013).
[2] H.C. Yong, L.Y. Ming, M.M.A.B. Abdullah, and K. Hussin, *Sci. Rep.*, **7**, 1-11 (2016).
[3] M.M.A.B. Abdullah, L. Jamaludin, H. Kamarudin, M. Binhussain, C.M.R. Ghazali, M,I. Ahmad, *Adv. Mater. Res.*,**686**, 227–233 (2013).
[4] L. Chen, Z. Wang, Y. Wang, J. Feng, *Materials (Basel).*, **9,** 1–12 (2016).
[5] N. A. Jaya, M.M.A.B. Abdullah, C.M.R. Ghazali, M. Binhussain, K. Hussin,R. Ahmad, *Mater. Sci. Forum*, **857**, 405–411 (2016).
[6] Muhammad Faheem, M.T., Mustafa Al Bakri, A.M., Kamarudin, H., Ruzaidi, C.M., Binhussain, M., Izzat, A.M. *International Review of Mechanical Engineering*, **7 (1)**, pp. 161-166 (2013)..
[7] L.Y. Ming., *Constr. Build. Mater.*, **156**, 9–18(2017).
[8] W.M.W. Ibrahim, K. Hussin, M. M.A.B. Abdullah, A.A. Kadir, M. Binhussain, *Appl. Mech. Mater.*,**2**,34 (2015).
[9] W. M. W. Ibrahim, K. Hussin, M. M. A. Abdullah, A. A. Kadir, and L. M. Deraman, *AIP Conf. Proc.*, **1885**, (2017).
[10] T. da S. Rocha, D. P. Dias, F. C. C. Franca, R. R. de S. Guerra, L. R. da C. de O. Marques, *Constr. Build. Mater.*, **178**, 453–461 (2018).
[11] N.A. Jaya, M. M.A.B. Abdullah, C.M.R. Ghazali, M. Binhussain, K. Hussin, R. Ahmad, *Key Eng. Mater.*, **700**, 3–11 (2016).
[12] R. Ahmad, M. M.A.B. Abdullah, J. Nur Ain *Mater. Sci. Forum*, **803**, 37–43 (2014).
[13] C. Y. Heah, H. Kamarudin, M.M.A.B. Abdullah, M. Luqman, I. Khairul Nizar, Y. M. Liew, *Aust. J. Basic Appl. Sci.*, **5**, 1026–1035 (2011).
[14] W. M. W. Ibrahim, K. Hussin, M. M. A. B. Abdullah, and A. A. Kadir, *AIP Conf. Proc.*, **4**, 56 (2017).
[15] R. Ahmad *et al.*, *Rev. Adv. Mater. Sci.*, **44**, 26–32 (2016).
[16] A. Romisuhani, M. M. Albakri, H. Kamarudin, and S. V. Andrei, *IOP Conf. Ser. Mater. Sci. Eng.*, **267**, 1 (2017).
[17] E. A. Aizat, M. M. A. Abdullah, Y. M. Liew, C. Y. Heah, *AIP Conf. Proc.*, **2030** (2018).

IOP Conf. Series: Materials Science and Engineering **743** (2020) 012024 doi:10.1088/1757-899X/743/1/012024

[18] D. A. Petrash *et al.*, *Earth-Science Rev.*, **171**, 558–582 (2017).
[19] C. K. Yip, J. L. Provis, G. C. Lukey, J. S. J. van Deventer, *Cem. Concr. Compos.*, **30**, 979–985 (2008).
[20] J. D. Rodriguez-Blanco, S. Shaw, L.G. Benning, *Am. Mineral.*, **100**, 1172–1181 (2015).
[21] E.A. Azimi, M.M.A.B. Abdullah, L.Y. Ming, H. C. Yong, K. Hussin, I.H. Aziz, *MATEC Web Conf.*, **78**, (2016).
[22] E. Cohen, A. Peled, G. Bar-nes, *J. Clean. Prod.*, **235**, 910–919 (2019).
[23] A. I. Casado, A. M. Alonso-Zarza, A. La Iglesia, *Sediment. Geol.*, **312**, 50–62 (2014).
[24] E.A. Aizat, M.M.A. Abdullah, Y.M. Liew, C.Y. Heah, AIP Conference Proceedings **2030**, 020274 (2018).

IOP Conf. Series: Materials Science and Engineering **743** (2020) 012025 doi:10.1088/1757-899X/743/1/012025

New High Strength Water Retaining Interlocking Pavers Block for High Mechanical Performing Pavement and Reducing Runoff

M A H Abdullah[1,*], N A Rashid[1], A L Abdul Rani[2] and M F Omar[1]

[1]Faculty of Engineering Technology, Department of Civil Engineering Technology, University Malaysia Perlis, Kampus UniCITI Alam Sungai Chuchuh, Padang Besar 02100 Perlis, Malaysia
[2]Faculty of Engineering Technology, Department of Chemical Engineering Technology, University Malaysia Perlis, Kampus UniCITI Alam Sungai Chuchuh, Padang Besar 02100 Perlis, Malaysia

Email: * afiqhizami@unimap.edu.my

Abstract. Interlocking paver blocks are used widely at low speed traffics road especially surrounding building complexes. Interlocking paver blocks are easy to install and provide aesthetic finishing to the landscape. However, usage of interlocking blocks in large areas reduce ground permeability that lead to higher runoff. New interlocking paver blocks were investigated to reduce runoff and the effect on its mechanical performance. Two types of paver block produced which is paver block with permeable concrete and paver block with void in the center. The results show that the compressive strength for paver block with permeable concrete is averagely higher than paver block with void by 31.5% while the flexural strength of all samples were between 1.0 to 1.7 MPa. Paver block with void recorded highest reduction of runoff compared to paver block with permeable concrete which is 25.5%. This is in line with water infiltration test result. It was concluded that the water retaining performance of paver block increase with increment of void volume and increase in void volume caused reduction on compressive strength of interlocking blocks.

1. Introduction

Interlocking paver block is commonly used as pavement in low speed and low volume traffic especially within building complexes. The main attraction of interlocking paver block would be its aesthetically pleasing colours and shapes besides its ability to bear traffic loading. Many researches have been conducted on increasing the adoption of recycle material in the production of paver block [3] [5] [10] [11] [12] [16]. This would produce a sustainable interlocking paver block. However, usage of interlocking pavement reduce the permeability of ground hence increasing the amount of runoff produced during rain. Few researches conducted on the potential of pavement to serve as drainage layer [6] [8]. This would provide larger drainage system without compromising area for traffic usage. However, the pavement design mostly involves thick pavement system that would be difficult to be implemented on existing pavement system. Hence, a newly designed pavement that could reduce runoff should be almost similar to current design of interlocking paver block which is 200 to 300 mm thickness to ease its adoption. Current research on high performance concrete that could achieve more than 100 MPa compressive strength provide the solution in reducing the thickness of pavement [7] [9]

Content from this work may be used under the terms of the Creative Commons Attribution 3.0 licence. Any further distribution of this work must maintain attribution to the author(s) and the title of the work, journal citation and DOI.
Published under licence by IOP Publishing Ltd

[15]. Pavement with high strength material could perform well in less volume that could be filled with void or porous material in order to reduce runoff. Researches on porous concrete with practical strength would enable the paver block with embedded porous concrete to absorb runoff while minimizing the loss in strength [1] [2] [4] [13] [14]. Hence, this study aimed to investigate the effect of paver block with varied void sizes and block with embedded porous concrete of varied porosity on their compressive capacity, flexural strength and runoff reduction through direct flow and slow infiltration.

2. Materials and methods

2.1 Mix design and Process.

Thirty Six numbers of interlocking paver blocks with dimension of 180 mm (length) x 70 mm (width) x 75 mm (height) were produced with different attributes (Figure 1). Table 1 shows the list of samples produced in triplicate. The mix constituents of base concrete block is as Table 2. The average slump flow measured using slump cone apparatus showed high workability of the mix which is 750mm (Figure 2). This high workability enabled the mix to form around designated spaces for void as per design. For IPB-2%PC, IPB-4%PC and IPB-6%PC; void with dimension of 126 x 50 x 50 mm was prepared from the surface of paver block (Figure 3). This void is to be filled with permeable concrete at later stage. Permeable concrete was made by adding air-entraining agent into high strength concrete to create pores. Mechanical properties concrete with varied air-entraining agent addition is given in Table 3. For IPB-20%V, IPB-40%V and IPB-60%V; paver block is cast until 60mm height with void introduced from surface based on the percentage assigned to each type of sample. Capping was placed prior to second casting of the paver block until full height of 75mm (Figure 1).

Table 1. Mix Design of CIB.

Sample designation	Properties	Numbers
IPB-2%PC	Contains permeable concrete of 2% air-entraining agent	6
IPB-4%PC	Contains permeable concrete of 4% air-entraining agent	6
IPB-6%PC	Contains permeable concrete of 6% air-entraining agent	6
IPB-20%V	Contains 20% void	6
IPB-40%V	Contains 40% void	6
IPB-60%V	Contains 60% void	6

Table 2. Mix Design of Interlocking Paver Block.

Dry component	Weight per block (g)	Total 18 nos
Cement	945.0	17.0
Water	189.0	3.4
Fine aggregate (600-1180micm)	945.0	17.0
High-range water reducer (HRWR)	28.4	0.5

Table 3. Mechanical Properties of Permeable Concrete.

Percentage by Cement Weight of Air-entraining Agent	Compressive Strength (MPa)	Water Absorption Percentage (%)
2%	95.4	2.81
4%	71.3	2.65
6%	56.8	2.19

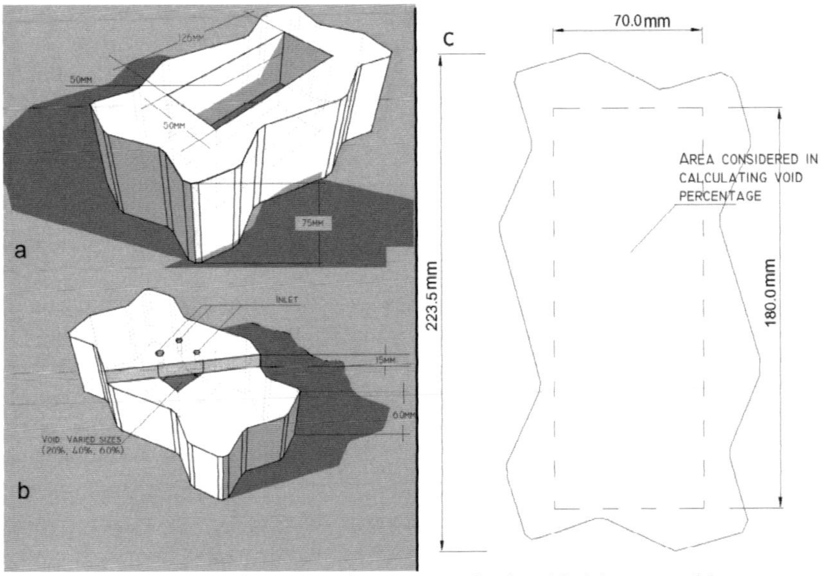

Figure 1. Overall view of Interlocking Paver Block with (a) permeable concrete and (b) void. Top view of Interlocking Paver Block (c).

Figure 2. Slump flow test for interlocking paver block mix.

2.2 Testing and Analyzing.

All of the samples were tested for runoff reduction followed by water infiltration test and then skid resistance test. Finally, three numbers of samples were tested for flexural strength based on ASTM C 293 and another three samples were tested for compressive strength after 28 days of curing process as per BS EN 1338:2003. The runoff reduction test were conducted using the rate of 200mm/hr to simulate maximum rainfall event. Oven-dried samples were placed in container of 600 x 400 x 300 mm and then sprinkled with water at designated rate. Then the paver block will be weighed to measure the water absorption. Samples were then oven-dried for 24 hours prior to water infiltration test that investigate water ponding effect on test sample. Oven-dried samples were placed in 600 x 400 x 300 mm contained and then submerged in water with 200 mm of excess water depth above the samples' surface level. Samples were submerged for 15 minutes before taken out for weighing. Skid resistance test was conducted as per ASTM E 303 on all samples after water infiltration test.

3. Results and discussion

3.1 Compressive and Flexural Strength.

Overall, the compressive strength of paver block samples are higher than 40MPa which is the requirement set by BS 6717 [18] (Table 4). Compressive strength of samples IPB-2%PC, IPB-4%PC and IPB-6%PC are higher by at least 92.3% compared to requirement. The highest value was recorded by sample IPB-2%PC which is 108.6 MPa. This may be due to the strength of paver block itself and contribution from permeable concrete that has quite substantial strength. Figure 3 shows that base of paver block cracked while the permeable concrete at the center is still intact which indicate its high strength. For samples with void, the compressive strengths are at least 5% higher than the recommended value. The highest value was recorded by sample IPB-20%V which is 106.5% higher compared to the referred value. In term of flexural strength, all samples recorded values within range of 1.5 to 1.7 MPa except for sample IPB-40%V and IPB-60%V. IPB-40%V and IPB-60%V recorded flexural strength of 1.0 and 1.2 MPa respectively which is quite large reduction compared to other samples. This may be due to void that reduce the volume resisting flexural deformation in the samples. This is congruent with result from sample with permeable concrete (IPB-2%PC – IPB-6%PC) that shows higher compressive and flexural strength due to placement of permeable concrete. The result for flexural strength of all samples are about 15 to 50% lower compared to flexural strength of block pavement with rubber [17].

Table 4. Result of compressive and flexural strength of concrete interlocking block (CIB).

Type of specimen	Average Compressive Strength at 28 days (MPa)	Average Flexural Strength at 28 days (MPa)
IPB-2%PC	108.6	1.5
IPB-4%PC	87.3	1.7
IPB-6%PC	76.9	1.6
IPB-20%V	82.6	1.5
IPB-40%V	62.5	1.2
IPB-60%V	42.1	1.0

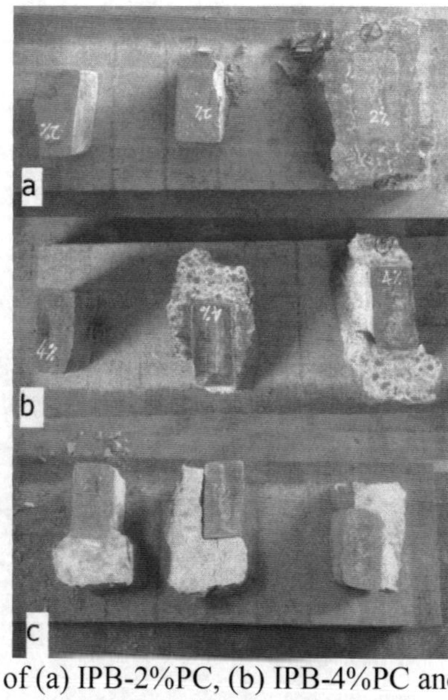

Figure 3. Compressed samples of (a) IPB-2%PC, (b) IPB-4%PC and (c) IPB-6%PC that show intact permeable concrete.

3.2 Runoff Reduction and Water Infiltration.

Percentage of runoff reduction of interlocking paver block samples show reduction of 6 to 25.5%. Highest runoff reduction by paver block with permeable concrete was recorded by IPB-6%PC with 18.7% (Figure 4). This may be due to the highest amount of air entraining agent in the permeable concrete. However, samples of paver block with void show higher percentage of runoff reduction compared to other samples. Highest percentage was recorded by IPB-60%V with 25.5% while the lowest percentage of 13.5 was recorded by IPB-20%V. This indicates higher efficiency in capturing flowing water in paver block with void compared to permeable concrete. In term of storing water through infiltration, paver block with void recorded highest percentage of water infiltration compared to paver block with permeable concrete (Figure 5). This is shown by sample IPB-60%V with 5.9% of water infiltration. This indicates an efficiency of this type of sample to absorb water using every part of the block as the percentage is the ratio of water absorbed to its dry weight. Paver block with void has lower mass of concrete compared to block with permeable concrete. However, paver block with permeable concrete such as IPB-2%PC only managed to record 1.32% of water absorption and the highest recorded by IPB-6%PC is 3.11%. This shows lower efficiency of these blocks to absorb water. It is worth to note the marginal difference in efficiency of water absorption between paver block with permeable concrete and paver block with void of less than 60. This shows that the paver blocks perform similarly under submerged condition but different in moving water such as runoff. Overall, it can be concluded type of paver block with void is more efficient in reducing runoff and absorbing water as the void is much larger compared to paver block with permeable concrete.

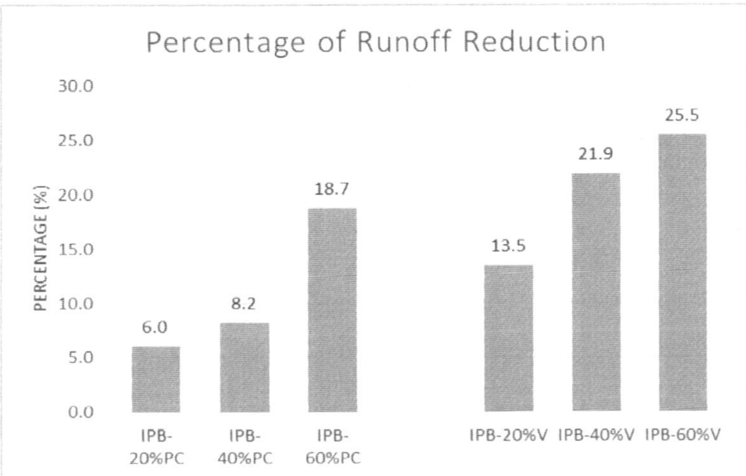

Figure 4. Result of water absorption.

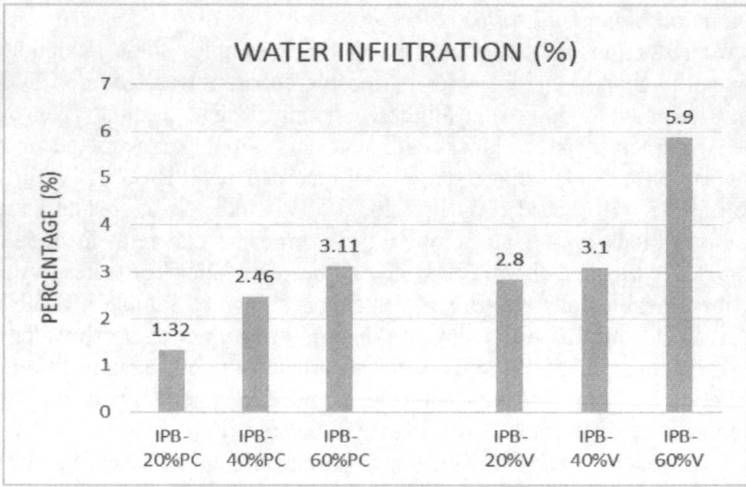

Figure 5. Result of water infiltration.

3.3 Skid Resistance.

BPN values for interlocking pavement with permeable concrete is within 34 to 56 which indicates roughness of permeable concrete that are not too varied (Figure 6). For samples of interlocking paver block with void, the BPN values varied from 45 to 78. This large variation is due to inlet holes allocated differently for each type of samples. Samples of IPB-20%V and IPB-40%V have a total of 4 inlet holes. However, distance between holes of IPB-40%V is greater compared to IPB-20%V due to greater dimension of void. For IPB-60%V, 6 inlet holes were provided at greater distance. It can be concluded that increase of inlet holes per area would increase the surface roughness of pavement block with void. It is also noted that inlet holes provide larger friction as indicated by BPN value of pavement block with void as compared to pavement block with permeable concrete. Overall, the BPN value for all samples except IPB-20%PC the safety value of road pavement in the research by Chou & Lee [19] which is 45.

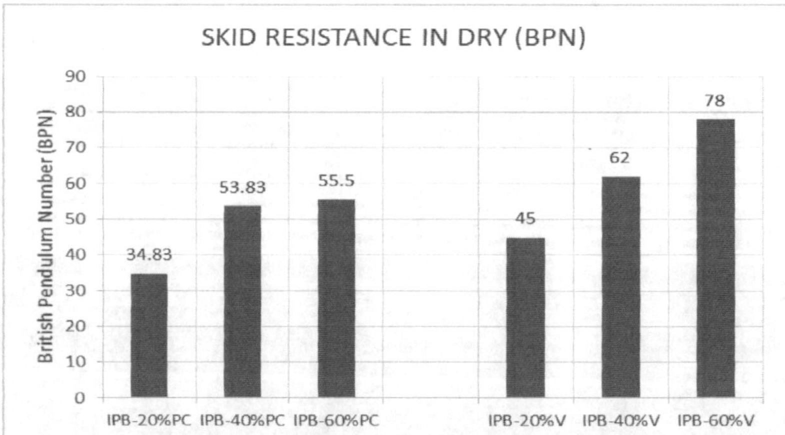

Figure 6. Result for the skid resistance for CIB.

4. Conclusions

Embedment of permeable concrete in interlocking paver block reduce compressive strength of block with increment of perviousness but enable the paver block to reduce runoff with acceptable skid resistance value. Introduction of void within the interlocking paver block greatly reduce runoff and increase skid resistance value while achieving acceptable compressive strength. Hence, development of paver block with runoff reduction capacity is viable in terms of mechanical performance and resistance for traffic users while having the ability to store water from precipitations. This would reduce the need for open drainages hence maximizing land usage. Future works could investigate the rate of water evaporation of saturated paver block and its long term effectiveness in reducing runoff especially due to clogging in extended usage.

Acknowledgement
This research was supported by the facilities provided by Fakulti Teknologi Kejuruteraan (FTK), Universiti Malaysia Perlis (UniMAP).

References
[1] A.R. Bhutta, M., Hasanah, N., Farhayu, N., Hussin, M. W., Tahir, M. B. M., Mirza, J. *Constr Build Mater*, **47**, 1243–1248 (2013).
[2] R. Abaeian, , H.P. Behbahani., S.J. Moslem, *Constr Build Mater*, **165**, 631–638. (2018).
[3] A. Abdelbary, A.R. Mohamed, *Conference*, **45**, 72, (2011).
[4] A.S. Agar-Ozbek, J. Weerheijm, E.K. Schlangen, K. Van Breugel., *Constr Build Mater*, **41**, 480–490 (2013).
[5] Dinesh.S, Dinesh.A, K. . (2016). *Int. J. Appl. Eng. Res,* **11**, 364–368 (2016).
[6] C.H. Elizabeth Pappas, *Watershed Management* **2010**, 364–375 (2011).
[7] Guo, Y. B., Gao, G. F., Jing, L., Shim, V. P. W., *Int J Impact Eng*, **108**, 114–135 (2017).
[8] Y. Liu, T. Li, H. Peng, *Science of the Total Environment*, **634**, 1119–1125 (2018).
[9] R. Kumar, S. Singh, L.P. Singh, *Constr Build Mater*, **153**, 506–513 (2017).
[10] Nikam, V. S., Rane, D. D., & Deshmukh, S. S. *IABSE Conference New Delhi, India* ***8*(3)**, 415–419 (2005).
[11] Patel Vivek V.1, Dr. Jayeshkumar Pitroda2, P. J. J. B. *International Journal of Current Trends in Engineering & Research (IJCTER)*, ***2*(5),** 33–42. (2016).
[12] Penteado, C. S. G., Viviani De Carvalho, E., & Lintz, R. C. C. *Journal of Cleaner Production*, **112**, 514–520 (2016).
[13] D. Sarwono, Djumari, Rochim, A. Setyawan, *Procedia Engineering*, **171,** 1429–1434 (2017).
[14] W Shen, Shan, L., Zhang, T., Ma, H., Cai, Z., & Shi, H. *Constr Build Mater*, **38**, 667–674 (2013).
[15] C.P. Sivathan., A.R. Santhakumar., S. Chandrasekaran, S. Viswanathan., R. Mathiyarasu, J.A. Kumar, R. Preetha, *Progress in Nuclear Energy*, **93,** 76–83. (2016).
[16] T. Uygunolu, I.B. Topcu, O. Gencel, W. Brostow, *Constr Build Mater*, **30**, 180–187. (2012).
[17] R.B. Murugan, C. Natarajan, S.E. Chen, *J. Traffic Transp. Eng. Engl. Ed.*, **3(5),** 483–491, (2016).
[18] BS 6717, Precast, unreinforced concrete paving blocks – Requirements and test methods. (2001).
[19] Chou, C., & Lee, N. *The 9th International Conference of Eastern Asia Society for Transportation Studies*. 12. (2011).

IOP Conf. Series: Materials Science and Engineering **743** (2020) 012026 doi:10.1088/1757-899X/743/1/012026

Influence of Various Sand Gradation on Mechanical Properties and Concrete Quality of High Performance Concrete (HPC)

K Muhamad[1], N F Isa[1], N H Hashim[1], Z Tusimin[2], N Yahya[1], N A Naamandadin[1], R A Razak[1] and P Y Tham[1]

[1]Faculty of Engineering Technology, University Malaysia Perlis, 02100 Sg Cuchoh, Perlis, Malaysia

[2]Department Construction Engineering Technology (Advanced System), Kolej Kemahiran Tinggi MARA Sri Gading, 86400 Parit Raja, Batu pahat, Johor, Malaysia

Email: khairunnisa@unimap.edu.my

Abstract. High Performance Concrete (HPC) has a few characteristics: high strength, high early strength, high modulus of elasticity, high durability and long life in severe environments, low permeability and diffusion, resistance to chemical attack, toughness and impact resistance and so on. Almost 70 to 80% of the volume of the concrete is occupied by aggregate which have a great impact on characteristics and properties of a concrete. Not only course aggregate but fine aggregate also contributes a lot toward the performance of HPC. Thus, this research studied on the influence of various sand gradations on mechanical properties and concrete quality of HPC. The sand gradation utilized in the study are normal size sand, sand passing through sieve size 1180µm, sand passing through sieve size 600µm and sand passing through sieve size 300µm. The mechanical properties of HPC being studied are compressive strength and splitting tensile strength while the concrete quality of HPC is determined by using water absorption and Ultrasonic Pulse Velocity (UPV) test. The results show that the mechanical properties and concrete quality increased as the size of sand particles decreased. The optimum sand gradation is the sand passing through sieve size 300µm which achieved highest mechanical properties and concrete quality showing the highest result of compressive strength (123.9MPa), splitting tensile strength (8.91N/mm2), lowest percentage of water absorption (0.55%) and highest UPV value (7743N/mm2).

1. Introduction

High Performance Concrete (HPC) can be defined as the concrete that reach special performance and have their own specific requirements that cannot be achieved by using conventional materials and normal mixing [1]. HPC is the concrete that contained mechanical properties which exceeds the normal strength concrete [2][3][4]. HPC has a few characteristics: high strength, high early strength, high modulus of elasticity, high durability and long life in severe environments, low permeability and diffusion, resistance to chemical attack, toughness and impact resistance and so on. Normally, the HPC is used to build bridge, tunnel and also high rise building [5].

Nowadays, earthquakes had become a common issue that might happen on all over the earth. During an earthquake, high horizontal and vertical forces will be subjected to the beam-column joint region. The force had magnitudes that much higher than those within adjacent beams and columns. In this case, the conventional design of Ordinary Portland Cement is no longer meet many functional

Content from this work may be used under the terms of the Creative Commons Attribution 3.0 licence. Any further distribution of this work must maintain attribution to the author(s) and the title of the work, journal citation and DOI.
Published under licence by IOP Publishing Ltd

requirements [3]. Therefore HPC which contains high mechanical and durability properties is developed to fulfill the requirement of the market especially for the heavy construction works.

The High Performance Concrete have its own specific properties. HPC can achieved high strength which is 70 to 140 MPa at 28 to 91 days; high early compressive strength which is 20 to 28 MPa at 1 to 3 days; low absorption which is 2% to 5% and an excellent concrete quality with UPV value above 3705 m/s [1]. Different sand gradation will affect the strength of the concrete. The concrete with finer sand particles will gain the higher compressive, flexural and splitting tensile strength. This is because the sand with finer particles able to hold the internal moisture content of the concrete better than the sand with coarser particles [5]. So the aim of this research is to study the effect of sand gradation in terms of passing sieve sized 1180μm, passing 600μm, passing 300μm and the normal sand size (i.e. combination of all sizes of sand)

Besides, the aggregate used in concrete have a great impact on characteristics and properties of a concrete [4]. Yet, the study of the effect of fine aggregate which is the sand grading of self-compacting concrete is very limited [5]. Hamiruddin N A et. al, studied a different sand gradation on Self-Consolidate Ultra-High Performance Concrete (UHPC) which consist of normal sand size, sand sized 600 μm -1180μm , sand sized 300μm - 600μm and sand sized 63μm - 300μm. Study shows that the used of 600 - 1180μm sand gradation represented the excellent UHPC performance due to the present of void. The reduction in voids is influenced by high cement content, very low (w/c), (HRWR) and fine particles [6].

2. Methodology

The mixture constituents (Table 1) consist of 4 different fine aggregate gradations, (1) normal sand, (2) sand passing through sieve size 1180μm, (3) sand passing through sieve size 600μm, and (4) sand passing through sieve size 300μm. High strength cement class 52.5N is used as a binder and high range water reducer type of superplasticizer that being used in concrete mix is Masterglenium Ace 8589 to reach a minimum flow of 750mm. The amount of superplasticizer added increased as the size of sand decrease. This is because the sand with smaller size will have more fine particle and hence the total surface area of the particle will increase therefore higher amount of superplasticizer is needed to react with every single particle of the concrete to achieved the workability. [6]

Table 1. Mixture constituents with different sand gradation.

Content	**HPC1**	**HPC2**	**HPC3**	**HPC4**
Cement (kg/m^3)	1000	1000	1000	1000
Water (kg/m^3)	200	200	200	200
Sand (kg/m^3)	669	669	669	669
Superplasticizer (kg/m^3)	22.0	23.6	23.9	29.9
Sand Gradation (μm)	Normal Sand	Passing 1180μm	Passing 600μm	Passing 300μm

The flow test was conducted is according to ASTM C 1611 [8] to determine the workability of HPC. Based on Karihaloo [7]. In this study, the flowability demonstrated in the range of 860mm - 1180 mm (Figure 1) which means that the amount of superplasticizer can be reduce to below than 2%. In this study the uses of superplastizer is to reduce the water cement ration usage in order to achieve high compressive strength of HPC [8].

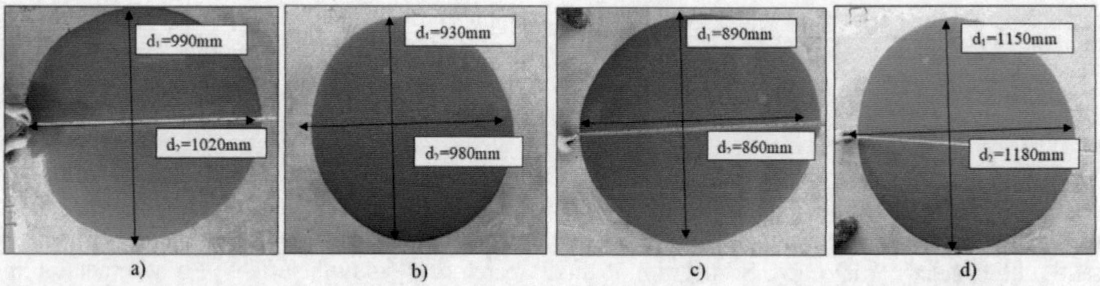

Figure 1. The flowability of HPC with different sand gradation; (a)HPC-Normal Sand (b)HPC-Passing 1180 μm (c)HPC- Passing 600 μm & (d)HPC- Passing 300 μm.

3. Results and Discussions

3.1 Compressive Strength

As indicated in figure 2, the compressive strength results show that all the specimen achieve high strength concrete which is above 100Mpa at 28 days [9]. These results also comply with the book of Design and Control of Concrete Mixtures which should in range of 70Mpa – 140 Mpa. Besides that, this mixture also achieve a high early strength concrete as it achieved above 20 to 28 MPa at 1 to 3 days as stated by Kosmatka, Kerkhoff, and Panarese, [5].

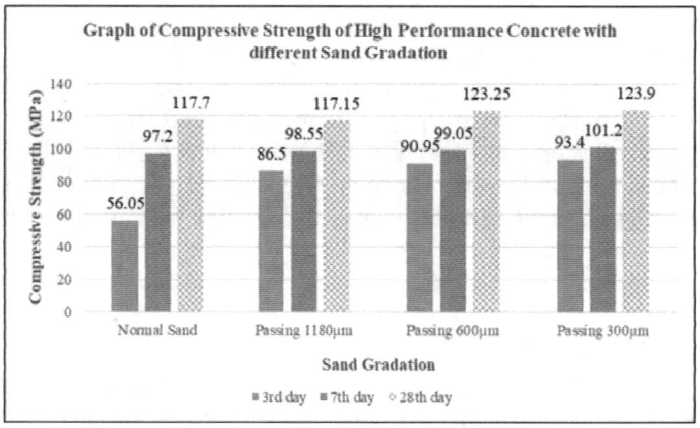

Figure 2. Compressive strength of HPC with different sand gradation.

In the study of different range of fine aggregate gradations, the result indicates that there is a slight different in compressive strength achievement due to the sample is taking as passing sieve sizes as stated above. It means that there are small particles of sand sized which will fill the void between larger particles size (Table 2). In this case it contributes a dense packing density between the particle sand inside the mixture proportion. A lesser air void contained in concrete mixed a higher compressive strength it will achieved [6]. These results also in line with Siong K L et. al [10] and Neville [11] which said that sand filler had contributed to the reduction of voids in mixture thus increase the strength of the concrete.

3.2 Splitting tensile strength

The splitting tensile strength of HPC increased as the size of sand particles decreased from 1180 to 300μm (Figure 3). The results showed the similar trend with the previous study done by Siong K L et al which the splitting tensile strength of concrete increase as the sand size decreased.They found that the increment might be happened as more internal moisture can be hold by the increment of finer particles. It's occurred when more water is needed to wet the surface of every particles, hence, the free water held by the fillers could be used to hydrate the other cement particles which are not hydrated yet.

IOP Conf. Series: Materials Science and Engineering **743** (2020) 012026 doi:10.1088/1757-899X/743/1/012026

This late hydration can reduced the strength loss caused by the tropical weathering, and hence increase in splitting tensile strength [10].

Table 2. Sieve Analysis of Normal Sand.

Sieve Size	Weight Retained (kg)	Cumulative Weight Retained (kg)	Cumulative % Retained	Cumulative Weight Passing (kg)	Cumulative % Passing
3.35 mm	0.0	0.0	0	0.5	100
2.0 mm	0.01	0.01	2	0.49	98
1.18 mm	0.01	0.02	4	0.48	96
600μm	0.08	0.1	20	0.4	80
300μm	0.31	0.41	82	0.09	18
150μm	0.08	0.49	98	0.01	2
Pan	0.01	0.5	100	0	0
Total	**0.50**				

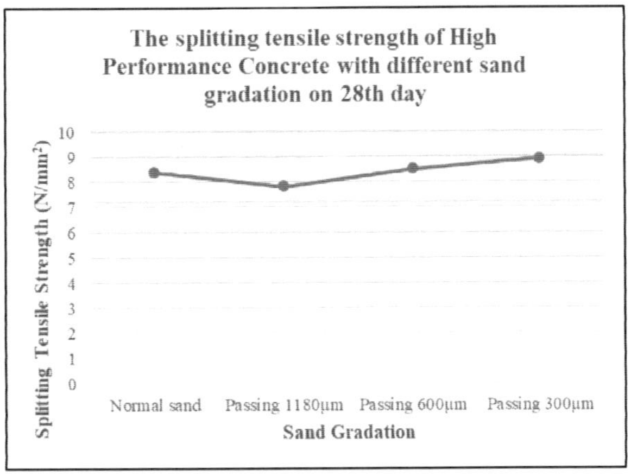

Figure 3. Split tensile strength.

3.3 Water Absorption

The percent water absorption of HPC (Figure 4) was influenced by the size of sand particles used for concrete mix. As the size of sand particles decreased, the percent water absorption will be decreased; it means that less water were absorbed by the concrete sample. The HPC with normal size sand had the highest percent of water absorption while that of HPC with sand passing through sieve size 300μm achieved the lowest percent water absorption. This is because the HPC with finer sand particles will filled up more space between the cement and the sand therefore less voids will be formed inside a concrete sample. When the number of voids in a concrete decreased, the ability of the concrete to absorb water will decreased as the same time.

The previous research by Gameiro [12] that study about the durability properties of structural concrete incorporating fine aggregates from the waste marble industry stated that the concrete that contained finer marble aggregates achieved lower water absorption by immersion. This is because the marble aggregates are smaller in size. The finer aggregates had increased the compactness of the particles inside the concrete and therefore reduce its ability to absorb water.

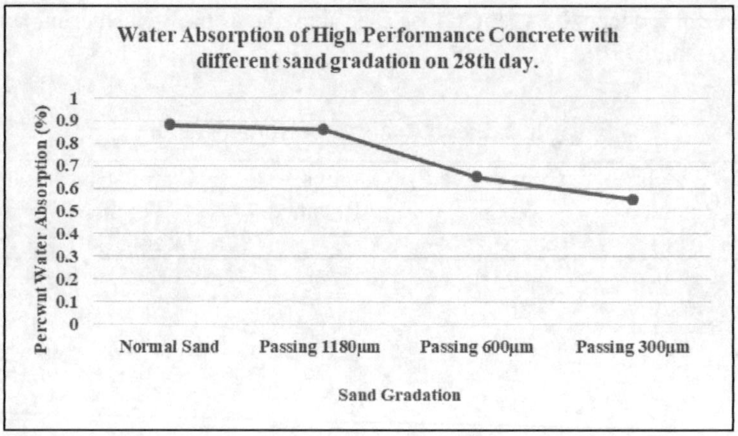

Figure 4. Water Absorption Capacity of HPC with Different Sand Gradation at 28th Day.

3.4 Ultrasonic Pulse Velocity (UPV)

The results of UPV (Table 3) shows the concrete quality of the HPC was in good and excellent quality. The concrete with excellent concrete quality mean that there are less voids being formed in the concrete. As compare to the results of UPV, the HPC-300 achieved the highest transition time because HPC-300 is the finest compared to others sand gradation. The sand particles can fill up more spaces between the cement and sand. Besides, the finer aggregates exhibit higher bonding strength between cement matrix and sand particles. As stated by G Kadir [13], the surface roughness of the aggregate grains has a significant effect on adhesion of the cement matrix – aggregate interface. The increase in compressive strength of aggregates with high surface roughness is caused by the stronger interfacial transition zone (ITZ). This situation is thought to be causing the increase in UPV values (Figure 5).

Table 3. Ultrasonic Pulse Velocity of High Performance Concrete with different sand gradation on 28th day.

		Ultrasonic Pulse Velocity (m/s)			
	Sand Gradation	Reading 1	Reading 2	Average	Concrete Quality
HPC-NS	Normal Sand	5100	5150	5125	Excellent
HPC-1180	Passing 1180μm	4545	4600	4573	Excellent
HPC-600	Passing 600μm	4445	4440	4443	Good
HPC-300	Passing 300μm	7810	7675	7743	Excellent

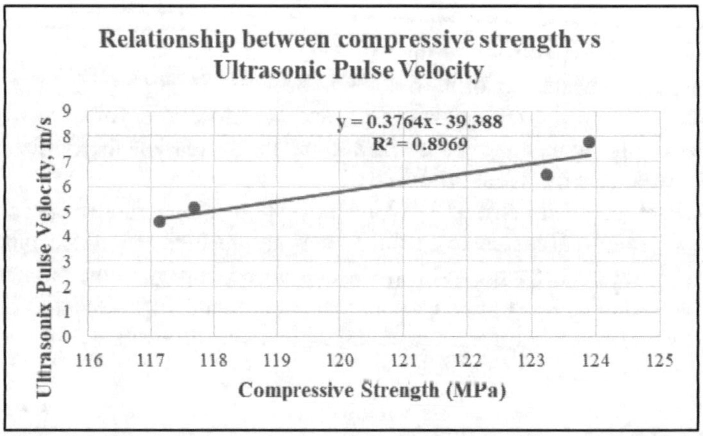

Figure 5. Relationship between compressive strength vs Ultrasonic Pulse Velocity.

4. Conclusion

In this study where the effects of fine aggregate gradation on mechanical and concrete quality were investigated, the following results were obtained:

- As the size of sand Particles decreased, both the mechanical properties and concrete quality of HPC increased.
- The optimum sand gradation of HPC to perform better mechanical properties and concrete quality is the sand passing through sieve size 300µm (HPC-300). The HPC-300 had achieved 93.40MPa early strength on the 3rd day, 101.20MPa compressive strength on the 7th day and 123.90MPa compressive strength on the 28th day which is higher than the other sand gradation. The splitting tensile strength of HPC-300 also achieved the highest value which equal to 8.91N/mm2 on 28th day. The percentage of water absorption for HPC-300 is 0.55% which is the lowest among the other sand gradation. Furthermore, the HPC-300 also showed the best concrete quality with highest UPV value which is 7743m/s.

Acknowledgement

The author would like to acknowledge the Faculty of Engineering Technology, University Malaysia Perlis (UniMAP) for providing equipment and laboratory in this project.

References

[1] Jalal M, Esmaeel M, Mohammad S, Ali R P, *Mater and Des* **34** 389-400 (2012).
[2] Jayesh S G and Awari U R, *Int J Res Eng Technol*, **5,** 1965 (2018).
[3] Kannan S U, *Int. J. Eng. Res and Dev*, **13**, 42-49 (2017).
[4] Muhamad K et al, *J. Adv. Res Eng*, **8** 23-28 (2019).
[5] Kosmatka S H, Kerkhoff B & Panarese W C (2002) Design and Control of Concrete Mixtures: *High Performance Concrete* 14th ed(Skokie: Portland Cement Association) **5420** 299-313.
[6] Hamiruddin N A, *Solid State Phenomena* **280** pp 476-480 (2018).
[7] Deeb R Ghanbari A, and Karihaloo B L , *Cement Concrete Comp*, **34**, 185-190 (2012).
[8] Al Tayeb M M and Hamouda H, *Civil and Environmental Research*, **7** 35-43 (2015).
[9] Mahure N V, Vijh G K, Sharma P, Sivakumar N and Murari R, *International Journal Of Earth Sciences and Engineering,* **4** , 871-874 (2011).
[10] Siong K L et. al ,*Construction and Building Materials* **38** 348-355 (2013).
[11] Neville A Properties of Concrete 4th ed 2009 Pearson Prentice Hall (Harlow, England).
[12] Gameiro F J G 2013 Durability properties of structural concrete incorporating fine aggregates from the waste marble industry Master's Dissertation at Civil Engieneering (Instituto Superior Técnico, Lisbon).
[13] Gucluer G, *Journal of Building Engineering* **27** , 100949 (2019).

Improving Indoor Environmental Quality (IEQ) and Comfortability of Lecture Rooms: Case Study – UniCITI Alam Campus, Padang Besar, Malaysia

Nurul Akmam Naamandadin[1], Siti Nur Aishah Mohd Noor[1], Norrazman Zaiha Zainol[1], Arita Hanim Awang[2], Syakir Amir Abdul Rahman[2], Khairunnisa Mohamad[1] and Muhammad Naim Ahmad[1]

[1]Faculty of Engineering Technology, Universiti Malaysia Perlis, Malaysia
[2]Kulliyyah of Architecture and Environmental Design, International Islamic University Malaysia, Malaysia

Email: nurulakmam@unimap.edu.my

Abstract. UniCITI Alam is a mixed development of commercial buildings and medium rise apartments. The whole development was designed based on the integration of student residential community with the combined commercial centre and temporary campus. The lecture rooms are located at the second floor of the triple-storey shop house. To improve the indoor environmental quality and comfortability of the teaching and learning activities, physical measurements of the indoor environment parameters (air temperature, relative humidity and carbon dioxide (CO_2) concentration) and light intensity are the main methods applied in the presented case study. Data collected will be compared with Malaysian Standard MS 1525: 2007 (Code of Practice on Energy Efficiency and Use of Renewable for Non Residential Building) and DOSH. The finding will be used for improving the environmental quality of the lecture rooms and enhancing the ecology circle.

1. Introduction and Background Information

UniCITI Alam is a mixed development of commercial buildings and medium rise apartments. It is located at the coordinate 6°38'58.4"N latitude and 100°15'23.6"E longitude. The whole development was designed based on the integration of student residential community with the combined commercial centre and temporary campus.

The design of lecture rooms in the UniCITI Alam is actually a refurbishment of the commercial building. The lecture rooms are located at the second floor of the triple-storey shop house (refer to Figure 1). There are sixteen (16) lecture rooms and two (2) lecture halls with a slightly different in sizes of the floor areas. One side of each lecture rooms is made of a curtain wall. All curtain walls are tinted with black films and covered by the white boards most of the wall area (refer to Figure 2 and Figure 3). Only four lecture rooms have windows at the corner side. All lecture rooms have been designed by not allowing natural ventilation and natural lighting to enter the rooms.

Content from this work may be used under the terms of the Creative Commons Attribution 3.0 licence. Any further distribution of this work must maintain attribution to the author(s) and the title of the work, journal citation and DOI.
Published under licence by IOP Publishing Ltd

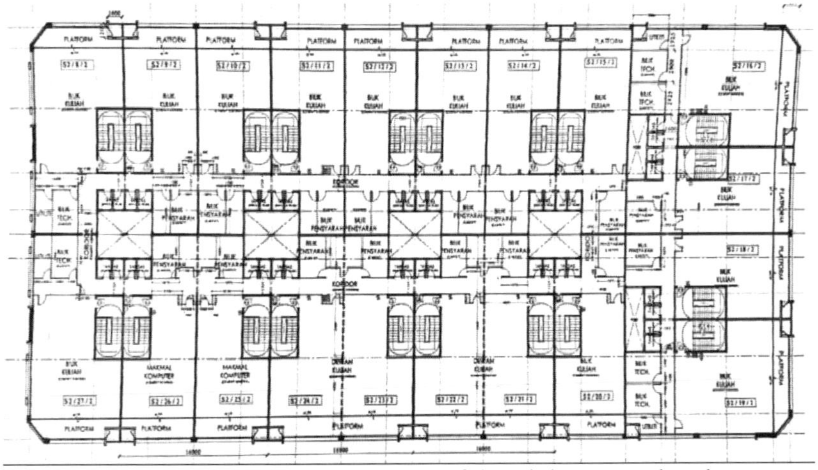

Figure 1. Floor plan of second floor of the triple-storey shop house.

Figure 2. Lecture room with artificial lights without occupants.

Figure 3. Lecture room with natural lighting without occupants.

2. Problem Statement

Many aspects shall be considered for the improvement of indoor environmental quality (IEQ) and comfortability of the lecture rooms such as indoor temperature, relative humidity, level of carbon dioxide and the light intensity. Stuffy and dark lecture rooms may not be suitable to create a comfort environment for teaching and learning activities. Since all the lecture rooms have been 'sealed', each of the lecture rooms have been equipped with air conditioners as a cooling system and artificial lighting as

IOP Conf. Series: Materials Science and Engineering **743** (2020) 012027 doi:10.1088/1757-899X/743/1/012027

a source of light. This approaches are not the best solutions and for the long term investment, high expenses of the electricity bills may not return the high profit to the university.

After a certain period of time, if the mechanical cooling technologies like air conditioning system cannot work properly, it will affect the temperature and air quality of the lecture rooms. Poor level of indoor environmental quality in the lecture rooms will create discomfort environment and the students feel difficulty to stay focused during lecture. It may also lead to Sick Building Syndrome (SBS).

3. Literature Review

According to Asere & Blumberga [1], spaces with high occupant density, e.g. in schools, conference rooms, offices, etc., are the main indoor pollutants with CO_2 and volatile organic compounds (VOCs), and air tight building envelopes contribute to a high degree of unacceptable IEQ. Mechanical ventilation systems have to be used to improve indoor environmental quality in buildings where adequate natural ventilation is not available. Such systems increase energy consumption with an assumption improving IEQ and thermal comfort but human productivity are not consistent with energy efficiency [1]. Studies show that the poorer the IEQ, the lower the human productivity (a 15% reduction in performance of schoolwork corresponds to about 1 year of teaching [2]. While Griffiths & Eftekhari [3] mentioned that the increased level of CO_2 has an impact on the students' ability to learn and it decreases this ability by 5%. Apart from IEQ, there have been articles studying a combination of thermal comfort along with the lighting factor [4].

3.1. Air Temperature

Refer to ASHRAE [5] air temperature is the average temperature of the air surrounding an occupant. MS1525:2007 recommended comfort temperature for Malaysia weather is 23°C - 26°C. Accordingly, room comfort conditions not only depend on air temperature but also mean radiant temperature, humidity, clothing, metabolic rate and air movement preference of the occupant [6].

3.2. Relative Humidity (RH)

DOSH [7] proposed the ideal comfort goes for relative dampness is 40-60%. While MS1525:2007 recommended RH 55% - 70%. Low humidity can bring about dryness of the eyes, nose and throat and may likewise expand the recurrence of friction based electricity stuns. High humidity, over 80% can be connected with weakness and report of "stuffiness" [7].

3.3. Carbon dioxide (CO_2)

Carbon dioxide (CO_2) is the most widely recognised indoor air contamination discharged by individuals. The levels of CO_2 indoor are reliant upon the quantity of individuals present and level of metabolic movement completed inside the air space. American Society of Heating, Refrigerating and Air-Conditioning Engineers (ASHRAE) Standard suggests greatest level of 1000 ppm for nonstop CO_2 presentation [8]. CO_2 is a key parameter for surveying indoor air quality and ventilation productivity. The ventilation that has deficient outside air admission can contribute to an abnormal state of CO_2 in certain region in the building. A ventilation rate of 10 Ls^{-1} to 20 Ls^{-1} for every individual will diminish the side effects of sick building syndrome (SBS) and achieve a superior air quality [9]. The concentration of carbon dioxide is strongly related with occupant density as well as the type of activity carried out [10]. Common effect of staying in a space with high concentration of carbon monoxide is feeling headache, sleepy and drowsy.

3.4. Light Intensity

Good lighting also creates a pleasant atmosphere and gives occupants a sense of well-being. This can improve their productivity and efficiency. Poor lighting can lead to visual fatigue and discomfort. It can affect the health of people at work causing symptoms like eyestrain, migraine and headaches [7]. According to DOSH [7], lecture hall and interior area for writing, typing, reading, and data processing, lighting must be 500 Lux. They also recommended ceilings and walls should be painted white or in a light colour.

4. Methodology

All experiments were conducted at lecture rooms in building S2-L2 in UniCITI Alam Campus at Padang Besar, Malaysia. There were four (4) experiments conducted; temperature, relative humidity, CO^2 and light intensity. The readings were recorded between 12.00 pm to 7.00 pm by using two instruments which were 5-in-1 Environmental Meter and Indoor Air Quality CO^2 Monitor.

5. Result and Finding

Table 1. Temperature, Relative Humidity, Carbon Dioxide and Light Intensity readings in lecture rooms with and without occupants.

Lecture Room	No of Occupants	Temperature (°C)		Relative Humidity (%)		Carbon dioxide (ppm)		Light Intensity (Lux)	
		with occupants	without occupants	with occupants	without occupants	with occupants	without occupants	with occupants	without occupants
Lecture Room 1	40	24.93	29.53	57.72	48.53	402.67	1591.67	203.67	269.33
Lecture Room 4	35	24.75	29.87	53.08	47.42	387.00	2707.33	189.00	258.00
Lecture Room 9	40	27.57	29.48	64.40	54.35	440.67	933.00	258.33	275.00
Lecture Room 10	60	26.33	29.37	45.93	48.53	504.67	3004.33	162.33	201.33
Lecture Room 12	60	28.03	31.56	55.32	40.85	500.33	2002.67	270.33	274.67
Lecture Room 14	60	27.20	28.43	56.28	43.22	415.33	2031.00	255.67	280.33
Lecture Hall	65	27.28	28.05	56.40	43.32	384.33	1223.00	130.33	277.67

Table 2. Temperature, Relative Humidity, Carbon Dioxide and Light Intensity readings suggested by DOSH and MS1525:2007.

Suggest by	Temperature (°C)	Relative Humidity (%)	Carbon dioxide (ppm)	Light Intensity (Lux)
DOSH	20°C - 26°C	40-60%	1000 ppm	500 Lux
MS1525:2007	23°C - 26°C	55% - 70%	-	300 – 500 Lux

Temperature readings from all lecture rooms show that they did not met the comfort temperature set by DOSH and MS1525:2007 when there were no occupant. Result without occupants temperatures were higher compared to with occupants because the air conditioners were turned off when they were left unoccupied. However, when the occupants more than 40, the temperature readings were not comfort to the occupants. The RH still in the acceptable range when refer to DOSH. All the data taken show that without occupants they were below the range suggested by MS1525:2007. The CO^2 readings taken during no occupant were reliable for all lecture rooms in the range between 384.33 ppm to 504.67 ppm. However, CO^2 readings taken during teaching and learning sessions mostly exceed the acceptable limit. The level of CO^2 depends on the number of occupants and also the activity carried out in that place. CO^2 is key parameter for indoor air quality and ventilation efficiency. Insufficient fresh air contributes to high level of CO^2 in certain rooms (Yau, Chew et. at., 2012). CO^2 cannot be removed by present air cleaning system, that's why ventilation very important for more comfortable indoor environment (Noh, Han, et. al., 2008). Based on the experiments conducted, they showed that there were big difference between CO^2 reading during with and without occupant. This is because the CO^2 produced by human beings. The average readings of light intensity were in the range 200-300 lux. The readings were taken in the conditions of artificial lightings on. The distance of the light sources to the instrument placement were same for all lecture rooms. The distance was in the same level of the ceiling high and the desk used by the students. The types and position of light sources were also same for each lecture room. The presence of occupants may affect the light intensity during teaching and learning sessions compared to without occupant. However, the source of artificial lighting still insufficient for writing, typing, reading, and data processing.

6. Conclusion

A study based on air temperature, relative humidity, level of carbon dioxide and level of light intensity had been made to improve indoor air quality and comfortability in the lecture rooms at UniCITI Alam Campus. Room temperature and RH in the lecture rooms can be improved. Frequent maintenance and services for air conditioners should be done in order to achieve the comfort temperature for the occupants. The level of carbon dioxide were in unacceptable condition and could affect the occupants and contribute the sick building syndrome. The level of light intensity in the lecture rooms were not sufficient and required an improvement for comfortability to the occupants. In conclusion, the lecture rooms at UniCITI Alam Campus need a major improvement in order to achieve good indoor environment quality and increase comfortability to the occupants. Staying long hours in the lecture room can lead to dryness of the eyes, nose and throat, eyestrain, migraine and headaches. Conducive teaching and learning environment can help the students to increase their performance.

Reference

[1] Asere L and Blumberga A 2018 Energy efficiency - Indoor air quality dilemma in public buildings *Energy Procedia.*

[2] Clausen G ., Høst J A, Sigsgaard;, Toftum G ., Bekö C . and Weschler, M.; Callesen, S.; Buhl, M. B.; Ladegaard, S.; Langer, B.; Andersen, J.; Sundell, C. G.; Bornehag T 2012 Indoor air policy for schools and day-care centres *Int. J. Indoor Environ. Heal.* **22** 467–75.

[3] Griffiths M and Eftekhari M 2008 Control of CO2 in a naturally ventilated classroom *Energy Build.*

[4] Han M, May R, Zhang X, Wang X, Pan S, Yan D, Jin Y and Xu L 2019 A review of reinforcement learning methodologies for controlling occupant comfort in buildings *Sustain. Cities Soc.*

[5] ASHRAE 2010 *Thermal Environmental Conditions for Human Occupancy* (United States of America: American Society of Heating, Refrigerating and Air-Conditioning Engineers, Inc.).

[6] Anon 2010 Ventilation for acceptable indoor air quality *ASHRAE Stand.*

[7] DOSH 2018 *Guidelines on Occupational Safety and Health for Lighting at Workplace* (Malaysia: Department of Occupational Safety and Health Malaysia).

[8] ANSI 2007 ANSI/ASHRAE Standard 62.1-2010, Ventilation for Acceptable Indoor Air Quality *Ashrae.*

[9] Yau Y H, Chew B T and Saifullah A Z A 2012 Studies on the indoor air quality of Pharmaceutical Laboratories in Malaysia *Int. J. Sustain. Built Environ.*

[10] Yau L C, Ding Y H and Chew B T 2011 Thermal comfort and Indoor Air Quality at Green Building in Malaysia *United Kingdom - Malaysia - Ireland Engineering Science Conference 2011* (Kuala Lumpur: UMIES 2011).

[11] Noh K C, Han C W and Oh M D 2008 Effect of the airflow rate of a ceiling type air-conditioner on ventilation effectiveness in a lecture room *Int. J. Refrig.* **31** 180–8.

Manufacturing of cast in situ ultra high performance fibre reinforced concrete (uhpfrc) - Workability, Tensile and Compressive Strength

M. Z. A. Mohd Zahid[1], B. H. Abu Bakar[2], F. Mohamed Nazri[2], A. Ayob[3], R. Abdul Razak[1]

[1]Department of Civil Engineering Technology, Faculty of Engineering Technology, Universiti Malaysia Perlis, Uniciti Alam Campus, 02100, Padang Besar, Perlis, Malaysia
[2]School of Civil Engineering, Universiti Sains Malaysia, Engineering Campus, 14300, Nibong Tebal, Pulai Pinang, Malaysia
[3]School of Environmental Engineering, Kompleks Pusat Pengajian Jejawi 3, Universiti Malaysia Perlis, 02600, Arau, Perlis, Malaysia

Email: mohdzulham@unimap.edu.my, cebad@usm.my, cefmn@usm.my, afizah@unimap.edu.my, rafizarazak@unimap.edu.my,

Abstract: The aim of this article is to share on the manufacturing process of Ultra High Performance Fibre Reinforced Concrete (UHPFRC) without heat curing or steam curing, hence, it makes UHPFRC possible to be used for in situ application. The common materials used in numerous previous studies were used in this work to produce UHPFRC: Cement type CEM I, fine sand, silica fume, water and third generation high range water reducer. Three tests were conducted, i.e. Compression test and slump flow test and tensile test. The targeted slump flow is in the range of 600 mm to 750 mm, minimum targeted compressive strength and tensile strength test is 115 MPa and 10 MPa, respectively. In addition, the UHPFRC must exhibit strain hardening properties. As a result, this study proposes the mix proportion of the above mentioned material used to produce UHPFRC which is suitable to be utilized for in situ application.

1. Introduction

Ultra High Performance Fibre Reinforced Concrete (UHPFRC) is relatively a new cementitious material. Its main features are flowable, high compressive strength (> 150 MPa), high tensile strength (> 10 MPa), high durability and exhibits strain hardening properties [1,2]. Basically, it consist of high content of cement, fine aggregate, superplasticizer, very low water content (water to cement ratio less than 2) and steel fibre. Initially it is used in manufacturing precast member, however, recently there is an attempt to use it repair and rehabilitation works [3,4,5,6,7,8]. When it comes to the in situ application like repair works, it is uneconomical to adopt heat or steam curing process and moreover, in producing UHPFRC, it is extremely difficult to obtain above mentioned UHPFRC feature just by replacing locally available material with the one established mixture proportions [8]. Therefore, this current study attempts to

Content from this work may be used under the terms of the Creative Commons Attribution 3.0 licence. Any further distribution of this work must maintain attribution to the author(s) and the title of the work, journal citation and DOI.
Published under licence by IOP Publishing Ltd

determine the most optimum UHPFRC mixture which has above mentioned properties without heat or steam curing process.

2. Material and method

In order to obtain appropriate UHPFRC mixture, the mixture optimization process was carried out in two stages as shown in Figure 1. In the first stage, several concrete mixtures without fibres were prepared and tested to find the best constituents combination with respect to the maximum compressive strength and targeted workability. This stage began with the thorough review work on the existing UHPFRC mixture and mixing procedures. Based on the various mix proportions from previous studies [8-14] there were a numbers of UHPFRC mix proportions have been prepared and tested. The key mix variables considered were:1) Silica fume-cement ratio 2) Water-cement ratio 3) Paste content 4) Type and content of superplasticizer 5) Aggregate size 6) Cement content.

There were various mixing procedures of UHPFRC adopted in previous studies. It is undeniable that mixing procedure greatly influences the quality of concrete to be produced. After a numbers of trials, this study found that the following mixing procedure of UHPFRC is suitable to be adopted:1)Place the dry material (except for steel fibre) in the mixing pan 2) Mix slowly for at least 2 minutes 3) Gradually add about 80% of water while mixing 4) Gradually introduce superplasticizer into the mixer while mixing. Continue until it shows dark in colour and moist 5) Add the remaining water and thoroughly mix with high speed for about 3 minutes 6) Reduce the speed of the mixer and gradually add 6 mm steel fibre and followed by 14 mm steel fibre 7) Thoroughly mix with high speed for about 2 minutes.

It is worth to mention that, the speed of the mixer plays a vital role to obtain high compressive strength concrete at 28th day (> 115 MPa) with self-compacting property. Low speed mixture requires more superplasticizer content to achieve targeted slump flow and this will affect the setting time as well as compressive strength of concrete. Therefore, it is strongly recommended that UHPFRC mixture to be mixed in mixer with various speeds. Moreover, based on various trial mix done in this study, it is also proposed that steel fibre to be introduced at the end or after all water and superplastizer were added and it must be added in gradually manner as it is greatly help to avoid coagulated of steel fibre in the mixture which will create more void in the mixture and eventually impair the compressive strength of concrete.

Note that, in an attempt to obtain the suitable UHPFRC mixture, this study has used cement from three different brands, namely Blue Lion, Blue Panda and Tasek as well as three different superplasticizers, namely MasterRheobuild 1100, Glenium, MasterGlenium 8589 and MasterGlenium Ace 8109 produced by BASF company. MasterGlenium 8589 and MasterGlenium Ace 8109 give almost similar performance in terms of rheology property and compressive strength, however, the retention time for MasterGlenium Ace 8109 is longer which is about 6 hours compared to 2 hours for MasterGlenium 8589, therefore, it is suggested to use MasterGlenium Ace 8109 for onsite application particularly for repair works. The combination of Blue Panda cement and MasterGlenium Ace 8109 was found to have the best rheology properties and compressive strength. The results of workability, compressive strength and tensile strength of trial samples will be discussed in chapter 4. Therefore, this study utilized Portland cement type CEM I 52.5 N, Blue Panda brand from Hume. This type of cement was recommended for UHPFRC as it has low C3A content which result in better compatibility with superplasticizer (Denarie, 2009) and widely used by other researchers i.e. Choe et al.,[9] , M Màca et al [10], Hassan and Jones [15], Habel et al., [16] and among others.

Prior to cast, the inverted slump flow test was performed following ASTM C1611[17] and the targeted slump spread was 600 mm to 760 mm and flow rate (T_{50}) should be greater than 2s following recommendation by Koehler & Fowler [18]. Noted that, in general, as slump flow increases the susceptibility to segregation also increases. In this method, a sample of freshly mixed concrete was placed in the cone in one lift without tamping or vibration. The mould was lifted and the concrete was allowed to spread. After spreading ceases, two orthogonal diameters of the concrete patty were measured as shown.

The slump flow was the average of the two diameters. This method also provides a procedure for a relative measurement of flow rate, viscosity, and stability. The flow rate, T_{50} was obtained by measuring the time taken by the fresh concrete achieve 500 mm diameter of spreading. The concrete viscosity influences the flow rate. The stability can be observed by visually examining the concrete patty for evidence of segregation.

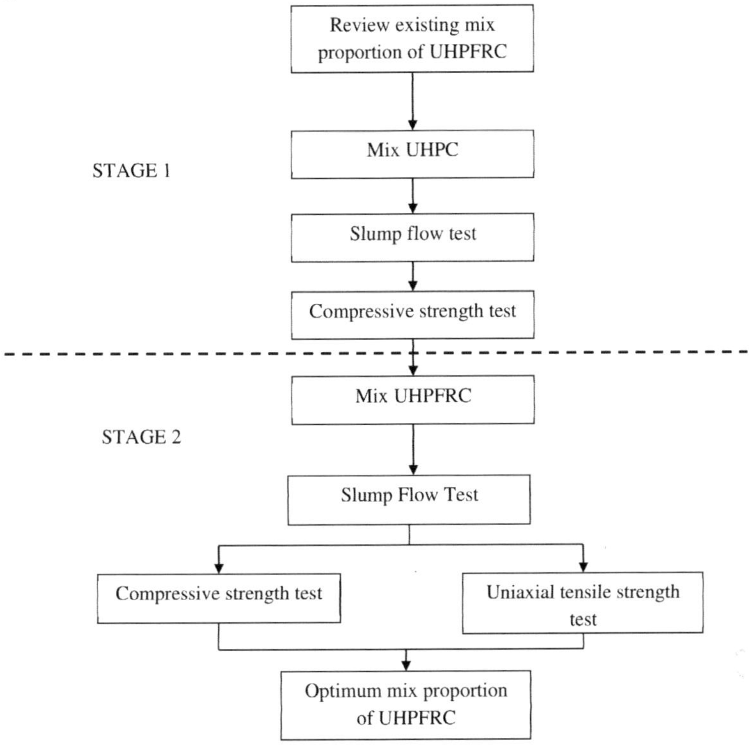

Figure 1. Mixing process flow of UHPFRC.

Besides inverted slump test, there were a numbers of test conducted by previous study to quantify the workability of UHPFRC. For instance, Benson and Karihaloo [19] adopted Cone Penetration Test (CPT) which widely used for soil to determine the workability properties of UHPFRC since it consists of graded sand and smaller particles in the form of cement and microsilica which make it like a sandy soil and a rough correlation between CPT value and slump are a CPT value of 20 mm is equivalent to a slump in excess of 250 mm and CPT value of 25 mm to a slump in excess of 300mm. Slump greater than 300mm considered very flowable. Kang & Kim (2011) used flow table test in accordance to ASTM C 1437 to measure the flowability. Meanwhile, Maca et al [10] and Wille, et al. [11] used standard flow table test in accordance to EN 1015-3 [20] and ASTM C230 [21], respectively. However this study choose inverted cone flow test to quantify the workability of UHPFRC as this test adopts standard slump cone apparatus which widely used in construction site.

The UHPFRC specimens were cast in 100mm x 100mm x 100mm cube and then cured for 28 days prior to test. Curing is a crucial process and it significantly affects the mechanical performance of cementitious material. Various curing method have been used by previous researchers in producing UHPFRC. For example Wu et al., [22] cured their sample in lime saturated water and the minimum compressive strength of UHPFRC was about 110 MPa at 28th day. According to Voo [23] and Voo & Foster [24] the specimens or members made of UHPFRC must be heat cured for 48 hours at a temperature of 90°C. Heat cure and steam cure are the most popular method and widely applied in producing UHPFRC as it may minimize the creep and long term shrinkage as well as smooth and good finishing [25]. A few studies have investigated the effect of curing process on the compressive strength of concrete. For instance, Persson [26] found that the compressive strength of water-cured, sealed and air-cured specimens were 110 MPa, 112 MPa and 119 MPa , respectively, after 28 days. Moreover, Lutfi Ay [14] showed that UHPC should not be water cured as it may create a partially open system which may harm the balance between inner and the outer part of the concrete. This can result in pre-stressing in the body and on the fibres. However, since this UHPFRC material to be used as repair material, neither heat nor steam curing method will be used as it is impractical to apply steam curing process at construction site for repair work, hence, only normal curing method was adopted in preparation of UHPFRC in this study. Furthermore, there were few studies have proved that the UHPFRC with high compressive strength of greater than 120 MPa were able to be attained with minimal curing i.e. ambient environment curing or air curing [9-14].

In this current study the UHPFRC specimens were cast in the 100mm x 100mm x 100mm cube and cured for twenty eight days in ambient environment prior to test. The compressive strength test was conducted with the loading rate of 2 kN/s in accordance to BS EN 12390 Part 3 [27]. Note that, BS EN 12390 Part 3 [27] suggested to apply compression load with loading rate ranging from 0.2 MPa/s up to 1.0 MPa/s which is equivalent to 2 kN/s up to 10kN/s for 100mm x 100mm x 100mm cube sample. The mix proportion by means of the best constituent combination with respect to the minimum compressive strength of 115 MPa and within the targeted workability (600 - 760mm) was chosen for the stage two.

In the stage two, the optimum UHPC mixture was added with steel fibre from 1% up to 2% by volume. Two different sizes were used (i.e. 6 mm and 14 mm long) and its mechanical properties based on supplier technical data sheet. According to Denarie [8], the typical fibre dosage in UHPFRC to exhibit the strain hardening characteristic is about 0.5% to 3%. Recently, Hoang et al. [28] reported that there were fibre balling appeared in UHPFRC paste with 3% (vol) of steel fibre and it degraded the compressive strength of UHPFRC. Therefore this current study utilized steel fibre content from 1% to maximum of 2% by volume. Then, the UHPFRC samples underwent similar process as stage one (i.e. slump flow test, curing and compressive strength test) in addition with uniaxial tensile strength test to examine the strain hardening behaviour of UHPFRC. According to Wille, et al, [11], tensile testing of materials with strain hardening behaviour accompanied by multiple cracking is particularly challenging. Currently, there are no testing standards available that define the test conditions, specimen geometry, and analytical procedures necessary to fully characterize the tensile properties of strain-hardening cementitious material. Predetermination of crack development by notching the specimen in bending or direct tension is commonly utilized to simplify the test setup, deformation control, measurement system and data analysis. However, doing so may not be appropriate for capturing the strain hardening behaviour of UHPFRC. Difficulties in obtaining evenly distributed stresses throughout the cross section and controlling a stable load versus displacement/crack opening response has limited the number of researchers performing direct tensile tests on cementitious materials and composites. Specimen shapes, which can be distinguished into dogbone shape [29-31] , unnotched [32] and notched prisms or cylinders [33] significantly influence the test results in all three stages of behaviour (i.e. Part I: strain based elastic part, Part II: strain based strain hardening part and Part III: crack opening based softening part).

In terms of uniaxial tensile test, this study followed the specimen preparation and testing procedure as proposed by Wille, et al. [11]. In accordance to Wille et al. [11], the UHPFRC was poured in layers in dogbone shaped specimens mould to full capacity, but without any vibration. After casting, the specimens were covered and stored at room temperature for 24 h. Afterward they were removed from their moulds and stored at room temperature for additional 25 days. All specimens are then tested at 28 days. About 24 h prior to testing, a spray coating was applied on the surface of the middle portion of each specimen for better crack detection. The uniaxial tensile test was conducted using Universal Testing Machine (UTM) Shidmadzu with capacity of 25 kN and loading rate was set to 0.01 mm/s. The dimension of dog bone shaped uniaxial tensile test specimen adopted in this study as per designed and proposed by Wille et al. [11] to full fill the following requirements:1) Accommodates small specimen in order to reduce the material usage 2)Employs specimens that are easy to cast and prepare 3)Use specimen with region of a constant area to capture multiple cracking

Wille et al [11] proposed that the largest size of material constituent must be less than one fifth of the smallest specimen dimension that is 3.4 mm. This allows testing paste and mortar like materials, such as the UHPC used in this current research, which has a maximum grain size of 0.8 mm. In comparison to the length of the fibers (13–30 mm), the height and width (17/30 mm) of the specimen moulds are relatively small. This is to ensure fibre alignment parallel with the load direction and it can be improved by placing the fresh material in layers. Finally, the mix proportion by means of the best constituent combination with respect to the maximum compressive strength, workability, tensile strength and strain hardening properties was obtained.

3. Result and Discussion

A series of test were conducted which involved plenty of specimens and it took two years to obtain the suitable mix of UHPFRC that full filled the requirement as repair material. The required properties of UHPFRC to be used as repair material are as follows [8]: 1) Self-compacting characteristic 2) High compressive strength and 3) High tensile strength with strain hardening property.

Even though there were many available UHPFRC mix in previous literature, very few mix satisfy all above mentioned properties at the same time. Moreover, most of the mix available in literature is not made from local available material and it is difficult to achieve satisfactory properties of UHPFRC just by replacing cement and superplasticizer from existing optimized UHPFRC mix with locally available material.

In this current study, the UHPFRC used was designed as self-compacting concrete and to achieve that purpose three different types of superplasticizer (MasterRheoBuild 1000, Glenium 6011, Glenium Ace 8109) have been tested and the one that achieved the targeted slump flow diameter around 600 to 760 mm was chosen. The inverted slump flow test was conducted on one mixture with those three different superplasticizers and it was observed that the flow diameter for Glenium 6011 and Glenium Ace 8109 were 590 mm and 780 mm, respectively, and no spread observed for MasterRheobuild 1000 specimens. Therefore, Glenium Ace 8109 type of superplasticzer was chosen.

It was found that silica fume also significantly affect the rheology property of UHPFRC. As shown in Figure 2, the slump flow decreases as the silica fume content increases from 5% to 20% and there was sharp reduction of slump flow diameter when 20% of silica fume was added into the mixture. It was also observed that the compressive strength of UHPFRC decreases as the silica fume content increases as illustrated in Figure 3. According to Vikan & Justnes [34], when silica fume content exceeds certain value, workability was reduced and greater viscosity lead to more void in fresh concrete. Therefore, this leads reduction of quality of concrete and negatively affects the compressive strength of concrete [22]. Based on these results, the silica fume content in UHPFRC mixture was limited to 10%.

Then the effect of water cement (W/C) ratio also has been examined and as tabulated in Table 1, it was found that reduction of W/C ratio to 0.175 significantly reduced the diameter of slump flow. Therefore, the w/c ratio was maintained to 0.2 as adopted by many researchers.

The influence of sand content also has been studied. Two different percentage of sand content has been considered i.e.: 18% and 88% as tabulated in Table 2. It clearly shows that the increase of sand content has resulted in reduction of slump flow diameter and increment of compressive strength. Therefore, the mix that rich with sand has been chosen in this study.

In this current study, amount of steel fibre in UHPFRC is one of the variables considered. Three different amount of steel fibre were taking into account, i.e. 1%, 1.5% and 2% by volume. The compression cube test and uniaxial tensile test were conducted to determine the compressive strength and tensile strength of UHPFRC, respectively. Results of both tests are shown in Figure 4 and 5. It can be seen that the tensile strength increase as amount of steel fibre increases. It is parallel with findings from previous studies [11].

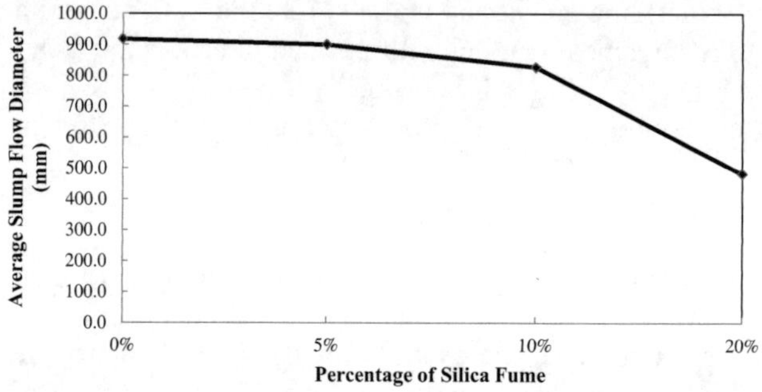

Figure 2. Effect of silica fume on slump flow diameter of UHPFRC.

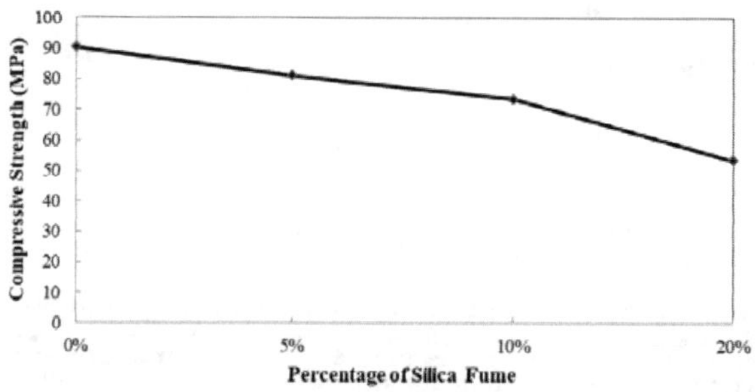

Figure 3. Relationship of silica fume and compressive strength.

Table 1. Effect of water-cement ratio on workability of UHPFRC.

	Water-Cement Ratio	D1 (mm)	D2 (mm)	Average Slump Diameter (mm)	T_{500} (s)
Specimen 1	0.200	820	835	827.5	3.80
Specimen 2	0.175	535	500	517.5	74.85

Table 2. Effect of sand content.

Sand Content (%)	Average Slump Diameter, D (mm)	Compressive Strength (MPa)
18	900	81
88	590	94

The average cube compressive strength of UHPFRC obtained from the compression test of three cube specimens from each concrete batch is shown in Figure 4. It can be seen that the compressive strength of UHPFRC is in the range of 119 MPa to 121MPa. Most of previous researchers defined minimum compressive strength of UHPFRC is 150 MPa [9-13]. However, for in situ application it is challenging to obtain minimum compressive strength of 150 MPa as it is difficult to apply heat or steam curing at site. Moreover, compressive strength is not the main feature for UHPFRC as tensile strength is more important. The tensile strength of UHPFRC is ranging from 5.92 MPa to 10.43 MPa as shown in Figure 6. It is crucial for cementitious repair material to have minimum tensile strength of more than 10 MPa as shrinkage deformations at early age are restrained to a more or less large extent by the existing structure, which gives rise to very high tensile stresses up to 10 MPa [8,35]. In this study, UHPFRC with 2% amount of fibre full fill this requirement.

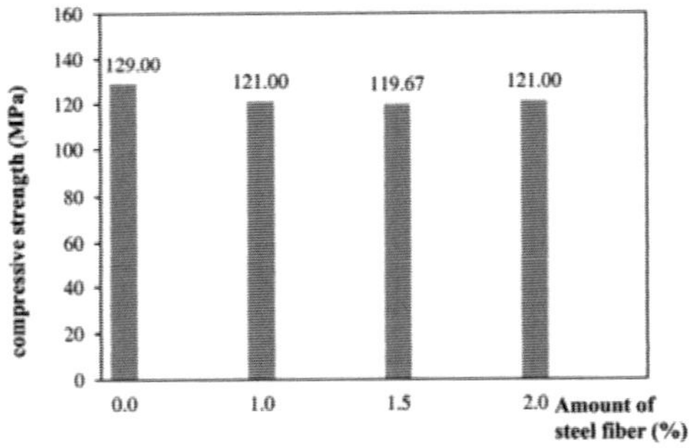

Figure 4. Average compressive strength of UHPFRC.

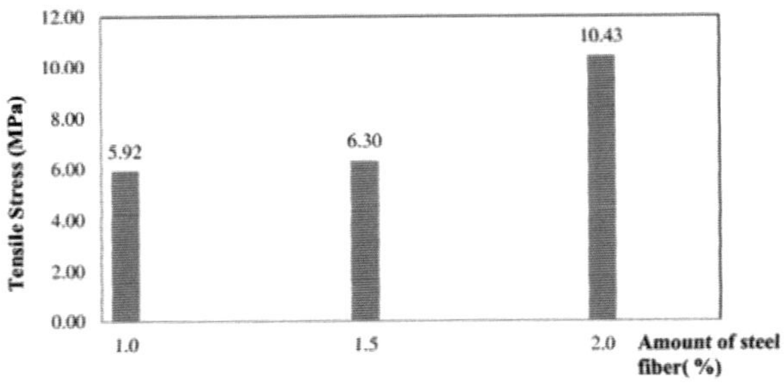

Figure 5. Tensile stress of UHPFRC.

Another important feature of UHPFRC is strain hardening property. Cementitious repair material which has this kind of property able to withstand the development of eigenstress due to restrained shrinkage by old concrete. Strain hardening material is when its tensile strength, σ_{pc}, is in access of its cracking stress, σ_{cc} [36]. Figure 6 shows the stress strain curves resulted from uniaxial tensile test. It clearly shows strain hardening behaviour for UHPFRC with 2% steel fibre amount and its σ_{pc} is higher than σ_{cc}. Meanwhile, in specimens with 1% and 1.5% volume of steel fibre, it was observed that there was strain hardening like trend after first cracking stress but the σ_{pc} was lower than σ_{cc}. Therefore, it was not considered as strain hardening property. This shows that the minimum amount of fibre to obtain UHPFRC with strain hardening property is 2% by volume. It is however, possible that these results may be substantially different if the type of steel fibre (e.g.: hooked and twisted type of steel fibre) were used or for greater size of specimens as discussed by Wille et al., [11]. Moreover, UHPFRC with 2% amount of fibre was extremely more ductile than the one with 1% and 1.5% as the maximum strain attained by 2.0% was greater than another two specimens. Finally, based on testing results discussed above, this study propose UHPFRC mixture as shown in Table 3 for cast in situ application like repair works.

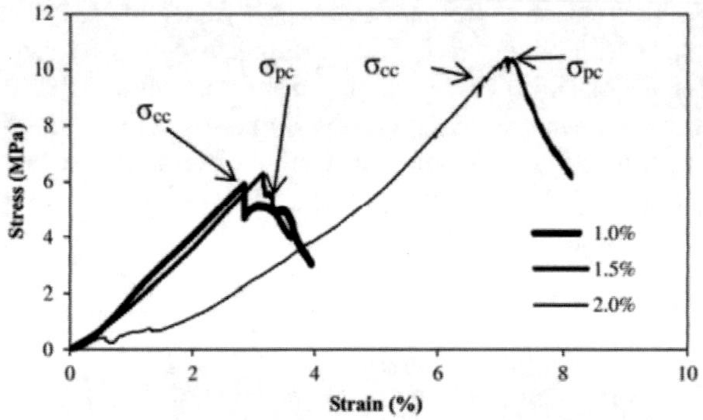

Figure 6. Tensile stress strain curve of UHPFRC.

Table 3. UHPFRC mixture .

Material	Amount (kg/m^3)	Ratio to cement
Cement	1000	1
Silica Fume	53	0.05
Fine Sand	887	0.89
Water	210	0.02
Superplasticizer	53	0.05
Steel Fibers	150	-

4. Conclusion

The aim of this study is to manufacture the optimum UHPFRC mixture for in situ application like repair works. The following key mix variables has been investigated 1) Silica fume-cement ratio 2) Water-cement ratio 3) Paste content 4) Type and content of superplasticizer 5) Aggregate size and 6) Cement content. Based on results from slump flow test, compression test and tensile test, the mix proportion that has following properties of flowable, high compressive strength (> 115 MPa), high tensile strength (> 10 MPa) and strain hardening were obtained as shown in Table 3.

Acknowledgement

The support provided by Malaysian Ministry of Higher Education and Universiti Sains Malaysia in the form of a research grant no.304/PAWAM/6050362/M154 for this study is highly appreciated. The authors also thank Universiti Malaysia Perlis for financial support in publishing this article.

References

[1] M Khairunnisa, MZAM Zahid, AR Rafiza, I Nurfitriah, NZ Zainol, MBHA Manaf, MM Ahmad, SNMN Aishah, NN Noh, MZ Johari, ASSM Zaidi 2018 *AIP Conference Proceedings* AIP Publishing 2030 020297.

[2] NA Hamiruddin, R Abd Razak, K Muhamad, MZA Zahid, CNS Aziz 2018 *Solid State Phenomena* Trans Tech Publications 280 476-480.

[3] Muhd Afiq Hizami Abdullah, Mohd Zulham Affandi Mohd Zahid, Afizah Ayob, Khairunnisa Muhamad 2019 *Journal of Structural Fire Engineering* Emerald Publishing Limited 10 56-75.

[4] NAAA Malek, K Muhamad, MZAM Zahid, NA Hamiruddin, NZ Zainol, N Yahya, N Bahaman, NM Ramli 2018 *MATEC Web of Conferences* EDO Sciences 195 01015.

[5] MZA Mohd Zahid, BH Abu Bakar, FM Nazri, MM Ahmad, K Muhamad *IOP Conference Series: Materials Science and Engineering* IOP Publishing 318 012023.

[6] MZA Mohd Zahid, K Muhamad 2017 *AIP Conference Procedings* AIP Publising 1885 020117 .

[7] Muhd Afiq Hizami Abdullah, Mohd Zulham Affandi Mohd Zahid, Badrol Abu Bakar, Fadzli Mohamed Nazri, Afizah Ayob 2015 *Applied Mechanics and Materials* Trans Tech Publications 802 283-289 .

[8] E Denarie 2009 Recommendation for the tailoring of UHPFRC recipes for rehabilitation. ARCHES D06. Available from: http://arches.fehrl.org.

[9] G Choe, G Kim, N Gucumski, S Lee 2015 *Construction and Building Materials* Elsevier Ltd 86 159-168.

[10] P Maca, R Sovjak, P Konvalinka 2014 *International Journal of Impact Engineering* Elsevier Ltd 63 158-163.

[11] K. Wille, S. El-Tawil, A.E. Naaman 2014 *Cement & Concrete Composites* Elsevier 46 53-66.

[12] R. Yu, P. Spiesz, H.J.H. Brouwers 2014 Cement and Concrete Research Elsevier 56 29-39.

[13] B Graybeal 2007 Technote FHWA Publication FHWA-HRT-11-038.

[14] Lutfi Ay 2004 *Proceedings of the International Symposium on Ultra High Performance Concrete* Kassel university press ISBN: 3-89958-089-9 695-701 .

[15] AMT Hassan, SW Jones 2012 *Construction and Building Materials* Elsevier 35 361-367.

[16] K Habel, M Viviani, E Denariė, E Brühwiller 2006 Cement and Concrete Research Elsevier 36 1362-1370.

[17] ASTM C1611/C 1611M-18, Standard Test Method for Slump Flow of Self-Consolidating Concrete, ASTM International, West Conshohocken, PA, 2018.

[18] Koehler, E.P. & Fowler, D.W. (2007) Inspection manual for self-consolidating concrete in precast members. Center for Transportation Research. The University of Texas at Austin.

[19] SDP Benson, BL Karihaloo 2005 *Magazine of Concrete Research* 57 347-352.

[20] EN 1015-3, Methods of test for mortar for masonry – Part 3: Determination of consistence of fresh mortar (by flow table), Comité Européen de Normalisation/British Standards Institution.

[21] ASTM C230 / C230M-14, Standard Specification for Flow Table for Use in Tests of Hydraulic Cement, ASTM International, West Conshohocken, PA, 2014.

[22] Wu, Z., Shi, C., Khayat, K.H. 2016 *Cement and Concrete Composites*. Elsevier Ltd. 71, 97-109.

[23] Voo, Y.L. 2010 *International Journal of Sustainable Construction Engineering & Technology*. Vol. 1, No. 2.

[24] Voo, Y.L., Foster, S.J. 2010 *The IES Journal Part A: Civil and Structural Engineering,* Vol. 3, No. 3, 168-187.

[25] Voo, Y.L., Nematollahi, B., Mohamed Said, A.B., Gopal, B.A., Yee, T.S. 2012 *International Journal of Sustainable Construction Engineering & Technology* Vol. 3, Issue 1. (ISSN: 2180-3242).

[26] Persson, B. 1996 *Advanced Cement Based Materials*, no. 3: p.107-123.

[27] EN 12390-3, Testing hardened concrete – Part3: Compressive strength of test specimens, Comité Européen de Normalisation/British Standards Institution.

[28] Hoang A.L., Fehling E. 2017 *Construction and Building Materials* Elsevier Ltd. 153, 790-806.

[29] Kamal, A., Kunieda, M. Ueda, N., Nakamura, H. 2008 *Cement & Concrete Composites* Elsevier Ltd. 30, 863-871.

[30] Kunieda, M., Kozawa, K., Ueda, N., Nakamura, H. 2010 *Proceedings of FraMCos-7. Fracture mechanic of concrete and concrete structures,* In: Jeju OH BH, editor Korea, May 23-28, 2010, p.501-508.

[31] Jun, P., Mechtcherine, V. 2010 *Cement and Concrete Composites.* Elsevier Ltd. Vol. 32, Issue 10, p.801-809.

[32] Graybeal, G.A., Baby, F. Marchand, P., Toutlemande, F. 2012 Proceedings of hipermat 2012, 3rd international symposium on UHPC and nanotechnology for high performance construction materials In: Schimdt, M., Fehling, E., editors., Kessel, Germany, March 7-9, p.395-402.

[33] Zhang, J., Stang, H., Li, V.C. 2000 *Journal Material Civil Engineering* ASCE 12(1), p.66-73.

[34] Vikan, H., Justnes, H. 2012 *Cement and Concrete Research.* Elsevier Ltd. 37(11), p.1512-1517.

[35] Eide, M. B., Hisdal J.M. 2012 Ultra-High Performance Fibre Reinforced Conrete (UHPFRC) – State of Art.COIN Project report 44. Available from World Wide Web: http://www.coinweb.no.

[36] Naaman, A.E. Reinhardt, H.W. 2003 *Proceedings of 4th RILEM symposium on high performance fibre reinforced cement composites HFFRCC4*. p.1-4.

Mechanical and physical properties of bottom ash/fly ash geopolymer for pavement brick application

Wan Mastura Wan Ibrahim[1,2*], Mohd Mustafa Al Bakri Abdullah[1], Romisuhani Ahmad[1,2], Muhammad Faheem Mohd Tahir[1], Nurul Aida Mohd Mortar[1] and Muhammad Amirul Aziem Noor Azli[2]

[1]Center of Excellence Geopolymer and Green Technology, School of Materials Engineering, Universiti Malaysia Perlis (UniMAP), P.O. Box 77, d/a Pejabat Pos Besar, 01007, Kangar, Perlis, Malaysia
[2]Faculty of Engineering Technology, Uniciti Alam Campus, Universiti Malaysia Perlis (UniMAP), 02100, Sungai Chuchuh, Padang Besar, Perlis, Malaysia

Email: * wanmastura@unimap.edu.my

Abstract. Geopolymers are amorphous to semi-crystalline with excellent physical and and mechanical properties. It has been used to become a potential binder to Ordinary Portland Cement (OPC) in certain applications due to its lower emission of carbon dioxide gases and low energy consumption sustainability criteria. Bottom Ash (BA) is one of the main industrial by-products and it is produced at the bottom of the furnace during the coal combustion process in electricity generation. The application of BA as a sustainable construction material in the building industry plays an important role in order to decrease the volume of residual waste and conserving existing natural fine aggregates. The objectives for this study is to study the effect of fly ash to bottom ash ratio and to determine the optimum ratio of fly ash to bottom ash geopolymer for pavement brick application. The chemical composition and morphology of geopolymer reinforcement was analysed by using X-ray Fluorescence and Scanning Electron Microscope. The molarity of the Sodium Hydroxide solution is fixed at 12M. The parameter used in this study are different weight percentage of fly ash geopolymer 0 wt%, 10 wt%, 20 wt%, 30 wt% and 40 wt%. The solid to liquid ratios for this study is 2.0. The curing temperature of this study is 80°C and the curing time is 24 hours. 100% of bottom ash geopolymer is used as a control variable for this study.

1. Introduction

Geopolymer is a new type of cementitious material which has received extensive attention internationally in recent years due to its use of industry waste source material and lower carbon dioxide (CO_2) emission during the material production [1]. The basic structure of geopolymer is silicon-oxygen tetrahedron and aluminium-oxygen tetrahedron that are connected by oxygen atom [2] [3]. Geopolymers are formed in high alkaline medium which releases Si and Al species through geopolymerization reaction involving the dissolution of aluminosilicates sources where the formation of small coagulates structures, gelation and lastly hardening to form hard solid. This reaction produced amorphous to semi-crystalline three dimensional polymeric structures containing Si-O-Al. This is suitable for materials that containing alumina and silica barrier phases such as natural rocks or secondary raw materials [4-8].

Content from this work may be used under the terms of the Creative Commons Attribution 3.0 licence. Any further distribution of this work must maintain attribution to the author(s) and the title of the work, journal citation and DOI.
Published under licence by IOP Publishing Ltd

IOP Conf. Series: Materials Science and Engineering **743** (2020) 012029 doi:10.1088/1757-899X/743/1/012029

Bottom ash is a part of non-combustible residue of combustion in a furnace or incinerator. It was produced in coal combustion and it sticks to hot side walls of coal-burning furnace and produces traces of combustible embedded in forming clinkers [9]. It contains ferrous (Fe) and non-ferrous (NFe) metals that can be recovered. It has high potential to be used in construction as lightweight aggregate as it has various properties such as high porosity, low shrinkage value, low density, good fire and chemical resistance and good immobilization of heavy metal [10]. The physical characteristics of bottom ash is determined by some factors which are type and quality of coal source, pulverized fineness and operating conditions of power plant. The particle of bottom ash is porous textures in angular shapes and dark colour [11].

Fly ash is a by-product of pulverized coal combustion in electric power generating plants. It reacts aggressively with calcium hydroxide to form compounds with cementitious value which increases concrete strength [12]. Selecting fly ash in concrete will attain environmental, economic and also structural benefits as increasing strength and fire resistance of concrete structures [13]. Adding of fly ash to concrete could improves mechanical properties of concrete at late ages but obtains lower strength up to 28 days. The reduction in early strength is a function of replacement percent. The mechanical properties and abrasion resistance showed continuous and significant improvement at the ages of 91 and 365 days due to the pozzolanic reaction of fly ash [14].

In this research, the effects on the addition of fly ash to the bottom ash-based geopolymer were studied based on the mechanical and physical properties of geopolymer mixtures. The properties of geopolymer was followed the American Society for Testing and Materials (ASTM) Standard for pavement brick application.

2. Experimental

2.1. Raw Materials

Raw materials used in this research were fly ash, bottom ash and alkaline activator. In this research, the alkaline activator used is combination of sodium hydroxide solution (NaOH) and sodium silicate solution (Na_2SiO_3). Fly ash and bottom ash are by-product that produced from coal combustion was supplied by Cement Industries of Malaysia Berhad (CIMA). Class C fly ash was used in this study. The sodium hydroxide in pallets form was used and dissolved in distilled water for one day to produce specific molarity concentration of sodium hydroxide solution. Then, the sodium hydroxide solution was added with sodium silicate solution to form an alkaline activator solution.

2.2. Mixing Process

First, bottom ash was mixed with fly ash at a different parameter (10 wt%, 20 wt%, 30 wt% and 40 wt%). Then, the mixture of bottom ash and fly ash were mixed with alkali activator solution for about 5 minutes or until achieved the homogeneity. The molarity of NaOH solution used is 12 M based on the previous study done by Wan Mastura et al., [15]. The mass ratio of Sodium Silicate to Sodium Hydroxide of 2.5, by mass was used and the ratio of solid to liquid ratio of 2.0, by mass was used in this research. The mixture was then poured into a 50 mm x 50 mm x 50 mm cube moulds in accordance with ASTM C109 [16] and covered with plastic film to avoid moisture evaporating during heat curing. The samples were cured in the oven for 24 hours at temperature 80° C.

2.3. Testing and Characterization

2.3.1. Chemical Composition Analysis

Chemical composition analysis has been done by using X-Ray Fluorescence Spectrometer (XRF). X-ray fluorescence spectrometer (XRF) is the analysis to get the information or study the element contained in raw materials. It is to determine the oxides content and percentage of the elements.

2.3.2. Microstructure Analysis
Scanning electron microscopy (SEM) model Hitachi TM 3000b is used in this research to analyse the microstructure on the surface of the geopolymer. This testing was undergo after the compressive test to determine the surface structure. The magnification used in this research are 100x, 500x and 1000x.

2.3.3. Compressive Strength Test
The compressive strength test was measured by using a Universal Testing Machine (UTM) in compliance with ASTM C 109 [16] on the (50×50×50) mm specimens. Compressive strength testing is the maximum compressive stress that under gradually applied load a given solid material will sustain without fracture. The samples were tested at 7th days of ageing. The three samples for each different percentage was compressed by using UTM machine and the results were recorded.

2.3.4. Water Absorption Analysis
Water absorption test was performed on 50×50×50 mm cube after 7 days of ageing according to ASTM C 140 [17]. Water absorption test is used to determine the amount of water absorbed under specified conditions. The formula in Equation 1 is used to measure the water absorption.

$$Water\ absorption = \left[\frac{(Wet\ weight - Dry\ weight)}{Dry\ weight} \times 100\right] \quad (1)$$

2.3.5. Density Analysis
Density measurement usually performed to determine the density of the samples. This measurement was performed by dividing the mass of the samples to the volume of the samples. The formula in Equation 2 is used to measure the density.

$$Density = \frac{mass\ of\ the\ sample\ (kg)}{volume\ of\ the\ sample\ (m^3)} \quad (2)$$

3. Results and Discussion
3.1. Chemical Composition Analysis of Raw Materials
In this research, X-Ray Flourescene (XRF) analysis was analysed to characterize the chemical composition of the raw material. Each composition of bottom ash and fly ash can be summarized in Table 1 below.

Table 1. Chemical composition of bottom ash and fly ash.

Compound	Bottom ash (wt. %)	Fly ash (wt. %)
SiO2	51.27	30.80
Al_2O_3	24.70	13.10
Fe_2O_3	4.31	22.99
TiO_2	1.24	0.90
CaO	7.46	22.30
MgO	2.19	4.00
P_2O_5	0.40	1.31
SO_3	5.57	0.49

The main constituents of bottom ash geopolymer and fly ash geopolymer are SiO2 and Al_2O_3 which is about 76% wt for bottom ash and 44% wt for fly ash. The variation in the principle constituents is about silica (25% wt-60% wt), Alumina (10% wt-30% wt), and Iron Oxide (5% wt-

25% wt). This is the expected composition since Silica and Alumina were compulsory for the geopolymerization process. Materials with high amount of Silica and Alumina content are good for binding action [18]. According to ASTM C618 [19], When the sum of silica and alumina of fly ash is less than 50 wt% and the CaO is more than 20 wt%, the fly ash was classified in class C.

Besides that, the percentage of Fe_2O_3 also higher for fly ash compared to bottom ash. Dakhane et al, [20] stated that Fe_2O_3 may effect on the strength development of the sample. Due to the both bottom ash and fly ash are rich in silica and alumina content, these materials are suitable as the source material to produce geopolymer for pavement brick application.

3.2. Microstructure Analysis

The Scanning Electron Microscopy (SEM) was used to analyse the microstructure of bottom ash, fly ash, and geopolymer to characterize the shape and surface texture of the mixture particles, and to acknowledge the behaviour of the materials. Figure 1 (a – e) present the microstructure of five different proportions of bottom ash/fly ash which were 0 wt%, 10 wt%, 20 wt%, 30 wt% and 40 wt%.

Figure 1. Microstructure images of five different proportions of bottom ash/fly ash geopolymer: a) 0 wt% b) 10 wt% c) 20 wt% d) 30 wt% e) 40 wt% under the magnification of x500.

Figure 1 (a) shows the 0 wt% of fly ash sample, indicating that it was 100 wt% of bottom ash, was used as the controlled sample. It showed that there were vast amount of cracks and pores existed. This is because, bottom ash had irregular shape, was rough and had gritty textural surface [10]. Figure 1 (b) shows the 10 wt% of fly ash geopolymer sample, it showed that the alkaline activator did not react completely with the mixture of ashes. It also showed the existence of voids and cracks but lesser than Figure 1 (a). Figure 1 (c) shows the 20 wt% of fly ash geopolymer sample, it showed that the number of porosity and crack still existed but in decreasing trend. Figure 1 (d) shows the 30 wt% of fly ash geopolymer sample. It showed that the number of cracks and pores decreased but still there were few small cracks can be seen. Figure 1 (e) shows the 40 wt% of fly ash geopolymer sample. It showed that there was less porosity found on the surface. The alkaline activator also reacted with the particle and there were still a few fine round particles which were not reacted or partially reacted that still can be seen.

The best microstructure that can be seen was at Figure 1 (e) which represented by proportion of fly ash geopolymer of 40% wt, shows denser and compact due to good reaction between the alkaline activator solution and the source material occurred. This was proved by the excellent results in compressive test.

3.3. Compressive Strength

The mechanical properties of bottom ash was evaluated through compression test. Several tests have been carried out to study the strength of the bottom ash at different weight percentage of fly ash added. In general, it has been observed that the compressive strength of the samples have been increased linearly with decrease of the percentage of fly ash added as shown in Figure 2.

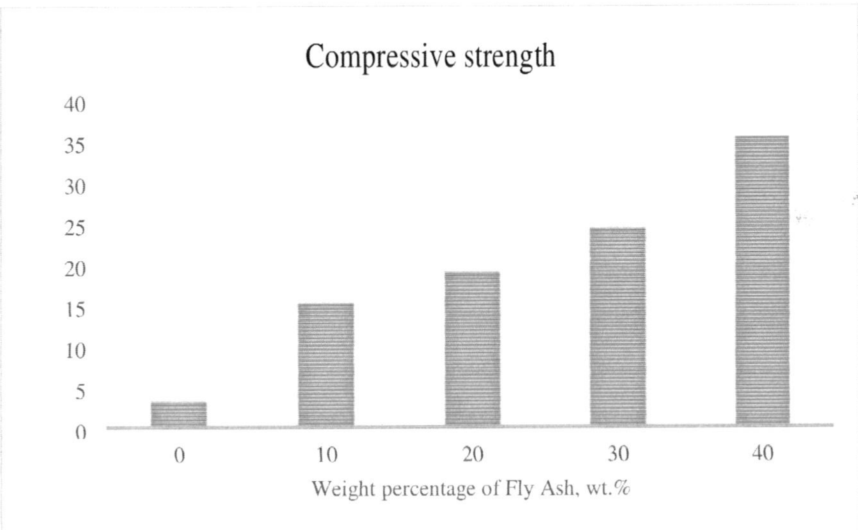

Figure 2. Compressive strength of five different proportions of bottom ash/fly ash geopolymer.

From the graph, the 0 wt% of fly ash geopolymer sample which act as the controlled sample presented lowest strength which is 3.136 MPa. This is due to the surface of the bottom ash was coarse and there were a lot of porosity, voids and cracks were found as shown in Figure 1 (a). Furthermore, the low strength produced also due to the non-completion of hydration between the alkaline activator and the bottom ash during the curing process. However, when the percentage of fly ash added with 10 wt%, 20 wt%, and 30 wt%, the strength of the sample increased to 15.250 MPa, 19.146 MPa and 24.531 MPa, respectively. The highest compressive strength obtained was 35.883 MPa when the weight percentage of fly ash added was 40 wt%. This can be proved by the microstructure of the geopolymer sample where the alkaline activator reacted completely with the mixture of bottom ash and fly ash. The cracks, voids and porosity has been decreased obviously as can be seen in the microstructure (Figure

1e). The addition of fly ash favor in increasing the strength during geopolymerization reaction process. It can be observed that the increase of fly ash increased the strength of all mixtures. This is because the reaction of fly ash with the bottom ash during hydration. The increase of fly ash content in the mixture resulted in chemical reaction and hence increased the strength [21].

3.4. Density Test

The physical properties of bottom ash also assessed through density test. Based on the observation, the density of the geopolymer sample shown the small increment from 0 wt% to 10 wt% which is 1446.56 kg/m^3 to 1526.56 kg/m^3. Then, the density increased slightly to 1830 kg/m^3 at 20 wt% of fly ash. However, the density of the bottom ash decreased drastically to 1553.6 kg/m^3 at 30 wt% of fly ash before it speeded to 1884.8 kg/m^3 at 40 wt% of fly ash added. The results of the density test have been plotted in Figure 3.

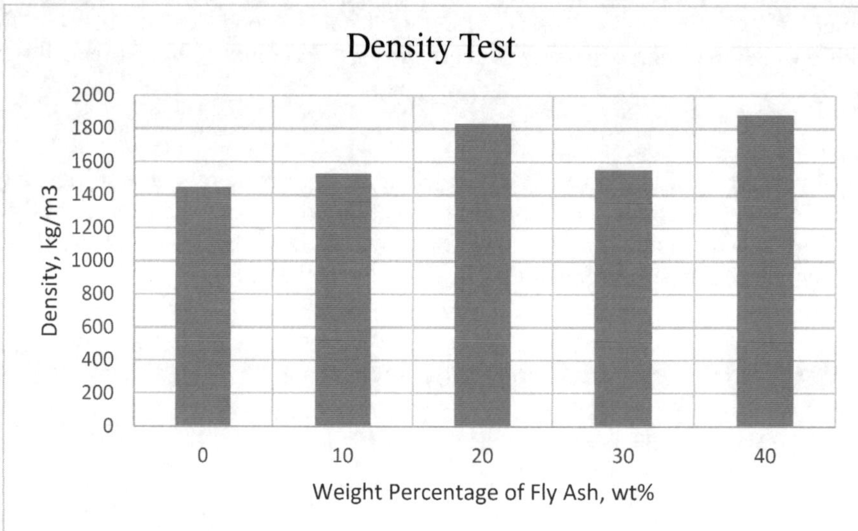

Figure 3. The relationship between the weight percentage of fly ash and the density test.

From the graph, it can be observed that the lowest density recorded is when the weight percentage of fly ash is at 0 wt% which is 1446.56 kg/m^3. The formation of large pores occurred at 0 wt% contributed to the lowest density. This results related to the lowest compression strength of the sample at 0 wt%. Besides, lowest density also due to the poor geopolymerization coalescence in pore formation results in larger pores [22]. The highest density has been recorded at the weight percentage of fly ash is 40 wt% which is 1884.8 kg/m^3. This is due to the uniform pore formation due to the effective geopolymerisation reaction process. The porosity of the sample is decreased, indicating to the geopolymer formed in hydration reaction gradually filled the pores and it became denser [23]. This can be proved from the compression strength where the strength of the sample was at highest at 40 wt%.

3.5. Water Absorption Test

The physical properties of bottom ash were evaluated through water absorption test. Based on the observation, the pattern of the water absorption had increased slightly at weight percentage of fly ash 0 wt% from 7.343% to 8.518% at weight percentage of fly ash 10 wt%. Then it declined drastically from 8.518% to 3.634% at 20 wt%. However, the trend increased drastically back to 8.228% at 30 wt% before it decreased slightly when the weight percentage of fly ash at 40 wt% which is 6.048%. The trend of the water absorption test have been summarized in Figure 4.

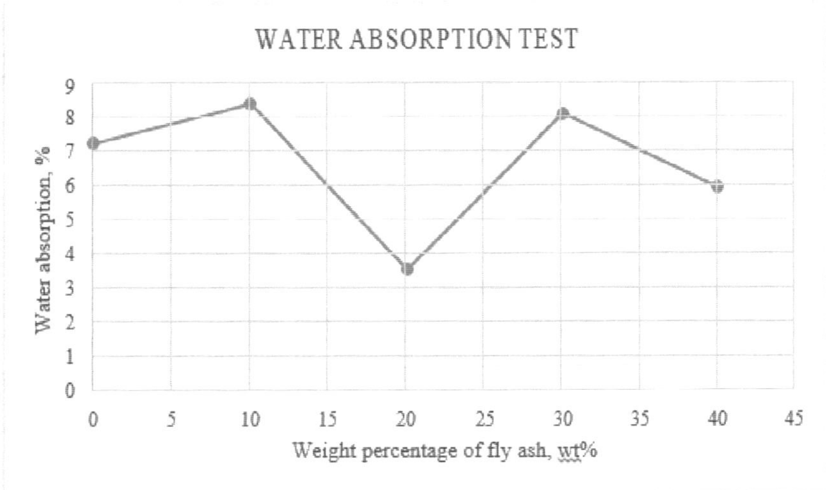

Figure 4. The relationship between the weight percentage of fly ash and the water absorption.

Based on the graph above, it can be concluded that the water absorption was at highest when the weight percentage of fly ash at 10 wt%. The water absorption rate is highest due to the increase in porosity. The water absorption rate received the maximum value when the porosities were found inside the sample. This can be proven from the morphology analysis where a vast amount of porous, voids and cracks can be found there. When the water absorption rate is increase, it will affect the density of the sample thus it will also affect the strength of the sample. This is why the compression test of the sample weight percentage of 0 wt%- 10 wt% were at lowest. However, during the water absorption rate at 30 wt% it had showed that the trend was drastically declined. This is due to the formation of pores became smaller and the water absorption rate became slower. Water absorption is a key factor affecting the durability of pavement brick [24]. The geopolymerisation process took place effectively during the reaction thus making the strength of the sample increased. The increase in density will attribute the particle size distributions and affected the compounds particles became denser and more reasonable in decreasing the porosity and the pore sizes inside the geopolymer [25].

4. Conclusion

The comparison between the microstructure, chemical composition, compressive test, density test and water absorption test have been successfully evaluated throughout this research. The influence of adding the fly ash into the bottom ash geopolymer can refine the microstructure and improve the strength of the geopolymer as long as the mixtures were homogenously mixed according to the parameter. The effects of fly ash to the bottom ash ratio can be seen in many aspects. Fly ash helped to increase the mechanical and physical properties of bottom ash geopolymer in term of strengths, microstructures, surface, density and the water absorption. The highest compressive strength that have been recorded in the research was 38.883 MPa compared to the controlled with 100 wt% of bottom ash geopolymer which was 3.136 MPa. Overall, it can be concluded that 40 wt% of weight percentage of fly ash geopolymer gives better properties and suitable to be used as pavement brick application.

Acknowledgments

The author would like to acknowledge the support from the International Social Project Via Research (INSPIRE) under a grant number of UniMAP/PPPI/9001-00581 from the Universiti Malaysia Perlis. This study was supported by Center of Excellence Geopolymer and Green Technology (CEGeoGTECH) UniMAP and Faculty of Technology Engineering, UniMAP.

References

[1] Fan F, Liu Z, Xu G, Peng H, and Cai C S 2018 *Construction and Building Materials* **160** 66-81.

[2] Yun-ming L, Cheng-yong H, Li L, Ain N, Mustafa M A A, Hussin K 2017 *Construction and Building Materials* **156** 9–18.

[3] Romisuhani A, Mustafa M A A, Kamarudin H, Andrei V S, Mohammed B, Nur Ain J 2016 *IOP Conference Series: Materials Science and Engineering* **133 (1)** 012023.

[4] Hefni Y, Zaher Y A E, and Wahab M A 2018 *Construction and Building Materials* **172** 728–734.

[5] Yahya Z, Abdullah M M A B, Hussin K., Ismail K N, Sandu A V, Vizureanu P, Razak R A 2013 *Revista de Chimie* ***64*(12)** 1408–1412.

[6] Abdullah M M A B, Hussin K, Bnhussain M, Ismail K N, Yahya Z, Razak R A 2012 *International Journal of Molecular Sciences* ***13*(6)** 7186–7198.

[7] Heah C Y, Kamarudin H, Bakri A M M Al, Luqman M, Nizar I K 2011 *Australian Journal of Basic and Applied Sciences* **5(7)** 1026–1035.

[8] Liew Y M, Heah C Y, Li L, Jaya N A, Abdullah M M A B, Tan S J, Hussin K 2017 *Construction and Building Materials* ***156*** 9–18.

[9] Kim H K, and Lee H K 2015 *KSCE Journal of Civil Engineering* **19** 1802–1818.

[10] Latifi N, Marto A, Safuan A, Jacky A R, and Jia L 2015 *Construction and Building Materials* **157** 1074–1083.

[11] Ngohpok C, Sata V, Satiennam T, and Klungboonkrong P 2017 00(0000) 1–8.

[12] Duan P, Yan C, Zhou W, and Luo W 2015 **3** 2261–2269.

[13] Abbas S, Saleem M A, Kazmi S M S, and Munir M J 2017 *Journal of Building Engineering* **14** 7–14.

[14] Ma C, Xie Y, Long G, Chen B, and Chen L 2017 *Construction and Building Materials* **157** 1074–1083.

[15] Wan Mastura W I, Mustafa M A A, Andrei V S , Kamarudin H, Ioan G S, Khairul N I, Aeslina A K, Mohammed B 2014 *Revista de Chimie –Bucharest* **65(11)** 1340-1345.

[16] ASTM C 109 2008 Compressive strength of hydraulic cement mortars *West Conshohocken PA* 1–9.

[17] ASTM C 140 2012 Standard test methods for sampling and testing concrete masonry units and related units *West Conshohocken PA* 1–18.

[18] Chindaprasirt P, Jaturapitakkul C, Chalee W, & Rattanasak U 2009 *Waste Management* **29(2)** 539–543.

[19] ASTM C 618 2012 Standard specification for coal fly ash and raw or calcined natural pozzolan for use *West Conshohocken PA* 44–46

[20] Dakhane A, Tweedley S, Kailas S, Marzke R, and Neithalath N 2017 *Materials and Design* **122** 236–246.

[21] Naganathan S, Yousef A, Mohamed O, and Mustapha K N 2015 *Construction and Building Materials* **96** 576–580.

[22] Wan Mastura W I, Kamarudin H, Mustafa M A A, Aeslina A K 2017 *AIP Conference Proceedings* **1835** 1-5.

[23] Guo X, Shi H, and Wei X, 2017 *Cement and Concrete Composites* **79** 53–61.

[24] Wan Mastura W I, Mustafa M A A, Romisuhani A, Amir N, Che Mohd Ruzaidi G, Masdiyana I 2019 *IOP Conference Series: Materials Science and Engineering* **551 (1)** 012082.

[25] Kumar E M, and Ramamurthy K 2017 *Construction and Building Materials* **156** 1137–1149.

IOP Conf. Series: Materials Science and Engineering **743** (2020) 012030 doi:10.1088/1757-899X/743/1/012030

Optimization of Pretreatment and Enzymatic Hydrolysis of Spent Coffee Ground for the Production of Fermentable Sugar

Lai Shan Jin[1], Midhat Nabil Salimi[1,*] and Syazni Zainul Kamal[1]

[1]School of Bioprocess Engineering, Universiti Malaysia Perlis, Perlis, Malaysia

Email: * nabil@unimap.edu.my

Abstract: The aim of this work was to optimize the condition of pretreatment and enzymatic hydrolysis for high yield of sugar production of spent coffee ground (SCG). Acid and alkaline pretreatment method were compared and the method with more sugar produced was selected. Response surface methodology was use for the analysis of conditions such as concentration of alkali, temperature and weight of SCG. The optimized condition obtained was 0.5% (v/v) of alkali, temperature of 100°C and 5% (w/v) of SCG. Enzymatic hydrolysis was carried out after the optimized condition of alkaline pretreatment. The conditions were pH, temperature and enzyme dosage. The optimized condition obtained was at pH 4.8, 0.01 ml of enzyme and temperature of 55°C.

1. Introduction

Coffee is one of the world's most consumed beverages, it is the second most profitable product in the world [1]. In 2015, 9.1 billion kg of coffee was being consumed by the people around the world [2]. The requirement for coffee is huge. Hence, it produces large quantity of residues called spent coffee ground (SCG) which will cause the environmental pollution. Brewing of coffee and preparation of instant coffee by treating coffee powder with hot water produce the spent coffee grounds. Six million tons of SCG produced annually from the coffee brewing and instant coffee production [3]. SCG contains various type of organic compounds which are carbohydrates, lipids, protein, lignin, ashes and polyphenols.

Pretreatment is an important step for the biochemical conversion of lignocelulosic materials into fermentable sugar. The structure of the lignocellulose which contains cellulose, hemicellulose and lignin is needed to change for the ease of enzyme to convert the carbohydrate polymers into sugar monomers [4]. Acid and alkaline pretreatment are the examples of chemical pretreatment. Acid pretreatment use concentrated or diluted acids as the reagents and it is usually carried out at high temperature [5]. Alkaline pretreatment usually involves calcium, sodium, ammonium and potassium hydroxide. This pretreatment method involves partial solvation of hemicellulose, changing in the structure of lignin and partial decrystallization of cellulose [6]. The other method for converting carbohydrate to fermentable sugar is enzymatic hydrolysis. The advantages of enzymatic hydrolysis are higher yield of fermentable sugar, milder working conditions higher selectivity and cheaper energy costs as compared to chemical pretreatment [7].

2. Materials

2.1 Raw material

Spent coffee ground (SCG) were obtained from McCafé of McDonald's at Kangar, Perlis. The material was dried in an oven at 60°C for 72 hours.

Content from this work may be used under the terms of the Creative Commons Attribution 3.0 licence. Any further distribution of this work must maintain attribution to the author(s) and the title of the work, journal citation and DOI.
Published under licence by IOP Publishing Ltd

2.2 Chemicals and Reagents

Sodium hydroxide, sodium potassium tartrate, glucose monohydrate, sodium hydrogen phosphate and sodium dihydrogen phosphate from HmbG Chemicals. Cellulase, citric acid and sodium citrate from Merck. Dinitrosalicylic acid (DNS) reagent from ALDRICH and sulphuric acid from QReC.

2.3 Apparatus and equipment

Erlenmeyer flasks (250 mL), test tubes, UV-Vis spectrophotometer, fume hood, oven, electronic balance, water bath, hot plate, micropipette, centrifuge, pH meter and incubator were used in the research work.

3. Procedures

3.1 Comparison of pretreatment method

Sulphuric acid and sodium hydroxide solution were common acid and alkali used in acid and alkali pretreatment. Working volume of 10ml was prepared by mixing 10% (w/v) of dried SCG with 1% (v/v) of sulphuric acid (H2SO4)/sodium hydroxide (NaOH) solution and put in the water bath of 60° C for 60 minutes [8]. Pretreated solid and liquid were separated by centrifuging at 8000 rpm for 5 minutes [9], whereas the pretreated SCG was washed with distilled water [10] to remove insoluble matter and the supernatant was used for sugar analysis by DNS method.

3.2 Reducing sugar analysis by DNS method

Sufficient amount of the mixtures obtained from the pretreatment and enzymatic hydrolysis were placed in test tubes. After that, 0.5 ml of DNS reagent was added and mixed well. Then, the test tubes were placed in boiling water bath for 5 minutes and was cooled to room temperature by washing through tap water. About 4.5 ml of distilled water was added into each test tube. Sufficient amount of solution was taken out from the test tubes and was put in cuvettes. The cuvettes were placed in UV-Vis spectrophotometer at absorbance of 540nm [11].

3.3 Optimization of alkaline pretreatment method

Alkaline pretreatment method was chose as it produced more fermentable sugar when compared to acid pretreatment method. Three parameters which were concentration of alkali, concentration of SCG and temperature were considered. For alkali pretreatment method, the range of studied parameters were 0.5-5% (v/v) of sodium hydroxide [12], 5-20% (w/v) of SCG weight [8] and temperature (50-100° C) [12]. For studying effect of temperature, water bath was used. Working volume of 10ml was prepared and put in the water bath of 60° C for 60 minutes [8].

The following steps were the same in the method of conducting alkaline pretreatment in comparing the pretreatment methods. These parameters was ran by using Central Composite Design (CCD). The three factors resulted experiment runs for the optimization of selected pretreatment method. Table 1 showed the experimental design matrix generated by DOE software. There was 20 runs of the experiments and the respond obtained was the sugar concentration.

Table 1. Optimization of alkaline pretreatment's parameters by using CCD method.

Run	Parameters			Sugar concentration (mg/ml)
	Concentration of alkali (%)	Temperature (°C)	Weight of SCG (%)	
1	4.09	60.1	8.04	
2	4.09	60.1	16.96	
3	4.09	89.9	8.04	
4	2.75	100.0	12.50	
5	2.75	75.0	12.50	
6	2.75	75.0	12.50	
7	2.75	75.0	20.00	
8	1.41	60.1	16.96	
9	1.41	89.9	16.96	
10	0.50	75.0	12.50	
11	2.75	75.0	12.50	
12	4.09	89.9	16.96	
13	2.75	75.0	12.50	
14	5.00	75.0	12.50	
15	2.75	75.0	12.50	
16	1.41	89.9	8.04	
17	2.75	50.0	12.50	
18	2.75	75.0	12.50	
19	2.75	75.0	5.00	
20	1.41	60.1	8.04	

3.4 Citrate buffer preparation

Concentration of 0.1M solution of citric acid and 0.1M solution of sodium citrate were prepared by dissolving 21.01g of citric acid and 29.41g of sodium citrate in 1L of distilled water respectively. The volume of solution needed to prepare citrate buffer with different pH was shown in Table 2. The total volume was made up to 100ml by adding of distilled water [13].

Table 2. Volume of solutions needed for preparing different pH of citrate buffer.

Citric acid solution (ml)	Sodium citrate solution (ml)	pH
33.0	17.0	4.0
31.5	18.5	4.2
28.0	22.0	4.4
25.5	24.5	4.6
23.0	27.0	4.8
20.5	29.5	5.0
18.0	32.0	5.2
16.0	34.0	5.4
13.7	36.3	5.6
11.8	38.2	5.8
11.8	38.2	5.8
9.5	41.5	6.0
7.2	42.8	6.2

3.5 Sodium phosphate buffer preparation

Concentration of 0.2M solution of sodium dihydrogen phosphate and 0.2M solution of sodium hydrogen phosphate were prepared by dissolving 27.8g of sodium dihydrogen phosphate and 71.7g of sodium hydrogen phosphate in 1L of distilled water respectively. The volume of solution needed to prepare sodium phosphate buffer with different pH was shown in Table 3. The total volume was made up to 200ml by adding of distilled water [14].

Table 3. Volume of solutions needed for preparing different pH of sodium phosphate buffer.

Sodium dihydrogen phosphate solution (ml)	Sodium hydrogen phosphate solution (ml)	pH
77.5	22.5	6.3
73.5	26.5	6.4
68.5	31.5	6.5
62.5	37.5	6.6
56.5	43.5	6.7
51.0	49.0	6.8
45.0	55.0	6.9
39.0	61.0	7.0
33.0	67.0	7.1
28.0	72.0	7.2
23.0	77.0	7.3
19.0	81.0	7.4
16.0	84.0	7.5
13.0	87.0	7.6
10.5	90.5	7.7
8.5	91.5	7.8
7.0	93.0	7.9
5.3	94.7	8.0

3.6 Optimization of enzymatic hydrolysis

The optimized condition from the selected pretreatment method was carried out and the mixture obtained was filtered and washed with distilled water to remove the alkali. The filtered SCG was dried in oven at 60° C for 24 hours. Working volume of 15ml was prepared by mixing dried SCG with 10ml of citrate buffer or sodium phosphate buffer and distilled water. Cellulase was used for enzymatic hydrolysis. The solutions was filled in test tubes and the test tubes were put in incubator for 24 hours [15]. The parameters were 0.01-0.03 ml of cellulase [16], pH (4-8) and temperature (35-55° C) [17]. The pH is adjusted during the mixing of the buffer with solutions and the cellulase was added after the pH was adjusted. These parameters were run by using Central Composite Design (CCD). Table 4 showed the experimental design matrix generated by DOE software. There was 20 runs of the experiments and the respond obtained was the sugar concentration.

Table 4. Optimization of enzymatic hydrolysis's parameters by CCD method.

Run	Parameters			Sugar concentration (mg/ml)
	pH	Enzyme dosage (ml)	Temperature (°C)	
1	7.2	0.026	50.9	
2	6.0	0.020	55.0	
3	6.0	0.020	45.0	
4	4.8	0.026	50.9	
5	6.0	0.020	45.0	
6	6.0	0.010	45.0	
7	4.0	0.020	45.0	
8	4.8	0.014	50.9	
9	6.0	0.020	45.0	
10	6.0	0.030	45.0	
11	6.0	0.020	35.0	
12	4.8	0.014	39.1	
13	7.2	0.014	50.9	
14	7.2	0.014	39.1	
15	7.2	0.026	39.1	
16	8.0	0.020	45.0	
17	6.0	0.020	45.0	
18	6.0	0.020	45.0	
19	6.0	0.020	45.0	
20	4.8	0.026	39.1	

4. Results and Discussion

4.1 Pretreatment Method Selection

Acid and pretreatment method were compared and Figure 1 showed the comparison of two pretreatment method investigated in the work. It was observed that alkaline pretreatment had a better result of sugar production which is 1.438 mg/ml while sugar production of acid pretreatment obtained was 0.453 mg/ml. The result obtained agreed with the previous work carried out by [18]. Alkaline pretreatment was better than acid pretreatment as it was more effective in breaking the bonds between the lignocellulose [19].

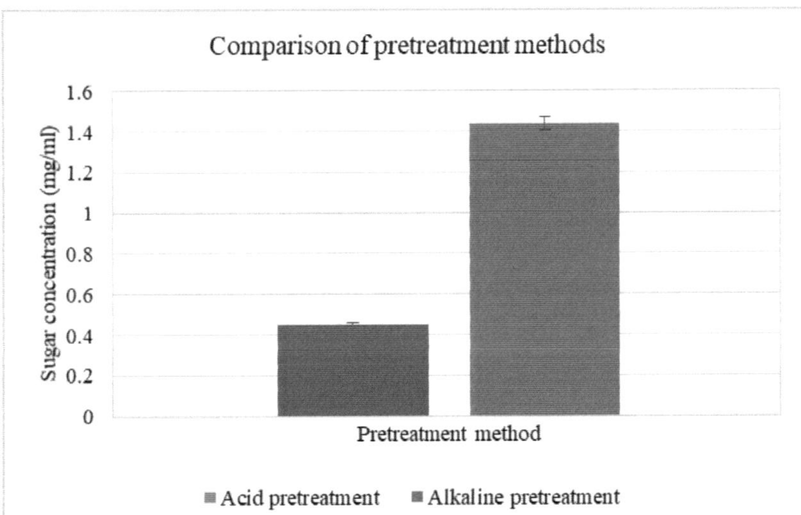

Figure 1. Bar chart of comparison of two pretreatment methods.

4.2 Optimization of parameters of alkaline pretreatment

CCD was used to find the optimum condition of the alkali pretreatment of SCG. Table 5 shows the sugar concentration obtained after alkaline pretreatment with different combination between parameters (concentration of alkali, temperature and weight of SCG) had been carried out. It can be seen that the highest sugar concentration obtained was 48.32 mg/ml at 4.09% (v/v) sodium hydroxide, temperature of 60.1˚ C and 16.96% (w/v) of weight of SCG. The calculation of sugar concentration was based on the glucose standard curve.

Table 5. Data obtained for the response of alkaline pretreatment based on the CCD matrix.

Run	**Parameters**			**Sugar concentration (mg/ml)**
	Concentration of sodium hydroxide (%)	Temperature (˚ C)	Weight of SCG (%)	
1	4.09	60.1	8.04	10.69
2	4.09	60.1	16.96	48.32
3	4.09	89.9	8.04	20.76
4	2.75	100.0	12.50	20.81
5	2.75	75.0	12.50	17.43
6	2.75	75.0	12.50	16.21
7	2.75	75.0	20.00	29.88
8	1.41	60.1	16.96	13.64
9	1.41	89.9	16.96	19.30
10	0.50	75.0	12.50	15.58
11	2.75	75.0	12.50	19.09
12	4.09	89.9	16.96	45.93
13	2.75	75.0	12.50	13.40
14	5.00	75.0	12.50	38.33
15	2.75	75.0	12.50	16.34
16	1.41	89.9	8.04	28.16
17	2.75	50.0	12.50	10.96
18	2.75	75.0	12.50	16.90
19	2.75	75.0	5.00	12.87
20	1.41	60.1	8.04	16.39

Figure 2a shows the 3D surface plot on the interaction of the parameters. This graph shows the interaction between concentration of alkali and temperature at the fixed 12.5% (w/v) weight of SCG. Based on the graph, the optimum region for concentration of alkali was 1.65% to 5% while for temperature was 50°C to 90°C. At concentration of 2.75%, the sugar concentration increased when temperature increased. It was due to removal of lignin increased when the temperature increased [20]. At temperature of 75°C, the sugar concentration increased when concentration of alkali increased as the solubilization of lignin was increased [20]. Figure 2b shows the interaction between concentration of alkali and weight of SCG at the fixed temperature of 75°C. At concentration of 2.75%, the sugar concentration increased when weight of SCG increased. At 12.5% weight of SCG, the sugar concentration increased when concentration of alkali increased as the solubilization of lignin was increased [20]. It could be observed that the sugar concentration increased when the concentration of alkali was increased from 0.5% (v/v) to 5% (w/v) and the weight of SCG was increased from 5% (w/v) to 20% (w/v). Figure 2c shows the interaction between temperature and weight of SCG at the fixed 2.75% (v/v) concentration of alkali. The optimum region for weight of SCG was 5% to 12.5% while for temperature was 50°C to 75°C. At temperature of 75°C, the sugar concentration increased when the weight of SCG decreased. On the other hand, at 12.5% weight of SCG, sugar concentration increased when the temperature decreased.

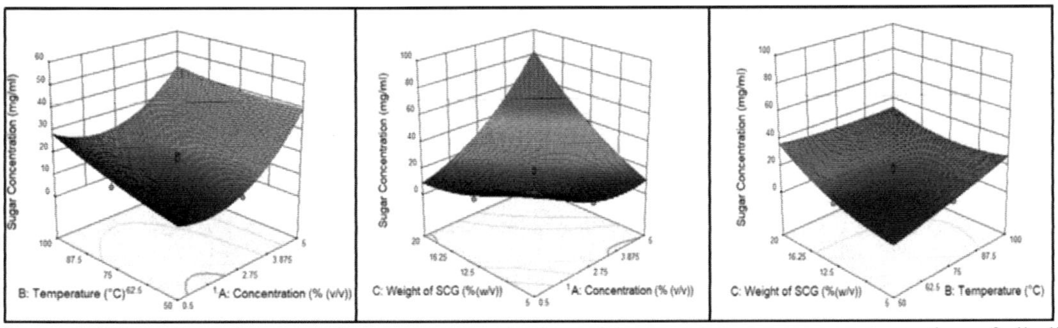

Figure 2. 3D surface plot towards sugar concentration: a) temperature versus concentration of alkali, b) weight of SCG versus concentration of alkali, c) weight of SCG versus temperature.

4.3 Validation of optimization of alkaline pretreatment

Various experiments were carried out on the given optimum condition to validate the model. The optimized condition for alkaline pretreatment was 0.5% (v/v) of alkali, temperature of 100°C and 5% (w/v) of SCG was carried out and performed in triplicate. The results were shown in Table 6. The average sugar concentration yields obtained was 57.0 mg/ml. The value obtained was near to the value predicted by the models which was 58.58 mg/ml of sugar concentration. According to [20], the optimum condition of alkali pretreatment was 90°C. This is near to the temperature of the optimum condition obtained.

Table 6. Validation of the model for optimization of alkaline pretreatment.

Run	Concentration of alkali (%)	Temperature (°C)	Weight of SCG (%)	Sugar concentration (mg/ml)		
				Actual	Predicted	% Error
1	0.5	100	5	56.50	58.58	3.55
2	0.5	100	5	54.78	58.58	6.49
3	0.5	100	5	59.78	58.58	2.04
			Average	57.02	58.58	4.03

4.4 Optimization of parameters of enzymatic hydrolysis

CCD was used to obtain the optimum condition of the enzymatic hydrolysis of SCG. Table 7 shows the sugar concentration obtained after enzymatic hydrolysis with different combination between parameters (pH, temperature and enzyme dosage) had been carried out. It can be seen that the highest sugar concentration obtained was 2.76 mg/ml at pH 4.8, temperature of 50.9° C and 0.026ml of enzyme while the lowest sugar concentration obtained was at the condition at pH 6, temperature of 35° C and 0.020 ml of enzyme. The calculation of sugar concentration was based on glucose standard curve.

Figure 3a shows the 3D surface plot on the interaction of the parameters. The graph shows the interaction between pH and enzyme dosage at the fixed temperature of 45°C. The optimum region for pH was 4.8 to 6 while for enzyme dosage was 0.01ml to 0.023ml. It could be observed that the sugar concentration increased when the enzyme dosage was increased from 0.01 ml to 0.03 ml. The increasing pH from 6 to 8 caused the decrease of sugar concentration. This was because the pH disrupt the charge of enzymes and the enzymes became denatured and unable to carry out chemical reactions [21]. Figure 3b shows the interaction between pH and temperature at the fixed 0.02 ml of enzyme used. The optimum region for pH was 5.4 to 6.4 while for temperature was 35°C to 45°C. At pH 6, sugar concentration increased when the temperature increased [17]. It could be observed that the highest sugar concentration had achieved with increasing of temperature from 35°C to 55°C as the enzyme activity increased when the temperature increased. Decreasing of pH from pH 7.2 to pH 4.8 obtained the highest sugar production as the enzyme became more active when it near to the optimum pH [22]. Figure 3c shows the interaction between enzyme dosage and temperature at the pH 6.0. It could be observed that the highest sugar concentration occurred when the temperature increased from 35°C to 55°C and enzyme dosage increased from 0.01 ml to 0.03 ml as the enzymatic hydrolysis required high enzyme dosage for the production of fermentable sugar [23].

Table 7. Data obtained for the response of enzymatic hydrolysis based on the CCD matrix.

Run	**Parameters**			**Sugar concentration (mg/ml)**
	pH	Enzyme dosage (ml)	Temperature (˚C)	
1	7.2	0.026	50.9	2.62
2	6.0	0.020	55.0	2.33
3	6.0	0.020	45.0	1.41
4	4.8	0.026	50.9	2.76
5	6.0	0.020	45.0	1.23
6	6.0	0.010	45.0	1.36
7	4.0	0.020	45.0	2.06
8	4.8	0.014	50.9	1.85
9	6.0	0.020	45.0	1.31
10	6.0	0.030	45.0	1.47
11	6.0	0.020	35.0	1.15
12	4.8	0.014	39.1	1.51
13	7.2	0.014	50.9	1.57
14	7.2	0.014	39.1	1.51
15	7.2	0.026	39.1	1.16
16	8.0	0.020	45.0	1.73
17	6.0	0.020	45.0	1.19
18	6.0	0.020	45.0	1.35
19	6.0	0.020	45.0	1.30
20	4.8	0.026	39.1	1.44

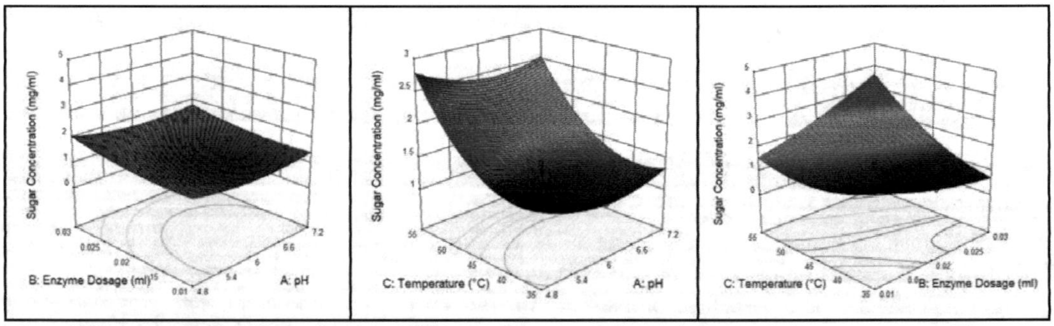

Figure 3. 3D surface plot towards sugar concentration: a) enzyme dosage versus pH, b) temperature versus pH, c) temperature versus enzyme dosage.

4.5 Validation of optimization of enzymatic hydrolysis

Various experiments were carried out on the given optimum condition to validate the model. The optimized condition for enzymatic hydrolysis was at pH 4.8, 0.01 ml of enzyme and temperature of 55°C was carried out and performed in triplicate. The results were shown in Table 8. The sugar concentration yields obtained was 1.824 mg/ml. The value obtained was near to the value predicted by the models which was 1.878 mg/ml of sugar concentration. From previous study, enzyme cellulase was found to be most active at temperature of 55°C and at pH 5.5 [17]. The optimum condition obtained was close to the result from the research.

Table 8. Validation of the model for optimization of enzymatic hydrolysis.

Run	pH	Enzyme dosage (ml)	Temperature (°C)	Sugar concentration (mg/ml)		
				Actual	Predicted	% Error
1	4.8	0.01	55	1.789	1.878	4.74
2	4.8	0.01	55	1.853	1.878	1.33
3	4.8	0.01	55	1.831	1.878	2.50
			Average	1.824	1.878	2.86

5. Conclusion

As a conclusion, the alkaline pretreatment method is better than acid pretreatment as it produced more fermentable sugar. RSM with CCD was successfully implemented to optimize the conditions of the process parameters (concentration of alkali, temperature and weight of SCG) for alkaline pretreatment and process parameters (temperature, pH and enzyme dosage) for enzymatic hydrolysis. The optimized condition obtained was 0.5% (v/v) of alkali, temperature of 100°C and 5% (v/v) of SCG for a maximum response of 57.02 mg/ml of sugar concentration while the optimized condition for enzymatic hydrolysis was at pH 4.8, 0.01 ml of enzyme and temperature of 55°C for a maximum response of 1.824 mg/ml of sugar concentration.

References

[1] Tucker, C. M., *Coffee Culture: Local Experiences, Global Connections*. 2nd edition, Routledge, New York. (2011).

[2] International Coffee Organization (2016). *Coffee Market Report*. Retrieved September 30, 2018, from https://icocoffeeorg.tumblr.com/post/148833167985/ coffee-market-consolidates-price-rises-but-robusta.

[3] Tokimoto, T., Kawasaki, N., Nakamura, T., Akutagawa, J., & Tanada, S. Removal of lead ions in drinking water by coffee grounds as vegetable biomass. *Journal of Colloid and Interface Science*, 281, 56–61. (2005).

[4] Mosier, N., Wyman, C., Dale, B., Elander, R., Lee, Y. Y., Holtzapple, M., & Ladisch, M.,. Features of promising technologies for pretreatment of lignocellulosic biomass. *Bioresource Technology*, 96, 673–686. (2005).

[5] Badiei, M., Asim, N., Jahim, J. M., & Sopian, K. Comparison of chemical pretreatment methods for cellulosic biomass, *International Conference on Chemical, Biological and Environmental Engineering*, 9, 170–174. (2014).

[6] Ibrahim, M. M., El-Zawawy, W. K., Abdel-Fattah, Y. R., Soliman, N. A., & Agblevor, F. A. Comparison of alkaline pulpine with steam explosion for glucose production from rice straw. *Carbohydrate Polymer*, 83, 720-726. (2011).

[7] Yang, B., Dai, Z.Y., Ding, S.Y., & Charles, E.W. (2011). *Enzymatic hydrolysis of cellulosic biomass*. Retrieved December 20, 2018, from https://scholars.opb.msu.edu/en/publications/enzymatic-hydrolysis-of-cellulosic-biomass.

[8] Kwon, E. E., Yi, H., & Jeon, Y. J. Sequential co-production of biodiesel and bioethanol with spent coffee grounds. *Bioresource Technology*, 136, 475–480. (2013).

[9] Foerster, S., Al-hamamre, Z., Foerster, S., Hartmann, F., Kröger, M., & Kaltschmitt, M. (2012). Oil extracted from spent coffee grounds as a renewable source for fatty acid methyl ester manufacturing Oil extracted from spent coffee grounds as a renewable source for fatty acid methyl ester manufacturing, *Fuel*, 96(June), 70–76.

[10] Bensah, E. C., & Mensah, M. Chemical pretreatment methods for the production of cellulosic ethanol : Technologies and innovations, *International Journal of Chemical Engineering*, 13, 1-21. (2013).

[11] Barnfeld, P. *Methods in Enzymology*, New York, NY: Elsevier. (1955).

[12] Juarez, G. F. Y., Pabilo~na, K. B. C., Manlangit, K. B. L., & Go, A. W Direct dilute acid hydrolysis of spent coffee grounds: a new approach in sugar and lipid recovery. *Waste Biomass Valorization*, 9(2) 235-246. (2017).
[13] Lillie, R. D. (1948). *Histopathologic Technique*. Blakiston. Philadelphia, Pennsylvania.
[14] Sorensen, S. P. L. Enzyme studies II. The measurement and meaning of hydrogen ion concentration in enzymatic processes. *Biochemische Zeitschrift*, 21, 131. (1909).
[15] Jooste, T., García-Aparicio, M. P., Brienzo, M., Van Zyl, W. H., & Görgens, J. F. Enzymatic hydrolysis of spent coffee ground. *Applied Biochemistry and Biotechnology*, 169(8), 2248–2262. (2013).
[16] Verhagen, S.. *Improving the Value of Spent Coffee Grounds by Converting Carbohydrates into Fermentable Sugars*. Retrieved January 13, 2019, from http://edepot.wur.nl/441154 (2018).
[17] Nataraja, S., Chetan, D. M., & Krishnappa, M. Effect of temperature on cellulose enzyme activity in crude extracts isolated from solid wastes microbes. *International Journal of Microbiology Research*, 2(2), 44–47. (2010).
[18] Chantorn, S., Aonkampa, T., Phannachet, K., & Apiraksakorm, J. Enhancement of the mannanase digestibility of various agricultural wastes by sodium hydroxide pretreatment. *In: International Conference on Microbial Taxonomy, Basic and Applied Microbiology*, 59. (2012).
[19] Taherzadeh, M. J., & Karimi, K. Pretreatment of lignocellulosic wastes to improve ethanol and biogas production: a review. *International Journal of Molecular Science*, 9, 1621–51. (2008).
[20] Ao, L. T., & En, J. R.. Effects of liquid-to-solid ratio and reaction temperature on NaOH pretreatment of *Achnatherum splendens*, *Asian Journal of Chemistry*, *25*(7), 3545–3548. (2013).
[21] Ariffin, H., Abdullah, N., Umi, M. S., Shirai, Y., & Hassan, M. A.. Production and characterization of cellulase by *Bacillus pumilus EB3*. *International Journal of Engineering and Technology*, 3, 47–53. (2006).
[22] Daniel, R. M., & Danson, M. J.. Temperature and the catalytic activity of enzymes: A fresh understanding. *Federation of European Biochemical Societies (FEBS) Journal*, 587(17), 2738–2743. (2013).
[23] Khaw, T. S., & Ariff, A. B. Optimization of enzymatic saccharification of palm oil mill effluent solid and oil palm fruit fibre to fermentable sugars. *Journal of Tropical Agriculture and Food Science*, 37(1), 85–94. (2009).

IOP Conf. Series: Materials Science and Engineering **743** (2020) 012031 doi:10.1088/1757-899X/743/1/012031

Evaluation and Optimization of Genomic DNA Extraction from Food Sample for Microfluidic Purpose

M. N. Afnan Uda[1, *], N. A. Parmin[2], Asral Bahari Jambek[1], U. Hashim[1, 2], M. N. A. Uda[3] and S. N. A. Shaharuddin[2]

[1]School of Microelectronic Engineering Group, Universiti Malaysia Perlis (UniMAP) Putra Campus, 02600 Pauh, Perlis Malaysia
[2]Institute of Nano Electronic Engineering (INEE), Universiti Malaysia Perlis (UniMAP), 01000 Kangar Perlis Malaysia
[3]School of Bioprocess Engineering Group, Universiti Malaysia Perlis (UniMAP) 02600 Arau Perlis

Email: * nurafnan92@yahoo.com

Abstract. Contamination of various food samples became one of the critical issues in food pathogen infection. Food pathogen can be detected by using digital polymerase chain reaction (PCR) and sequencing. These methods were reliable but consuming and take a longer time for detection. The present work describes the innovation to develop a technology to extract double-stranded deoxyribonucleic acid (dsDNA) from food samples and then denatured dsDNA into and single-strand DNA (ssDNA) for further use on the chip using microfluidic device. Microfluidic device is a lab-on-chip device that consist of microfluidic channels that provide paths for biomolecules to flow to individual point of care. DNA extraction is the process by which DNA is separated from proteins, membranes, and other cellular material contained in the cell from which it is recovered. Lysis solution is used in the process of extraction the DNA to break up the cells containing DNA from protein and other cellular materials. This extraction firstly be done in the most labour-intensive in obtaining the DNA biomolecules. Extraction methods may require an overnight incubation, may be a protocol that can be completed in minutes or a couple of hours by using a commercial kit. The disadvantages of the laboratory and commercial kit is due to time-consuming, poor cost-effectiveness, the need to use big laboratory and a complicated process which need an expertise to conduct the experiment and interpret the data. This research is proposed to design and fabricate a microfluidic device that has DNA extraction capabilities. In this research DNA extraction using a commercial kit will be used as a comparison for the quality of the result. The microfluidic device can be used in health care delivery system and will help the doctors in diagnostic process to identify disease of a patient rapidly. Other than that, the output extracted from microfluidic device will be used for DNA probe target interaction for diagnostic kit. The major advantage of microfluidic device is that it consumes less time compared to the conventional chemical methods.

1. Introduction

Nowadays, there are many pressing need in science fields especially in the medical fields and forensics community that seek for a sensitive, robust, and reliable integrated method of DNA extraction that is rapid, cost effective, and neither labour nor space intensive [1]. DNA extraction is a process of purification of DNA from sample through chemical and physical methods [2-6]. The DNA extraction purposes is to separate it from proteins, membrane and other cellular materials. The purification of DNA

Content from this work may be used under the terms of the Creative Commons Attribution 3.0 licence. Any further distribution of this work must maintain attribution to the author(s) and the title of the work, journal citation and DOI.
Published under licence by IOP Publishing Ltd

is to make sure that a pure DNA biomolecule is obtained during this process. In the past, the process of extraction and purification of nucleic acids used to be complicated, time-consuming, labour-intensive and limited in terms of overall throughput [7]. The first DNA extraction process was done in the chemical laboratory where they needed to use a lot of equipment and go through complex procedures to get DNA biomolecule. This method takes three days for the result to be obtained including the culturing process. Then, the DNA extraction kit was manufactured and takes one hour for the doctor to obtain the result. DNA extraction method was improved to modify approach that takes less those 20 minutes [8]–[10]. Due to time-consuming problem in obtaining the result, this optimization was used to accelerate the time of result attainment by creating a lab-on-chip device [11]. The lab-on-chip device is an idea of a microfluidic chip with a pattern of micro channel which is moulded or engraved which is shown in figure 1. On the chip, there are pathways that allows liquids to flow through. This microfluidic chip provide paths for biomolecules to flow to the point of care that will detect the result of the fluids [12]. Microfluidic device is only taking a few minutes for the users to attain the result.

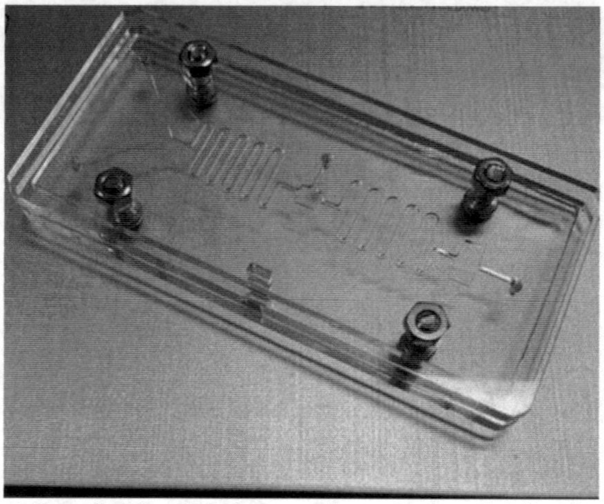

Figure 1. Microfluidic chip that consists of microfluidic channels.

In this study, developing a PDMS microfluidic device with DNA extraction capabilities to extract a single-stranded DNA to obtain rapid and quality result of DNA biomolecules is the main existential of this research. In order to attain the single-stranded DNA, process of mixing, homogenising, incubation, centrifuge and precipitation is being repeated until the outcome was obtained. Single-stranded DNA is obtained from double-stranded DNA. The process is done when using the commercial kit DNA extraction to obtain the single-stranded DNA. However, as micro and nanotechnology have been developed, integration of a single micro device such as the lab-on-chip device are suitable approach to be used for sampling and monitoring as well as mixing, reaction, incubation and analysis of small volumes of fluids which will give a rapid result [4]. The design of the microfluidic is done by AutoCAD software and fabrication of microfluidic by using PDMS.

2. Material and methods

2.1 Chemicals, Reagents and Equipment

DNA extraction process using DNA extraction kit is done by using micropipette, mixer, shaker, centrifuge, nanodrop spectrophotometer and UV-Vis Spectrophotometer. The chemicals that been used are lysis buffer, buffer AW1 and AW2 and Elution Buffer AE. Buffer AW1 and AW2 serve as a washing agent for the DNA which will make sure that it is clean from other bimolecular materials. Meanwhile lysis buffer separate protein from the sample DNA. Elution Buffer AE act as a binding agent to make sure that the DNA is bind together. For fabrication process of microfluidic, there are synthetic materials that been used which are acetone, SU-8 photoresist, Isopropanol (IPA) and glass. Acetone and IPA serve in substrate cleaning to eliminate contaminants and particles from the surface of the glass. SU-8 is an average utilized epoxy-based negative photoresist which is used to make design with high aspect ratio. It is utilized as a high-resolution mask in fabrication process where it used most in fabrication of microfluidic and MEMS parts.

2.2 DNA Extraction Methods

DNA extraction process can be done using multiple types of kits. The kits that are used during extracting the DNA is commercial extraction kit. This kit takes about one to two hours to obtain the single-stranded DNA. The main function of DNA extraction is to obtain the single-stranded DNA. The samples that been used such as milk, yogurt drink, yakult and other liquid food samples. The DNA that have been extracted is then been tested the purity by using NanoDrop spectrophotometer and UV-Vis spectrophotometer. DNA was extracted by preparing the sample be. If the sample is in liquid form, it would be easy to extract DNA from the sample. However, if it is solid forms such as rice or chicken, the sample needs to be ground until the liquid comes out from the food. Next, is the mixing process between lysis and sample. The lysis buffer is used to isolate protein from the molecular cells. The lysis buffer that is used in the mixing process with the sample is NaOH. Then quickly incubate the sample at 65°C with 500 rpm. The main purpose of this process is to break up connective tissue holding the cell together. This will make them to be easier to lyse. It will also inactivate DNAs and enzymes that would degrade and inhibit PCR. After that, the sample will undergo cleaning process which use washing buffer that helps in purification of DNA and elimination of cellular debris. Then, the extraction process continues with elution process. Elution is a process of extracting material for other material by washing it with a solvent. The washing agent that is used in the extraction process is AE Buffer solution. The solution will help in removing the DNA from the matrix. This process is done by using the spin column tube which the tube will be span in the centrifuge that will eventually pull the elution buffer through the matrix. After elution, the double-stranded is converted to single-stranded by adding a ratio of 1:1 of the double-stranded DNA of the sample and 0.01M of NaOH. The single-stranded DNA of the sample is then stored inside a fridge at -20°C.

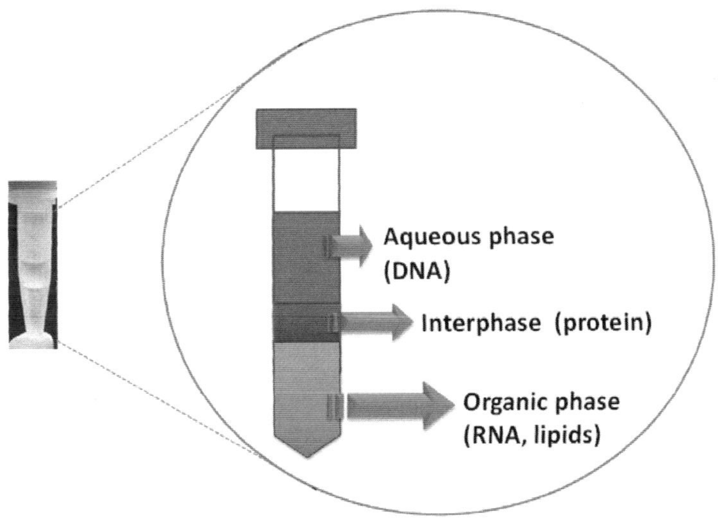

Figure 2. DNA extraction and separation process.

2.3 Microfluidic Device Fabrication

A few steps are included in the production of the microfluidic gadgets. Sample preparation is the first step in the production of microfluidic [8-10]. The result and technique used in this production process will be influenced by this development. The manufacturing process is carried out with a microscopic glass in this project. The 75 mm x 25 mm x1 mm microscopic glasses are used as shown in Table 1.

IOP Conf. Series: Materials Science and Engineering **743** (2020) 012031 doi:10.1088/1757-899X/743/1/012031

Glass was used as the foundation of the microfluidic device. The glass must be microscopically cleaned. The glass is cleaned from the glass surface with Isopropanol (IPA) and distilled waters (DI water).

Table. 1 Specification of Microfluidic Chip.

Properties	Specification
Microfluidic glass	23.00 mm x 31.67 mm
Channel 1 width	0.5 mm
Channel 2 width	0.8 mm
Inlet radius	1 mm
Outlet width	1.6234 mm
Point of care width	5.3202 mm

After the cleaning process is completed, the next step is resistant coating. In this process, the thin SU-8 layer is spread across the glass surface and the grip of the entire SU-8 laying onto the glass surface is enhanced. On the spinning coater, the glass plate is attached. In the center of the glass plate, about 1 ml SU-8 is then dropped. Spin velocity is set for 30 seconds to 1000rpm and for 10 seconds to 50rpm. The sample is then softly baked at temperature of 95°C for 45 minutes using a hot plate after spinning coating with SU-8 3050. During the soft baking process, the solvent level of SU-8 layer is reduced, and thus diminishes the risk of exposed resist loss, swelling and the adhesion defects. The sample remains on the cold plate for 20 minutes. Then, the designation mask was exposed to UV light. The exposed region remained after the development since negative photo resistance is employed in our cases. The exposure time is 55 sec . The glass slide is then moved to the warm plate for 45 minutes to bake hard at 95°C. The glass diaphragm is then developed. The sample is developed using the developer solution that is SU-8 developer. The purpose of the process is to remove the undesirable resistance and shape the pattern. The rinsing and drying process is followed by SU-8 developers, Isopropanol (IPA) and Distilled Water so that the process is not developed after removal from the glass surface. The rest of the pattern on the glass slide is referred to as the microfluidic mold used in the following process. The form SU-8 received from the last step is used as a form in soft lithography following the development procedure. A beaker combines a PDMS solution with a PDMS cure agent. In order to dissolve completely, the PDMS blend is blended together for 10 minutes. The PDMS is cured during the blending of PDMS and cure agent by organometallic cross linkage process. The PDMS and the cure operator should be set to 10:1. The expansion of the PDMS elastomer will result in higher and low polyvalence. The solution is then fed into the SU-8 mold inside the self-designed box of the case as seeded in Figure 2. The procedure is to obtain the desired microfluidic instrument shape and measurement. The next step is to vacuum the self-designed box inside the vacuum chamber to evacuate the increases within the mix. When the mixture is vacuum, the vacuum chamber temperature is increased by 65°C and the mixture remains within the chamber for 30 minutes. The PDMS elastomer layer after soft lithography is deliberately removed from the container and cut into pieces.

2.4 DNA Extraction using Fabricated Microfluidic Device

DNA extraction process by using microfluidic device does have a mixing process that is same with the DNA extraction process using the DNA extraction commercial kit. The difference is the time taken to obtain the double-stranded DNA and reducing the equipment that needed to be used during the DNA extraction process. Mixing the lysis buffer to samples of food is the first mixing process. This procedure is performed for cell disruption and cellular content extraction. The second process to obtain a single beam DNA is to mix NaOH solution into a double beamed DNA. This mixing process is done by simultaneously drop the sample and lysis solution into the first and the second inlet of the Microfluidic device. The double-stranded DNA is obtained at the first output and been tested using UV-Vis and NanoDrop. Then, the double-stranded DNA is converted to single-stranded DNA by dropping a 0.01M of NaOH at the third inlet of the Microfluidic device.

IOP Conf. Series: Materials Science and Engineering **743** (2020) 012031 doi:10.1088/1757-899X/743/1/012031

3. Result and Discussion

3.1 DNA Extraction using Commercial Kit

DNA quantification by using NanoDrop spectrophotometer was used to quantify the concentration the specimen of the nucleic acid since the efficiency of downstream reactions rely on the concentration of these nucleic acids. DNA can be quantified from several types of specimens such as genomic DNA, PCR product, plasmid DNA. There are two main methods which the extracted DNA amount can be determined. DNA yield can be assessed using four different methods which are absorbance (optical density), agarose gel electrophoresis, fluorescent DNA binding dyes and a luciferase-pyro phosphorylation-coupled quantitation system. In this research, absorbance method was used to obtain the results.

After the entire extraction process was completed, double-stranded DNA was obtained. It took between one and a half hours to obtain the double stranded DNA. However, the ssDNA only took five minutes processing where the dsDNA was mixed with a 0.01μM of NaOH solution with 1μL:1μL ratio to denature the dsDNA. There are 24 samples of dsDNA and ssDNA that been obtained from DNA extraction process using commercial Kit. The 24 samples is obtained from 6 fluid samples which one fluid sample have 4 output which is a pure sample double-stranded DNA, a sample mix with synthetic E.coli double-stranded DNA, a sample mix with synthetic Listeria double-stranded DNA and a sample mix with synthetic Salmonella double-stranded DNA. The sample is tested using NanoDrop Spectrophotometer. The result was plotted in a graph as shown in Figure 3.

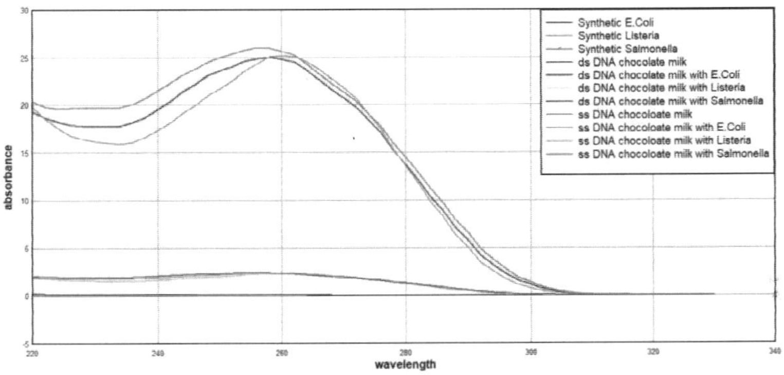

Figure 3. Synthetic samples, double-stranded DNA sample and single-stranded DNA sample.

Figure 3 shows that there are huge different between the sample that is not added with synthetic samples and the samples that is added with synthetic samples. DNA concentration and purity both can be determined by measuring the absorption of ultraviolet light. DNA absorbs UV more or less strongly depending upon the wavelength. NanoDrop spectrophotometer, will takes measurements at wavelengths of 260 and 280 nm, and in addition determines an absorption spectrum from 220 – 330 nm. The graph shows that all the synthetic samples have a high peak on the 260 wavelength. This indicate that there are high reading on DNA being detected in the synthetic samples. 260/280 - ratio of absorbance at 260 nm and 280 nm. The ratio of absorbance at 260 and 280 nm is used to assess the purity of DNA and RNA. A ratio of ~1.8 is generally accepted as "pure" for DNA. If the ratio is appreciably lower in either case, it may indicate the presence of protein, phenol or other contaminants that absorb strongly at or near 280 nm.

Table 2. Double-stranded and single-stranded DNA purity and concentration.

Samples	Double-Stranded DNA		Single-Stranded DNA	
	Concentration(ng/μL)	Purity	Concentration(ng/μL)	Purity
Synthetic *E.coli*	-	-	819.763	1.81
Synthetic Listeria	-	-	832.265	1.85
Synthetic Salmonella	-	-	847.502	1.77
Chocolate Milk	3.927	2.07	1.966	2.28
Chocolate Milk with *E.coli*	3.628	2.12	75.767	1.83
Chocolate Milk with Listeria	5.704	2.02	75.984	1.81
Chocolate Milk with Salmonella	3.696	2.26	77.837	1.78

Table 2 shows the concentration and purity of double-stranded DNA and single-stranded DNA. Based on the sample, the purity of the DNA could be determine. This is to make sure that the DNA is totally free from other biomolecule such as protein. In the table, most of the samples range between 1.8-2.0 which was still in the range of a clear DNA.

3.2 DNA Purity and Concentration Calculation

The concentration of the extracted DNA was estimated spectrophotometrically by using UV/Visible spectrophotometer Ultraspec 2100 pro [16]. Spectrophotometric assay was used to determine the quantity and purity of the DNA. The quantitation of DNA was done by taking absorbance at 260 nm, 280nm and 320 nm[17]. The ratio of absorbance at 260 nm and 280 nm were calculated using formula 2. The value of pure DNA sample should be in the range of 1.8 to 2.0. A lower ratio is an indication of protein contamination. DNA solution with A260 nm contains approximately 50 μg/mL of single-stranded DNA. Thus, the concentration of DNA was calculated according to the formula 1.

$$\text{Concentration of DNA } \mu g/mL = (A_{260} - A_{320}) \times dilution\ factor\ (df) \times 50\frac{\mu g}{ml} \quad (1)$$

$$\text{DNA Purity} = \frac{A_{260} - A_{320}}{A_{280} - A_{320}} \quad (2)$$

4. Conclusion

This research goal has been successfully achieved by developing a double-stranded and single-stranded chip microfluidic device. The effort is made by using PDSS with a lab glass slide as the base for mouth to design and manufacture a Microfluidic Chip for a lab on chip devices. The device functional test shows that the sample can be extracted using the Microfluidic Chip according to the path designed. Therefore, as set out in Chapter 1, the research has effectively met three objectives. The first goal is achieved by successfully designing and producing the Microfluidic device using the PDMS technology in soft lithography. Evaluation of single-stranded DNA and double-stranded DNA is done using NanoDrop spectrophotometer and UV-Vis spectroscopy which lead to the achievement to the second and third objectives.

Acknowledgement

The author would like to thank all staff members of the Institute of Nanoelectronic Engineering in University Malaysia Perlis (UniMAP) for their technical advice and contributions, directly and indirectly. Collaborative Research in Science and Technology Centre (CREST) is acknowledged for providing grant for this research program.

Reference

[1] Azizah N., Hashim U., S. C. B. Gopinath, and S. Nadzirah, "A direct detection of human papillomavirus 16 genomic DNA using gold nanoprobes," *Int. J. Biol. Macromol.*, vol. 94, pp. 571–575, 2017.

[2] A. Ayoib, U. Hashim, S. C. B. Gopinath, and M. K. Md Arshad, "DNA extraction on bio-chip: history and preeminence over conventional and solid-phase extraction methods," *Applied Microbiology and Biotechnology*. 2017.

[3] D. M. Prendergast *et al.*, "Application of PCR for rapid detection and serotyping of Salmonella spp. from porcine carcass swabs following enrichment in semi-solid agar," *Food Res. Int.*, 2012.

[4] N. Azizah, U. Hashim, S. C. B. Gopinath, and S. Nadzirah, "Gold nanoparticle mediated method for spatially resolved deposition of DNA on nano-gapped interdigitated electrodes , and its application to the detection of the human Papillomavirus," 2016.

[5] N. Z. Natasha, R. D. A. A. Rajapaksha, M. N. A. Uda, and U. Hashim, "Electrical DNA Biosensor using Aluminium Interdigitated Electrode for E . Coli O157 : H7 detection," vol. 020235, 2017.

[6] N. A. Parmin *et al.*, "Voltammetric determination of human papillomavirus 16 DNA by using interdigitated electrodes modified with titanium dioxide nanoparticles," *Microchim. Acta*, vol. 186, no. 6, 2019.

[7] S. C. Tan and B. C. Yiap, "DNA, RNA, and Protein Extraction: The Past and The Present," *J. Biomed. Biotechnol.*, vol. 2009, pp. 1–10, 2009.

[8] M. N. A. Uda, R. D. A. A. Rajapaksha, M. N. A. Uda, U. Hashim, and A. B. Jambek, "Selective detection of E.coli O157:H7 bacteria DNA using electrical based aluminium interdigitated electrode biosensor," *AIP Conf. Proc.*, vol. 2045, no. December, 2018.

[9] R. D. A. A. Rajapaksha, M. N. A. Uda, U. Hashim, S. C. B. Gopinath, and C. A. N. Fernando, "Impedance based Aluminium Interdigitated Electrode (Al-IDE) biosensor on silicon substrate for salmonella detection," *IEEE Int. Conf. Semicond. Electron. Proceedings, ICSE*, vol. 2018-Augus, pp. 93–96, 2018.

[10] R. D. A. A. Rajapaksha, U. Hashim, M. N. Afnan Uda, C. A. N. Fernando, and S. N. T. De Silva, "Target ssDNA detection of E.coli O157:H7 through electrical based DNA biosensor," *Microsyst. Technol.*, 2017.

[11] M. G. Mauk, C. Liu, M. Sadik, and H. H. Bau, "Microfluidic devices for nucleic acid (NA) isolation, isothermal NA amplification, and real-time detection," *Methods Mol. Biol.*, 2015.

[12] G. Luka *et al.*, "Microfluidics integrated biosensors: A leading technology towards lab-on-A-chip and sensing applications," *Sensors (Switzerland)*. 2015.

[13] A. K. Au, W. Lee, and A. Folch, "Mail-order microfluidics: Evaluation of stereolithography for the production of microfluidic devices," *Lab Chip*, 2014.

[14] T. Adam, U. Hashim, T. S. Dhahi, and M.N.A.Uda, "Fabrication of PDMS Based Micro Fluidic Devices Fabrication of PDMS Based Micro Fluidic Devices," in *2014 Fifth International Conference on Intelligent Systems, Modelling and Simulation*, 2014, no. January, pp. 717–721.

[15] R.D.A.A Rajapaksha, N. A. N. A. M. N. A. Uda, U.hashim, S.C.B.Gopinath, and C.A.N.Fernando, "Multichannel PDMS microfluidic based nano- biolab-on-a-chip for medical diagnostics," in *AIP Conference Proceedings 2045*, 2018, vol. 020020, no. December, pp. 1–7.

[16] N. Azizah, U. Hashim, S. Nadzirah, M. K. M. Arshad, A. R. Ruslinda, and S. C. B. Gopinath, "Comparison of different methods for extraction and purification of human Papillomavirus (HPV) DNA from serum samples," in *AIP Conference Proceedings*, 2017, vol. 1808.

[17] F. H. Stephenson, "Nucleic Acid Quantification," in *Calculations for Molecular Biology and Biotechnology*, 2016.

IOP Conf. Series: Materials Science and Engineering **743** (2020) 012032 doi:10.1088/1757-899X/743/1/012032

Aluminium Interdigitated Electrode Based Biosensor for Specific ssDNA Target Listeria Detection

M. N. Afnan Uda[1,*], Asral Bahari Jambek[1], U. Hashim[1, 2], M. N. A. UDA[3] and M. A. F. Bahrin[1]

[1]School of Microelectronic Engineering Group, Universiti Malaysia Perlis (UniMAP) Putra Campus, 02600 Pauh, Perlis Malaysia
[2]Institute of Nano Electronic Engineering (INEE), Universiti Malaysia Perlis (UniMAP), 01000 Kangar Perlis Malaysia
[3]School of Bioprocess Engineering Group, Universiti Malaysia Perlis (UniMAP) 02600 Arau Perlis

Email: * nurafnan92@yahoo.com

Abstract. Nowadays interdigitated electrode (IDE) based sensor have stimulated increasing interest in the application of biosensor filed. A large number of finger electrodes as comb structure gain high sensitivity through electrical measurements. In this paper, we have investigated Listeria bacteria detection through the electrical based IDE. Listeria monocytogenes is a food borne pathogen-based bacterium that can cause dangerous disease to human, some infection may result in death. The AutoCAD software was used to design the chrome mask of IDE sensor and the fabrication process was done using conventional photolithography method. The fabricated Al IDE morphologically analyzed using a low power microscope (LPM), a high-power microscope (HPM) and 3D profiler. Functionalization step of the Al IDE, silanization process was done using (3-Aminopropyl) triethoxysilane (APTES), immobilization process was done using carboxylic probe Listeria and Tween-20 as a blocking agent for nonspecific binding on the non-immobilized area of the biosensor surface. The biosensor was validated with complementary, non-complementary and single base mismatch ssDNA targets. Different concentration of complementary ssDNA target from 1 fM to 1 μM was done for the sensitivity detection.

1. Introduction

Listeria monocytogenes is a facultative intracellular pathogen that can causes listeriosis, particularly in pregnant women, new born and adult with weak immune systems. This bacterium is widespread in the environment and infections occur through ingestion of contaminated food [1],[2]. Fever, muscle aches and diarrhea are the symptoms caused by Listeria. If the infections spread to the nervous system, symptom such as headache, stiff neck, confusion, loss balance can occur [3]–[6]. Based on the International Journal of Food Microbiology reported that the prevalence of *Listeria* in raw chicken (60%), raw beef (50%), fresh prawns (44%) and leafy vegetables (22%). It is found that the occurrence of *Listeria* in imported frozen beef was 75%, 30.4% in local beef samples and 12% in fermented fish. Therefore, higher percentage of *Listeria* in food will give dangerous illness to people. Recently, conventional method has been introduced to detect foodborne pathogens illness. Conventional method like Polymerize Chain Reaction (PCR) has already been developed but it required expensive equipment, skilled personal and labor intensive gel-based detection that shows feeble sensitivity and specificity [7]–

Content from this work may be used under the terms of the Creative Commons Attribution 3.0 licence. Any further distribution of this work must maintain attribution to the author(s) and the title of the work, journal citation and DOI.
Published under licence by IOP Publishing Ltd

IOP Conf. Series: Materials Science and Engineering **743** (2020) 012032 doi:10.1088/1757-899X/743/1/012032

[10]. Previously a lot of methods in developing biosensor for modifying sensor surface have been introduced [11]–[13]. However to detect foodborne pathogens disease there is some problem in detecting very small concentration down to fM range concentration of ssDNA target [14]–[16]. In this research, we introduce and develop a sensitive, selective and rapid biosensor using Al IDE to detect ssDNA Listeria bacteria. The major steps of functionalization as biosensors are, silanization using APTES, immobilization using carboxylic ssDNA Listeria probes and blocking the non-immobilized area using Tween-20 [17], [18]. Then, the functionalized biosensor was hybridized with complementary, non-complementary and single base mismatch ssDNA. Fabricated Al IDE is physically characterized using a low power microscope (LPM), high power microscope (HPM) and 3D profiler. The I-V characteristics were performed for every functionalization steps, selectivity, and sensitivity measurements.

2. Material and methods

2.1 Chemical, Reagents and Instruments

The materials Ethanol (C2H5OH), Sodium Hydroxide (NaOH), (3-Aminopropyl) triethoxysilane (APTES) and Deionized water (DI Water) and they were obtained from Sigma Aldrich USA. All of the other chemicals were analytical reagent grade and purchased commercially. Deionized water (Di Water) was used throughout this experiment. The 30-base synthetic oligonucleotides were purchased from AIT Biotech Pvt. Ltd (Singapore). Table 1 shows the Carboxylic modified synthetic probes and target oligonucleotides sequences.

Table 1. 30-mer oligonucleotides sequences for listeria probe, complementary, non-complementary and single base mismatch.

Oligonucleotide	Sequences
30-mer probe	5'-(COOH) CGC GCG GCA TCC GCA TCA ATA ATA CCG GCC –3'
30-mer complementary	(5'- GGC CGG TAT TAT TGA TGC GGA TGC CGC GCG –3')
30-mer non-complementary	(5'- CCG GCC ATA ATA ACT ACG CCT ACG GCG CGC –3')
30-mer single base mismatch	(5'- GGC CGG TAT TAT TGT TGC GGA TGC CGC GCG –3')

The Al IDE mask was designed using AutoCAD Software. After that, the design is transferred to a commercial chrome mask, Silterra (M) Sdn Bhd to fabricate the Al IDE. LPM, HPM, and 3D profiler were used for physical and morphological characterization of bare Al IDE. Lastly, the I-V characterization measurements were completed by using (Keithley 2450), Kickstart software and Probe Station.

2.2 Silanization, Immobilization and Hybridization

First, the active area of Al IDE was washed with NaOH to make hydroxyl layer on the SiO_2 surface. After that, the bare Al IDE surface was functionalized with APTES. 2 µl APTES was dropped on the Al IDE active area. It will build an 'active' layer on the Al IDE. It contains hydroxyl group which attack and displace the alkoxy group on the saline and thus forming a covalent of -Si-O-Si- bond [14-15]. Then, the IDE was incubated for 1 hour inside the dry cabinet and rinsed with DI Water. After that, 2 µl of DNA probe was immobilized on the surface. Form the reaction, it will form a recognition layer of the covalent amide bond between carboxyl group [16-18]. Then, let it to dry for 1 hour inside the dry cabinet. Then, rinsed with DI water to remove unbounded probe DNA on the surface. After that, 2 µl of tween-20 was dropped on to the functionalized area and kept 15 minutes inside the dry cabinet. Remove the Tween-20 by using DI water and left for 5 minutes to dry. Tween-20 was performed as a blocking agent, to repeal unspecific ssDNA target binding on the biosensor surface. Thus, the target ssDNA unable to direct bind with APTES layer. Then, the Al IDE biosensor prepared to be hybridized.The functionalized IDE biosensor was validated using the hybridization process. It was accomplished using different target ssDNA Listeria. The hybridization process was done by synthetic sample of complement, non-complement and mismatch target on the active area of Al IDE's surfaces. 2 µl of DNA targets from 1 fM to 1 µM were dropped on Al IDE biosensor to hybridize with immobilized ssDNA probes for sensitivity measurements. Then, left the hybridization process for 1hour. After that, rinsed the Al IDE surface with DI water to remove unbounded DNA target. The I-V measurement measured nano ampere range changes through the silanization, immobilization, and hybridization process. Figure 1 shows the process preparation step of sensing surface Al IDE.

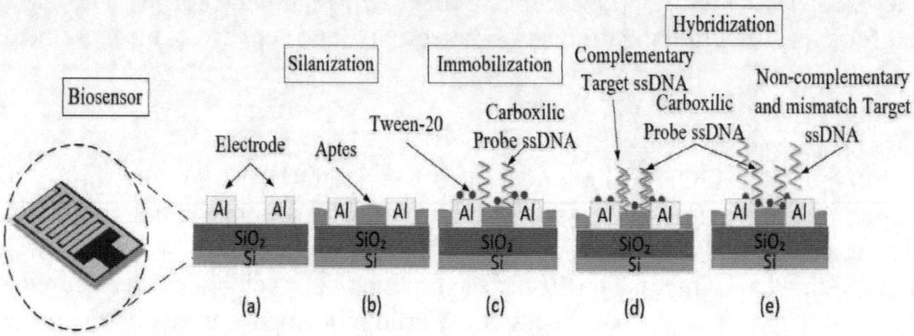

Figure 1. Functionalization steps of the sensing surface. (a) Bare Al IDE,
(b) Silanization with APTES, (c) Immobilization with *Listeria* ssDNA Probe and tween-20,
and (d) Hybridization with *Listeria* complementary target ssDNA
(e) Hybridization with *Listeria* non-complementary and mismatch target ssDNA

3. Result and Discussion

3.1 Surface Characterization

The physical characterization of Al IDE surface is displayed in Figure 2. Physical characterization is one of the most essential steps in making sure that there were no shortage on the Al IDE surface. A part from that, it is also to observe the material composition on Al IDE surfaces and to examine the existence of particles or squalor on top of IDE surface. It is significant to make sure the fabrication process is done with circumspect to ensure the measurement is not affected during the characterization process. All images have been taken using LPM and HPM with 5x, 10x, 20x and 50x resolution power. The surface topography shows uniform structure without shortage and any contaminants on the surface.

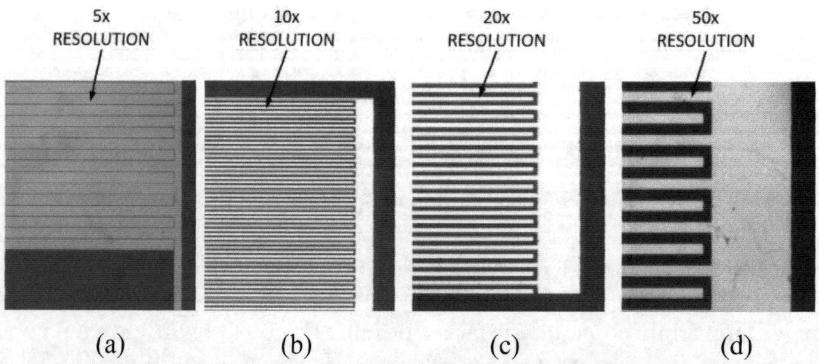

Figure 2. Al IDE physical characterization using LPM and HPM images
(a) 5x resolution, (b) 10x resolution, (c) 20x resolution and (d) 50x resolution.

Figure 3 shows the analysis of the IDE structure with 3D profiler. The IDE structure with 3D profiler depicts some nearly even electrode edges, which may be further confirmed that etching has reached to the maximum level of the development process. The maximum height of the Al surface from the grain with 532 nm and the average height is 366 nm, were confirmed by the uniformity of the surface. It is confirmed that the etching has reached the maximum developed design.

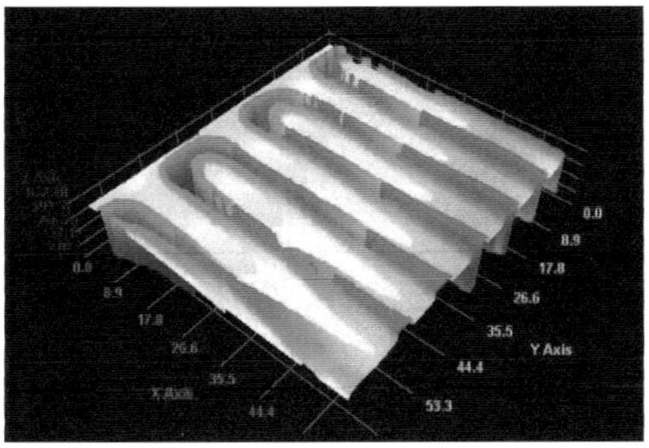

Item	Value
Start Pos (um)	X: 2.96 Y: 2.66
Area Size (um)	58.90 x 65.71
Center Pos (um)	X: 32.26 Y: 35.52
Pitch (um)	0.296
Minimum Height (nm)	0.000
Maximum Height (nm)	532.500
Ave Height (nm)	366.583

Figure 3. 3D profile of the IDE surface and specific parameters.

3.2 Electrical Characterization for Listeria Detection

Figure 4 shows the I-V characterization was performed by using (Keithley 2450), probe station and Kickstart software with Al IDE biosensor. The voltage between the two electrodes was being prescribed from 0 V to 1 V. Electrical measurement of Al IDE's structured is proposed and the result can be determined by electrical characterization current voltage I-V characterization for more accurate measurement. From the I-V measurement, the bare Al IDE current at 1 V was 2.24 x 10^{-11} A. I-V characteristics of biosensor preparation process of different steps, after salination, after immobilization and test with complementary ssDNA which is hybridization process.

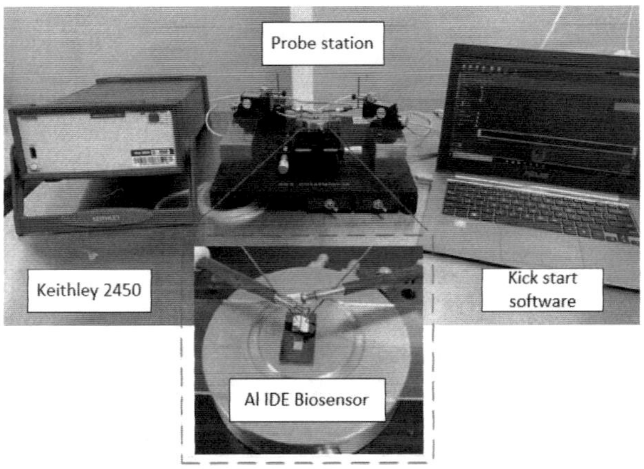

Figure 4. Electrical Characterization measurement using (Keithley 2450), Kick start software and probe station with Al-IDE biosensor.

The current keep increasing to nA range after silanization with APTES. APTES to provide contact between the organic and inorganic surfaces of a single-stranded DNA probe. The current captured at 1V shows the value which was 1.2 × 10^{-9} A. The current value then continuous increase after ssDNA probe was immobilized on the top of APTES layer. The current captured at 1V shows the value which was 3.35 × 10^{-9} A. Finally, the 2 µl of 1 nM concentrated Listeria ssDNA target was hybridized on the probe Listeria. The current captured of the target at 1 V is 5.4 × 10^{-9} A. DNA is negatively charged and after hybridization with complementary target ssDNA, free total positive charge carrier in APTES layer is increased because of the increment of the total negative charge of DNA probe and target. This sensing component changes the resistivity in the APTES layer according to the concentration of target ssDNA which are captured by ssDNA probes. Hybridization of synthetic target ssDNA can cause electric field increase. So current will decrease when electric field is increase. The current generation mechanism can be explained using following equations.

$$Conductivity\,,\ \sigma = \frac{l}{p} \tag{1}$$

$$Resistivity,\ R = \frac{pl}{A} \tag{2}$$

$$Electric\ field,\ E = \frac{k.Q}{d^2} \tag{3}$$

Figure 5 shows the selectivity measurement that was tested for complementary, non-complementary and single base mismatches target of Listeria separately. 1 nM concentrated targets and 2 μl volume was used for selective measurements. Curve 3 until 6 shows the Listeria Probe, complement, non-complementary and mismatch measurement captured at 1V are 3.3×10^{-9} A, 5.4×10^{-9} A, 1.8×10^{-9} A and 2.1×10^{-9} A respectively. The accessibility of immobilized probes to complementary target sequences can be enhanced by treating the surface with a small molecule blocking agents. Moreover, carboxylic group rapidly displaces the weaker adsorptive contact between oligonucleotides leaving the probes was moored probe through carboxylic end group. While, for probe and non-complement current curve shows closely same value because non-complementary ssDNA cannot be bind with immobilized ssDNA probes. As well as they cannot bind with APTES layer due to the blocking agents cover the area which was unbounded by immobilized ssDNA. Thus, non-complementary target ssDNA thoroughly removed after washing with deionized water.

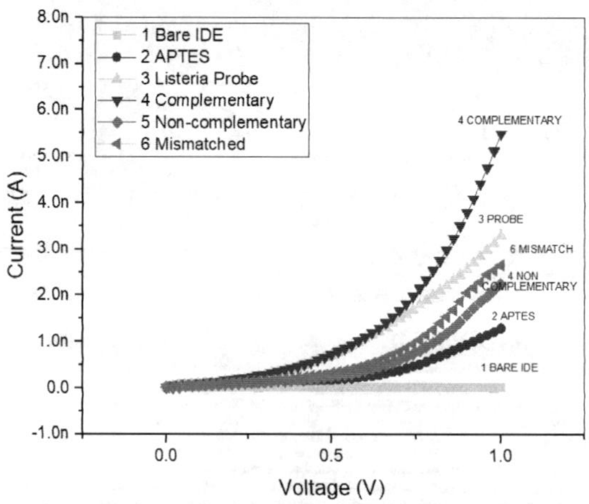

Figure 5. Graph of Current Vs Voltage for Al IDE with Complementary, Non-complementary and Mismatched ssDNA Listeria

Figure 6 shows the sensitivity measurement using different concentration target Listeria. The current variations were steadily increased based on the increment the target ssDNA concentration as 1 fM, 1 pM, 1 nM, 1 μM and current value at 1 V are 3.6×10^{-9} A, 4.1×10^{-9} A, 5.4×10^{-9} A and 6.1×10^{-9} A respectively. From the graph, shows that the current will continue to increase when the concentration of target *Listeria* increase.

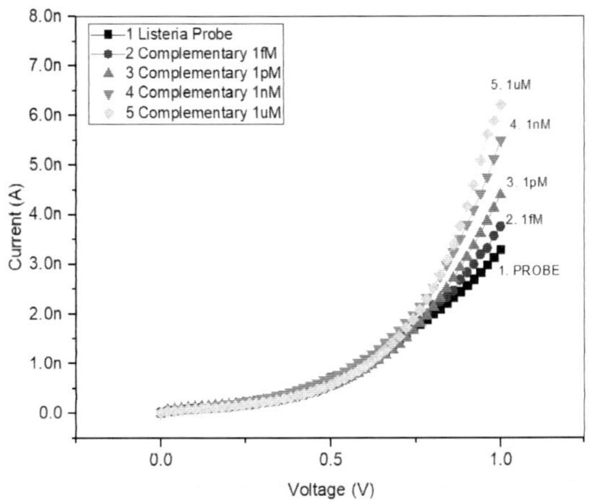

Figure 6. Graph of I-V for Al IDE with functionalized Al IDE surface and different concentration target Listeria.

4. Conclusion

In this paper, we described an approach for Listeria target ssDNA via electrical Al IDE based biosensor. The developments of biosensor based on Al IDE are very efficient. This IDE can directly detect current from very small range up to large range of current with different concentration of Listeria targets. Thus, the result shows that it is high in stability, selectivity and sensitivity in detecting complement, non-complement and mismatch target of Listeria. The sandwich layer of Al IDE which includes the salinization of APTES, immobilization of Listeria probe and the hybridization of Listeria target was also designed to overcome the selectivity problem for conventional methods like the PCR. Thus, we hope the conception of the mechanism and biosensor to be beneficial for commercial diagnostic applications.

Acknowledgement

The author would like to thank all staff members of the Institute of Nanoelectronic Engineering in Universiti Malaysia Perlis (UniMAP) for their technical advice and contributions, directly and indirectly. Collaborative Research in Science and Technology Center (CREST) is acknowledged for providing grant for this research program.

References

[1] R. D. A. A. Rajapaksha, U. Hashim, N. Z. Natasha, M. N. A. Uda, V. Thivina, and C. A. N. Fernando, "Gold nano-particle based Al interdigitated electrode electrical biosensor for specific ssDNA target detection," *Proc. 2017 IEEE Reg. Symp. Micro Nanoelectron. RSM 2017*, pp. 191–194, 2017.

[2] R. D. A. A. Rajapaksha, U. Hashim, M. N. Afnan Uda, C. A. N. Fernando, and S. N. T. De Silva, "Target ssDNA detection of E.coli O157:H7 through electrical based DNA biosensor," *Microsyst. Technol.*, 2017.

[3] M. Xu, R. Wang, and Y. Li, "Electrochemical biosensors for rapid detection of Escherichia coli O157:H7," *Talanta*, vol. 162, no. October 2016, pp. 511–522, 2017.

[4] S. R. Balakrishnan *et al.*, "Development of highly sensitive polysilicon nanogap with APTES/GOx based lab-on-chip biosensor to determine low levels of salivary glucose," *Sensors Actuators, A Phys.*, vol. 220, pp. 101–111, 2014.

[5] P. Poltronieri, V. Mezzolla, E. Primiceri, and G. Maruccio, "Biosensors for the Detection of Food Pathogens," *Foods*, vol. 3, no. 3, pp. 511–526, 2014.

[6] S. Nadzirah, N. Azizah, U. Hashim, S. C. B. Gopinath, and M. Kashif, "Titanium dioxide nanoparticle-based interdigitated electrodes: A novel current to voltage DNA biosensor recognizes E. coli O157:H7," *PLoS One*, vol. 10, no. 10, pp. 1–15, 2015.

[7] M. N. A. Uda, R. D. A. A. Rajapaksha, M. N. A. Uda, U. Hashim, and A. B. Jambek, "Selective detection of E.coli O157:H7 bacteria DNA using electrical based aluminium interdigitated electrode biosensor," *AIP Conf. Proc.*, vol. 2045, no. December, 2018.

[8] N. Z. Natasha, R. D. A. A. Rajapaksha, M. N. A. Uda, and U. Hashim, "Electrical DNA Biosensor using Aluminium Interdigitated Electrode for E . Coli O157 : H7 detection," vol. 020235, 2017.

[9] N. A. N. Yahaya, R. D. A. A. Rajapaksha, M. N. A. Uda, and U. Hashim, "Ultra-low current biosensor output detection using portable electronic reader," *AIP Conf. Proc.*, vol. 1885, 2017.

[10] R. D. A. A. Rajapaksha, N. A. N. Yahaya, M. N. A. Uda, and U. Hashim, "Development of portable electronic reader for picoampere detection for two-electrode based amperometric biosensor applications," *AIP Conf. Proc.*, vol. 2045, no. December, pp. 1–7, 2018.

[11] M. N. A. Uda *et al.*, "Immunosensor development formatting for tungro disease detection using nano-gold antibody particles application," *AIP Conf. Proc.*, vol. 1808, pp. 10–14, 2017.

[12] N. Z. Natasha, R. D. A. A. Rajapaksha, M. N. A. Uda, and U. Hashim, "Electrical DNA biosensor using aluminium interdigitated electrode for E.Coli O157:H7 detection," *AIP Conf. Proc.*, vol. 1885, 2017.

[13] S. A. B. Ariffin, T. Adam, U. Hashim, S. Faridah, I. Zamri, and M. N. A. Uda, "Plant diseases detection using nanowire as biosensor transducer," *Adv. Mater. Res.*, vol. 832, pp. 113–117, 2014.

[14] R. D. A. A. Rajapaksha, N. A. N. Azman, M. N. A. Uda, U. Hashim, S. C. B. Gopinath, and C. A. N. Fernando, "Multichannel PDMS microfluidic based nano-biolab-on-a-chip for medical diagnostics," *AIP Conf. Proc.*, vol. 2045, no. December, 2018.

[15] S. C. B. Gopinath, U. Hashim, and M. N. A. Uda, "Evaluation of factor IX deficiency by interdigitated electrode (IDE)," *AIP Conf. Proc.*, vol. 1808, pp. 1–5, 2017.

[16] M. N. A. Uda *et al.*, "Immunosensor development for rice tungro bacilliform virus (RTBV) detection using antibody nano-gold conjugate," *AIP Conf. Proc.*, vol. 1808, 2017.

[17] R. D. A. A. Rajapaksha, U. Hashim, M. N. A. Uda, and C. A. N. Fernando, "High-performance Electrical Variable Resistor Sensor for E . coli O157 : H7 Detection," vol. 10, no. 1, pp. 61–64, 2017.

[18] R. D. A. A. Rajapaksha, M. N. A. Uda, U. Hashim, S. C. B. Gopinath, and C. A. N. Fernando, "Impedance based Aluminium Interdigitated Electrode (Al-IDE) biosensor on silicon substrate for salmonella detection," *IEEE Int. Conf. Semicond. Electron. Proceedings, ICSE*, vol. 2018-Augus, pp. 93–96, 2018.

[19] R. D. A. A. Rajapaksha, U. Hashim, and C. A. N. Fernando, "Design, fabrication and characterization of 1.0 μm Gap Al based interdigitated electrode for biosensors," *Microsyst. Technol.*, vol. 23, no. 10, pp. 4501–4507, 2017.

[20] U. Hashim *et al.*, "Development of electronic reader for formaldehyde detection sensor," *Proc. - 2015 2nd Int. Conf. Biomed. Eng. ICoBE 2015*, no. March, pp. 30–31, 2015.

[21] S. Nadzirah, U. Hashim, and N. Malihah, "Structural and Electrical Properties of Titania Thin Films," vol. 925, pp. 300–303, 2014.

[22] S. Nadzirah and U. Hashim, "Annealing Effects on Titanium Dioxide Films by Sol – Gel Spin Coating Method," pp. 159–162, 2013.

[23] R. Singh, M. Das, G. Sumana, R. K. Gupta, S. Sood, and B. D. Malhotra, "Sensors and Actuators B : Chemical Biosensors for pathogen detection : A smart approach towards clinical diagnosis," *Sensors Actuators B. Chem.*, vol. 197, pp. 385–404, 2014.

Mechanical and Rheological Properties of Concrete with Ceramic Tile Waste as Partial Replacement of Fine Aggregate

N Yahya[1*], S M S Razak[1], M K Othman[2], S N A M Noor[1], K Muhamad[1] and M K M Jaih[1]

[1]Faculty of Engineering Technology, Universiti Malaysia Perlis (UniMAP), 02100 Padang Besar, Perlis, Malaysia
[2]Muda Agricultural Development Authority (MADA), MADA Headquarters, 05990, Alor Setar, Kedah, Malaysia

E-mail: * norhaizura@unimap.edu.my

Abstract. The rapidly increasing in population has led to the higher of construction, repairing and renovation activity that lead to produce large amount of construction material waste. The disposal of broken ceramic tiles during construction is one of the factors which contribute to this matter and can lead to land pollution. On the other hand, the natural resource in construction such as fine aggregate also facing depletion in order to cater the current and future demand. Therefore, this paper explores the properties of concrete with ceramic tile waste used as a partial replacement for fine aggregate. About 45 cube samples, 30 prism samples and 15 control samples were casted. Various percentage of ceramic tile waste has been introduced as partial replacement for fine aggregate with proportion of 10%, 15%, 20% and 30%. Tests for mechanical and rheological properties which have been done to identify the concrete performance are compressive strength, flexural, water absorption and slump test. From the results obtained, the sample of concrete contain of 15% ceramic tile waste as fine aggregate replacement has reached the optimum strength in both compressive strength and flexural strength. However, by using 20% of ceramic tile waste as fine aggregate replacement does show higher workability and water absorption.

1. Introduction

Construction activity such as renovation, demolition and repairing work have increased the amount of solid waste. Construction waste contribute the highest percentage of solid waste in the world with various of ceramic materials such as ceramic tile, brick tile, roof tile and the other ceramic products. Malaysia has recorded an approximately 8 million ton of construction waste per year which stated by Solid Waste and Public Cleansing Management Corporation [1].

It is crucial for a developing country to has a growing construction industry with a good waste management system so that the rapid development will not affect the environment. Another option is to reduce the waste from construction industry by reuse or recycling the waste. In this study, the focus is to reuse the ceramic tiles waste obtained from the construction site as a partial replacement for fine aggregate in concrete in terms of its mechanical and rheological properties performance.

Improper management of ceramic tiles waste will not only give impact to land pollution but also to human health if ceramic tiles waste were disposed in ash form. It can affect the lung function, cardiovascular disease and chronic abnormal [1].

Content from this work may be used under the terms of the Creative Commons Attribution 3.0 licence. Any further distribution of this work must maintain attribution to the author(s) and the title of the work, journal citation and DOI.
Published under licence by IOP Publishing Ltd

Furthermore, with this study, it also contributes to less use of fine aggregate which cater the issue of insufficient of natural resources. The problem arose from continuously mining the sand which effect the ecosystem and cause the degradation of the river, as well as affect the water quality [1] can be reduce.

2. Literature Review

Previously, a study on ceramic tiles as a replacement for both fine and coarse aggregate in concrete mixture has been done [2]. About 10% of ceramic tile used as fine aggregate replacement while for coarse aggregate replacement contain of three percentage which are 30%, 60% and 100%. The mix concrete is for M30 grade and from the results obtained, with 10% of ceramic tiles as fine aggregate replacement and 30% of ceramic tiles replace as coarse aggregate, the compressive strength has achieved 31.70N/mm^2. The compressive strength increased to 34.61N/mm^2 when the percentage of ceramic tiles as coarse aggregate replacement increased to 60%. However, with the used of 100% ceramic tiles as coarse aggregate has shown the decrease in compressive strength value which it only reaches until 28.34N/mm^2.

Another study shows replacement of both fine and coarse aggregate with amount of 22.1% will increase the compressive strength value [3]. On the other hand, 20% replacement of tile ceramic waste and ceramic sanitary for natural limestone aggregate and fine aggregate also achieved higher value of compressive strength compared to the conventional concrete [4].

The study of ceramic aggregate as alternative material in replacing fine aggregate show the flexural strength has increased with the increment of ceramic aggregate percentage from 25% to 75% content [5]. Other than that, one study shows that the flexural strength in M30 concrete at 28 days are higher with the increment content of crushed ceramic tiles as fine aggregate replacement due to the reaction of the crushed ceramic tiles with the Portland-pozzolana cement grade 43 which used in the mix design [6].

In terms of workability, previous study shows that the slump test is similar with the control sample if the ceramic tile was replaced as coarse aggregate and it shows higher value compared to control sample if the ceramic tile waste is replaced as fine aggregate [7]. The optimum content which was obtained from the study is 20% ceramic tile for fine aggregate replacement. Another study on influence of water absorption using ceramic tiles as aggregate replacement shows the higher value of water absorption compare to the normal sample [8].

3. Experimental Procedure

Experimental procedure is done starting from the preparation of raw material, mixing and conducting related testing on the samples prepared.

3.1. Raw Material

Type of ceramic tile waste used in this study as raw material is porcelain tile which has been collected from the construction site area. This raw material is first having to be cleaned from any other substance or mud. It then will be break into small pieces before crushed in order to get the 4.75mm size aggregate. Then, it will be dry for 24 hours in oven. The procedure stated is shown in Figure 1.

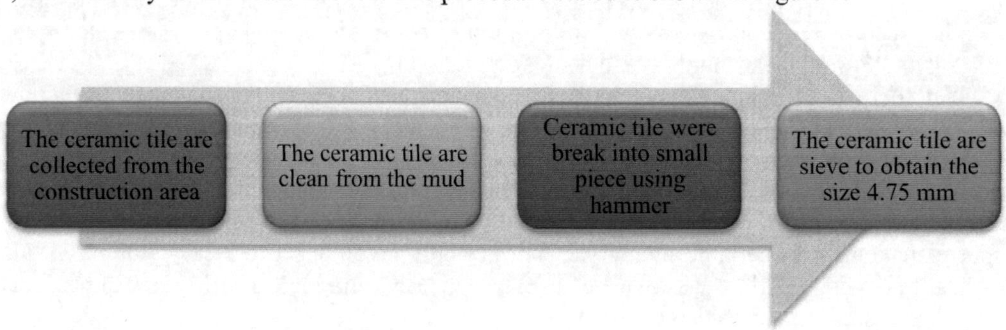

Figure 1. Preparation of ceramic tile waste as raw material.

IOP Conf. Series: Materials Science and Engineering **743** (2020) 012033 doi:10.1088/1757-899X/743/1/012033

3.1.1. Mix Design and Process.

Mix design of concrete grade 25 for control samples (CS) and samples with ceramic tile waste (CTWC) are as shown in Table 1. Ordinary Portland Cement (OPC) grade 42.5 has been used in this study with 0.5 ratio of water-cement content. The coarse aggregate with the range size of 16mm to10mm will be mixed with fine aggregate with the range size are 4.75mm to 1.18mm. Four different proportion of CTWC are used to replace fine aggregate which are from 10%, 15%, 20% and 30% of ceramic tile fine aggregate will be replaced with fine aggregate. The mix design is then will placed into cube and beam size samples. The samples will be cured in the water before testing for 7 and 28 days.

Table 1. Design concrete mix.

Percentage replacement of fine ceramic tile (%)	W/C ratio	Cement content (kg/m^3)	Water (kg/m^3)	Fine aggregate (kg/m^3)	Coarse aggregate (kg/m^3)	Ceramic tile waste (kg/m^3)
0% (CS)	0.5	380	190	558	1302	NA
10% (CTWC-10)	0.5	380	190	502.2	1302	55.8
15% (CTWC-15)	0.5	380	190	474.3	1302	83.7
20% (CTWC-20)	0.5	380	190	446.4	1302	111.6
30% (CTWC-30)	0.5	380	190	390.6	1302	167.4

3.2. Testing and Analysing

Testing made to identify the performance of concrete with CTW replacement for fine aggregate in terms of mechanical and rheological properties. For mechanical properties, compressive and flexural strength test has been done to the samples. The compressive strength test of concrete was carried out according to BS 1881: Part 116: 1983 while flexural strength test of concrete was carried out in accordance with BS 1881: Part 118: 1983. Rheological properties for this study focused on the workability of the concrete samples by the standard slump test according to the BS 1881 Part 102: 1983 and the water absorption test according to BS 1881-122. Figure 2 shown the process for water absorption which involved water curing process, oven dry process and weight.

Figure 2. Water absorption test process.

4. Results and Discussion

4.1. Compressive Strength

The compressive strength test done on the cube samples after 7 and 28 days. Based on Figure 3, compressive strength for CS at 7 days has reached 37.1MPa and 33.4MPa at 28 days. On the other hand, the compressive strength value for CTWC-10 and CTWC-15 samples has reached higher compared to CS samples with value of 42.3 MPa and 44.5MPa each at 7 days while at 28 days the value are 36.9MPa and 40.5MPa each. For CTWC-20 and CTWC-30samples at 7 days, the values are lower than CS samples but still higher than the design of normal strength concrete in this study which is 25MPa. However, the values are higher compared to CS samples at 28 days which are 35.4MPa and 33.9MPa each. The positive results by replacing fine aggregate with ceramic tile waste in concrete may due to the pozzolanic activity of ceramic micro particle when combined with cement compound. Other than that, the type of fine aggregate used in these samples that are very fine may cause the increase of compressive strength value.

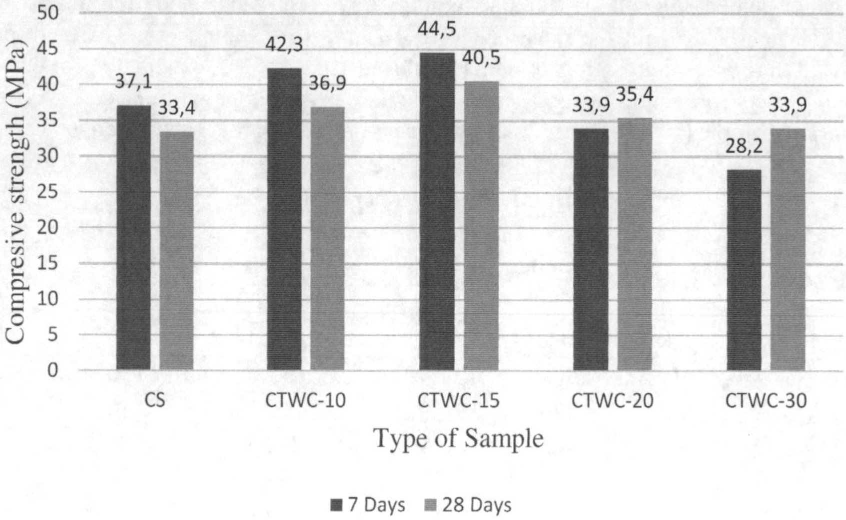

Figure 3. Compressive Strength Test.

4.2. Flexural Strength

The flexural strength test done on the prism samples shows that flexural strength for CS at 7 days has reached 4.1MPa and 5.1MPa at 28 days. On the other hand, the flexural strength value for CTWC-10 and CTWC-15 samples has increment value compared to CS samples with value of 4.6MPa and 4.8MPa each at 7 days while at 28 days the value are 5.6MPa and 5.8MPa each. For CTWC-20 and CTWC-30 samples at 7 and 28 days, the values are lower than CS samples. The reduction of flexural strength in replacement of ceramic tile as fine aggregate may be due to the ceramic material did not have good adhesion between the cement paste and the aggregate surface. In this study, the ceramic tile surface that been used are smooth on both top and bottom surface of ceramic tile that can affect the low cohesion between cement paste and ceramic tile fine aggregate accounts the reduce of flexural strength.

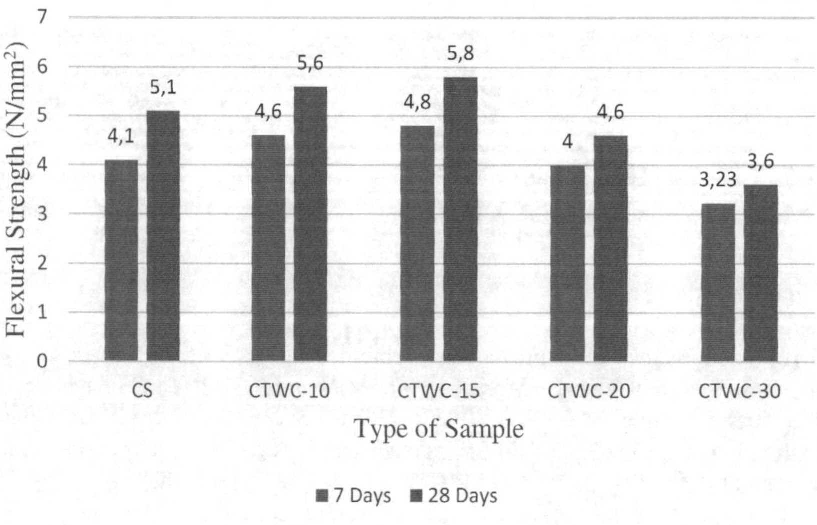

Figure 4. Flexural strength test.

4.3. Workability of Concrete

Figure 5 show the result of slump test. There is similar result for CS, CTWC-10, CTWC-15 and CTWC-20 samples which is 30mm. Meanwhile, CTWC-30 samples shown result of decreasing value which is 20mm. This is in line with previous study which stated that the replacement of fine aggregate with ceramic tile in the concrete inhibit to the decrement of slump test [6]. This finding can be explained as when the surface of ceramic tile is smooth, it might reduce the workability of concrete. Other than that, the finer particle requires more water to wet their larger surface, whilst the irregular shape and rougher texture of an angular demand more water [3].

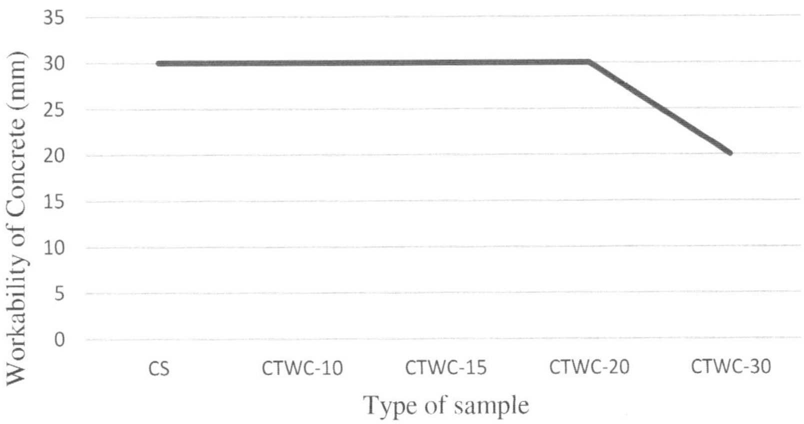

Figure 5. Workability of concrete.

4.4. Water Absorption of Concrete

Based on the Figure 6, it shows that the percentage of water absorption is increase with the increase of ceramic tile waste content used as fine aggregate replacement. The percentage of improvement for water absorption is from 8.76% for CTWC-10 up until 15.65% for CTWC-30. This result is similar with previous study which stated that the ceramic waste has water absorption is 0.18% compare with natural aggregate [8]. The higher percentage of ceramic tile used; the higher porosity of the concrete preserved in the air. Other than that, the surface area, pore structure and clay content play important role on the higher of water absorption. However, the higher of water absorption give the effect of reduction of compressive strength of concrete since it can give the improper bonding in the interfacial transition zone [4].

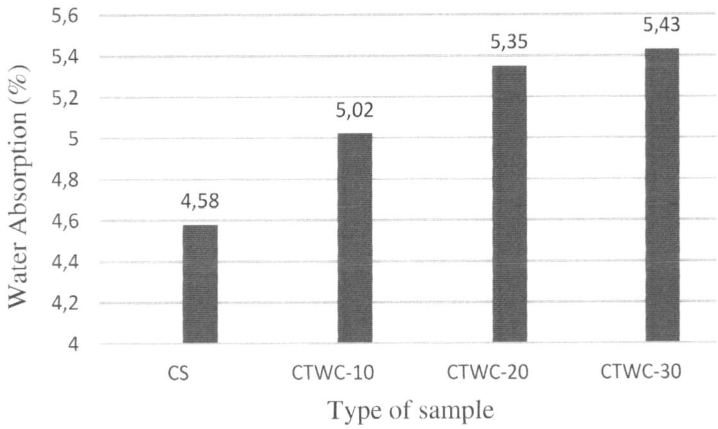

Figure 6. Water absorption of CS and CTWC.

Table 2. Results of water absorption and percentage of improvement.

Type of mixture	Water absorption (%)	Percentage of improvement on water absorption (%)
CS	4.58	-
CTWC-10	5.02	8.76
CTWC-15	5.12	10.55
CTWC-20	5.35	14.39
CTWC-30	5.43	15.65

5. Conclusion

This study focusses on the mechanical and rheological properties of concrete by partial replaced of fine aggregate with ceramic tile waste. From the results obtained, the sample of concrete contain of 15% ceramic tile waste as fine aggregate replacement has reached the optimum strength in both compressive strength and flexural strength. However, by using 20% of ceramic tile waste as fine aggregate replacement does show higher workability and water absorption.

Acknowledgement

The author would like to acknowledge the support from Faculty of Engineering Technology, University Malaysia Perlis (UniMAP) in this research.

References

[1] Saadi, N., Ismail, Z., & Alias, Z. (2016). A review of construction waste management and initiatives in malaysia. *Journal of Sustainability Science and Management*, **11**(2), 101–114.

[2] P.Rajalakshmi, Dr.D.Suji, M. Perarasan, E. N. (2016). Studies on Strength Characteristics on Utilization of Waste Ceramic Tiles as Aggregate in Concrete, *International Journal of Engineering Science and Technology (IJEST),* **4**(1), 114–125.

[3] Awoyera, P. O., Ndambuki, J. M., Akinmusuru, J. O., & Omole, D. O. (2016). Characterization of ceramic waste aggregate concrete. *HBRC Journal*, 1–6.

[4] Pitarch, A. M., Reig, L., Tomás, A. E., & López, F. J. (2017). Effect of Tiles, Bricks and Ceramic Sanitary-Ware Recycled Aggregates on Structural Concrete Properties. *Waste and Biomass Valorization*, **0**(0), 1–15.

[5] Ramadevi, K. (2017). a Study on Properties of Concrete With Ceramic Waste Replaced for Fine Aggregate. *International Journal of Civil Engineering and Technology*, **8**(88), 1730–1737.

[6] Chandel, U. S. (2017). Investigation on Fine Aggregate by Broken Tiles in Concrete. *International Journal of Advance Research, Ideas and Innovations in Techology*, **3**(6), 150–154.

[7] Wattanasiriwech, D., Saiton, A., Wattanasiriwech, S., Kamala, R., Rao, B. K., Kumar, N. S. T., Fern, F. (2015). Effect of Waste Ceramic Tiles in Partial Replacement of Coarse and Fine Aggregate of Concrete. *Journal of Cleaner Production*, **2**(3), 390–394.

[8] Rao, S., Valikala, G., & Ghorpade, V. G. (2013). Influence of Water Absorption of the Ceramic Aggregate on Strength Properties of Ceramic Aggregate Concrete. *International Journal of Innovative Research in Science, Engineering and Technology*, **2**(11), 6329–6335.

IOP Conf. Series: Materials Science and Engineering **743** (2020) 012034 doi:10.1088/1757-899X/743/1/012034

A Conceptual Paper on Application of Energy Efficiency Measures for Reducing Carbon Footprint in Zinc Phosphate Coating Process of Automotive Industries

Abuzaar Mohd Zulkifli[1], Fazdliel Aswad Ibrahim[1], Norrazman Zaiha Zainol[1], Mardhiah Farhana Omar[1] and Lim Chun Keat[1]

[1]Department of Civil Engineering Technology, Faculty of Engineering Technology, Universiti Malaysia Perlis, Malaysia

Email: norrazman@unimap.edu.my

Abstract. Automotive manufacturing consumes a huge amount of energy to operate its production lines. Several processes involve including zinc phosphate coating process, which is a process that commonly used in automotive industries to improve the corrosion resistance of the metal. The process required plenty of hot water supplied by the boiler. Liquefied petroleum gas (LPG) is used to boil the water to produce thermal energy for the process. Although the LPG is crucial in the production of thermal energy but its release by-products such as carbon dioxide and other carbon compounds emitted during the coating process due to the burning of the fossil fuel. The by-products contribute to global warming issues and produce waste to the environment. This research is present to design the energy efficiency best practices on the reduction of carbon footprint in automotive industries. A status quo to be referred to as baseline is determined, i.e. energy consumption and wastage by assessing the thermal requirement of the said manufacturing process. The data obtained is analysed to identify the best measures of energy efficiency. The thermal assessment method is used for data analysis to determine the potential saving of fossil fuels. Besides, pinch analysis assist in the determination of the process optimisation and system optimisation which lead to the reduction of operational cost and wastes as well as efficiently applying the heat during the coating process. Hence, the decreases in energy consumption reduce dependency on fossil fuels and wastes. Eventually, the application of best practices of energy-efficient measures diminishes the carbon footprint for the coating process in automotive industries.

1. Introduction

In this 21st century, the growth of the automotive industry is considered as 'industry of industries' which potentially drive the industrialisation ahead due to its linkage and spilt-over effects on other manufacturing industries. The automotive industry is an organisation involved in design, manufacturing, development, marketing and selling of motor vehicles. The automotive manufacturing process involved chassis production unit, coating process, surface finish centre, pre-assembly and the final assembly line. Coating process plays a significant role in the manufacturing process such as zinc phosphate coating, as it is widely used for corrosion resistance, the primer for painting and decoration. In the coating process, steam boiler is commonly used to supply hot water which requires a large amount of fossil fuel to heat the water before transferring to the other stages. Therefore, boiler consumes high energy, commonly fossil fuels such as liquefied petroleum gas (LPG) to provide thermal energy. However, burning fossils fuels caused a tremendous amount of carbon dioxide

Content from this work may be used under the terms of the Creative Commons Attribution 3.0 licence. Any further distribution of this work must maintain attribution to the author(s) and the title of the work, journal citation and DOI.
Published under licence by IOP Publishing Ltd

IOP Conf. Series: Materials Science and Engineering **743** (2020) 012034 doi:10.1088/1757-899X/743/1/012034

emissions [1] as well as other by-products to the environment. By improving energy efficiency, which is the critical components of sustainability may reduce the energy consumption of the manufacturing process [2]. Therefore, the determination of the best measure of energy efficiency that aims to reduce the burning of fossil fuels should be the primary concern.

In this research, the energy consumption of the zinc phosphate coating process is studied based on the manufacturing process flow of the automotive industry, i.e. motorcycle manufacturer located at Penang, Malaysia. The research focuses on the reduction of fuel usage and optimising energy consumption in zinc phosphate coating process by applying energy efficiency measures. The ultimate objective is to reduce the carbon footprint. The parameters under study are temperature, pressure and flow rate to set the current status, i.e. baseline of the process. By evaluating the process flow, stream list and Sankey diagram are established. Then, pinch analysis assists in the determination of the process and system optimisation. The best practices energy efficiency measures are recommended to the stages that have the potential for energy saving. The conceptual of energy recovery is verified by three analysis methods which are economics, energy and environmental analysis.

1.1. Problem statement
The automotive industry nowadays is highly desired by many countries due to its contribution toward economic growth, job creation, and technology development [3]. However, a new challenge is faced by the industries in this current environmental era, i.e. global warming and climate change. In Malaysia, the primary cause of greenhouse gas (GHG) emissions in manufacturing firms can be traced to three sources which are electricity consumption, the direct combustion of fuels in the manufacturing process and the fuel consumed in logistical activities [4]. Hence, the automotive industry faces immense pressures in the environmental arena [5]. In the automotive industry, zinc phosphate coating process is one of the processes that commonly used for corrosion protection, the primer for painting and decoration due to low processing cost [6]. The process required thermal energy which provided by heating the water or air in a boiler through the burning of fossil fuels, i.e. LPG [7] to the required temperature for each stage in the process. Even though LPG provides high thermal energy, but it's polluting the environment as well [8]. Hence, the zinc phosphate coating process indirectly causes environmental issues significantly compared to other stages of vehicle manufacturing [9].

2. Literature Review
Climate changes are driven by human activities such as the burning of fossil fuels in the industry sector, in particular, the production of greenhouse gas emissions that directly impact on environment. In Malaysia, industrial sectors in one of the contributors of carbon dioxide emission. According to statistics from Asian Pacific Energy Center (APEC), the outlook period for carbon dioxide emission from energy consumers in Malaysia is projected to grow about 4.2% annually reaching 414 million tons of carbon dioxide in 2030 [4]. Hence, the industrial sector, like the automotive industry, should aware of the need to address environmental issues in their products and processes.

The automotive industry is defined as the manufacture of motor vehicles, including most components, such as engines and bodies but excluding tires, fuels and batteries [10]. In this era, automotive industries are facing many challenges due to the recent significant increase in overall vehicle production with a rise in energy cost and environmental responsibilities [11]. The manufacturing process of the automotive industry is complex and intense, and it consumes a large number of raw materials as well as utilities, i.e. water and electricity [11]. Besides, the automotive industries responsible for more than one-third of worldwide primary energy use because they rely heavily on energy in its processes [12]. Therefore, an evolution of the manufacturing process in the automotive industry is required to move towards low-carbon emissions and increased sustainability of the process [11] which due to many thermal processes in the automotive industry such as zinc phosphate coating which required burning of fossil fuels to produce thermal energy.

IOP Conf. Series: Materials Science and Engineering **743** (2020) 012034 doi:10.1088/1757-899X/743/1/012034

2.1. *Zinc Phosphate Coating Process*

Zinc phosphate coating is one of the critical processes in the automotive industry. It is generally used for the purposes of corrosion protection, the primer for painting, metal forming lubricants, wear reduction, electrical insulation and even decoration. In this process, many parameters influence the thickness, morphology, protective properties and chemical composition of the coating [13]. For instance, if the temperature of the solution is low, the heat is insufficient for phosphating reaction and reduces the coverage rate of the surface [14]. Therefore, a massive amount of thermal energy is required by the process which conventionally supplies by the steam boiler.

2.2. *Thermal Energy and its effect on Environment*

In a thermal process that comprises of the heat exchanger to provide required temperature for the manufacturing process, the first law of thermodynamic is typically used to measure the efficiency of energy utilisation by a boiler in actual operation [15]. Moreover, heat power is calculated using the first law of thermodynamic for control volume as well as obtains the energy consumption for the manufacturing process. The basic parameters, i.e. pressure, temperature, and volume, are identified as an input to assist in the determination of optimisation measures by the process. Besides, the Hagen-Poiseuille equation is used to ascertain the flow rate of fluid in the piping system.

Energy consumption is defined as the amount of energy consumed in a process or system. In the automotive industry, around 70% of thermal energy of all automotive industry's total energy consumption is required due to the manufacturing process required thermal energy in the form of steam, hot air or water [16]. Therefore, the reduction of energy consumption is very vital because it can minimise the environmental issues caused by the manufacturing process [12] in which as the energy consumption decreases, less carbon dioxide emissions due to lesser burning of fossil fuels to generate the required heat in the manufacturing process. Ultimately, environmental issues such as global warming effect and greenhouse gases emissions will be lessened.

A recent report, Carbon Risks and Opportunities in the S&P 500, commissioned by the Investor Responsibility Research Center Institute for Corporate Responsibility (IRRCi), found that more than 80% of companies of automotive faced carbon risks resulting from emissions in their supply chain [17]. The burning of fuels released carbon dioxide, which then, absorbed partly by the world's oceans resulting in acidification of those oceans, which leads to ecological and biological changes. Continuously increasing concentration of atmospheric leads to global warming and climate change [1]. Hence, assessing carbon dioxide emissions in the automotive industry using carbon footprint measurement is recommended. Assessing the carbon footprint in the manufacturing process perhaps provides a better understanding, quantify and rigorously analyse the impact of carbon emissions in the manufacturing process [18].

2.3. *Efficient Management of Energy*

Energy management is the strategy or method of meeting energy demand when and where it is needed. Energy management contributes to the monitoring and controlling of energy and has become a concerning issue recently [19]. Nowadays, the role of energy management is crucial in minimising energy costs and wastes without affecting production and quality as well as to reduce the environmental effects. In the automotive industry, energy management is a central strategy to leverage the present excess heat for the realisation of an efficient manufacturing process [11]. Hence, the waste can be minimised, and the energy consumption can be reduced once the energy is optimised by effective and efficient energy management in the process and production lines. Energy management is positively related to energy efficiency [20]. Therefore, energy efficiency is vital in the manufacturing process of the automotive industry and considered as one of the critical components of sustainability [2]. The reduction in energy consumption by improving the energy efficiency of a system can be achieved by using energy efficiency evaluation which is an effective management strategy [21]. The energy efficiency becomes a driver for the manufacturing industry as it is one of the most significant consumers and carbon emitters in the world [22]. The possible strategies are explored for energy consumption reduction in the processes, a systematic overview of energy consumption classification and prediction methods for different levels of energy consumption have conducted to investigate methods could be used for energy efficiency evaluation in the processes [21]. The methods such as thermal insulation and heat recovery are commonly applied as a solution for the EE measures.

Thermal insulation is defined as the decrease of heat transfer from the surface of the materials [22]. By improving the insulation of the steam piping and steam boilers, potential savings of 6% to 26% can be achieved [23]. Heat recovery is known as the high temperature is used for one purpose or the waste heat from one process step applied to another process step [24]. The heat recovery in industrial processes has been one of the significant areas of research to reduce fuel consumption, lower harmful emissions and improve production efficiency [25]. The waste heat can be recovered through various waste heat technologies to provide valuable energy sources and significantly reduce energy consumption. There are several types of heat recovery technologies, such as heat exchanger and heat pumps. A heat exchanger is an equipment used for the transfer of heat between two or more fluids at different temperatures [26]. In the manufacturing process, the heat exchanger is installed to preheat the feedwater, so that it can reduce the steam boiler fuel requirements. Two types of heat exchangers can be used to recover heat from boiler exhaust which is gas-gas heat exchanger and gas-liquid heat exchanger [23]. Heat pumps are commonly used to change the energy from a low-temperature level to high-temperature level. It has been used to good effect for waste heat recovery, which creates a temperature lift in the heat source. An excess of 40°C of temperature lifts can be achieved by using the heat pump. Commonly, two types of heat pump mostly use in the industry, which is an open-cycle heat pump and closed-cycle vapour compression heat pump can be considered for this purpose. However, a thorough study is recommended before opting for the heat pump since the overall efficiency of energy recovery is decreased due to the energy consumption of the heat pump [27].

System optimisation is defined as a methodology, act or process of obtaining entirely perfect, effective and best result to a system [28]. A methodology such as energy efficiency can optimise the system to reduce energy consumption and energy waste. Process optimisation is a popular tool by operating complex processes technically to achieve the best improvement and asset utilisation [29]. Process optimisation can find the optimal parameters for multiple machines and adjust them simultaneously to minimise energy consumption. In the manufacturing process, energy efficiency can be through integrating process optimisation with energy efficiency scheduling [30]. Therefore, the manufacturing process in the automotive industry can be optimised by applying the best practices of energy efficiency.

3. Methodology

The baseline, i.e. current state of the energy usage for the process is determined by analysing the utility bills and existing energy consumption of the target process. Then, the flow sheet is drawn based on the process flow of the zinc phosphate coating section in an automotive factory, complete with essential information collected upon a site visit. The flow sheet of zinc phosphate coating process is in the form of a block diagram. A block diagram is the simplest form of presentation of the overall process. It provides an overview of the process involved. Each block can represent a single piece of equipment or a complete stage in a process. The numbers in the flow sheet indicate the stream number for each flow in the stream list.

The data of zinc phosphate coating process, which is flow rate, humidity, temperature and pressure are recorded for a specified period. Data logger and measuring instruments such as environmental meter, flow meter, thermometer gun and a digital pressure gauge are used to obtain the required data of the process. The flow rate, which measured in kg/s, of fluid in the zinc phosphate coating process, is collected by using a digital ultrasonic flow meter. The temperature of the fluid in the zinc phosphate coating process is collected by using a thermometer gun. The unit of the temperature is Celsius, °C.

Stream no	1	2	3	4	5	6	7	8	9	10
	Predegreasing 1					Predegreasing 2				Water Rinse 1
	Frame body in	Hot water supply from Hex 1	Hot water return to Hex 1	Make-Up Water Supply	Frame body in	Hot water supply from Hex 2	Hot water return to Hex 2	Make-Up Water Supply	Frame body in	Make-Up Water Supply
Water, kg/h										
Air, kg/h										
LPG, kg/h										
Stack kg/h										
Temp, °C										
S/L/G										
cp, kJ/kg°C										
Enthalpy, kW										

Figure 1. A stream list of a process.

A stream list as shown in Figure 1 is evaluated which consists of temperature (°C), mass flow (kg/s), heat capacity (kJ/kg°C), enthalpy (kW), energy (kW) and operating hour (h) for each stage in the zinc phosphate coating process. The stream list is generated in a table form. The recorded data such as temperature, mass flow, humidity, pressure and operating hour categorised as raw data is input to the stream list, then, enthalpy and energy consumption is calculated by using the equations.
The formula of the energy consumption is

- Energy [kWh] = Q (kW) * Operating hour (h).

The equation of enthalpy is

- Enthalpy, Q [kW] = $m * q$ [kW] where m = mass flow, q = specific heat capacity, c * temperature change, ΔT.

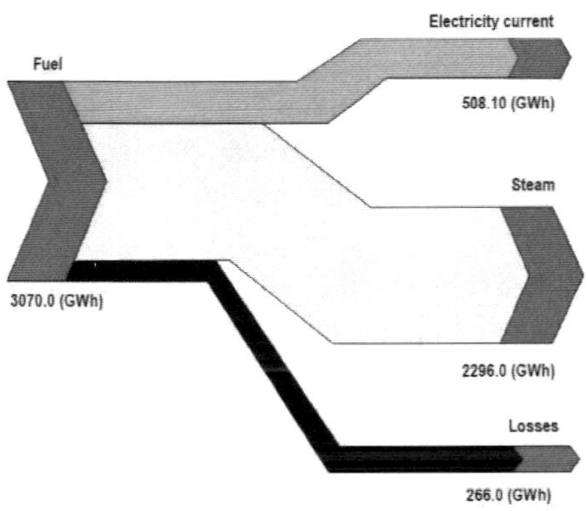

Figure 2. A typical Sankey diagram.

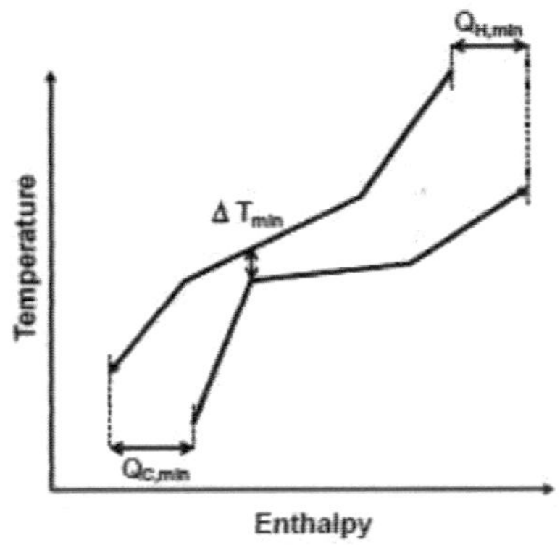

Figure 3. Pinch analysis uses the hot and cold composite curve to determine the energy targets.

The result of the stream list will be the input for Sankey diagram, which is a flow diagram in which the width of the arrows proportional to the flow quantity. The Sankey diagram represents the flows such as energy consumption in a process system or between processes. The Sankey diagram is an essential aid in identifying inefficiencies and potential for savings when dealing with resources [31]. The energy calculated in the stream list is used to generate the Sankey diagram of zinc phosphate coating process. The Sankey diagram is generated by using SankeyMATIC software. Therefore, the different use of energy and energy sources can be identified.

Pinch analysis is a method for reducing the energy consumption in a process by calculating thermodynamically feasible energy targets and achieving them by optimising heat recovery systems, applying energy efficiency and process operating conditions. Besides, pinch analysis which developed by Bodo Linnhoff in 1977 is required to identify energy target and determines the minimum driving force across any of the heat exchanger's inlet/outlet streams [32]. Pinch analysis can identify the potential to apply heat recovery to the process. The result of pinch analysis will be used to determine the specification of heat exchanger fit for heat recovery purposes.

3.1. Verification of Baseline and Proposed EE Measures

Three analyses which are economics, energy and environmental analysis is carried out to compare with the baseline and verify the finding which is recommended practical measures of energy efficiency that feasible to be integrated into the zinc phosphate coating process. Firstly, an economic analysis is carried out to test the method of energy efficiency suggested. The analysis is all about analysing the economic aspects of the recommended measures. It is used to verify that the method is operating effectively and profitable. The total investment of heat recovery in zinc phosphate coating process, annual maintenance cost and other costs are calculated. Then, the pay-back period is calculated to verify that profit is making. Besides, it is also compared to the previous electricity bill with the current electricity bill after the applications of energy efficiency. The energy analysis is used to analyse the process which the energy consumption after the energy efficiency is applied. The energy consumption of zinc phosphate is calculated after the technique or method of energy efficiency is applied. The thermal energy generated is also calculated after the heat recovery technique applied. An environmental analysis is carried out to the method of energy efficiency suggested. It used to analyse the process, whether it bring impacts to the environment. The carbon footprint is calculated after the technique or method of energy efficiency is applied.

4. Conclusion and Recommendation

In the automotive industry, zinc phosphate coating process plays an essential role in the manufacturing processes; however, it caused GHG, i.e. carbon dioxide emissions to the environment. Conventionally, the majority of manufacturers burnt of a large amount of fossil fuel to heat the fluid, i.e. gas or water in the steam boiler to provide thermal energy for the process. Therefore, finding the best measures of effective and efficient energy efficiency to minimise the fuel consumption in the automotive industry and those measures are feasible to be applied is the primary concern. By using the thermal assessment to analyse the data from the existing state, the status quo can be determined. The status quo is evaluated, which include of stream list of the process, Sankey diagram and pinch analysis. The result iswill be used as the input to determine the practical method feasible for the process, i.e. zinc phosphate coating. The application of energy efficiency is significant to the automotive industry in reducing energy consumption and energy waste as well as optimising the process and system.

Hence, the emission of carbon dioxide can be reduced because of the reduction of burning fossil fuels. The economic, energy and environmental analysis are carried out for this research to counter check the measures. Theses analyses are vital because it helps to determine how feasible the measures in term of economy, energy and environmental performance. Hence, the application of the proposed energy efficiency measures s not only reduce the operation cost and time, but increasing productivity as well as promotes a pleasant environment due to the carbon footprint is reduced.

Acknowledgment

The author would like to acknowledge the support from the Fundamental Research Grant Scheme (FRGS) under a grant number of FRGS/1/2017/SS03/UNIMAP/03/1 from the Ministry of Higher Education Malaysia.

References

[1] Fernando, Y., & Lin, W. (2017). Resources , Conservation & Recycling Impacts of energy management practices on energy efficiency and carbon emissions reduction : A survey of Malaysian manufacturing firms. **126**(July), 62–73.

[2] Unver, U., & Kara, O. (2019). Energy efficiency by determining the production process with the lowest energy consumption in a steel forging facility. *Journal of Cleaner Production*, **215**, 1362–1370.

[3] Fadly, N. H., Anis, F. M. Z., Juriab, C., Nurzatul, A. S. L. J., & Suzaituladwini, H. (2012). The Development of Sustaining Lean Improvements and Sustainable Performance in Malaysian Automotive Industry. *International Journal of Lean Thinking*, **3**(2), December 2012.
[4] Hosseini, S.E., Wahid, M.A., Aghili, N., 2013. The scenario of greenhouse gases reduction in Malaysia. *Renew. Sustain. Energy Rev.* **28**, 400–409.
[5] Geffen, C. A., & Rothenberg, S. (2002). Suppliers and environmental innovation. *International Journal of Operations & Production Management*, **20**(2), 166–186.
[6] Li, G. Y., Lian, J. S., Niu, L. Y., Jiang, Z. H., & Jiang, Q. (2006). Growth of zinc phosphate coatings on AZ91D magnesium alloy. *Surface and Coatings Technology*, **201**(3–4).
[7] Abishek, S., & Elangovan, R. K. (2018). Design and Implementation of On-Site Emergency Planning for Liquefied Petroleum Gas Utilizing Automobile Industries. *Advances in Fire and Process Safety Springer Transactions in Civil and Environmental Engineering*, 253-262.
[8] Bansal, P., Pandya, A., & Ahuja, A. (2017). Experimental Performance Evaluation on S.I. Engine with Gasoline and Liquefied Petroleum Gas as Fuel. *Techno-Societal 2016*, 183-191.
[9] Prendi, L., Henshaw, P., & Tam, E. K. L. (2006). Automotive coatings with improved environmental performance. *International Journal of Environmental Studies*, ***63***(4), 463–471.
[10] Binder, A. K., & Rae, J. B. (n.d.). Automotive industry. Retrieved from https://www.britannica.com/technology/automotive-industry.
[11] Giampieri, A., Ling-Chin, J., Taylor, W., Smallbone, A., & Roskilly, A. P. (2019). Moving towards low-carbon manufacturing in the UK automotive industry. *Energy Procedia*, **158**, 3381–3386.
[12] Ghazanfari, B. (2015). Modeling Energy Consumption in Automotive Manufacturing. *Electronic Theses and Dissertations*, **156**.
[13] Fathyunes, L., Azadbeh, M., Tanhaei, M., & Sheykholeslami, S. O. R. (2017). Study on an elaborated method to improve corrosion resistance of zinc phosphate coating. *Journal of Coatings Technology and Research*, **14**(3), 709–720.
[14] Popić, J. P., Jegdić, B. V., Bajat, J. B., Veljović, D., Stevanović, S. I., & Mišković-Stanković, V. B. (2011). The effect of deposition temperature on the surface coverage and morphology of iron-phosphate coatings on low carbon steel. *Applied Surface Science*, **257**(24), 10855–10862.
[15] Wang, X., Shen, B., Lin, H., Zhang, Q., Gao, J., Yu, Z., & Yi, H. (2018). Energy-exergy analysis and energy efficiency improvement of coal-fired industrial boilers based on thermal test data. *Applied Thermal Engineering*, **144**, 614–627.
[16] Uppal, A., Kesari, J. P., & Zunaid, M. (2016). Designing of solar process heating system for indian automobile industry. *International Journal of Renewable Energy Research*, **6**(4).
[17] Lee, K. H. (2012). Carbon accounting for supply chain management in the automobile industry. *Journal of Cleaner Production*, **36**, 83–93.
[18] Lee, K. H. (2011). Integrating carbon footprint into supply chain management: The case of Hyundai Motor Company (HMC) in the automobile industry. *Journal of Cleaner Production*, **19**(11), 1216–1223.
[19] L.A. Tiago, C.M. António, A.F. José, Strategies to make renewable energy sources compatible with economic growth, *Energy Strat. Rev.* **18** (2017) 121–126.
[20] Antunes, P., 2014. A Maturity Model for Energy Management.
[21] Tuo, J., Liu, F., Liu, P., Zhang, H., & Cai, W. (2018). Energy efficiency evaluation for machining systems through virtual part. *Energy*, **159**, 172–183.
[22] Apostolos, F., Alexios, P., Georgios, P., Panagiotis, S., & George, C. (2013). Energy efficiency of manufacturing processes: A critical review. *Procedia CIRP*, **7**, 628–633.
[23] Barma, M.C., Saidur, R., Rahman, S.M.A., Allouhi, A., Akash, B.A., Sait, S.M., 2017. A review on boilers energy use, energy savings, and emissions reductions. *Renew. Sustain. Energy Rev.* **79**, 970–983.
[24] Smith, C. B., & Parmenter, K. E. (2016). General Principles of Energy Management. *Energy Management Principles*, 35–44.

[25] Jouhara, H., Khordehgah, N., Almahmoud, S., Delpech, B., Chauhan, A., & Tassou, S. A. (2018). Waste heat recovery technologies and applications. *Thermal Science and Engineering Progress*, **6**, 268–289.
[26] Sundén B. Introduction to heat transfer. *Southampton: WIT Press*; 2012.
[27] Law, R., Harvey, A., & Reay, D. (2013). Opportunities for low-grade heat recovery in the UK food processing industry. *Applied Thermal Engineering*, **53**(2), 188–196.
[28] Frangopoulos, C. A. (2018). Recent developments and trends in optimization of energy systems. *Energy*, **164**, 1011–1020.
[29] Poe, W. A., & Mokhatab, S. (2017). Process Optimization. In Modeling, Control, and Optimization of Natural Gas Processing Plants.
[30] Zeng, Z., Hong, M., Li, J., Man, Y., Liu, H., & Li, Z. (2018). Integrating process optimization with energy-e ffi ciency scheduling to save energy for paper mills. *Applied Energy*, **225**(January), 542–558.
[31] Schmidt, M. (2008). The Sankey Diagram in Energy and Material Flow Management Part I : History. **12**(1).
[32] Ebrahim, M., & Kawari, A.-. (2000). Pinch technology: an efficient tool for chemical-plant energy and capital-cost saving. *Applied Energy*, **65**(1-4), 45–49.

IOP Conf. Series: Materials Science and Engineering **743** (2020) 012035 doi:10.1088/1757-899X/743/1/012035

Partially Replacement of Cement By Sawdust And Fly Ash in Lightweight Foam Concrete

M F Omar[1*], M A H Abdullah[1], N A Rashid[1] and A L Abdul Rani[2]

[1]Faculty of Engineering Technology, Department of Civil Engineering Technology, University Malaysia Perlis, Kampus UniCITI Alam Sungai Chuchuh, Padang Besar 02100 Perlis, Malaysia
[2]Faculty of Engineering Technology, Department of Chemical Engineering Technology, University Malaysia Perlis, Kampus UniCITI Alam Sungai Chuchuh, Padang Besar 02100 Perlis, Malaysia

Email: * mardhiah@unimap.edu.my

Abstract. The rapid growth of population has led to increased demand for fast, affordable and quality housing development. Today, the construction industry in Malaysia has shifted from conventional methods to Industrial Building Systems (IBS). The most commonly used IBS component is precasat concrete with lightweight foam concrete. This study focuses on the main component of foam lighweight concrete, which is a partially replacement of cement by sawdust and fly ash. Among the features of lightweight concrete is density below 1800 kg /m3. Therefore, the objectives of this study is to determine the effects of sawdust and fly ash as part of cement replacement in terms of mechanical properties (compressive strenght) and physical properties (water absorption). In addition, this study also determine the optimum percentage of cement replacement by sawdust and fly ash in building material. The percentage of saw dust and fly ash used in this study as a partial replacement cement are 5%, 10%, 15% and 20%. The results show that increasing the percentage of mix propotion will increase the water absorption rate as well as decrease the compressive strenght of strength. Also, the density and compressive strength of lightweight foam concrete will decrease as the percentage of partial replacement cement increases. According to JKR Standard Specification for Building Works that referred in Malaysia, the minimum compression strength of lightweight foam concrete allowed for hollow blocks is 2.8 N / mm2. The results obtained from this study show lightweight concrete blocks using saw dust and fly ash as part of the cement replacement meet the standards and can be commercialized in the industrial building system development.

1. Introduction

Lightweight concrete is a type of concrete made up of Ordinary Portland Cement, sand, water and voids. The voids are usually imported into the system by adding the suitable foaming agent to the water at the required mix ratio in a foam generator to produce manufactured foam [1]. Lightweight concrete which density between 400 kg/m3 and 1850 kg/m3 with high flowability, low cement usage, low aggregate usage and good thermal insulation [2]. The foaming agent can be used with all types of OPC and blended into cement only, or sand-cement mortar slurries to produce foamed concrete when used it. There are two types of foaming agents that can be added into the mix design which categorized into the protein-based foaming agents and synthetic foaming agents. Protein-based foaming agents can be obtained from animal proteins out of the horn, blood, bones of cows, pigs and other remainders of

Content from this work may be used under the terms of the Creative Commons Attribution 3.0 licence. Any further distribution of this work must maintain attribution to the author(s) and the title of the work, journal citation and DOI.
Published under licence by IOP Publishing Ltd

animal carcasses [3]. Besides that, synthetic foaming agents are made up of purely chemical products. Synthetic foaming agent which is used widely in the world due to its easier production and availability in every country [4]. It has permanent properties and working life much longer compared to a protein-based foaming agent [5, 6]. The synthetic foaming agent is very stable at concrete with densities above 1000 kg/m3 and it can give more strength to the concrete [7]. In addition, sawdust is produced during sawing of logs of timber into different sizes like a small irregular shape or small fragments of wood [8]. Sawdust has many advantages such as low bulk density, improved in heat preservation and heat insulation property, and reduce the pollution to the environmental [9]. Therefore, sawdust has been identified as one type of suitable filler material in which the industry can be used to produce hollow concrete blocks [9, 10]. Another material is fly ash as a type of waste produced by burning coal in an electrical generating station. It can help to improve workability, reduce permeability, and increase the ultimate strength. It also reduced bleeding, had a better surface and reduced heat of hydration [11]. The suitable replacement in the construction of new road networks was by applied the usage of brown coal fly ash [12].

2. Experimental Procedure

2.1 Raw Material

Sawdust used in this research was collected from the wood factory around the Kangar and Kaki Bukit, Perlis, Malaysia. The collected sawdust was put into the furnace for 1 hour at 600°C [13]. Next, the burnt sawdust ash was cooled under normal conditions for 24 hours. After that, the sawdust ash was ground into powder form by using mortar. The ground sawdust ash was sieved through 600 μm sieve (sieve no. 30). The foaming agent is formed through the Sodium Lauryl Sulphate (SLS) along with Sodium hydroxide (NaOH) at an SLS concentration of 2.5% (the ratio of SLS to water is 1:40 by volume) and NaOH concentration of 4% (the ratio of NaOH to water is 1:25 by volume) with a pressure of 130 KPa. The foam is generated by using a foam generator [14].

2.2 Mix Design and Process

Three types of materials will be mixed, which are cement together with 5%, 10%, 15% and 20% combination of sawdust and fly ash as a partial cement replacement material, sand as fine aggregate and water. The water to solid ratio used in the trial was fixed at 0.38. Besides that, the amount of foam used was fixed at 10 kg/m3. In addition, for the concrete mixing process the amount of Ordinary Portland Cement (OPC), sand, sawdust and fly ash were weighted according to the trail mix. All the weighted materials are poured into a mixing pot for mixing until evenly mixed. After that, the water was weighed and added into the dry mix slowly. At the same time, the foam was generated by using the foam generator. The foam was then weighted and added into the wet mix. The wet mix was mixed until it was uniformly mixed. Then, the inverted slump test will be carried out to determine the properties of lightweight concrete. Lastly, the fresh concrete mixes and poured into the mold for casting.

2.3 Testing and Analyzing

The fresh density test was carried out based on ASTM C796 (2004) [19] to ensure the densities of lightweight concrete is below 1800 kg/m3. The inverted slump test was carried out to determine its workability per ASTM C1611 (2005) [20]. As complied with BS 1881-116 (1983) [21], the compressive strength test was conducted by using the Compressive Strength Testing Machine. The load is applied continuously at a constant rate within the range 0.2 N/(mm2·s) to 0.4 N/(mm2·s) until the specimen failed. A water absorption test was carried out in this research to calculate the water absorption capacity of hardened lightweight concrete in accordance with BS 1881-122 (1983) [22].

3. Results and Discussion

3.1 Fresh Density Test

Fresh density test is one of the tests which can use to determine the unit weight of the concrete admixture before it is casting. The density of lightweight concrete is below 1800 kg/m3. Table 1 to Table 4 shows the differences between fresh densities of various concrete samples before and after adding the foam were in the range of ± 50 kg/m3. The decrease of the densities can be due to the foam volume as it will increase the voids inside the concrete [6, 15]. The manufacturing of foamed concrete which had been done in the industry was controlled in the range of ± 50 kg/m3 [16].

Table 1. The fresh density test for 5% of cement replacement.

Description	Density of sample (kg/m^3)					
	1	2	3	4	5	6
Before adding foam	1755	1705	1710	1690	1700	1725
After adding foam	1725	1730	1750	1720	1650	1690
Differences	-30	25	40	30	-50	-35

Table 2. The fresh density test for 10% of cement replacement.

Description	Density of sample (kg/m^3)					
	1	2	3	4	5	6
Before adding foam	1570	1590	1575	1605	1610	1620
After adding foam	1605	1620	1605	1580	1580	1590
Differences	35	30	30	-25	-30	-30

Table 3. The fresh density test for 15% of cement replacement.

Description	Density of sample (kg/m^3)					
	1	2	3	4	5	6
Before adding foam	1520	1480	1470	1450	1490	1500
After adding foam	1480	1530	1510	1490	1475	1465
Differences	-40	50	40	40	-15	-45

Table 4. The fresh density test for 20% of cement replacement.

Description	Density of sample (kg/m^3)					
	1	2	3	4	5	6
Before adding foam	1370	1320	1330	1375	1390	1310
After adding foam	1350	1370	1380	1320	1340	1340
Differences	-20	50	50	-45	-50	30

3.2 Slump Value

The slump test value must between 500mm to 700mm for self-consolidating concrete [15]. The slump test value obtained for 5%, 10%, 15% and 20% of cement replacement were over 500mm which can be categorized as self-consolidating concrete. Table 5 shows the slump test value for each percentage of cement replacement of the sample.

Table 5. Slump test value for each percentage of cement replacement.

Percentage of cement replacement (%)	Slump test value (mm)
5	520
10	510
15	525

3.3 Density and Compressive Strength

28 days compressive strength of each percentage of cement replacement of the samples were studied in order to determine its suitability to be used as a building material. The density of the lightweight concrete is decreased when the percentage of cement replacement is increased as shown in Figure 1. This is because the density of sawdust and fly ash used to replacement the cement had reduced the overall density of the lightweight concrete [16]. In addition, the decrease in density will also decrease the compressive strength. Based on British Concrete Association (BCA), the dry density of 1600 kg/m3 of foamed concrete will have compressive strength between 7.5 – 10.0 N/mm2. Based on Figure 1 and Figure2, 5% of cement replacement achieved the highest compressive strength as compared with others for 28 days of age and had achieved the standard set by BCA. The compressive strength is decreasing linearly when the cement replacement is increasing. For the sample of 10%, and 15% cement replacement by sawdust and fly ash, the compressive strength set by the BCA should be at the range between 6.0 – 8.0 N/mm2. The result obtained was in the range of the standard by BCA. In addition, the compressive strength for the sample of 20% cement replacement by sawdust and fly ash had achieved the BCA as the average density is 1341.7 kg/m3. As BCA standard of compressive strength for density between 1200 kg/m3 to 1400 kg/m3 was 4.5 – 5.5 N/mm2.

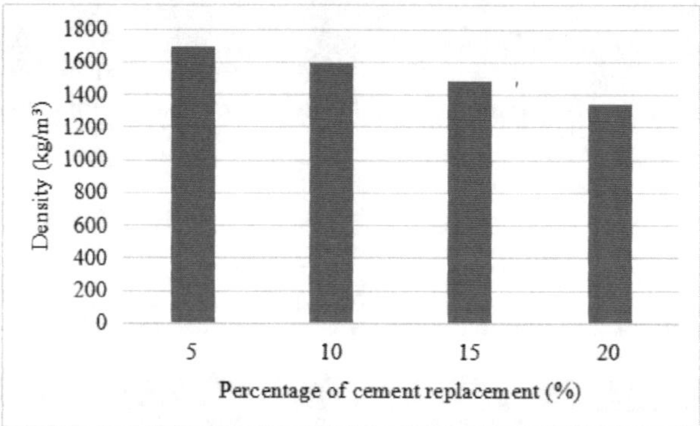

Figure 1. The density of the each percentage of cement replacement of the samples at 28 days.

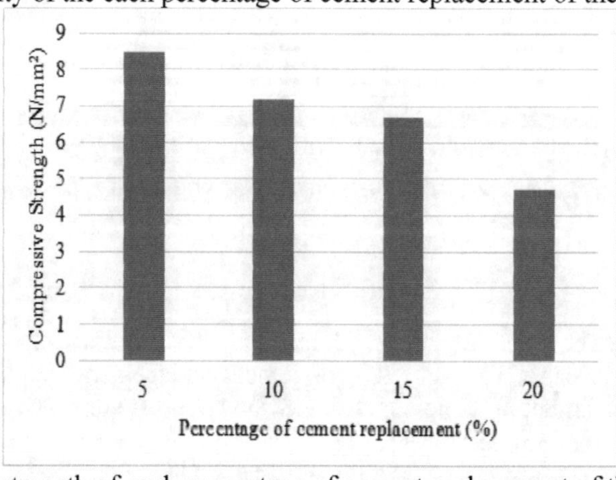

Figure 2. Compressive strength of each percentage of cement replacement of the samples at 28 days.

3.4 Water Absorption

Water absorption is a test that can be classifying the durability of the concrete. Normally, concrete with low water absorption was better as it can protect the reinforcement within the concrete. Based on Figure 3 and Figure 4, the water absorption increases when the weight of the concrete becomes lighter. The concrete with lower density can produce large amount of air-voids between them. The air voids can be filled up with the water which causes high water absorption. Besides that, the higher percentage of cement replacement had a significant increase in the percentage of water absorption. The percentage of water absorption was high can be caused by the total voids distributed inside the sample that had been increased due to the replacement cement by sawdust and fly ash [17]. In addition, the maximum water absorption for reinforced masonry units under category lightweight concrete is 288 kg/m3 which is stated in the ASTM (C90-01) [23]. Based on the result obtained, the concrete with 20% cement replacement is not suitable to be used as masonry units as the water absorption is more than 288 kg/m3.

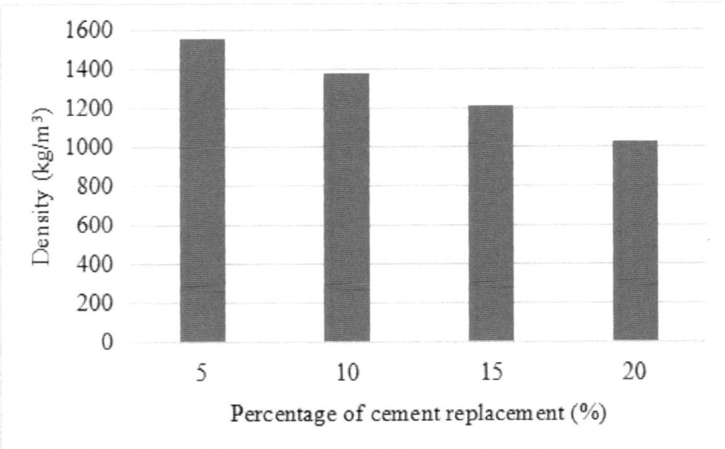

Figure 3. Density of the samples with different percentage of cement replacement.

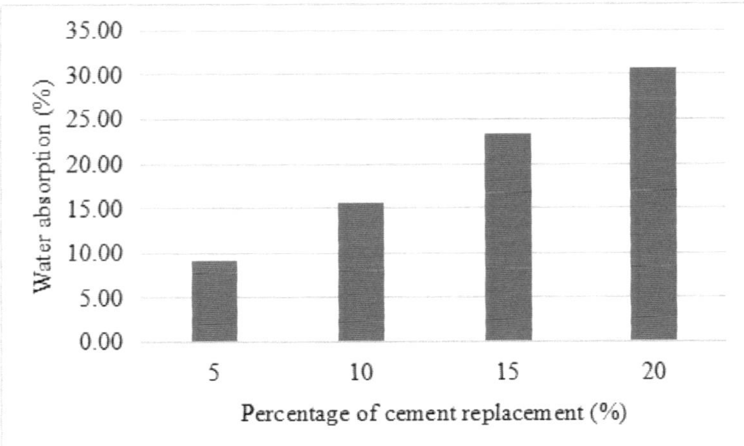

Figure 4. Water absorption of the samples with different percentage of cement replacement.

3.5 Application as Building Material

The eco-friendly building material had a high demand in the market to reduce environmental issues. Based on the result obtained the sample with 5%, 10% and 15% of cement replacement by sawdust and fly ash can be chosen to be used as a building material. This is because of the compressive strength of 5%, 10%, and 15% had achieved the requirement with an acceptable range of water absorption. The density between 1600 kg/m3 and 1800 kg/m3 was recommended to be used as slabs

[18]. In addition, the concrete sample with 15% cement replacement can be used as housing applications.

4. Conclusions
The density of lightweight concrete will affect the compressive strength. The highest compressive strength of lightweight concrete with partially cement replacement is at 5%. The water absorption for the sample 5%, 10% and 15% of cement replacement was accepted in the range of water absorption which was 288 kg/m^3. Furthermore, 5%, 10% and 15% cement replacement by sawdust and fly ash can be used as building material such as slabs and housing applications.

Acknowledgement
This research was supported by the facilities provided by Fakulti Teknologi Kejuruteraan (FTK), Universiti Malaysia Perlis (UniMAP).

References
[1] E.B. Coker, S. Sadiku, J.I. Aguwa, M. Abdullahi, *Niger Technol J*, **35**, 699–706 (2016).
[2] Y.H.M. Amran, N. Farzadnia, A.A.A. Ali, *Constr. Build. Mater,* **101**, 990–1005 (2015).
[3] A.J. Hamad, *Int. J. Mater. Sci. Eng,* **2** no 2 pp 152–157 (2014).
[4] A.S. Sharma, G. Suhil, L. Visali, *Int. J. Res. Appl. Sci. Eng. Technol*, **5**, 1078 (2017).
[5] S.F. Ahemed, M. Hashmath, *Int. J. Adv. Technol. Innov. Res.*, **9**, 1385–138 (2017).
[6] M. Jalal, A. Tanveer, K. Jagdeesh, F. Ahmed, *Int. J. Civ. Eng. Res* ,**8**, 2278–36(2017).
[7] M.A. Yahya, M.A. Yusof, *Journal of Engineering and Applied Sciences*, **12**, 4063–4067 (2017).
[8] Kumar D, Singh S, Kumar N, and Gupta A, *Glob. J. Res. Eng.*, **14**, 33–36 (2014).
[9] Cheng Y, You W, Zhang C, Li H, Hu J, *Engineering*, **5** , 943–947 (2013).
[10] A.S.M. Awal, A.A.K.Mariyana, M.Z. Hossain, *Int. J. GEOMATE*, **10**, 1918–1923 (2016).
[11] V. Goud, N. Soni, *IOSR J. Eng.,* **6(10),** 69–75 (2016).
[12] M. Ondova, N. Stevulova, L. Meciarova, *Procedia Eng.* **65**, 45–5 (2013).
[13] S. Chowdhury, M. Mishra, O. Suganya, *An overview Ain Shams Eng. J.*, **6**, 429–43(2015)
[14] Rin G, Anand K B *IOP Conf. Ser. Mater. Sci. Eng.*, **310**, 1–10 (2018).
[15] E. Mahmoud , A. Ibrahim, H. E. Chabib, V.C. Patibandla, *Int. J. Concr. Struct. Mater.* **7(2)**, 155–163, (2013).
[16] J. Kamau, A. Ahmed, F. Hyndman, P. Hirst, J. Kangwa J, *EJERS, European Journal of Engineering Research and Science,* **2(3)**, 36–43 (2017).
[17] N.M.I.S. Salehuddin, R.C. Amat, N.L. Rahim, T.N.T. Izhar, *Int. Conf. Environ. Sci. Dev.*, **5**, 497–501 (2013).
[18] Mohd Sari K A and Mohammed Sani A R, Applications of Foamed Lightweight Concrete *MATEC Web Conf.*, **01097**, 1–5(2017).
[19] ASTM C796 2004 *American Society for Testing and Materials* Standard Test Method for Foaming Agents for Use in Producing Cellular Concrete Using Preformed Foam .
[20] ASTM C1611 2005 *American Society for Testing and Materials* Standard Test Method for Slump Flow of Self-Consolidating Concrete ASTM International BS.
[21] BS EN 1881-116 1983 *British Standards Institution* Method for determination of compressive strength of concrete cubes (London).
[22] BS EN 1881-122 1983 *British Standards Institution* Testing Concrete Method for Making Test Cubes from Fresh Concrete (London).
[23] ASTM C90-01 2011 *American Society for Testing and Materials* Standard Specification for Loadbearing Concrete Masonry Units.

IOP Conf. Series: Materials Science and Engineering **743** (2020) 012036 doi:10.1088/1757-899X/743/1/012036

Production of Iron Oxide Nanoparticles by Co-Precipitation method with Optimization Studies of Processing Temperature, pH and Stirring Rate

Beh Hui Hui[1] and Midhat Nabil Salimi[1]

[1]School of Bioprocess Engineering, Universiti Malaysia Perlis, Perlis, Malaysia

Email: nabil@unimap.edu.my

Abstract. Iron Oxide Nanoparticle, maghemite (γ-Fe_2O_3) has received great interest and extensively used in biomedical field. Optimization studies were carried out in the production of γ-Fe_2O_3 nanoparticles by using co-precipitation method. Iron (II) chloride and iron (III) chloride were used as precursors which are dissolved in distilled water followed by centrifugation, drying and grinding process in order to obtain dried dark brown precipitated γ-Fe_2O_3 powder. The effect of different processing temperature (30 to 70°C), pH (10 to 12) and stirring rate (300 to 700rpm) towards crystallite size of γ-Fe_2O_3 were investigated by using Response Surface Methodology (RSM) and Central Composite Design (CCD). Based on analysis of variance (ANOVA), the determination coefficient, R^2 obtained was 0.9890 where stirring rate was the parameter that affected the most on the crystallite size. Optimization processing condition that produce smallest crystallite size of 7.3657 nm was 50 °C, pH 11.40 and 550 rpm by using Design of Expert software (DOE). Characterization of γ-Fe_2O_3 powder samples were evaluated by using different analytical tools such as Fourier Transform Infrared (FTIR), X-Ray Diffractometer (XRD) and Scanning Electron Microscope (SEM). Iron oxide group (Fe-O), hydroxyl (OH-) group and carbon dioxide (CO_2) were identified in FTIR spectrum. The characteristic peak occurring at $2\theta = 35.4°$ indicated presence of γ-Fe_2O_3 in the samples. The γ-Fe_2O_3 particles appeared generally in spherical shape in SEM analysis.

1. Introduction

Recently, magnetic nanoparticles received a great interest and being extensively used around the globe. The superparamagnetic properties of iron oxide (γ-Fe_2O_3) nanoparticles have gained significant attention due to its attractive feature which contribute in nanomedical application. The concept of utilizing magnetic nanoparticles in gene and drug delivery was introduced in the late 1970s [1]. The first success of gene therapy with the used of γ-Fe_2O_3 nanoparticles was achieved in Italy in which modified gamma retroviral vector was utilized to transfer therapeutic gene to targeted stem cell in order to cure the life threatening disease [2]. The fatal disease such as cancer is treated with traditional approach such as surgery, immunotherapy, radiation therapy and chemotherapy. However, these approach have shortcoming of damaging normal cell [3]. Therefore, emergence of γ-Fe_2O_3 nanoparticles as non-viral vector have overcome the mentioned issue.

There are various type of magnetic nanoparticles available nowadays. Among them, iron oxide nanoparticles, IONPs maghemite (γ-Fe_2O_3) owns the unique properties such as nanoscale size, large ratio of surface-to-volume, low curie temperature and high coercivity [4]. Size and shape of IONPs significant influences the application in various field. γ-Fe_2O_3 nanoparticles has high surface activity with the size less than 100 nm are more efficient in biomedical application [5]. In spite of the progression of technology, it is still challenging to govern the size and shape of IONPs which will influence their magnetic properties [6].

Content from this work may be used under the terms of the Creative Commons Attribution 3.0 licence. Any further distribution of this work must maintain attribution to the author(s) and the title of the work, journal citation and DOI.
Published under licence by IOP Publishing Ltd

In the last decades, different types of efficient method to synthesis iron oxide nanoparticles are developed by many researchers as to synthesis shape-controlled, monodispersed and biocompatible iron oxide nanoparticles. Numerous chemical, physical and biological route have been utilized in order to produce appropriate surface chemistry of magnetic nanoparticles [7]. The common chemical method including co-precipitation, sol-gel, hydrothermal, microemulsion and thermal decomposition. Among these methods, co-precipitation method is the most promising and cost effective method to produce γ-Fe2O3 nanoparticles due to its simplicity and high productivity. The characteristics of final product such as surface response, particle size and shape of the IONPs depends upon preparation technique. Only the nanosized γ-Fe_2O_3 nanoparticles with appropriate size, shape and purity has noteworthy impact in diverse biomedical applications and different scientific research areas.

In this research, co-precipitation method is applied to produce IONPs maghemite (γ-Fe_2O_3) for optimization and characterization studies with variable processing parameter such as processing temperature, pH and stirring rate. Therefore, γ-Fe_2O_3 nanoparticles can be produced by decisive control of these parameters. The γ-Fe2O3 nanoparticles formed from all the processing conditions will be systematically characterized and compared by using Scanning Electron Microscopy (SEM), X-Ray Diffraction (XRD) as well as Fourier Transform Infrared (FTIR).

2. Materials

2.1 Chemicals

Iron (II) chloride ($FeCl_2.4H_2O$, 99% pure), Iron (III) chloride ($FeCl_3$, 98% pure, anhydrous), sodium hydroxide (NaOH, 98% pure) were purchased from Acros Organics and were used without further purification.

3. Procedures

3.1 Synthesis of maghemite nanoparticles

In this experiment, γ-Fe_2O_3 nanoparticles have been produced by co-precipitation method. 0.7954 g of iron (II) chloride, $FeCl_2.4H_2O$ and 1.2974 g of iron (III) chloride, FeCl3 were each dissolved in distilled water. The mole ratio of iron (II) chloride and iron (III) chloride were fixed at 1:2 throughout the experiment. The solution was then mixed for 30 minutes by using hot plate with magnetic stirrer. In mixing steps, different processing temperatures, pH and stirring rates were regulated. After that, the mixed solution was separated by using centrifuge at 4000 rpm for 15 minutes. The separated precipitate was dried in the oven at temperature of 100 °C for 24 hours. Finally, the dried dark brown precipitated sample was collected and crushed into powder form by using a pestle and mortar. The chemical balanced equation for the reaction $FeCl_2.4H_2O$ and FeCl3 is shown in Equation 1-3 where γ-Fe_2O_3 was the final product and water was the by-product:

$$Fe^{2+} + 2Fe^{3+} + 8OH^- \rightarrow Fe_3O_4 + 4H_2O \quad (1)$$

$$Fe_3O_4 + 0.25O_2 + 4.5H_2O \rightarrow 3Fe(OH)_3 \quad (2)$$

$$2Fe(OH)_3 \rightarrow \gamma Fe_2O_3 + 3H_2O \quad (3)$$

3.2 Particle Characterization Technique

γ-Fe_2O_3 nanoparticles formed from all the processing conditions can be systematically characterized in terms of functional groups, phase composition and crystallite size as well as morphology structure by using Fourier Transform Infrared Spectroscopy (FTIR), X- Ray Diffraction (XRD) and Scanning Electron Microscopy (SEM), respectively.

3.2.1 Fourier Transform Infrared Spectroscopy (FTIR)

Fourier Transform Infrared (FTIR) spectrometer (Perkin Elmer, Spectrum Version 10.02.00) was performed in order to investigate the functional groups in γ-Fe_2O_3 with the resolution of 4000-450 cm-1. According to Nazari et al. [8], the most characteristic functional groups in the FTIR spectrum of synthesized maghemite nanoparticles are the vibration of iron oxide group (Fe-O), hydroxyl group (OH-) and atmospheric carbon dioxide, CO_2. The powder sample was grounded into fine powder before being placed onto the plate in order to obtained more accurate result.

3.2.2 X-ray Diffraction (XRD)
The phase composition of maghemite (γ-Fe2O3) was undertaken by using an X-Ray Diffractometer (XRD-6000 Shimadzu). The analysis of phase composition and crystallite size will be generated by using wavelength at λ=0.15406 nm over the 2θ range of 20° to 70° and proceed with scanning rate of 2°/min [9]. The peak broadening of XRD will be used to evaluate the crystallite size of IONPs based on the Scherrer formula.

3.2.3 Scanning Electron Microscopy (SEM)
SEM (JSM-6460LA) was used to examine the surface of γ-Fe2O3 nanopowder. A focused electron beam was scanned over a surface of sample to create an image when detection of secondary electrons. Cheng et al. [10] reported maghemite nanoparticle observed using SEM to be spherical shape.

3.3 Optimization study of γ-Fe2O3 nanoparticles by using Design-Expert® Software
DOE software (Design Expert 7.1.5) was applied to determine the number of experimental runs with the used of Response Surface Methodology (RSM) and Central Composite Design (CCD). The application of DOE able to obtain more information from fewer experiment. Response Surface Methodology (RSM) and Central Composite Design (CCD) were used to study the effect of parameters towards the response which is crystallite size of γ-Fe2O3. The parameters chosen were shown in Table 1.

Table 1. Selected processing parameters for optimization experiment.

Factors	Name	Units	Low (-α)	High (+α)
A	Processing temperature	°C	30	70
B	pH	-	10	12
C	Stirring rate	rpm	300	700

4. Result and Discussion
4.1 Maghemite nanoparticles characterization analysis
4.1.1 Fourier Transform Infrared Spectroscopy (FTIR)
FTIR spectrum analysis is an important tool to investigate the biomolecules present in IONPs. The spectra generated by FTIR was investigated to identify the functional groups and organic compounds contained in nanoparticles. The most characteristic functional groups in the FTIR spectrum of synthesized maghemite nanoparticles, γ-Fe2O3 are vibration of iron oxide group (Fe-O), hydroxyl group and atmospheric carbon dioxide [8]. Five samples of dark brown powder form of γ-Fe2O3 nanoparticles with different processing temperature, pH and stirring rate is selected and investigated by using Perkin Elmer Spectrum Version FTIR equipment with the resolution of 4000-450 cm-1. The tested ATR-FTIR spectrum of γ-Fe2O3 nanoparticles result are shown in the Figure 1.

The tested sample powder of IONPs, maghemite (γ-Fe2O3) shows infrared absorption spectra as in Figure 1. The result obtained for five samples shows the usual band of IR absorption. The absorption band assigned to the stretching mode of hydroxyl group of water molecules can be observed at around 3391-3403 cm-1. Alternatively, for O-H stretch vibration of hydrogen-bonded hydroxyl group, the adsorbed water band obtained was in the range of 1607-1624 cm-1. The stretching mode of hydroxyl O-H group occurred at around 3100-3600 cm-1 was the major evidence of presence of OH group of water [11]. Two absorption bands appeared at 3,170 and 3,408 cm−1 represented the stretching band of O-H group [8]. Chakrabarti et al. [12] reported that γ-Fe2O3 nanoparticle at 1621 cm-1 and 3387 cm-1 absorption peak corresponded to the adsorbed water. Presence of hydroxyl group indicated crystallization of water in the powder. Hence, the result of the absorption band for O-H group was successfully achieved.

Chauhan et al. [13] stated presence of iron oxide functional group occurred at region of low frequency. Octahedral and tetrahedral sites of maghemite occurred vibration of Fe-O at absorption bands of 630 cm-1 to

550 cm-1 range [14]. From the Figure 1, the strong and obvious absorption peak that obtained was in the range of 631-634 cm-1 indicated the vibration of Fe-O functional group of maghemite nanoparticles. As the maghemite molecule contained main elements of iron and oxygen, existence of Fe-O group is necessary. Absorption peak for Fe-O group can be observed for the five tested sample powder. According to Aliahmad and Nasiri Moghaddam [15], the intense peak for Fe-O group at 586 cm-1. On the other hand, the absorption peaks at 636 cm-1 and 636 cm-1 assigned to Fe-O group of maghemite [16].

According to Aliahmad and Nasiri Moghaddam [15], atmospheric carbon dioxide can be observed at 2361 cm-1. From Figure 1, IR absorption band at 2922 cm-1 shows the presence of carbon dioxide as maghemite nanoparticle exposed to atmosphere that contained carbon dioxide. Besides, 2337 cm-1 and 2361 cm−1 wavelength in absorption spectrum also indicated existence of atmospheric carbon dioxide [12]. Formation of IR band occurred only in the presence of dipole moment changing in the molecules. In short, dipole moment changes during the molecular vibration [17].

Nazari et al. [8] stated that pure maghemite contained other peaks such as at 892 cm-1 and 795 cm-1. The result of the five samples presented the remaining peaks of maghemite in the range of 804-898 cm-1. Furthermore, the absorption peak at 1330-1337 cm-1 can be considered as vibration of ferrophase complex. However, there is one of the absorption spectrum which is overlap with the others sample, probably due to the polarization effect that affects the light reflection. In addition, crystalline sample considered a nuisance in ATR-FTIR analysis as will reflect the beam and lead to wrong spectra. Hence, crushing of crystalline sample into fine powder is necessary due to the sensitivity of the instrument.

FTIR result obtained clearly indicated three main functional groups, Fe-O, OH- and CO2 were observed in their respective IR region. The formation of spectra for all the tested IONPs powder were followed the same trend regardless the effect of processing factors such as pH, temperature and stirring rate. Hence, the IONPs tested are free from impurities.

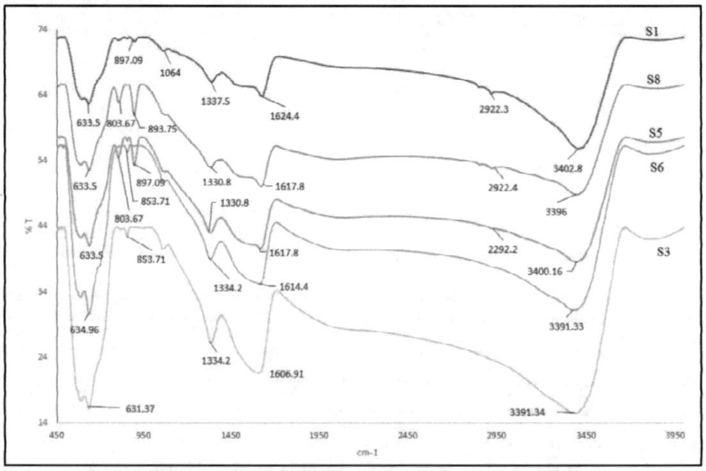

Figure 1. ATR-FTIR spectrum of maghemite.

4.1.2 X-ray Diffraction (XRD)

The phase composition of IONPs maghemite (γ-Fe2O3) was undertaken by using an X-Ray Diffractometer (XRD-6000 Shimadzu). Determination of crystallite size and phase composition was generated by using wavelength of Cu-Kα radiation (λ = 0.15406 nm) over 2θ range of 20 to 70° proceed with scanning rate of 2°/min. Figure 2 shows the diffraction pattern of maghemite (S4) that synthesized at a processing temperature of 50°C, pH 11 and stirring rate of 500 rpm. There are three different diffraction peaks confirming to the crystal planes of crystalline IONPs [18]. Moreover, diffraction pattern correspond to the formation of maghemite has been confirmed by indicating the characterization peak occurring at 2θ = 35.4° with (311) plane appeared in all of the tested samples. Database International Centre for Diffraction Data (ICDD) was applied during the analysis and the characteristic of maghemite was verified. The X-ray diffractogram of all the tested samples that synthesized from different processing condition were almost similar.

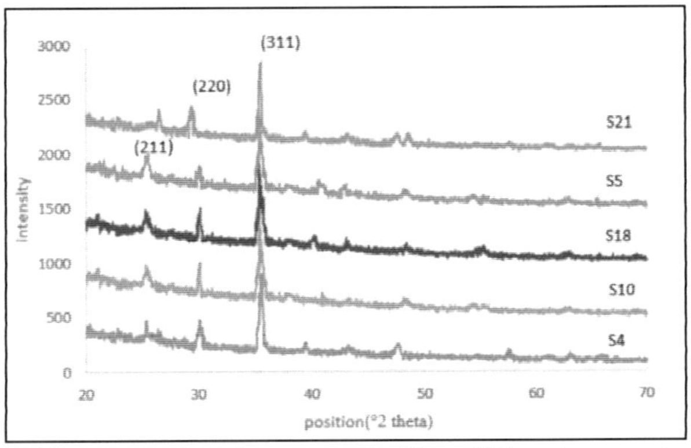

Figure 2. XRD pattern of maghemite, γ-Fe_2O_3 at different processing condition.

By using Scherrer equation 4, crystallite size of synthesized IONPs was calculated with the use of full width at half maximum at the peak broadening of diffraction reflection which occurred at 2θ = 35.4°. The result of crystallite size was tabulated in Table 2.

$$X_s = \frac{0.94\,\lambda}{FWHM\ cos\theta} \tag{4}$$

Where Scherrer constant 0.94, X_s is the crystallite size (nm), FWHM is full width at half maximum of diffraction peak (rad), λ is the wavelength radiation for Cu-Kα and θ is diffraction angle (°).

4.1.3 Scanning Electron Microscopy (SEM)

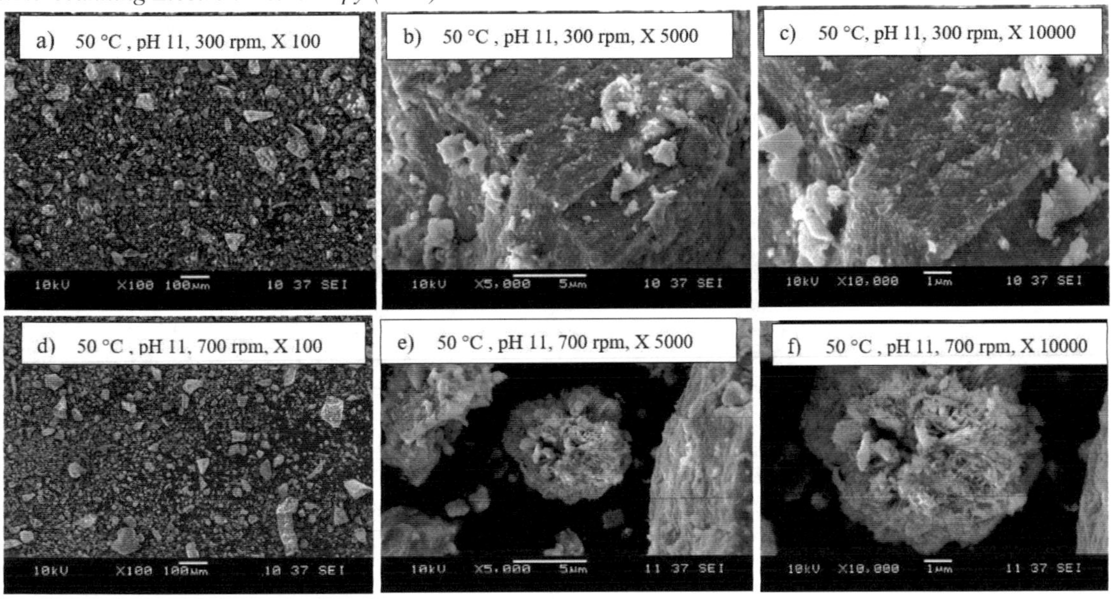

Figure 3. SEM image of IONPs at different stirring rate with constant temperature and pH.

IOP Conf. Series: Materials Science and Engineering **743** (2020) 012036 doi:10.1088/1757-899X/743/1/012036

The morphology study of IONPs maghemite was studied by using JSM-6460LA scanning electron microscope (SEM). Figure 3 depicted the SEM pictures of IONPs maghemite at the same processing temperature (50 °C) and pH 11 with different stirring rates (300 rpm and 700 rpm). The particles were generally spherical shape with some is cuboid-like formed at a lower stirring rate while the particles with spherical shape with some elongated ellipse-like were observed at higher stirring rate. The agglomerated nanoparticles may be due to magnetic characteristic of IONPs [19]. The agglomeration of IONPs also indicated the small surface to volume ratio of the particles. Moreover, the non-spherical shape formation can be explained by the shear produced during stirring process [20]. High stirring rate possess induced shear in the mixing zone that contribute to the vortex within the system which built up aggressive flow behavior and caused some of the formation of elongated particles. In addition, there is high energy dissipation at high stirring rate which enhance the nucleation rate, crystal growth mechanism and subsequent morphology is affected [21]. Therefore, IONPs showed a high tendency to agglomerate where they tended to stick together. Based on Table 2, it can be seen that 27.6080 nm and 9.9870 nm of IONPs were formed at 300 rpm and 700 rpm respectively with the constant processing temperature of 50 °C and pH 11.

4.2 Optimization study of maghemite nanoparticles by using Design-Expert ® Software

In this model design, Design Of Experiment software (Design Expert 7.1.5) was applied to determine the amount of experimental runs to optimize the synthesis of IONPs with the used of Response Surface Methodology (RSM). Central Composite Design (CCD) is an experimental model design which is useful to study the effect of processing temperature, pH and stirring rate towards the crystallite size of IONPs. There are three parameters were set at low (-α) and high (+α) level while the response set as crystallite size in quadratic model. This model consists of 20 experimental runs with 6 central points of the parameters in order to determine the crystalline size of IONPs under different combination of conditions. Table 2 shows the outcome of crystallite size of IONPs which is a responses that calculated by using Scherrer's formula based on the full width half maximum of X-ray Diffraction (XRD) spectrum diffraction peak.

Table 2. Crystallite size (nm) of IONPs for different process conditions.

Run	Factor 1 A: Temperature (°C)	Factor 2 B: pH	Factor 3 C: Stirring rate (rpm)	Response: crystallite size of IONPs (nm)
1	50.00	11.00	500.00	9.2472
2	50.00	11.00	500.00	9.5321
3	70.00	11.00	500.00	26.2270
4	50.00	11.00	500.00	9.0642
5	38.00	11.60	380.00	25.5622
6	62.00	11.60	620.00	11.0010
7	50.00	12.00	500.00	12.6532
8	38.00	11.00	620.00	25.1235
9	50.00	11.00	500.00	9.5121
10	50.00	11.00	700.00	9.9870
11	50.00	11.00	500.00	9.6421
12	62.00	11.60	380.00	25.7895
13	38.00	10.40	380.00	25.0312
14	50.00	11.00	500.00	11.1102
15	62.00	10.40	380.00	22.0933
16	62.00	10.40	620.00	18.1211
17	38.00	11.60	620.00	14.3334
18	30.00	11.00	500.00	31.5491
19	50.00	10.00	500.00	18.4072
20	50.00	11.00	300.00	27.6080

4.2.1 Analysis of Anova

Analysis of variance (ANOVA) can be used to determine the reliability of a design model by reducing random viability [22]. ANOVA is a structural system design by taking account on the presence of uncertainties involved in the reliability of the structure of a model. Structural reliability is a theoretical basis for determining the likelihood of failure, which is also known as reliability index of structural systems. The ANOVA test result was accomplished in order to validate the quadratic polynomial model's significance. Table 3 below tabulated the ANOVA for optimization study of quadratic model whereas R-Squared value for response parameter is shown in the Table 4.

Table 3. Analysis of variance (ANOVA) for optimization study quadratic model.

Source	Sum of squares	Df	Mean Square	F value	p-value	
Model	1135.13	9	126.13	99.69	< 0.0001	significant
A: Temperature	35.43	1	35.43	28.00	0.0004	
B: pH	39.96	1	39.96	31.58	0.0002	
C: Stirring rate	259.51	1	259.51	205.11	< 0.0001	
AB	5.84	1	5.84	4.62	0.057	
AC	7.27	1	7.27	5.74	0.0375	
BC	61.26	1	61.26	48.42	< 0.0001	
A^2	628.43	1	628.43	496.69	< 0.0001	
B^2	50.98	1	50.98	40.29	< 0.0001	
C^2	132.83	1	132.83	104.99	< 0.0001	
Residual	12.65	10	1.27			
Lack of fit	9.99	5	2.00	3.75	0.0866	Not significant
Pure error	2.66	5	0.53			
Cor total	1147.78	19				

*df – degree of freedom

Table 4. R-Squared value for response parameter.

Parameters	Value
Standard deviation	1.12
R-Squared	0.9890
Adjusted R-Squared	0.9791
Predicted R-Squared	0.9306
Mean	17.58
C.V. %	6.40
PRESS	79.64
Adequate precision	26.889

Based on the Table 3, the F-value of model obtained is 99.69 whereas p-value of the model is below 0.0001. This implies that the individual terms in the response model generated was statistically significant on interaction effect and only 0.01% chance of noise occurrence. The p-value expressed the significance of each factor, if the factor has a p-value below 0.05 indicate that the model terms are significant with a confidence level of 0.95 [23].

In this ANOVA analysis, the variable A, B, C, AC, BC, A2, B2, and C2 had significant effect on the response which is crystallite size of IONPs due to the p-value less than 0.05. However, the variables AB showed insignificant effect towards response due to the p-value 0.057 is greater than 0.05. Those insignificant variables were retained because there is no influence of model reduction on insignificant variables. "Lack of fit F-value" have to be insignificant so that the model fit well in the experimental design. In this analysis, 3.75 lack of fit was obtained and there is 8.66% chance that F-value of Lack of Fit this large might occur due to the noise. The pure error implied the variability of the observations of each treatment. In order to reduce large variance of experimental data, more replicates of experiments shall be conducted since 3 replication of experiments generated by DOE have been carried out.

In order to assess the model performance, analysis of determinates coefficients R-squared (R2), predicted R-squared, adjusted R squared and adequate precision were tabulated in Table 4. Based on Table 4, the R2 obtained was 0.9890 which close to value of 1.0 indicated the model was highly reliable and fit to actual data. Predicted R2 is a measurement of predictive capability of the regression model. The difference between "predicted R2" (0.9306) and "adjusted R2" (0.9791) is 0.05. This indicated that the "predicted R2" is in a reasonable agreement with "adjusted R2". If there is difference greater than 0.2 will cause data problem or model distribution. "Adequate precision" compares the range of the values predicted at design points with the prediction of average error. In other words, it is a measure of the experimental signal to noise ratio. From Table 4, "adequate precision" of 26.889 greater than 4 indicated that the performance of model is in an adequate signal situation to circumnavigate the space of design [24].

The coefficient of variation (C.V.) is a measure of relative variability on the ratio of standard deviation to average mean. The higher the C.V 6.4%, the higher the dispersion level around the mean and less reliable of this optimization study.

4.2.2 Analysis of regression

Table 5 shows coded parameters for crystallite size of IONPs, maghemite (γ-Fe_2O_3). The final equation of model in terms of coded parameters was developed by using the experimental data that listed as below.

Table 5. Analysis of regression

Factor	Coefficient estimate
Intercept	9.71
A: Temperature	-1.61
B: pH	-1.71
C: Stirring rate	-4.36
AB	0.85
AC	-0.95
BC	-2.77
A^2	6.60
B^2	1.88
C^2	3.04

The relationship between the factors and response has been developed by using Central Composite Design (CCD). A mathematical regressive model in terms of coded parameters which was linearly correlation related to variables that influenced the crystallite size of IONPs is shown in Equation 5.

$$\text{Crystallite size} = 9.71 - 1.61\,A - 1.71\,B - 4.36\,C + 0.85\,AB - 0.95\,AC - 2.77\,BC + 6.60\,A2 + 1.88\,B2 + 3.04\,C2 \quad (5)$$

Internally studentized residual is the residual (difference between actual value and predicted value) divided by that residual estimated standard deviation. Based on Figure 4, normal probability plot of internally studentized residuals was scattered closely along the positive slope. This showed an accurate model with good agreement throughout the operating variable ranges as the residuals distributed in a linear line. Besides, Figure 5 shows there was a great correspondence between the actual value (experimental data) and predicted value (evaluated by DOE). In short, the result from both graph considered as a good fit to the quadratic model.

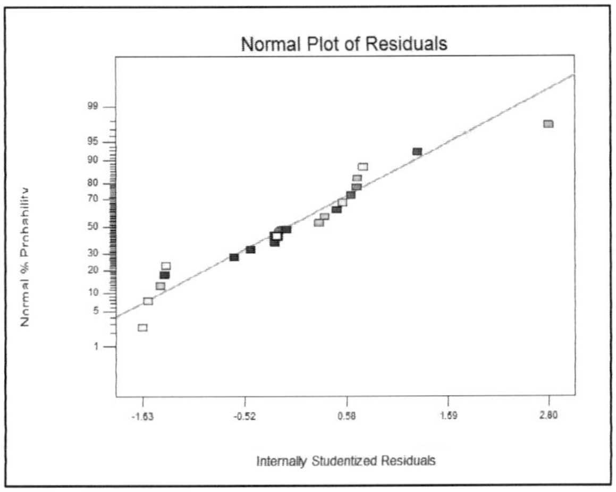

Figure 4. Normal plot of internally studentized residuals.

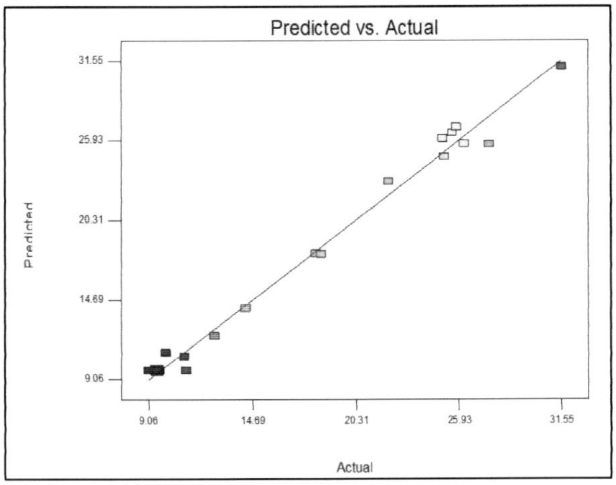

Figure 5. Predicted value vs. Actual value.

4.2.3 Analysis of model

Response Surface Methodology (RSM) is a powerful optimization tool used to determine the response of that crystallite size of IONPs, maghemite (γ-Fe2O3) at optimum condition for each parameters. Three parameters were employed in this study such as pH, temperature and stirring rate. Figure 6, 8 and 10 showed an egg-shaped 2D contour plot which indicated the relationship between the related study parameters. Furthermore, Figure 7, 9 and 11 illustrate 3D surface of hyperbola minimum hill with a minimum point located at the bottom of experimental design. The minimum point represented the minimum value of particle size. All the factors exerted positive effect towards the minimization of crystallite size as the result exhibit downward quadratic curve line.

4.2.3.1 Effect of processing temperature and pH towards IONPs crystallite size

The optimum region for processing temperature (A) was in the range of 44.3 °C to 58 °C whereas for pH (B) was in the range of 10.60 to 11.60. The particle size of γ-Fe2O3 decreased when the pH increased when temperature is fixed at 50 °C. For all the pH value, it shows that particle size decreased when processing temperature increased from 38 °C to 50 °C. However, the particle size started to increase when the processing temperature above 50 °C. The increased particle size at higher temperature can be explained by the Ostwald-ripening phenomenon where normal mechanism of growth is distorted and particle is redeposited onto the stable nuclei [9]. Besides, gradual decreased in size of particle with increased temperature due to the accelerated nuclei collision rate in which extent of aggregation within the particle is reduced by the large energy [25]. Yusoff et al. [9] agreed with this result where the particle size decreased when temperature and pH increased.

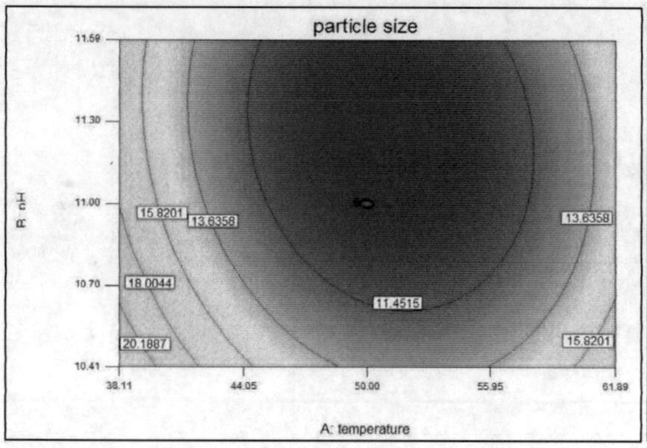

Figure 6. 2D contour plot on the interaction of A (processing temperature) and B (pH) towards particle size of γ-Fe_2O_3.

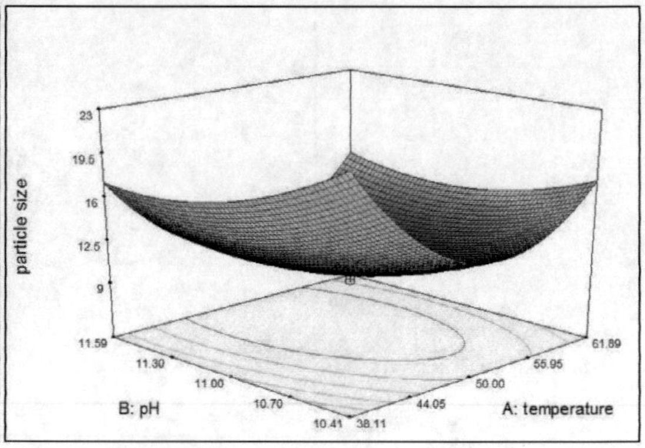

Figure 7. 3D surface plot on the interaction of A (processing temperature) and B (pH) towards particle size of γ-Fe_2O_3.

4.2.3.2 Effect of processing temperature and stirring rate towards IONPs crystallite size

Based Figure 8 and Figure 9 below, the optimum region for processing temperature (A) was 44.30 °C to 60.30 °C while for stirring rate (C) was 470 rpm to 619 rpm. At temperature of 50 °C, the particle size decreased when the stirring rate increased. This is due to high degree of agitation, the contribution of high energy to the suspension medium resulted dispersion of solution into smaller precipitate. At 500 rpm, the particle size of γ-Fe2O3 decreasing from 38 °C to 50 °C but particle size started to increase when the temperature above 50 °C. Larger size of particles are formed at low stirring velocity because the uniformity of the reaction solution are distorted [9]. The nucleating species tends to aggregate when it is not equally dispersed throughout the solution [26]. In contrast, high stirring rate resulted smaller particle size as the mixture contain greater seed formation which led to smaller particle size. Therefore, stirring rate will affect nucleation and aggregation which in turn affect the size of particles [27].

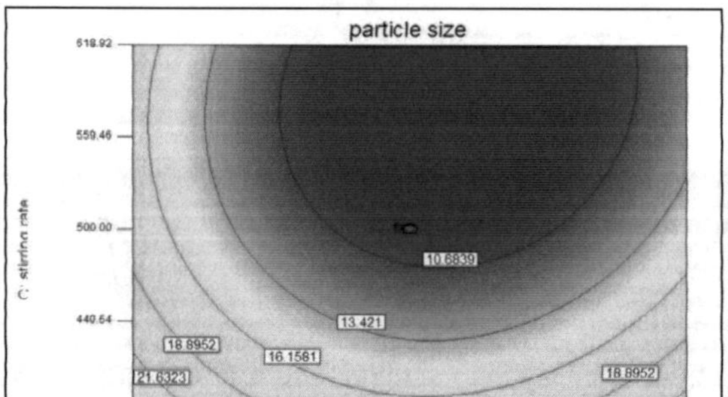

Figure 8. 2D contour surface plot on the interaction of A (processing temperature) and C (stirring rate) towards particle size of γ-Fe_2O_3.

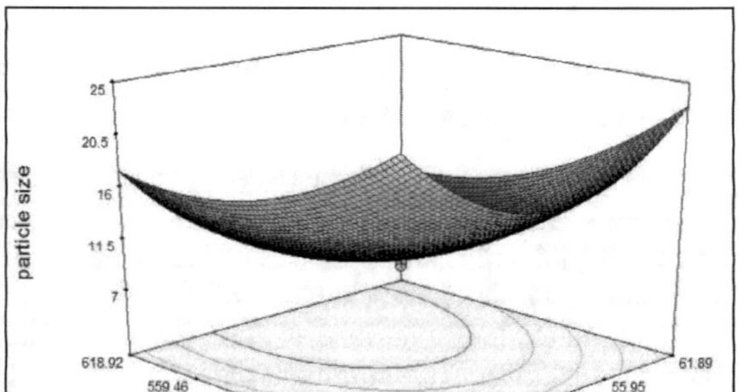

Figure 9. 3D surface plot on the interaction of A (processing temperature) and C (stirring rate) towards particle size of γ-Fe_2O_3.

4.2.3.3 Effect of pH and stirring rate towards towards IONPs crystallite size

Figure 10 and 11 clearly showed that in the range of pH 10.80 to pH 11.60 and 490 rpm to 619 rpm was the optimum region for the parameter pH (B) and stirring rate (C) respectively. Besides, particle size decreased when the stirring rate increased. This is due to the homogeneity of reaction solution is enhanced at high mechanical stirring. On the other hand, particle size decreased when pH increased from 10.40 to 11 but particle size started to increase after pH 11. There is excess ion present in the growth phase when pH of solution is increased which lead to low interfacial tension. Hence, the smaller particle size is formed at this state as the reaction rate is faster and more initial nuclei capable to precipitate. It is also observed that smaller particle size can be obtained at higher stirring rate as in agreement in the study of Yusoff et al. [9].

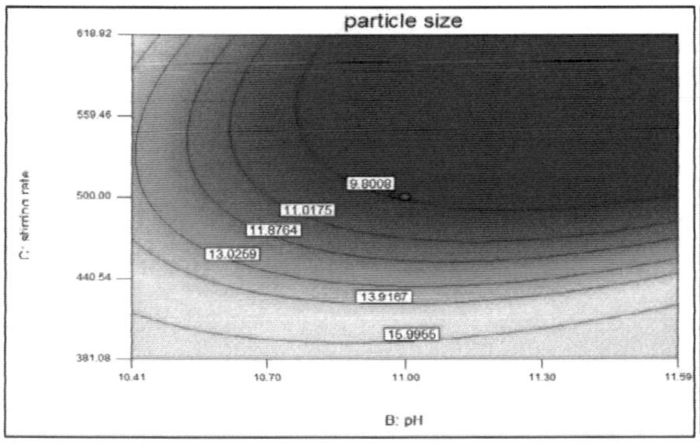

Figure 10. 2D contour plot on the interaction of B (pH) and C (stirring rate) towards particle size.

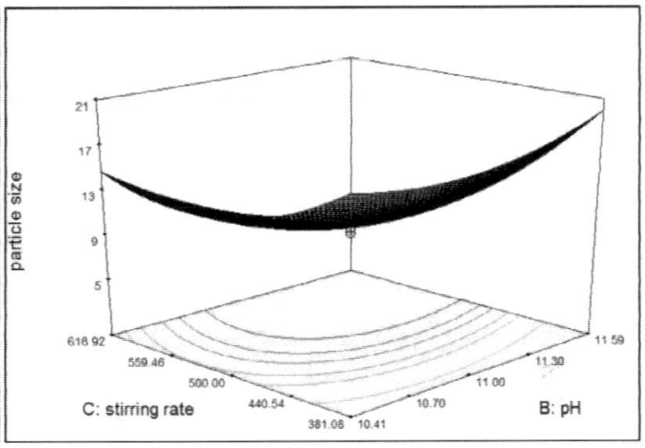

Figure 11. 3D surface plot on the interaction of B (pH) and C (stirring rate) towards particle size.

4.2.3.4 Validation of experimental model

Table 6 demonstrate the desired goals for the factors and response of particle size γ-Fe2O3 whereas Table 7 show the validation test result for optimization verification. Based on Table 7, the selected solution generated by Design Expert Software represented the optimum condition to produce the smallest particle size of γ-Fe2O3. The smallest particle size obtained was 7.3657 nm at 50 °C, pH 11.40 and 550 rpm. The difference between the value of predicted size (7.4965 nm) and actual experimental value was only small percentage error of 1.75 %. Since the percentage error was less than 5 %, it can be concluded that selected solution condition has been validated.

Table 6. Desired goals for factors and response of particle size γ-Fe_2O_3.

Factor	Goal	Lower limit	Upper limit
Processing temperature	In range	30 °C	70 °C
pH	In range	10	12
Stirring rate	In range	300 rpm	700 rpm
Particle size	minimize	9.0642 nm	31.5491 nm

Table 7. Validation test result for optimization verification.

Run	A: Processing temperature	B: pH	C: Stirring Rate	Current response of IONPs crystallite size (nm)		
				Predicted	experimental	Error (%)
1	50	11.40	550.00	7.4965	7.3003	2.62
2	50	11.40	550.00	7.4965	7.4941	0.03
3	50	11.40	550.00	7.4965	7.3027	2.59
Average					7.3657	1.75

5. Conclusion

In conclusion, the overall goal of this project has been achieved since the characterization and optimization studies on the synthesis of IONPs, maghemite (γ-Fe_2O_3) by using co-precipitation has been successfully carried out. The investigation effect of processing parameters towards the smallest particle size of γ-Fe_2O_3 has been synthesized at higher temperature, higher pH and higher stirring rate. The method that has been utilized is co-precipitation owing to high convenience, high production and low cost. The statistical study was performed by Response Surface Methodology (RSM) with the use of CCD for the optimization of γ-Fe_2O_3 production. The determination coefficient (R^2) obtained was 0.9890. The model F-value obtained was 99.69 and p-value $<$ 0.0001 indicated this model was statistically significant. The optimization of γ-Fe_2O_3 production was completed at 50 °C, pH 11.40 and 550 rpm and 7.3657 nm of γ-Fe_2O_3 produced as the smallest crystallite size. The result of Fourier Transform Infrared Spectroscopy (FTIR) analysis showed the production of γ-Fe_2O_3 by using co-precipitation revealed three main functional group, OH^- , CO_2 and Fe-O group. X-ray Diffraction (XRD) analysis has confirmed the presence of γ-Fe_2O_3 occurred at the peak of $2\theta = 35.4°$. Besides, Scanning Electron Microscopy (SEM) analysis showed agglomerated nanoparticles at lower stirring rate while spherical shaped were obtained at higher stirring rate.

Acknowledgement

Acknowledge with much appreciation to my Final Year Project (FYP) supervisor, Assoc. Prof. Dr. Midhat Nabil Ahmad Salimi, who gave continuous guidance and contributed stimulating ideas, knowledge and encouragement throughout this project.

References

[1] Senyei, A., Widder, K., & Czerlinski, G. (1978). Magnetic guidance of drug-carrying microspheres. *Journal of Applied Physics*, *49*(6), 3578–3583.

[2] Herranz, F., Almarza, E., Rodríguez, I., Salinas, B., Rosell, Y., Desco, M., Ruiz-Cabello, J. (2011). The Application of Nanoparticles in Gene Therapy and Magnetic Resonance Imaging. *Microscopy Research and Technique*, *74*(7), 577.

[3] Stephen, Z. R., Kievit, F. M., & Zhang, M. (2011). Magnetite Nanoparticles for Medical MR Imaging. *Materials Today (Kidlington, England)*, *14*(7–8), 330–338.

[4] Ali, A., Zafar, H., Zia, M., Ul Haq, I., Phull, A. R., Ali, J. S., & Hussain, A. (2016). Synthesis, characterization, applications, and challenges of iron oxide nanoparticles. *Nanotechnology, Science and Applications*, 9, 49–67.

[5] Lu, A.H., Salabas, E. L., & Schüth, F. (2007). Magnetic Nanoparticles: Synthesis, Protection, Functionalization, and Application. *Angewandte Chemie International Edition*, *46*(8), 1222–1244.

[6] Andrade, A. L., Valente, M. A., Ferreira, J. M. F., & Fabris, J. D. (2012). Preparation of size-controlled nanoparticles of magnetite. *Journal of Magnetism and Magnetic Materials*, *324*(10), 1753–1757.

[7] Ali, K., Sarfraz, A. K., Mirza, I. M., Bahadur, A., Iqbal, S., & Ul Haq, A. (2015). Preparation of superparamagnetic maghemite nanoparticles by wet chemical route and investigation of their magnetic and dielectric properties. *Current Applied Physics*, *15*(8), 925–929.

[8] Nazari, M., Ghasemi, N., Heydar, Mohammad, M., & Motlagh, M. (2014). Synthesis and characterization of maghemite nanopowders by chemical precipitation method. *Journal of Nanostruct Chem*, 2-5.

[9] Yusoff, A. H. M., Salimi, M. N., & Jamlos, M. F. (2018). Critical Parametric Study on Final Size of Magnetite Nanoparticles. *IOP Conference Series: Materials Science and Engineering*, *318*, 012020.

[10] Cheng, Z., Tan, A. L. K., Tao, Y., Shan, D., Ting, K. E., & Yin, X. J. (2012). Synthesis and Characterization of Iron Oxide Nanoparticles and Applications in the Removal of Heavy Metals from Industrial Wastewater. *International Journal of Photoenergy*, 2, 1–5.

[11] Namdeo, M., & Bajpai, S. K. (2009). Investigation of Hexavalent Chromium Uptake By. *Journal of Environmental, Agricultural and Food Chemistry*, *7*(7), 3082-3094.

[12] Chakrabarti, S., Ganguli, D., & Chaudhuri, S. (2004). Optical properties of γ-Fe_2O_3 nanoparticles dispersed on sol–gel silica spheres. *Physica E: Low-Dimensional Systems and Nanostructures*, *24*(3–4), 333–342.

[13] Chauhan, R. P., Mathur, R., Singh, G., Kaul, A., Bag, N., Singh, S., Mishra, A. K. (2013). Evaluation of Folate Conjugated Superparamagnetic Iron Oxide Nanoparticles for Scintigraphic/Magnetic Resonance Imaging. *Journal of Biomedical Nanotechnology*, *9*(3), 323–334.
[14] Iyengar, S. J., Joy, M., Ghosh, C. K., Dey, S., Kotnala, R. K., & Ghosh, S. (2014). Magnetic, X-ray and Mossbauer studies on magnetite/maghemite core-shell nanostructures fabricated through an aqueous route. *Journal of Royal Society of Chemistry*, 4, 64919–64929.
[15] Aliahmad, M., & Nasiri Moghaddam, N. (2013). Synthesis of maghemite (γ-Fe_2O_3) nanoparticles by thermal-decomposition of magnetite (Fe_3O_4) nanoparticles. *Materials Science-Poland*, *31*(2), 264–268.
[16] Zhu, H.Y., Jiang, R., Xiao, L., & Li, W. (2010). A novel magnetically separable γ-Fe_2O_3/crosslinked chitosan adsorbent: Preparation, characterization and adsorption application for removal of hazardous azo dye. *Journal of Hazardous Materials*, *179*(1–3), 251–257.
[17] Hartman, S., & Robertson, R. E. (1960). Nucleophilic Displacement Of Methyl Sulphonic Esters By Hydroxide Ion In Water. Canadian Journal of Chemistry, *38*(11), 2033-2038.
[18] Ghadi, A., Mahjoub, S., Tabandeh, F., & Talebnia, F. (2014). Synthesis and optimization of chitosan nanoparticles: Potential applications in nanomedicine and biomedical engineering. *Caspian Journal of Internal Medicine, 5(3), 156-61.*
[19] Chuan, L., Eldin, W., & Feng, R. (2012). Agglomeration of magnetic nanoparticles. The *Journal of Chemical Physics*, 136(12), 124109.
[20] Genes, D. Z., Scirocco, R., Mewis, J., & Vermant, J. (2008). Flow-induced orientation of non-spherical particles: Effect of aspect ratio and medium rheology. *Journal of Non-Newtonian Fluid Mechanics*, *155*(1–2), 39–50.
[21] Salimi, M.N. (2013). Systematic Investigation of Calcium Phosphate Produced by Wet Chemistry Method and Supercritical Processing Techniques. Degree of Doctor of Philosophy Thesis, School Chemical Engineering University of Birmingham. *Journal of Powder Technology*, 109-118.
[22] Chakraborty, S., & Chowdhury, R. (2016). Sequential experimental design based generalised ANOVA. *Journal of Computational Physics,* 317, 15-32.
[23] Shalil Miraz Mohamed, R., & Darman, N. (2015). Synthesis and Characterization of Hydroxyapatite. *Malaysia Journal of Analytical Sciences, 21*(1), 136-148.
[24] Regti, A., Laamari, M. R., Stiriba, S. E., & El Haddad, M. (2017). Use of response factorial design for process optimization of basic dye adsorption onto activated carbon derived from Persea species. *Microchemical Journal*, 130, 129-136.
[25] Grover, L., Salimi, M. N., Bridson, R. H., Grover, L. M., & Leeke, G. A. (2013). Effect of Processing Conditions on the Formation of Hydroxyapatite Nanoparticles. *Powder Technology*, *218*(12), 109–118.
[26] Prabu, S. B., Karunamoorthy, L., Kathiresan, S., & Mohan, B. (2006). Influence of stirring speed and stirring time on distribution of particles in cast metal matrix composite. *Journal of Materials Processing Technology*, 171, 268-273.
[27] Li, Y. H., Xu, B. W., Lu, J. J., & Kang, D. Z. (2006). On computational complexity of the extended fuzzy description logic with numerical restriction. *Journal of Software*, *17*(5), 968–975.

Glucose Production from Sugarcane Bagasse by Two Stages Chemical Pretreatment & Hydrolysis

Nurul Syiffa Husna Che Roslan[1] and Midhat Nabil Salimi[1]

[1]School of Bioprocess Engineering, Universiti Malaysia Perlis, Perlis, Malaysia.

Email: nabil@unimap.edu.my

Abstract. The aim of this work was to produce glucose from sugarcane waste. It consists of high cellulose, hemicellulose and lignin which enables to be converted into a glucose production. Three conditions of acid hydrolysis was measured by using One Factor at a Time (OFAT) which consist of sulphuric acid concentration, retention time and temperature. The highest yield and optimum conditions from OFAT parameters were further continue with the enzymatic hydrolysis by using Central Composite Design (CCD). The optimization of enzymatic hydrolysis for the conversion of cellulose to glucose was conducted by using three independent variables which were pH, temperature and enzyme dosage. Based on the data of Design of Expert (DOE) software from optimization of enzymatic hydrolysis, the optimum glucose production was determined by using Dinitrosalicylic Acid Reagent (DNS) method.

1. Introduction

For many years, increasing in interest was happened in a form of agricultural wastes as feedstock for biofuels production as it was considered in second generation. In Brazil, sugarcane was the major bioenergy crops where it used for the production of bioethanol. It is the most abundant co-products in the Brazilian sugar and ethanol industry where one ton of sugarcane was able to produce 280kg of bags [1]. Bagasse can be burned as a fuel about 50% to combined heat and power systems to meet energy which needs of the mills and the remainder is stored. Hence, the new development of technologies for ethanol production from sugarcane bagasse will increased the efficiency of ethanol production without expand the size of agricultural areas and can reduced the current conflict which effect the change in land use to meet growing energy demand [5].

Furthermore, sugarcane bagasse is a lignocellulose of a biomass. The cell wall mainly consists of three different components which were lignin, hemicellulose and cellulose. The high percentage of glucose and xylose from fraction of hemicellulose and cellulose in bagasse and the ability of microorganisms can convert the sugars into ethanol were the main factor of using this co-product for the biofuels production, one of the example of biofuel was ethanol production [2].

Sugarcane bagasse consist of mixture of hard fibre with smooth and soft parenchyma (pith) with high hygroscopic property and also consist of 35% moisture, it also have 60% of depithing efficiency and it will be removed about 20% as a pith either as a sugar or mill of pill premises of the production tonnage [5]. Thus, it have high tendency to produced bioethanol because of high carbohydrate contents which were 39.8% cellulose and 32.2% of hemicellulose.

In addition, for production of glucose it involved two steps which consist of chemical pretreatment and hydrolysis process. The removal of lignin and hemicellulose was done by pretreatment process of biomass to make the porosity of materials is reduced and also to reduce in

Content from this work may be used under the terms of the Creative Commons Attribution 3.0 licence. Any further distribution of this work must maintain attribution to the author(s) and the title of the work, journal citation and DOI.
Published under licence by IOP Publishing Ltd

cellulose crystallization and rise. Fraction of hemicellulose in the biomass can be removed through process of dilute acid hydrolysis by using sulphuric acid as the catalyst [3]. To breaks down the lignin bond that exists in sugarcane bagasse, biomass pretreatment can be applied which can prevents the accessibility of cellulase enzyme. In addition, the hydrolysis of enzymatic rate can be enhanced by using the pretreatment process.

Then, the cellulose of enzymatic process is performed by higher specific of cellulase enzyme. The product that can be obtained after enzymatic hydrolysis is reducing sugar such as glucose. The features that influenced the cellulose of enzymatic hydrolysis of are temperature, retention time, pH, substrate concentration, enzyme dosage and also agitation speed [6].

2. Materials

2.1 Chemicals

Concentrated sulphuric acid (95%) was purchased from Fisher Pte Ltd, while for glucose, sodium citrate, citrate acid, potassium tartrate, Dinitrosalicylic acid reagent were purchased from Sigma Aldrich.

2.2 Plant materials

Sugarcane bagasse were collected at Taman Muhibbah, Perlis. The sugarcane bagasse was washed by using distilled water in order to remove dust and dry at 70°C overnight by putting the sugarcane bagasse into the oven. The dried sugar cane bagasse were ground into powder form and sieved by passing through a 1mm screen. Then, the sample were kept in a tightly closed container at room temperature to prevent contamination.

2.3 Procedures

2.3.1 Chemical preparation

The concentration of sulphuric acid (H2SO4) required in this pretreatment process was diluted sulphuric acid. The preparation of the solution by the dilution of concentrated sulphuric acid where the proportion of the concentrated acid is diluted with distilled water by using Equation 3.1.

$$V \times 1.84\ g/mL \times 0.96 = [(VF - V) + 1.84\ V]\ (\%\ of\ H2SO4) \quad (1)$$

Where VF = Final volume of diluted sulphuric acid
V =Volume of sulphuric acid in mL

The purity of the sulphuric acid were accounted in the dilution procedure.1.0% w/v of sulphuric acid were prepared from the concentrated sulphuric acid 96% w/v with the density of 1.84 g/mL by adding 5.69 mL of concentrated sulphuric acid solution into volumetric flask and dilute to 1.0% w/v with the final volume 1000mL with distilled water [4].

2.3.2 Citrate buffer

The citrate buffer were prepared from the stock solution of citric acid and conjugate base by mixing 10.5g of citric acid and 14.7g of sodium citrate in 500mL of distilled water respectively. Then, 82 mL of the citric acid solution were mixed with 18mL of sodium citrate solution in a volumetric flask. The total volume of the mixture were made up to 1 litre with distilled water. The solution were stirred with magnetic stirrer, 1M of sodium peroxide were added to adjust the pH to 4.5 with the help of a pH probe [4].

2.3.3 Dinitrosalicyclic Acid Reagent (DNS)

The DNS reagent were prepared by using conical flask with total volume of 250mL and the conical flask were wrapped with aluminium foil. First, 2.0g of sodium hydroxide were dissolved in 100mL of distilled water which produced 0.5mol/L sodium hydroxide solution. Then, the solution flask were

IOP Conf. Series: Materials Science and Engineering **743** (2020) 012037 doi:10.1088/1757-899X/743/1/012037

placed on a hot plate at 60°C and stirred for the entire preparation steps. Next, 75 g of potassium tartrate were added to the heated sodium hydroxide solution. After that, 2.5 g of 3, 5-dinitrosalicylic acid were added to mixture. The mixture then diluted to a final volume of 250ml by distilled water. The DNS solution were allowed to cool down to room temperature and kept in the refrigerator during the whole experiment [5].

2.3.4 Acid hydrolysis

For acid hydrolysis, 3.0 g sugarcane bagasse powder were added with 20mL of sulphuric acid in a 100mL conical flask, then the flask was wrapped with aluminium foil and heated in a water bath. The sample after the acid hydrolysis was filtered by the filter paper, the liquid fraction was tested by reducing sugar concentration.

For the first OFAT, the concentration of the sulphuric acid was the factor, then the temperature was set as 90°C and the retention time of 60 minutes in the water bath then the glucose concentration was tested using DNS method. The best sulphuric acid concentration is brought to next OFAT as a constant factor.

For the second OFAT, the retention time was the factor, then the sulphuric acid concentration based on the previous OFAT and the temperature was set to 90°C as constant in water bath then the reducing sugar concentration was tested. The best retention time is carry forward to the third OFAT as a constant factor.

For the third OFAT, the temperature is the factor, then the best concentration of sulphuric acid and retention time from the first and second OFAT was set as constant in a water bath then, the reducing sugar concentration was tested.

3. Results and Discussion

3.1 Acid Hydrolysis

One Factor At a Time (OFAT) was done by continuing manipulated three factors which were sulphuric acid concentration, retention time and temperature.

Figure 1 shows the effect of acid concentration on reducing sugar production (g/L). The first factor used to manipulate was the acid sulphuric concentration, the temperature and retention time were set as a constant which was 90°C and 60 minutes respectively. The highest reducing sugar concentration was 0.65 g/L produced at 0.5 % of sulphuric acid concentration. The reducing sugar concentration increased from 0 % to 0.5 % but when sulphuric acid concentration was between 1.25% and 3%, the reducing concentration decreased. This was because of formation of a xylose found as the main sugar in sugarcane bagasse of a hemicellulose hydrolysate[4].

Figure 2 shows the effect of retention time on reducing sugar concentration (g/L). 90 minutes produced the highest reducing sugar concentration which was 0.29 g/L. Based on the graph, after the optimum condition of the retention time, the increased value of retention time made the yield of reducing sugar concentration decreased because of the possibility of formation of other side products instead of glucose due to the extreme temperature. This is due to the highest chances that the pentose was not fully decomposed to furfural and disturbed the acid hydrolysis process [4].

Figure 3 shows the effect of temperature on reducing sugar concentration (g/L). It shows the increasing of reducing sugar production and 90°C was the optimum temperature which obtained 0.365 g/L of reducing sugar concentration as the sugarcane bagasse was decomposed by the sulphuric acid. However at 100°C it showed the decreasing of reducing sugar concentration. Glucose. The reducing sugar concentration was reduced because a portion of the sugar was degraded into enzyme-inhibiting by-products such as furfural (2-furaldehyde), 5-hydroxymethylfurfural (5-HMF), acetic acid, gypsum, vanillin, and aldehydes (4-hydroxybenzaldehyde, syringaldehyde) as a result of the conditions caused the cellulose to rapidly break into sugars [4].

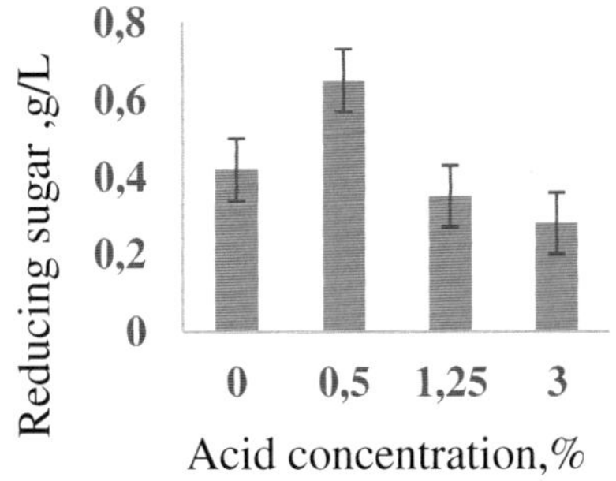

Figure 1. Effect of acid concentration on reducing sugar (g/L).

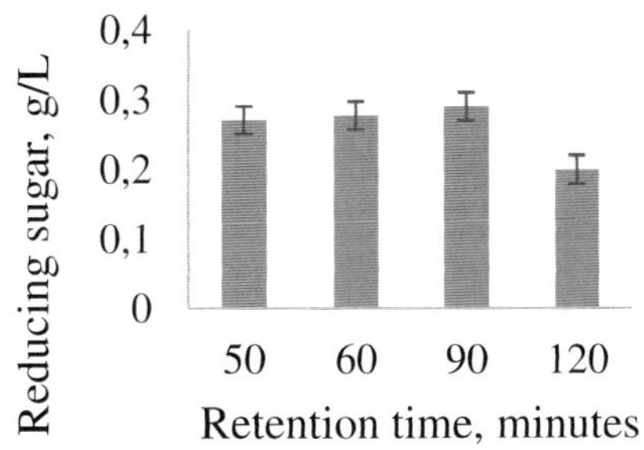

Figure 2. Effect of retention time on reducing sugar (g/L).

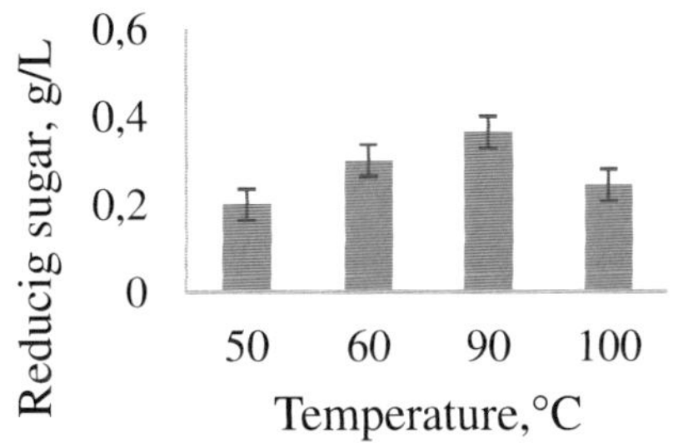

Figure 3. Effect of temperature on reducing sugar (g/L).

3.2 Optimization of Enzymatic Hydrolysis

The optimization of enzymatic hydrolysis of glucose production was conducted by using Design of Expert (DOE) on the Response Surface Methodology (RSM) via Central Composite Design (CCD) by using 5 level for each numeric factor. There were three independent variables selected to optimize the glucose production, which were temperature, pH and enzyme dosage by using cellulase enzyme while the response variables was glucose concentration. The effect on temperature, pH and enzyme dosage were determined in the range of 30°C to 50°C, pH 4 to pH 7 and 0.10mL to 0.20 mL respectively. The constant variables was agitation speed at 150 rpm. Based on Table 1, the optimum conditions for enzymatic hydrolysis was determined from the highest value of glucose concentration which at run 10, consists of pH 6.39, temperature 34°C and enzyme dosage 0.15mL.

IOP Conf. Series: Materials Science and Engineering **743** (2020) 012037 doi:10.1088/1757-899X/743/1/012037

Table 1. Design matrix of CCD and the experimental result.

Std	Run	Factor 1 : pH	Factor 2 : Temperature, (°C)	Factor 3 : Enzyme dosage, (mL)	Response: Glucose concentration, (g/L)
15	1	5.50	40.0	0.10	0.234
14	2	5.50	40.0	0.15	0.683
7	3	4.61	46.00	0.20	0.468
4	4	6.39	46.00	0.18	0.772
16	5	5.50	40.00	0.12	0.585
13	6	5.50	40.00	0.15	0.670
1	7	4.61	34.00	0.10	0.543
5	8	4.61	34.00	0.12	0.740
18	9	5.50	40.00	0.18	0.640
2	10	6.39	34.00	0.15	0.783
3	11	4.61	46.00	0.12	0.369
17	12	5.50	40.00	0.12	0.570
19	13	5.50	40.00	0.15	0.756
20	14	5.50	40.00	0.15	0.720
11	15	5.50	30.00	0.15	0.630
9	16	4.0	40.00	0.15	0.622
10	17	7.0	40.00	0.15	0.774
8	18	6.39	46.00	0.18	0.745
6	19	6.39	34.00	0.18	0.540
12	20	5.50	50.00	0.15	0.603

3.3 Regression Analysis

The result obtained from CCD were fitted in a quadratic or second-order equation to explain the dominance effect of substrate concentration and the enzyme dosage in terms of coded as A, B and C respectively. The purpose of the quadratic model was the mathematical explanation of the variables. The final empirical models in terms of coded and actual factors were shown in the following equation:

Final equation in terms of coded factors:
Glucose Concentration
$= 0.73 + 0.038A - 0.021B + 0.024C - 0.036AB + 0.061AC + 0.17BC - 8.401E\text{-}003A^2 - 0.037B2 - 0.15C^2$

Where the alphabetical was represents as:
A: pH
B: Temperature
C: Enzyme dosage

Final equation in terms of actual factors:
Glucose concentration
= 1.01252 + 0.087903 pH - 0.029585 temperature – 0.56996 enzyme dosage – 0.0105612 pH – 1.052622 temperature – 168.778722 enzyme dosage

3.3.1 Analysis of Variance (ANOVA)

The analysis of variance (ANOVA) was a series of statistical hypothesis testing which was used to investigate whether statistical relationship between the means of independent variables. To determine the model satisfies the assumption of ANOVA, which was shown in Table 2 and Table 3.

Table 2. Analysis of Variance (ANOVA) for surface quadratic model.

Source	Sum of Squares	df	Mean Square	F Value	p-value Prob > F	
Model	0.38	9	0.042	27.39	< 0.0001	Significant
A-pH	0.014	1	0.014	9.33	0.0122	
B-temperature	4.866E-003	1	4.866E-003	3.20	0.1040	
C-enzyme dosage	4.860E-003	1	4.860E-003	3.20	0.1041	
AB	2.392E-003	1	2.392E-003	1.57	0.2384	
AC	0.013	1	0.013	8.80	0.0141	
BC	0.061	1	0.061	39.94	< 0.0001	
A^2	8.222E-004	1	8.222E-004	0.54	0.4791	
B^2	0.016	1	0.016	10.61	0.0086	
C^2	0.17	1	0.17	111.05	< 0.0001	
Residual	0.015	10	1.521E-003			
Lack of Fit	0.011	6	1.764E-003	1.52	0.3557	Not significant
Pure Error	4.627E-003	4	1.157E-003			
Cor Total	0.39	19				

Table 3. R-Squared value for model developed from CCD.

Parameters	Value
Standard Deviation	0.039
Mean	0.62
C.V%	6.27
R-Squared	0.9610
Adjusted R-Squared	0.9259
Predicted R-Squared	0.8493
Adequate Precision	18.517

3.3.2 Graphical Analysis

In the previous discussed, regression analysis was the residual difference between the observed value of the dependent variable and predicted variable. The normal plot with residue and the plots of residual versus predicted were analysed as shown in Figure 4 and Figure 5.

In the normal probability plot of the raw data, the analysis of variance shown a straightforward relationship with the residual. Figure 4 was the quadratic polynomial model satisfies the assumptions analysis of variance (ANOVA) was near to the normal. If the model was correct and the assumption was satisfied, the residual might without structure. In fact, they should be no relation to any other variable including plotting the residuals versus the fitted predicted response. A simple check would include plotting the residuals versus the fitted predicted values. A plot of residuals versus the rising predicted response value test the assumption of the constant variance. Based on figure 5, the plot was a random scatter which was justified the need for no alternation to minimize personal error.

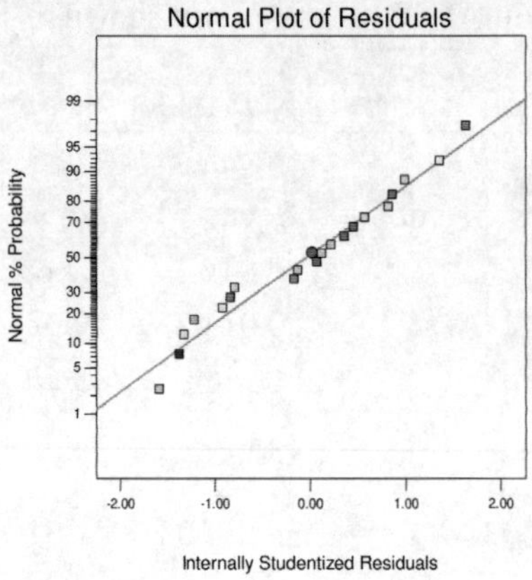

Figure 4. Normal probability plot of standardized residuals.

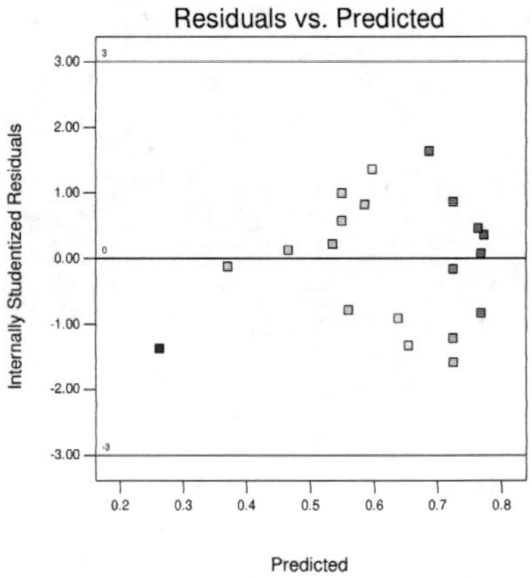

Figure 5. Studentized residual versus predicted values.

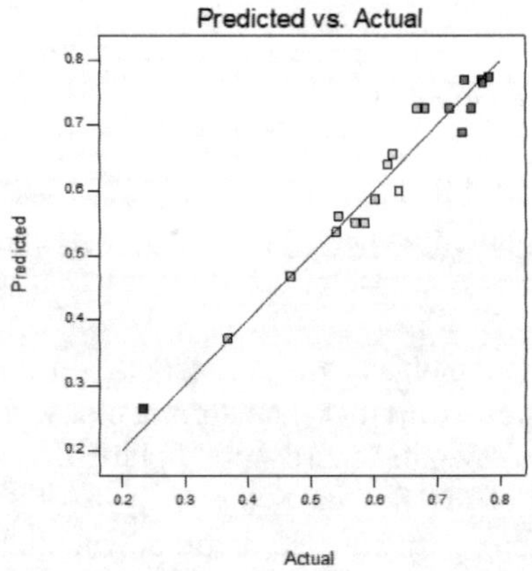

Figure 6. Predicted versus actual glucose concentration.

Figure 6 shows the predicted versus actual glucose concentration graph. The predicted versus the actual plot reveals how well the models can be fit into the data. The actual data means the experimental data from the experiment where the predicted data was evaluated by the DOE software. Based on Figure 6, there was a distance results between the actual and predicted data. Points was a distant from the straight line indicate that the existence of outliers. The result improved in order to fit well in the quadratic model.

3.3.3 Effects of temperature, pH and enzyme Dosting

The response surface was used to obtain the maximum response which was the total glucose concentration at the optimum conditions at each design variables. Variables temperature, pH and enzyme dosing were considered in this study for the interaction and the effect the response variable which was the glucose concentration produced.

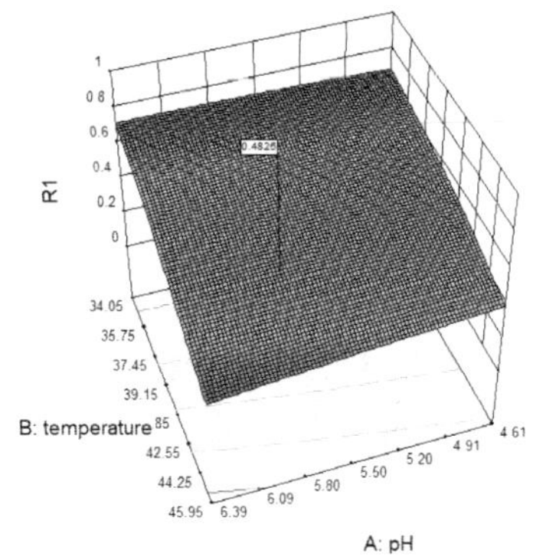

Figure 7. 3D plot interaction between temperature, pH and enzyme dosing.

The two dimensional (2D) contour plot and three dimensional (3D) response surface were used as a graphical tool to display the interaction between the temperature, pH and enzyme dosing on glucose production. Figure 7 and figure 8 shown the relationship between three variables. Figure 7 demonstrated that a surface with a maximum point located inside the experimental region. Figure 8 shown a flat shape contour plot with the related variables were vital.

Based on the 2D and 3D figures shown the glucose increased when the temperature increased but it started to decline when the temperature and pH after the optimum value. Temperature and pH were main factors governing the saccharification. The variation in temperature and pH activity in the enzyme cellulase might due to the active side chain that were fully occupied at this concentration and thus produced the optimum glucose. Temperature and pH value must be optimized to indicate the good hydrolysis on the specified maximum concentration leading to high glucose yield and decreases overall cost.

In addition, the effect of enzyme dosing was a plot in the range of 0.10ml to 0.20ml, thus the optimum result was 0.15ml. Biomass which required the lower enzyme dosage must be utilized to continuous development in order to decrease the cost of enzyme.

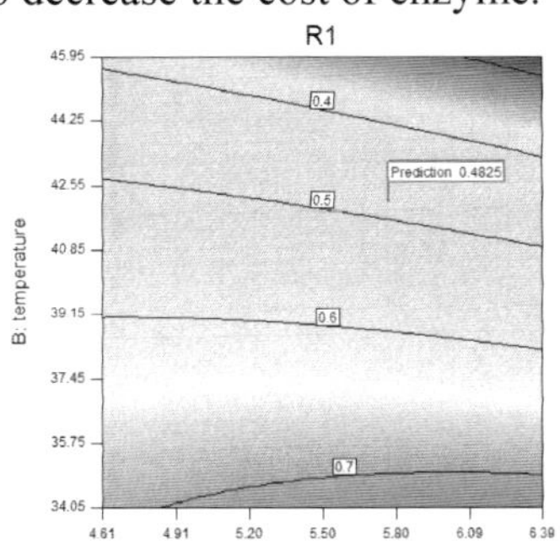

Figure 8. 2D of Contour plot of interaction between temperature, pH and enzyme dosage.

4. Conclusion

The sugarcane bagasse was a suitable raw material for the glucose production. Method in One Factor at a Time (OFAT) was able to determine the best conditions of acid hydrolysis which consisted of three parameters, namely acid concentration, retention time and temperature which were 0.5 %, 90 minutes and 90°C respectively. By using Surface Response Methodology (RSM) through Central Composite Design (CCD), the optimum conditions between the relationship of temperature, pH, and enzyme dosage were carried out by 20 experiments for glucose concentration determination. The experimental results showed that at optimum parameters of 34°C, pH of 6.39 and 0.15mL of enzyme dosage produced 0.783 g/L of glucose. Predicted R-Squared in Analysis of variance (ANOVA) correlation coefficients was 0.8493 and adjusted R-Squared was 0.9259 indicated moderate of the experimental data evaluation by a quadratic regression model. The determination of the coefficient of R-Squared was 0.9610 which corresponds to the standard deviation of 0.039. The magnitude of the R-Squared indicated there was a close relationship between the value of actual and also with the predicted value and the value for model F was 27.39. Last but not least the high yield of glucose production was 0.783 g/L.

References

[1] Aminudin, E., Khalid, N. H. A., Azman, N. A., Bakri, K., Din, M. F. M., Zakaria, R., & Zainuddin, N. A. (2017). Utilization of Baggase Waste Based Materials as Improvement for Thermal Insulation of Cement Brick. *MATEC Web of Conferences*, *103*, 01019.

[2] Curtis, M. J., Donald, D. A., & Ridd, M. (2012). Determination Of Sucrose And Glucose In Sugarcane Using A Glucose Biosensor, *34*.

[3] Dussán, K. J., Silva, D. D. V, Moraes, E. J. C., Priscila, V., & Felipe, M. G. A. (2014). Dilute-acid Hydrolysis of Cellulose to Glucose from Sugarcane Bagasse, *38*, 433–438.

[4] Lavarack, B. P., Griffin, G. J., & Rodman, D. (2002). The acid hydrolysis of sugarcane bagasse hemicellulose to produce xylose, arabinose, glucose and other products. *Biomass and Bioenergy*, *23*(5), 367–380.

[5] Lopes, E. S., Dominices, K. M. C., Lopes, M. S., & Tovar, L. P. (2017). Enzymatic Hydrolysis Exploration and Fermentation : Acid Pretreatment and Delignification in Sugarcane Bagasse for 2G Ethanol Production, *57*, 151–156.

[6] Philippini, R. R. (2017). Pretreatment of Sugarcane Bagasse from Cane Hybrids : Effects on Chemical Composition and 2G Sugars Recovery Pretreatment of Sugarcane Bagasse from Cane Hybrids : Effects on Chemical Composition and 2G Sugars Recovery. *Waste and Biomass Valorization*.

IOP Conf. Series: Materials Science and Engineering **743** (2020) 012038 doi:10.1088/1757-899X/743/1/012038

Fabrication of electrolyte-gate nanocrystalline diamond- based field effect transistor (NCD-EGFET) for HIV-1 Tat protein detection

Nurul Atiqah Ahmad[1], Ruslinda A Rahim[1*], Nur Syakimah Ismail[2], Bohuslav Rezek[3]

[1]Institute of Nano Electronic Engineering, Universiti Malaysia Perlis, 01000 Seriab Perlis, Malaysia.
[2]School of Microelectronic Engineering, Universiti Malaysia Perlis, 02600 Arau Perlis, Malaysia.
[3]Institute of Physics, Academy of Science, Na Slovance 1999/2, 18200 Prague, Czech Republic.

Email: * ruslinda@unimap.edu.my

Abstract. In this paper, we reported on the fabrication process of electrolyte-gate field effect transistor using nanocrystalline diamond as a sensing transducer. The fabrication procedure was begin with the growth of nanocrystalline diamond thin film on silicon/silicon dioxide (Si/SiO_2) substrate using microwave plasma-enhanced chemical vapour deposition (CVD). Then the photolithography process was performed in order to design and pattern the field effect transistor device with the active gate channel of 60 µm length and 20 µm width. Each device consists of three active gate channel which connecting to three different pairs of source and drain contact. The surface morphology of fabricated NCD-EGFET was characterized using Scanning Electron Microscope to clarify the active gate channel of the device and the grain size of nanocrystaline diamond. The current-voltage (*I-V*) measurement of the device were carried out to study the electrical behaviour for HIV-1 Tat protein detection via RNA aptamer as sensing probe.

1. Introduction

Biosensor is defined as a device that uses a specific biochemical reaction facilitated by isolated organelles, enzymes, tissues, or the whole cell to detect chemical compounds usually by optical, electrical or thermal signals. It is the combination between the component that recognizes the analyte (bioreceptor) and produces a signal, a signal transducer, and a signal reader.

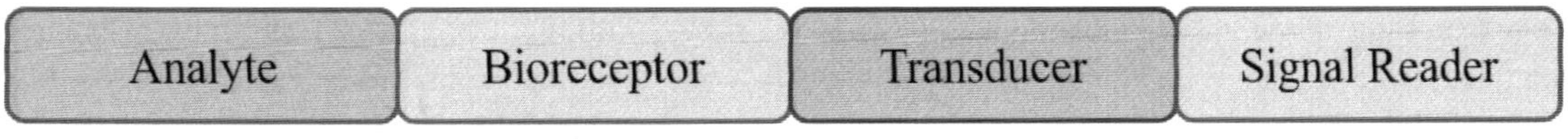

Figure 1. Components of biosensor.

There are several types of biosensor have been developed by researchers such as potentiometric, amperometric, impedimetric, and voltammetric biosensor [1]–[3]. Potentiometric biosensor including field effect transistor (FET) is one of a well-known sensing device for biological detection. Compared to other types of biosensor, FET is more favourable as it needs low production cost, high possibility of

Content from this work may be used under the terms of the Creative Commons Attribution 3.0 licence. Any further distribution of this work must maintain attribution to the author(s) and the title of the work, journal citation and DOI.
Published under licence by IOP Publishing Ltd

IOP Conf. Series: Materials Science and Engineering **743** (2020) 012038 doi:10.1088/1757-899X/743/1/012038

miniaturization, flexible configuration for simple measurement, and most importantly biocompatible especially for bioanalytical applications [4], [5].

Several types of FET have been reported previously such as electrolyte-gate FET (EGFET)[6]–[9], back-gate FET (BGFET) [10]–[12], and top-gate FET [13], [14]. EGFET is a type of field effect transistor that work in wet condition. The working principle of EGFET is based on the charge modulation from the biomolecules on the active gate area. The EGFET is different from BGFET in terms of the gate development. For BGFET, metal gate is used as gate while for the EGFET, metal gate is replaced by electrolyte and Ag/AgCl reference electrode. For biosensing application, it is better to used EGFET as biomolecules will be more stable in wet condition.

Herein, we demonstrated the fabrication of electrolyte-gate field effect transistor on silicon substrate with nanocrystalline diamond as transducing material. Nanocrystalline diamond was chose as transducer as it promises stability, surface-wettability, non-toxicity, large volume to surface ratio, electron affinity, and high electrical conductivity.

2. Materials and Methods

2.1. Growth of nanocrystalline diamond thin film

The nanocrystalline diamond thin films were growth on silicon substrates. Silicon substrates in size of 10 mm × 10 mm × 1 mm were primarily oxidized in oxidation furnace to gain the oxide layer on top of the substrates. Next, all the substrates were ultrasonically cleaned in isopropyl alcohol for 10 min and deionized water (DIW) was used to rinse the substrates. Subsequently, substrates are immersed for 30-40 min into an ultrasonic bath with a colloidal suspension of diamond nanopowder for nucleation process. For the nucleation solution, 200 ml of DIW was mixed with 5 ml of NanoAmando solution. NanoAmando solution consists of 5 nm dispersed buckydiamond with concentration of 5.0 w/v/% and 4.8 nm ± 0. 6 nm diamond size. From this process, the formation of 5-25 nm thin layer of diamond will occur necessary to initiate the diamond growth in a thin film.

Once the nucleation process done, nanocrystalline diamond (NCD) thin films were grown in a microwave ellipsoidal cavity reactor by chemical vapor deposition (CVD) process. The process takes 4.5 h with the gas pressure of 30 mbar, gas mixture 1% methane (CH_4) in hydrogen gas (H_2) and microwave power of 1000W. The deposition temperature was in the range of 550-600 °C. These deposition parameters led to the growth of ~450 nm thick diamond film with the grain sizes ~250 nm. Nanocrystalline diamond thin films were further hydrogenated in the same microwave plasma reactor at 600°C in hydrogen plasma (H_2) for 10 min to induce the surface conductivity.

2.2. Fabrication of nanocrystalline diamond-based electrolyte-gate field effect transistor (NCD-EGFET)

Fabrication of NCD-EGFET were divided into four steps; mark aligning, channel aligning, contact aligning, and final cover. In the primary step, the samples were cleaned using nitrogen gas (N_2) to remove the dust from the samples. Next, the sample was put into the spin coater and dropped with Hexamethyldisaline (HDMS) 1:4 which work as the adherent layer. The sample was spin coated for 10 sec at 3000 rpm. Subsequently, the sample was dropped with positive photoresist MA 15 and spin coated at 3000 rpm for 30 sec. The sample was then soft baked at 100 °C for 2 min. Further, the sample was aligned under the mask aligner and exposed under UV for 6 sec to build the crosses on the sample. The sample was developed in the developer for 45 sec and rinsed with distilled water. The sample was then exposed to the oxygen gas in the oxygen plasma for 1 min. The next process was 50 nm gold deposition on the sample with the adherent layer of 10 nm titanium (Ti). After the metal deposition process, the gold was removed by lift off procedure in acetone so that the gold only remained on the crosses of the sample.

The channel aligning step was started with putting the sample into the spin coater and dropped with HDMS 1:4 with condition of 3000 rpm. The sample was then dropped with positive photoresist MA 15 and spin coated at 3000 rpm for 30 sec followed by soft baked at 100 °C for 2 min. Next, the sample was aligned under the mask aligner for 10 sec. The sample was exposed under UV light for 6 sec to

IOP Conf. Series: Materials Science and Engineering **743** (2020) 012038 doi:10.1088/1757-899X/743/1/012038

build three active gate channels with size 20 μm x 60 μm on the sample substrate. After the aligning process, the sample was developed in the developer for 45 sec and rinsed with distilled water. Subsequently, the sample was then exposed to the oxygen gas in the oxygen plasma for 1 minute followed by removing the photoresist.

For the contact aligning procedure, the same process as channel aligning were done starting from spin coating the sample with HDMS 1:4 up to developing the contact pads for 45 sec in the developer. The next process was metal coating with 50 nm gold on the sample with 10 nm Ti as adherent layer. After the deposition process, the gold was lifted off in acetone so that the gold only remained on the arm and contact pad of the sample. The sample was then checked its conductivity which should be good at 10^{-8} A.

As the conductivity of the sample was fine, the final cover of the sample was done. The sample was put into the spin coater and dropped with HDMS 1:4 and spin coated for 15 sec at 4000 rpm. The sample was then dropped with positive photoresist AR-P-3220 and spin coated at 4000 rpm for 60 seconds followed by soft baked at 95 °C for 150 seconds. Next, the sample was aligned under the mask aligner with UV for 13 sec to build the final cover on the sample so that only contact pads and active gate channel are free from photoresist. After the aligning process, the sample was developed in the developer for 6 min and rinsed with distilled water. The sample was then post-bake in the oven at 100 °C for 45 min. The final photolithographic step created openings of 60 μm × 20 μm to define the active gate area.

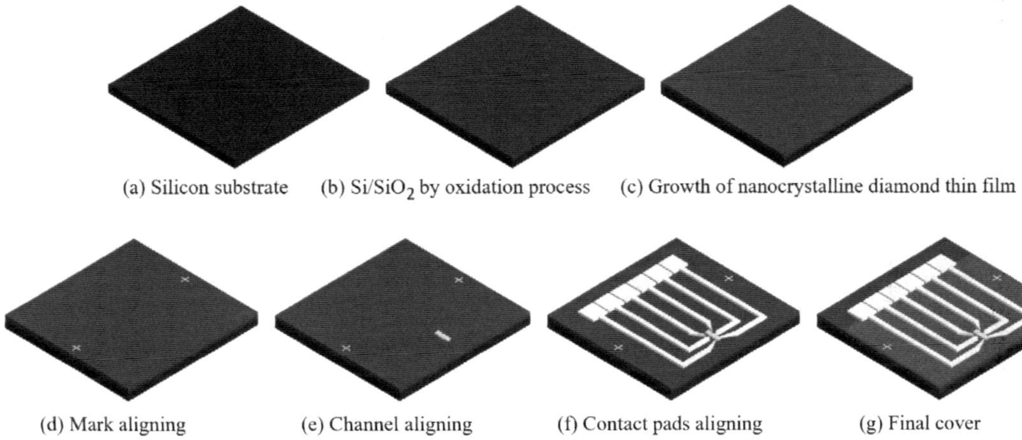

Figure 2. Steps for EGFET fabrication.

2.3. *HIV-1 Tat detection*

Primarily, RNA aptamer was immobilized on the active gate channel of the NCD-EGFET. Every sample was treated with a mixture of 0.4 M 1-ethyl-3-(3-dimethylaminopropyl) carbodiimide hydrochloride (EDC) and 0.1 M N-hydroxysuccinimide (NHS) for 1 h to trigger the carboxylic functional groups of the sample. Then, the first strand of RNA aptamer was manually drop-casted on the surface of the active gate channel and incubated at 38 °C for 2 hours in a humidified chamber, followed by rinsing with PBS. The binding of 100 pM HIV-1 Tat with the probe RNA aptamer was done at room temperature for 1 h. The concentration of the aptamer derived second strand used was also 100 pM. After that, Tris–HCl buffer solution was used as rinsing agent to avoid the nonspecific binding. An 8.3 M urea solution was used to remove the tested HIV-1 Tat from the active gate channel to study the repeatability of the measurement.

IOP Conf. Series: Materials Science and Engineering **743** (2020) 012038 doi:10.1088/1757-899X/743/1/012038

3. Results and Discussion

3.1. Surface morphological studies

Nanocrystalline diamond thin film was analyse using scanning electron microscopy (SEM). Fig 3 shows the image of grain particle of nanocrystalline diamond with average size of ~250 nm, observed at 100,000× magnification. Fig 4 shows the arrangement of active gate channel of fabricated device. The active gate channels showed are light coloured determining the conductivity of the channels. The gap that joining the electrodes were 60 µm in length and 20 µm width. All the gaps were terminated with hydrogen bonding.

Figure 3. The grain particle of NCD.

Figure 4. The arrangement of active gate channel of the fabricated device.

3.2. Electrical characterization

Once the fabrication process was done, the device was electrically measured using Keithley 2400 Source Meter and 6487 Picoammeter. The I_{ds}-V_{gs} measurement was conducted using an Ag/AgCl as reference electrode in 1 mM PBS at pH 7.4 for HIV-1 Tat protein detection. Based on fig 5, the drain-source current, I_{ds} reduced after the introduction of HIV-1 Tat to the RNA aptamer. This is due to the reduction of electron transfer from source to drain [15]. Moreover, the reaction between positively charge ions (HIV-1 Tat) the positive charge holes from the NCD surfaces also affect the current reduction. Based on previous study, it has been discovered that HIV-1 Tat rich in positively charge ions with the isoelectric point (pI) of 9.88 [16]. Since, the binding of HIV-1 Tat was done at pH 7.4, it is confirmed that HIV-1 Tat is positively charged. The interaction between the RNA aptamer with HIV-1 Tat causing the gate potential to shift towards the negative direction by 174.8 mV for the I_{ds}–V_{gs} transfer characteristic, with respect to the potential charge changes created by the biomolecular interaction. This result validates that the NCD-EGFET responses were only identified when specific interaction between recognition molecules (RNA aptamer and HIV-1 Tat) occurred on the active gate channel. To confirm the binding of HIV-1 Tat on RNA aptamer, the experiment was repeat on the same device. From fig 6, the shifting in gate potential were recorded at 174.3 mV and 173.9 mV, subsequently. From this result, we can estimate that the binding of 100 pM HIV-1 on RNA aptamer will cause the shifting in gate potential by approximately 174 mV. This can be the reference line for further studies of HIV-1 Tat detection on NCD-EGFET via RNA aptamer as sensing probe.

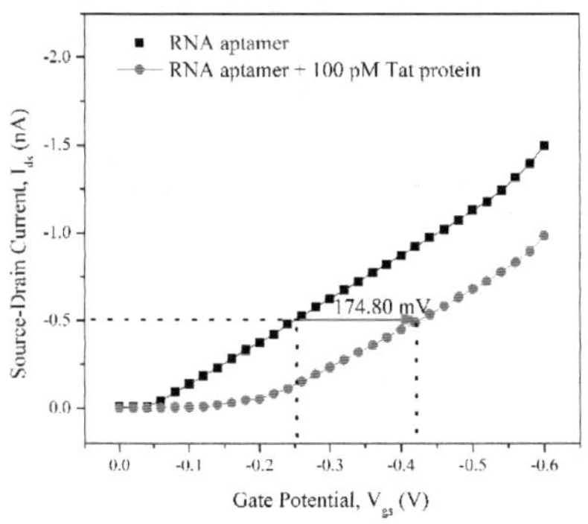

Figure 5. The I_{ds}-V_{gs} measurement of NCD-EGFET during the detection of 100 pM HIV-1 Tat, conducted in phosphate buffer solution at pH 7.4. The drain-source voltage, V_{ds} was biased at -0.1 V and the gate potential, V_{gs} was swept from 0 to -0.6 V.

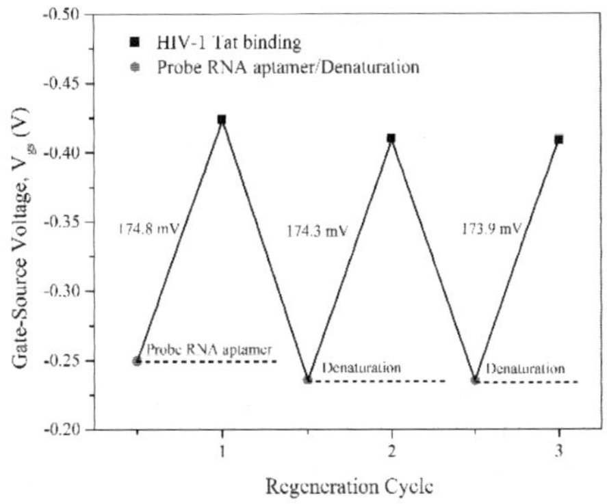

Figure 6. The repeatability measurement of 100 pM HIV-1 Tat detection performed on the same device.

4. Conclusion

In summary, we have successfully fabricated nanocrystalline diamond-based electrolyte-gate field effect transistor for biosensing application. The grain size of the NCD were successfully proved to be in nanometres range. Later, the detection of HIV-1 Tat on NCD-EGFET was successfully conducted with the RNA aptamer as sensing probe. In the future, studies on the HIV-1 Tat detection will be conducted on NCD-EGFET in the terms of sensitivity, specificity, and selectivity in order to develop a highly precise biosensor for biological diagnosis.

References

[1] R. Monošík, M. Streďanský, and E. Šturdík 2012 *Acta Chim. Slovaca*, vol. **5** 109–120.
[2] A. P. F. Turner 2013 *Chem. Soc. Rev.* **42** 3184–3196.
[3] V. S. P. K. S. A. Jayanthi, A. B. Das, and U. Saxena 2017 *Biosens. Bioelectron.* **91** 15–23.
[4] E. Stelmach *et al.* 2019 *Electrochim. Acta* **309** 65–73.
[5] A. Rahim Ruslinda, K. Tanabe, S. Ibori, X. Wang, and H. Kawarada 2013 *Biosens. Bioelectron.* **40** 277–282.
[6] V. Procházka, M. Cifra, P. Kulha, T. Ižák, B. Rezek, and A. Kromka 2017 *Appl. Surf. Sci.* **395** 214–219.
[7] B. Rezek, H. Watanabe, D. Shin, T. Yamamoto, and C. E. Nebel 2006 *Diam. Relat. Mater.* **15** 673–677.
[8] K. Song, T. Hiraki, H. Umezawa, and H. Kawarada 2012 *Applied Physics Letter* **90** 22–25.
[9] G. D. Tabi, B. Nketia-Yawson, S. H. Kang, C. Yang, and Y. Y. Noh 2018 *Org. Electron. physics, Mater. Appl.* **54** 255–260.
[10] R. Adzhri *et al.*2016 *Anal. Chim. Acta* **917** 1–18.
[11] M. F. M. Fathil *et al.* 2016 *Anal. Chim. Acta* **935** 30–43.
[12] A. R. *et al.*2017 *Sensors and Actuators A: Physical* **259** 57–67.
[13] K. Takagi, T. Nagase, T. Kobayashi, and H. Naito 2016 *Org. Electron. physics, Mater. Appl.* **32** 65–69.
[14] P. Bolshakov, P. Zhao, A. Azcatl, P. K. Hurley, R. M. Wallace, and C. D. Young 2017 *Microelectron. Eng.* **178** 190–193.
[15] F. M.F. *et al.* 2019 *Process Biochem.* **8** 55.
[16] S. Shojania and J. D. O'Neil 2006 *J. Biol. Chem.* **281** 8347–8356.

Mechanical Properties of Coconut Shell Reinforced Unsaturated Polyester Composites with the Presence of Titanium Dioxide as Additional Filler

Nurzarihah Shahidan[1], Mohamad Bashree Abu Bakar[1], Mazlan Mohamed[1]*, Azwan Mat Lazim[2], Ayub Ahmed Janvekar[3]

[1]Advances Material Research Cluster, Faculty of Bioengineering and Technology, Universiti Malaysia Kelantan, Jeli Campus, 17600 Jeli, Kelantan, Malaysia
[2]School of Chemical Sciences and Food Technology, Faculty of Science and Technology, University Kebangsaan Malaysia, 43600 Bangi, Selangor, Malaysia [3]School of Mechanical Engineering, Universiti Sains Malaysia, Engineering Campus, 14300 Nibong Tebal, Penang, Malaysia

Email: * mazlan.m@umk.edu.my

Abstract. Unsaturated polyester (UPE) composites filled with coconut shell (CS) powder was prepared by compression molding technique. To improve the mechanical and physical properties of CS-UPE composites, titanium dioxide (TiO_2) filler was added. CS also undergo two types of chemical treatment which were alkaline (NaOH) and silane treatment. Fourier transform infrared (FTIR) was used to investigate the changes of fibers chemical constituents after the two chemical treatments have been done. CS-TiO_2-UPE composites was characterized with physical testing (density, water absorption and thickness swelling), mechanical properties (tensile and flexural test) and scanning electron microscopy (SEM). Overall, the result shows that, the presence of CS and TiO_2 filler in UPE composites demonstrated the highest density. For water absorption and thickness swelling study, it was observed that the treated CS-UPE composites showed lower water absorption properties in comparison to those of untreated CS-UPE composites. The presence of TiO_2 filler showed an increased in those water absorption properties but still low compared to CS filled UPE. Both chemical treatment of CS caused a significant increase in the tensile and flexural properties. The presence of TiO_2 filler demonstrated a significant improvement in modulus properties but at the expense of strength properties. SEM investigation has shown that the surface modification of CS has better fiber-matrix interaction. Thus, chemical treatments on the CS improve fiber/matrix adhesion of UPE composites.

1. Introduction

Composite materials are one of the most influence materials used in industry applications such as automotive industry, building and construction industry and packaging industry besides metals, ceramics and polymers. Composites are known as the combination of two or more types of materials to give balance and at the same time enhance the properties such as physical and mechanical properties of the materials.

Content from this work may be used under the terms of the Creative Commons Attribution 3.0 licence. Any further distribution of this work must maintain attribution to the author(s) and the title of the work, journal citation and DOI.
Published under licence by IOP Publishing Ltd

Biocomposite is a types of materials which involve the combination of reinforcement and matrix which one or both are from biological origins [1]. For examples, fibers such as rice straw, kenaf, banana, pineapple, oil palm, jute and sisal reinforced to polymer matrix. The use of biocomposites in industries are quite popular among the manufacturers as the reinforcement/fiber is easy to find, can be renewable, eco-friendly and save the operating cost. In this study, biocomposite will be fabricated by using a combination of coconut shell (CS) as reinforcement, titanium dioxide (TiO_2) as the additional filler and unsaturated polyester (UPE) as the polymer matrix.

Coconut (*Cocos nucifera*) are came from the *Arecaceae* of family (palm family) which is known as one and the only species of genus *Cocos* [2]. In making the biocomposite materials, there are several parts of coconut that can be used as reinforcement such as shell and coir. Coconut shell is known as one of the most important reinforcement produced in tropical countries such as Malaysia, Indonesia, Thailand, Sri Lanka and India and it is widely used in making biocomposite. This is because of high specific strength and modulus properties [3,4]. According to [5] CS reinforced composites showed 80% better elongation at break and 20% better Charpy impact strength than soft wood composites [6,7].

Due to hard-wearing quality and high hardness which is not fragile like glass fiber, good acoustic resistance, moth-proof, not toxic, resistant to microbial and fungi degradation, and not easily combustible, the coconut particles are remarkable interest in the automotive industry [8,9]. Coconut shells reinforced matrix composites are classified in structural biocomposite which can be defined as one that need pressure to carry load in use because of its high compressive strength can be used in broad range applications of load bearing [10]. Titanium dioxide (TiO_2) is one of the additional filler besides calcium carbonate ($CaCO_3$) and aluminum oxide (AlO_3).

TiO_2 has been used since decades due to the speciality which can increase the stiffness and the modulus of the composite. TiO_2 will be added into the combination of CS and UPE as an additional filler. UPE resins are used widely in the composites industry besides phenolic resins and epoxy resins since UPE are not only good in mechanical properties, they are also low cost and easily use [11]. Combinations of the UPE with natural fillers, such as coconut shells, are believed to improve the physical and mechanical properties of the composites.

2. Experimental setup

2.1 Mechanical Testing

Tensile test was conducted towards the composite material samples to measure the length of the test specimens. The specimen was cut into suitable size according to ASTMD-638-I for tensile test. This experiment was conducted by using Instron universal testing machine with its cross of speed of 5 mm/min. From this testing, the volume fraction and the average value of the composite material samples was taken for analysis.

$$\text{Tensile strength} = \frac{\text{maximum load force (F)}}{\text{surface area (A)}} \tag{1}$$

$$\text{Tensile modulus} = \frac{\text{stress}}{\text{strain}} \tag{2}$$

Three-points bending flexural test was known as the test which conducts according to the ASTMD-790 standard to produce flexural strength and flexural modulus. The composite samples was tested at a cross head speed of 3 mm/min with the temperature of 25 $\pm$ 3 °C and the humidity of 50 %. Average value of the samples was recorded. The flexural strength in a three-point bending was calculated by using equation as follows.

$$\sigma_{max} = \frac{3P_{max}L}{bh^2} \quad (3)$$

Where, P_{max} is the maximum load at failure (N), L is the span (mm), b and h is the width and thickness of the specimen (mm) respectively. The flexural modulus was calculated from the slope of the initial portion of the load-deflection curve by using equation as follows.

$$E = \frac{mL^3}{4bh^3} \quad (4)$$

Where, m is the initial slope of the load deflection curve for each stacking sequence, the samples was tested and average result was obtained.

2.2 Scanning Electron Microscopy

The morphology of the tensile fracture surface of the sample composites was carried out by using a scanning electron microscope (SEM) model Jeol JSM-IT100. SEM was used to analyst the interfacial adhesion between matrix and reinforcement of the damaged composites after conducted the tensile testing.

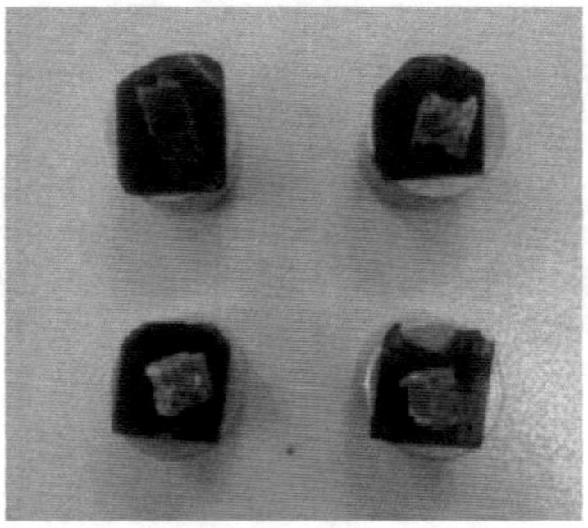

Figure 1. Samples for Scanning Electron Microscopy (SEM).

The composite sample was cut into a small pieces as shown in Figure 1 so that can be placed onto the pellet. SEM was operated at an acceleration voltage of 10 kV. To collect SEM images of the fracture surfaces of the tensile samples, a magnification ranged from ×25 to ×350 was used. The overall process of the preparation and characterization of CS-TiO_2-UPE composite.

3. Results

3.1 Mechanical Properties of CS-UPE Composites

3.1.1. Tensile and Flexural Strength

According to [12], the tensile of polymer composite increased with the addition of CS content. This is due to CS contain hard lignocellulose which give better strength to the composite [13]. Same goes to the flexural strength, which increasing with the addition of CS content.

With the same amount of CS content, the silane treated CS composite, UCS and alkaline treated CS composite, UCN show increase in tensile strength compared to untreated CS composite, UC. The reason is due to both of alkaline and silane treatments give better strength to the reinforcement. Alkaline treatment removed the impurities such as wax, oil, lignin and improves the rough surface. Meanwhile, silane treatment give smooth surface of the composite. For flexural strength, UCS exhibits lower flexural strength compared to UC while UCN exhibits higher flexural strength compared to UC.

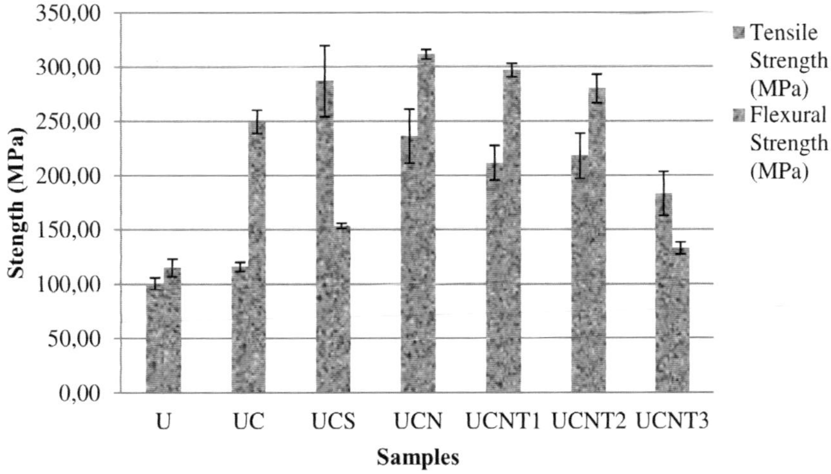

Figure 2. Tensile and flexural strength of UPE composites

With the same amount of CS content and the same treatment method which is alkaline treatment, the tensile strength of UCNT1, UCNT2 and UCNT3 shows different. UCNT2, which contain 1 wt.% TiO_2 filler become the highest tensile strength among the three followed by UCNT1, 0.5 wt.% TiO_2 and UCNT3, 1.5 wt.% TiO_2 respectively. This indicates that UCNT3 is the best percentage of filler used because the tensile strength decreased as more filler are added. Meanwhile, the flexural strength of UCNT1, UCNT2 and UCNT3 are decrease respectively when TiO_2 filler are added. UCNT1 which contain 0.5 wt.% of TiO_2 is the highest flexural strength, followed by UCNT2 which contain 1 wt.% TiO_2 and UCNT3 which contain 1.5 wt.% TiO. This indicates that the flexural strength decreased as the filler are added.

3.1.2. Tensile and Flexural Modulus

Figure 3 shows the graph of tensile and flexural modulus of the 7 composite specimens. The tensile modulus of U, UC, UCS, UCN, UCNT1, UCNT2 and UCNT3 are 0.51 GPa, 0.60 GPa, 0.96 GPa, 0.77 GPa, 1.21 GPa, 1.25 GPa and 1.38 GPa respectively.

Based on the graph, UC exhibits higher tensile modulus compared to U. This is due to the addition of CS as the reinforcement in the UC composite compared to U which only contain neat UPE. However, with the same amount of CS content but with different types of treatments, both UCS and UCN which undergo silane treatment and alkaline treatment respectively show lowered tensile modulus compared to UC which contain untreated CS. By comparing the two composites with the treatment, UCS exhibits higher tensile modulus compared to UCN. The graph also indicates that the tensile modulus of CS composite become higher as more filler is added. The flexural modulus of U, UC, UCS, UCN, UCNT1, UCNT2 and UCNT3 are 0.54 GPa, 4.28 GPa, 1.41 GPa, 8.03 GPa, 6.09 GPa, 3.58 GPa and 1.45 GPa respectively. Based on the graph, flexural modulus become higher as the CS was added. This is due to UC exhibits higher flexural modulus compared to UPE. UCS shows lower flexural modulus compared to UC but UCN exhibits higher flexural modulus compared to UC. With the same amount of CS content and also same treatment to CS, the graph shows the tensile modulus of CS composite increase as more TiO_2 filler are added.

IOP Conf. Series: Materials Science and Engineering **743** (2020) 012039 doi:10.1088/1757-899X/743/1/012039

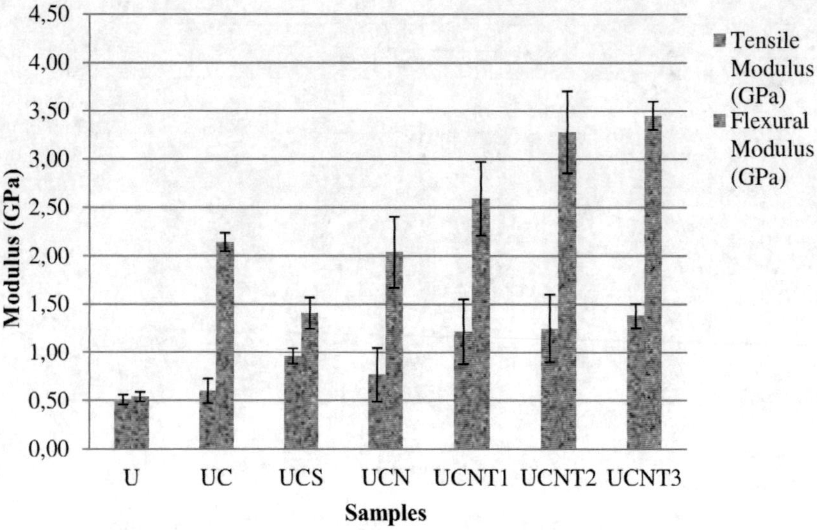

Figure 3. Tensile and flexural modulus of UPE composites.

3.2 Scanning Electron Microscopy

3.2.1. Morphological analysis of CS-TiO_2-UPE Biocomposites

The Scanning electron microscope was used to examine the tensile fracture surface of neat UPE, CS-UPE and CS-UPE-TiO_2 composites. The SEM micrograph of the tensile fracture surface of neat UPE is shown in Figure 4.7. The neat UPE composite has been represented as the control sample. The micrograph exhibits the homogenous surface and matrix tearing of UPE.

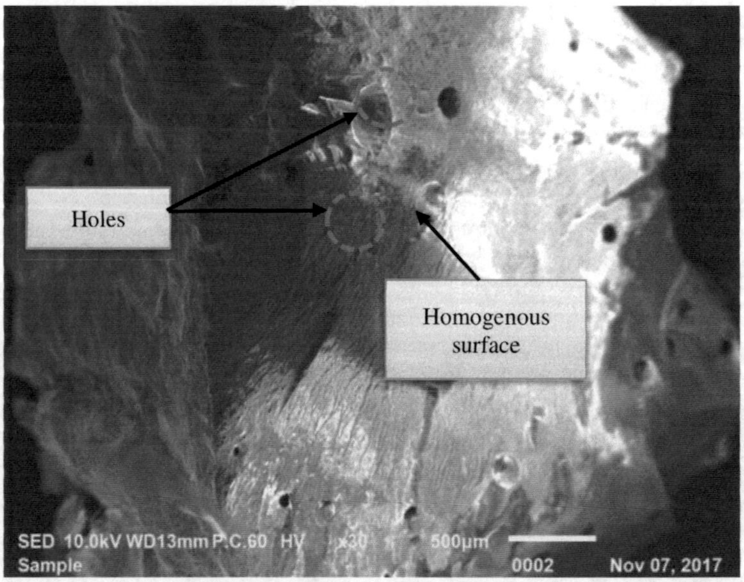

Figure 4. Scanning electron micrograph of tensile fracture surface of neat UPE composite.

As 30 wt.% CS is being added, there show the different between these two types of composites. SEM shows rough fracture surface and detachment of CS from matrix. This indicates the poor wetting and interfacial adhesion between CS and UPE matrix. The holes are the evidence weak bonding between CS and UPE matrix. The impurities observed on the surface of UC composite can act as crack initiators, thereby lessening the strength of the UPE matrix.

4. Conclusions

Mechanical properties consist of tensile and flexural strength and also tensile and flexural modulus. Based on the graph, tensile and flexural modulus increased as the CS was added and also after the chemical treatment. However, tensile and flexural modulus slowly down as TiO_2 filler addition increased. For tensile and flexural modulus, the modulus increased when CS content was added. Nevertheless, after the chemical treatment, the tensile and flexural modulus slightly decreased but increased back after the addition of TiO_2 filler. Morphology studies demonstrated the defects such as holes, crack and agglomerations which contributed to the reduction of mechanical properties such as tensile and flexural strength.

References

[1] Durowaye, S. I., Lawal, G. I., Akande, M. A., & Durowaye, V. O. *Mechanical properties of particulate coconut shell and palm fruit polyester composites*. International Journal of Materials Engineering, 4(44), 141-147. (2014).

[2] Ali, M. (2011). *Coconut fibre: A versatile material and its applications in engineering*. Journal of Civil Engineering and Construction Technology, 2(9), 189-197.

[3] Bhaskar, J., & Singh, V. K. *Physical and Mechanical Properties of Coconut Shell Particle* Reinforced-Epoxy Composite. J. Mater. Environ. Sci, 4(2), 227-232 (2013).

[4] N. Aziz, M. Mohamed, M. Mohamad, M. A. Mohd Hazim, M. S. A. Aziz, H. Yusoff, and Z. I. Rizman. *Influence of Activated Carbon Filler on the Mechanical Properties of Wood Composites*. ARPN J. Eng. Appl. Sci., vol. 10, no. 1, pp. 376–386, (2015).

[5] Agunsoye, J. O., Isaac, T. S., & Samuel, S. O. *Study of mechanical behaviour of coconut shell reinforced polymer matrix composite*. Journal of Minerals and Materials Characterization and Engineering, 11(8), 774-779. (2012).

[6] M. Mazlan, A. Rahim, A. M. M. A. I. Bakri, A. F. Zubair, Y. M. Najib, and A. B. Azman. *Thermal Management of Electronic Components by Using Computational Fluid Dynamic (CFD) Software , FLUENT ™ in Several Material Applications (Epoxy , Composite Material & Nano-silver)*. Adv. Mater. Res., vol. 795, no. 1982, pp. 141–147 (2013).

[7] M. Mazlan, A. Rahim, A. M. Mustafa Al Bakri, M. a. Iqbal, W. Razak, and M. S. Salim. *A New Invention of Thermal Pad Using Sol-Gel Nanosilver Doped Silica Film in Plastic Leaded Chip Carrier (PLCC) Application by Using Computational Fluid Dynamic Sofrware, CFD Analysis*. Adv. Mater. Res., vol. 795, pp. 158–163 (2013).

[8] Bhaskar, J., & Singh, V. K.. *Physical and Mechanical Properties of Coconut Shell Particle Reinforced-Epoxy Composite*. J. Mater. Environ. Sci, 4(2), 227-232 (2013).

[9] Ibrahim, S. A., Mohamed, M., Ramle, S. F. M., Amini, M. H. M., Abdul Aziz, M. S., & Rizman, Z. I. *Biocomposite material to enhance heat transfer of wood (shorea faguetiana and palaquim sp) for green building in malaysia*. ARPN Journal of Engineering and Applied Sciences, 10(1), 301-312 (2015).

[10] Rozali, N. A., Abu Bakar, M. B., Masri, M. N., Sulaiman, M. A., Mohamed, M., & Mohd Thirmizir, M. Z. *Mechanical and water absorption properties of hybrid kenaf/glass fibre mat reinforced unsaturated polyester composites*. Materials Science Forum, Vol. 888, pp. 228-233, (2017).

[11] Ibrahim, M. S., Sapuan, S. M., & Faieza, A. A. *Mechanical and thermal properties of composites from unsaturated polyester filled with oil palm ash*. Journal of Mechanical Engineering and Sciences, 2, 133-147. (2012).
[12] Ting, T. L., Jaya, R. P., Hassan, N. A., Yaacob, H., Jayanti, D. S., & Ariffin, M. A. M. *A Review of Chemical and Physical Properties of Coconut Shell in Asphalt Mixture*. Carbon, *7*(15), 85-89. (2016).
[13] Bisanda, E. T. N., & Ansell, M. P. *The effect of silane treatment on the mechanical and physical properties of sisal-epoxy composites*. Composites Science and Technology, 41(2), 165-178 (1991).

IOP Conf. Series: Materials Science and Engineering **743** (2020) 012040 doi:10.1088/1757-899X/743/1/012040

Application of Ionic Nano Particles in Fish Processing Wastewater by using Biochemical Treatment: A Short Review

Mohamad Anuar Kamaruddin[1,2,a], Lim Ewc Teik[3,b] and Muhd Nazmi Ismail[4,c]

[1,3]Environmental Technology Division, School of Industrial Technology, Universiti Sains Malaysia, 11800, Penang, Malaysia
[2]Center of Excellence Geopolymer and Green Technology (CEGeoGTech), School of Materials Engineering, Universiti Malaysia Perlis (UniMAP), Perlis, Malaysia
[4]Civil Engineering Department, Politeknik Sultan Abdul Halim Mu'adzam Shah (POLIMAS), Bandar Darulaman, 06000 Jitra, Kedah, Malaysia

Email: [a] anuarkamaruddin@usm.my, [b] anuar197@gmail.com, [c] nazmiphd@gmail.com

Abstract: Concerns of microorganisms growth in fish processing plants are crucial to the operation of the fish processing facilities. The presence of toxic microorganisms in the effluent could bring detrimental effects to the environmental and public health as several disease outbreaks could possibly take place with the release of infectious wastewater from the facilities. Environmental Quality (Industrial Effluent) Regulations 2009 published by Department of Environment Malaysia has clearly stated that the discharge limit of total coliform into water bodies that are safe for marine life, fisheries and human consumption must be not exceeded 600MPN/100 mL while for *E.coli* is not more than 400MPN/100mL.To cope with the laws and regulations, the industry has practiced several methods on disinfecting the effluent such as chlorination or ultraviolet treatment. However, the excessive use of such methods can cause deterioration to the environment as the by-products get released from the effluent are usually harmful to the natural systems of the environment. This paper aims to summarize recent progress on disinfectant methods that have been reported and published in the literature.

1. Introduction

Malaysia fish processing industry is considered as one of the important contributor to the national gross development. Based on the data recorded by Department of Fisheries Malaysia, fish processing industries alone, which includes the making of fish fillets, fresh, chilled or frozen products contributed about 14 thousand tonnes of yield which equal to USD 44 million. A trend of large industry development for these purposes are being supported by the present Malaysian government. The trend is expected to continue with the growing numbers of small and medium enterprises for fish processing plant coming out in the coming years.

Fish processing activity in Malaysia has been established like other manufacturing industry that releases large volume of wastewater. Fish processing industry produces a high volume of wastewater daily. In order for this industry to be recognized and well accepted by the authority, there are several stringent regulations to be adhered with. For example, operation license from the authority must be obtained prior operating the facility. Meanwhile, strict requirement by Department of Fishery Malaysia has been imposed for fish processing activity and its products need to follow enacted regulations and complying to international standards. The discharge from processing facility is regarded as industrial has wastewater needs. The discharge needs to adhere to Environmental Quality (1974) Industrial

Content from this work may be used under the terms of the Creative Commons Attribution 3.0 licence. Any further distribution of this work must maintain attribution to the author(s) and the title of the work, journal citation and DOI.
Published under licence by IOP Publishing Ltd

Effluent (Regulations) 2009 for various parameters that covers physical, chemical and biological parameters.

1.1 Wastewater Generation

In general, there are a few fragmented procedures involves during fish processing activity. Typical activity and their byproduct is shown in Fig. 1. The wastewater generated for whole activity can be as much as 30 m3 per batch. Throughout the process, water is commonly used for cleaning, beheading and washing generate large amount of wastewater. The constituents in the wastewater typically with oil and grease, chemical oxygen demand (COD), biological oxygen demand (BOD), total suspended solids (TSS) and

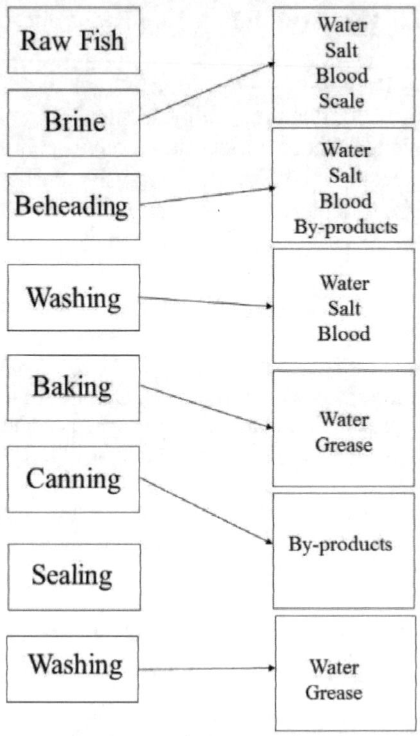

Figure 1. Fish processing and its byproducts (Source: (Chowdhury et al., 2010)).

1.2 Fish Processing and its Wastewater

Typically, large volume of water during processing of fish is needed as the manufacturer required to cleanse the fish product for every process of scaling, washing, beheading, gutting and filleting [1]Those downstream processing industries and fish processing carries wastewater laden with various kind of organic contaminants, salts and oils (proteins, lipids and enzymes) which are quite difficult to treat. In fact, large volume of wastewater is also produced after the canning process as they usually contain the raw leftover materials that constitutes large portion of the organic contaminants [2]. The major issue associates with the fish processing industry is due to excessive organic loading discharge to the receiving water body without effective treatment in place. Albeit, there have been significant progress of efficient treatment of the wastewater by selected promising method, however, the excessive use of such methods can cause deterioration to the environment as the by-products get released from the effluent and carries harmful substances to the natural systems of the environment.

The COD of effluent are commonly found to be higher than BOD for fish processing wastewater due to high organic load production which leads to a higher amount of bacteria coliform produced [2]. It is important to find a sustainable treatment method for fish-processing wastewater that allows treating of wastewater that fulfills the discharge limit or available for reuse in other sectors. Typical wastewater treatment can be ranging from physical, chemical and combination of physico-chemical treatment. Each of the treatment has their advantages and drawbacks depending to the extent of discharge limit allowable by the authority. One of many methods is known as disinfection which is considered as important procedure in a wastewater treatment system which serves as a preventive method that ensuring biological strains which causes severe diseases are not present prior to release as effluent [3]. The disinfectant particles commonly interact with bacteria cells by interrupting the transmission of electron or causes damage to the DNA of the bacteria by disrupting its cell membrane [4]. According to Lee et al. [3], the low operation cost of chlorination method is the main cause for chlorine overuse in industrial sectors.

2. Disinfectant Technology

There have been a lot of chemical agents that have been synthesized and commercialized in the recent years. These chemicals include chlorine dioxide, ozone, organic acids, peroxyacetic acid, electrolyzed oxidizing water and hydrogen peroxide [5]. The properties of these disinfectants are listed in Table 1 as summarized by [6]. Accordingly, there have been a substantial evidence of disinfectant agents application due to their inactivation capabilities against various pathogenic and spoilage microorganisms.

Table 1. List of disinfectant chemicals and their properties [7].

Parameters	Chlorine	Chlorine dioxide	Ozone	Peroxyacetic acid	H_2O_2
Boiling point, °C	-34.04	11	-112	107	-0.43
Melting point, °C	-101.5	-59	-192	-	150.2
Solubility in water	0.007 g/mL	8 g/L (at 20 °C)	0.570 g/L (at 20 °C)	Completely soluble	Miscible
Density	3.2 g/L (at 0 °C)	2.757 g/L	2.144 g/L (at 0 C)	1.0375 g/mL	1.450 g/cm^3
Acidity	11.7	3.0	7/5	8.2	11.75
Oxidation potential (eV)	1.36	1.57	2.1	1.81	1.8

2.1 Chlorination

The use of chlorine is a very well known method to disinfect effluent before discharging them into any streams, river or ocean [5, 8]. The ability of chlorine to breakdown nucleic acids and also the cell membranes of microorganisms makes it a very good choice of disinfectant for treating the wastewater in industrial sector [7]. It is commonly known for its ease of application, disinfecting effectiveness and low costing in the treatment of municipal sewage [6].However, the formation of carcinogenic chloramine are one of the main problem from overdosing chlorine in wastewater disinfection [9]. Chloramine are formed by mixing chlorine with ammonia as shown in equation, which is very toxic to fishes and could decimate certain fish species [10].

$$NH_3 + HOCl \Leftrightarrow NH_2Cl + H_2O \quad \text{Monochloramine}$$

$$NH_2Cl + HOCl \Leftrightarrow NHCl_2 + H_2O \quad \text{Dichloramine}$$

The current emphasis on stricter enforcement in effluent standards causes exponential growth in the use of chlorination for disinfecting effluent [11]. In contrast, not much have been done by those who proposed and supported such methods to find out the subsequent effect of the increased usage of chlorine [12].

2.2 Ozonation
Ozone may act as a primary disinfectant which provides an alternative to the use of chlorine acting as powerful oxidizing agent. Ozonation provides a higher efficiency in removing several types of organic micropollutants especially pesticides in downstream water system [13].The residual time for ozone in water bodies are shorter than the other type of disinfectants as ozone decays rapidly compared to other chemical additives. Besides that, ozonation will also be able to improve the removal rate of algae in water bodies and eliminate other micropollutants that gives out bad smell and color.

2.3 Ultraviolet
Another popular choice for industrial application in deactivating pathogens contained in wastewater is the Ultraviolet (UV) disinfection [14]. Compared to conventional methods like chlorination and ozonation, ultraviolet disinfection does not require any addition of chemicals and no toxic by-products are produced from the use of it [15]. It is found that the use of ultraviolet disinfection will not lead to generation of disinfectant-resistant bacteria as the rays does not have direct contact with the water bodies and pathogens, but the disadvantage of this method will have much higher implementing cost [14]. Fig. 2 shows typical ultraviolet and ozone set up for fish processing wastewater.

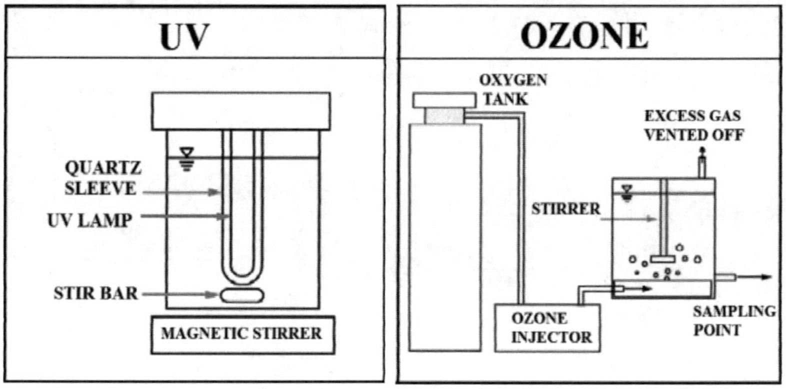

Figure 2. UV and ozone set up.

2.4 Environmental Concerns of Disinfectant Application
Although several popular disinfection methods for fish processing wastewater such as the use of chlorine, ozone and ultraviolet may be able to reduce bacteria growth in the effluent, with the exception that all of them does in fact have side effects that may just transferred the pollution from water to another medium without solving the root problems. Table 1 summarizes the disadvantages and limitations of the few common disinfection methods obtained by some researchers in disinfecting industrial wastewater.

Type of disinfectant	Limitation	Source
Chlorine	◆ Toxic by-products ◆ Promotes growth of disinfectant-resistant bacteria	[16] [9] [5] [17]
Ozone	◆ Toxic by-products ◆ High cost in industrial scale ◆ Toxic gas and corrosive upon in contact with moisture	[3]
Ultraviolet	◆ Very high cost in industrial scale ◆ May be radioactive to human	[13] [14]

3. Biochemical Treatment

There are various processes available for the treatment of wastewater with the most common method such as chemical destabilization and membrane processes [18]. Compliance with the discharge limits might not be achieved completely even when some of these techniques are used. The major economic advantage is the main reason that most industry treat their effluent with biochemical treatment system as it is cheaper, produces a cleaner effluent that fulfil the discharge limit and emit lesser toxic by-products. Biochemical treatment uses natural processes which relies on the help of bacteria in the decomposition of organic complex substances. The method is also more environmental friendly comparing to physical or chemical treatment [2].

Some researcher conducted studies and found that the copper(II) ions was affecting the cell metabolism of the microorganisms present in wastewater [19]. Therefore, copper compounds are usually present in pesticides and algae-killing solution in swimming pools or natural lakes. The ability of copper ions in interfering the microbe's cell metabolism shows the antimicrobial effect clearly. A study done by European researchers stated that the removal of copper ions from biological wastewater treatment systems can be easily done by adsorption and precipitation of copper ions [20]. The potential impact of the copper ions towards the environment are also much lower compared to the chlorine ions. A review paper in the same journal showed that about 33 to 98% copper ions can be removed from activated sludge treatment which are definitely easier to remove compared to *E.coli* and *Salmonella sp.* removal or even the toxic compounds formed from chlorine overdosing. Thus, we can opt to use copper ions in treating wastewater instead of using the costly ultraviolet treatment or chlorination which may cause detrimental effect to the environment and hard to remove the excess ions.

Quality of wastewater effluents can be monitored by relying on the indicator bacteria namely the coliforms, *E.coli* and *Pseudomonas* [21]. Thus, several bacteriological tests such as enumeration of *E.coli* and coliforms are developed in modern world to cater for the needs of industries in monitoring the microbiological quality of their wastewater effluents. Solid medium plating method are commonly used to test the presence of coliforms and E.coli due to its low costing and quick analysis [12].

One of the main issues with membrane filtering application in wastewater treatment are the formation of biofilm [16]. It would cost a lot to change the membrane filter periodically so it is important to control bio-fouling on the membrane surfaces [22].It is found that having copper ions on the membrane surface would show a significant anti-biofouling activity and extends the periodic changing of membrane filters by 30% [19].

4. Conclusion

Although many efforts concerted on disinfectant technology have been devoted especially in the materials science field, there is growing interest for the application of antimicrobial products for diverse field in the recent times. Albeit, knowledge of the frontier of the bacterial.

Acknowledgement

Authors gratefully acknowledge the grant received from Universiti Sains Malaysia under Innovation Seed Fund Project (AUP100248).

References

[1] Cristóvão, R.O., et al., *Fish canning industry wastewater treatment for water reuse–a case study.* Journal of Cleaner Production, **87**: p. 603-612 (2015).
[2] Chowdhury, P., T. Viraraghavan, and A. Srinivasan, *Biological treatment processes for fish processing wastewater–A review.* Bioresource technology,. **101**(2): p. 439-449 (2010).
[3] Lee, O.-M., et al., *A comparative study of disinfection efficiency and regrowth control of microorganism in secondary wastewater effluent using UV, ozone, and ionizing irradiation process.* Journal of hazardous materials,. **295**: p. 201-208 (2015).
[4] Li, Q., et al., *Antimicrobial nanomaterials for water disinfection and microbial control: potential applications and implications.* Water research, **42**(18): p. 4591-4602 (2008).

[5] Joshi, K., et al., *Novel disinfectants for fresh produce.* Trends in Food Science & Technology, 2013. **34**(1): p. 54-61.
[6] Li, Y., et al., *Three-step effluent chlorination increases disinfection efficiency and reduces DBP formation and toxicity.* Chemosphere, **168**: p. 1302-1308 (2017).
[7] Zhang, Y., et al., *Inactivation of antibiotic resistance genes in municipal wastewater effluent by chlorination and sequential UV/chlorination disinfection.* Science of the total environment, **512**: p. 125-132. (2015).
[8] Yang, X., C. Shang, and J.-C. Huang, *DBP formation in breakpoint chlorination of wastewater.* Water research,. **39**(19): p. 4755-4767 (2005).
[9] Zheng, J., et al., *Effects and mechanisms of ultraviolet, chlorination, and ozone disinfection on antibiotic resistance genes in secondary effluents of municipal wastewater treatment plants.* Chemical Engineering Journal, **317**: p. 309-316 (2017).
[10] Dunnick, J.K. and R.L. Melnick, *Assessment of the carcinogenic potential of chlorinated water: experimental studies of chlorine, chloramine, and trihalomethanes.* JNCI: Journal of the National Cancer Institute,. **85**(10): p. 817-822. (1993).
[11] Oliver, J.D., M. Dagher, and K. Linden, *Induction of Escherichia coli and Salmonella typhimurium into the viable but nonculturable state following chlorination of wastewater.* Journal of water and health, **3**(3): p. 249-257 (2005).
[12] Sangadkit, W., et al., *Practical coliforms and Escherichia coli detection and enumeration for industrial food samples using low-cost digital microscopy.* Procedia Engineering, **32**: p. 126-133. (2012).
[13] Mehrjouei, M., S. Müller, and D. Möller, *A review on photocatalytic ozonation used for the treatment of water and wastewater.* Chemical Engineering Journal,. **263**: p. 209-219 (2015).
[14] Song, K., M. Mohseni, and F. Taghipour, *Application of ultraviolet light-emitting diodes (UV-LEDs) for water disinfection: A review.* Water research,. **94**: p. 341-349 (2016).
[15] Bourgin, M., et al., *Evaluation of a full-scale wastewater treatment plant upgraded with ozonation and biological post-treatments: Abatement of micropollutants, formation of transformation products and oxidation by-products.* Water research, **129**: p. 486-498 (2018).
[16] Du, Y., et al., *Increase of cytotoxicity during wastewater chlorination: impact factors and surrogates.* Journal of hazardous materials,. **324**: p. 681-690 (2017).
[17] Rajnochova, M., L. Tuhovcak, and J. Rucka. *Drinking Water Supply without Use of a Disinfectant.* in *E3S Web of Conferences*.. EDP Sciences (2018).
[18] Delee, W., et al., *Anaerobic treatment of textile effluents: a review.* Journal of Chemical Technology & Biotechnology: International Research in Process, Environmental AND Clean Technology, 1998. **73**(4): p. 323-335.
[19] Ochoa-Herrera, V., et al., *Toxicity of copper (II) ions to microorganisms in biological wastewater treatment systems.* Science of the total environment, **412**: p. 380-385 (2011).
[20] Crane, R., et al., *Fate and behaviour of copper and zinc in secondary biological wastewater treatment processes: I Evaluation of biomass adsorption capacity.* Environmental technology, **31**(7): p. 705-723 (2010).
[21] Odonkor, S.T. and J.K. Ampofo, *Escherichia coli as an indicator of bacteriological quality of water: an overview.* Microbiology research,. **4**(1): p. e2-e2 (2013).
[22] Zhao, C., et al., *Highly antifouling and antibacterial performance of poly (vinylidene fluoride) ultrafiltration membranes blending with copper oxide and graphene oxide nanofillers for effective wastewater treatment.* Journal of colloid and interface science,. **505**: p. 341-351 (2017).

IOP Conf. Series: Materials Science and Engineering **743** (2020) 012041 doi:10.1088/1757-899X/743/1/012041

Food Waste Quantification and Characterization as a Measure towards Effective Food Waste Management in University

Mohamad Anuar Kamaruddin[1,2,a], Nurul Nadia Jantira[3,b] and Rasyidah Alrozi[4,c]

[1,3]Environmental Technology Division, School of Industrial Technology, Universiti Sains Malaysia, 11800, Penang, Malaysia
[2]Center of Excellence Geopolymer and Green Technology (CEGeoGTech), School of Materials Engineering, Universiti Malaysia Perlis (UniMAP), Perlis, Malaysia
[4]Faculty of Chemical Engineering, Universiti Teknologi MARA Cawangan Pulau Pinang, Kampus Permatang Pauh, 13500, Permatang Pauh, Pulau Pinang

Email: [a] anuarkamaruddin@usm.my, [b] anuar197@gmail.com,
[c] rasyidah.alrozi@ppinang.uitm.edu.my

Abstract. Globally, one-third of the food produced is wasted or underutilized which accounted for about 1.3 billion tonnes of food. If the food waste (FW) is disposed of in the landfill, it decomposes under the anaerobic condition that releases methane gas, a harmful greenhouse gas. FW is also referring as biodegradable materials where it has a higher moisture content that easily breakdown by the microbe and could produce odor problems. Thus, a study on FW generation rate and its characteristics were carried out. Physical and chemical characteristics of the FW which include the density, food waste composition, the generation rate, pH, moisture content, ash content, C/N ratio and the calorific value were determined based on the Malaysian Standard MS 2505:2012. Based on the field and laboratory work analysis, it was observed that the generation rate for the café A and B was between 0.04 kg/cap/day to 0.112 kg/cap/day and 0.01 kg/cap/day to 0.119 kg/cap/day. Meanwhile, the density of the FW was in the range between 0.035 kg/m^3 to 0.077 kg/m^3. Based on the standard method recommendation, 5 compositions were tabulated which include FW, plastic film, plastic rigid, paper and others (aluminum, steel and etc.). Food waste demarcated was further divided into five categories such as rice, noodles, vegetable, shells and meat & bones. Rice and noodles produced the largest amount which was 52.50% and 55.52%. Moisture content and an ash content of the FW was recorded at 71.52% and 25.31% where it was slightly acidic (6.16) in nature. The calorific value for wet and dried FW was 4539.914 kcal/kg and 2091.3 kcal/kg. Based on the result obtained, the suitable method to treat food waste from the cafeteria was proposed.

1. Introduction

Malaysia is one of the countries that have rapid population growth, urbanization, economic levels and rising of community that generates an enormous rate of MSW [1]. The increasing of waste generation per capita can be influenced by the continuous economic growth and development. For example, there are many rural areas have been transformed into industrial areas and provide many job opportunities to rural communities. Thus, the job opportunities intriguing the citizen from another country like Bangladesh, Nepal, Myanmar and many more to migrate in these areas. The rising of population growth and migration of the citizen into this country contributing to increasing of average solid waste generation [2, 3]. The quantity of solid waste generated in Malaysia is exponentially increasing year by year. The solid waste generated was increased as much as 2,900 tons/day from 2001 to 2005. The solid

Content from this work may be used under the terms of the Creative Commons Attribution 3.0 licence. Any further distribution of this work must maintain attribution to the author(s) and the title of the work, journal citation and DOI.
Published under licence by IOP Publishing Ltd

IOP Conf. Series: Materials Science and Engineering **743** (2020) 012041 doi:10.1088/1757-899X/743/1/012041

waste generated was estimated will reach 30,000 tons/day in 2020. For example, a metropolitan city like Kuala Lumpur generates a lot of solid waste because of the population growth. The solid waste generated was estimated to reach 7,713.6 tons/day in the year 2020 [4].

The composition of MSW in Malaysia consists of organic or FW, paper, plastic, glass, metal, textile and so on. But, organic or FW is predominant components in MSW [5, 6]. Besides, there are 15,000 tons of daily FW generated including 3,000 tons of food that can be consumed and should not be discarded in Malaysia.

However, there is a variety of adverse environmental impacts involving FW. Global climate change on FW resulting in unnecessary greenhouse gas emission that compromising water and land. This will cause natural ecosystems becoming lesser than it should [7, 8]. Besides, untreated FW can increase of CO2 emissions. According to the European Commission, the importance of household waste such as FW must be taken care of. Therefore, by divesting FW from municipal solid waste composition, can increase the lifespan of a landfill. Furthermore, landfilling is one of the options for MSW disposal in Malaysia [9].

Until now, there is no documented report on the FW generation rate, characteristics and their potential has been established or available especially in USM. Therefore, there is a need to address these issues so that the result of the proposed study will be useful for future research work. Apart from that, there is urgency for this study to be carried out because, in line of a new implementation of separation at source campaign by the state government, no results have been put available to the stakeholders. Thus, this study will provide some baseline data potential of diverting FW from the waste stream in order to reduce the dependency on landfill.

2. Material and method

2.1 Sampling site

The selected location for this study was the University Sains Malaysia on Penang Island, Malaysia located at coordinates 5.3559 North and 100.3025 East. The study involved 2 cafes at Cafeteria Bakti Permai within the university compound. These cafes located near to student hostel and academic building which was School of Industrial Technology. Both cafes were coded as Café A and B. Café A stands for Kak Ani's Café and Café B stands for Koay Teow Th'ng's Café as shown in Figure 1.

Figure 1. Sampling site.

2.2 Food Waste Generation Rate

The data collected from the FW generation was chosen to be from food preparation and uneaten food. The calculation of daily FW generation was computed by weighing the FW generated by each café every day. The FW collected was fixed at 3 pm and weighed by using a weighing balance. Data obtained were recorded in the datasheet that provided in Appendix A. Next, the density of FW generate was determined by the total volume of FW in the tray multiply with the volume of the tray. The data was recorded into the datasheet in Appendix B. Other than that, the estimation of FW per capita rate per day was determined by the number of eateries recorded from Monday to Friday. The FW generation rate per day was calculated by using the following equation:

$$\text{Rate} = \frac{\text{Food Waste Weight (kg)}}{\text{Total Eateries}} \quad (1)$$

2.3 Sample Processing and Characterization

The dry matter (DM) determination was carried out following a method as recommended by Adhikari [10]. The FW sample was dried at the temperature 105°C for 24 hours in the oven (Memmert, Germany). After that, the dried sample was cooled by using desiccators. The size of the dried sample was reduced by using a grinder (Waring, USA)). Then, the sample was sifted into small size of 0.25 mm by using sieve shaker (Retsch, Germany). DM was used to arbitrate both characteristics which are physical and chemical. The dried sample was placed in a plastic bag and sealed so that it was not exposed to the air. pH value was measured by soaking 1 part dried sample with 10 part of distilled water for 60 minutes. Then, the pH value was measured by using a HACH Session 3 pH Meter (HACH, USA). The size of the sample is less than 180 µm [11]. Moisture content is the amount of water in the FW which is referred to the percentage of the weight loss during drying. The sample of raw FW was placed in the evaporating dish and weighed by using analytical balance, W1. After that, the sample of raw FW has been dried in the oven about 24 hours at 105°C. Then, the dried sample was cooled in the desiccators and weighed, W2. The percentage of the moisture content was calculated by using this formula which is taken from the guidelines from United Nation for Education Program published in 2002. C/N ratio of the FW was determined by using CHNS elemental analyzer (Elementar, Italy). The sample size is less than 180 µm. The calorific value of the FW sample was analyzed based on the method recommended by UNEP. In general, raw FW and dried FW were ground into a small particle (<1mm). The bomb calorimeter (IKA C200) was used to ignite the sample. Nickel wire was used for ignition. Then, the oxygen was filled to the cylinder which accommodates sample with ignition wire. The sample was weighted using combustion bag (0.5 to 0.8 g). The temperature of the equipment should not greater than 25°C for a half-hour. The value of calorific value reading was recorded in unit kcal/kg.

3. Result and Discussion

3.1 Food waste management system

Based on observation, the USM compound did not provide appropriate food waste management. The generation, collection, and transportation of FW were managed by the owners of the cafeterias. All the plates used by the eateries were collected by the representative of each café. The uneaten FW was placed together with the kitchen waste in the primary bucket. As we know, FW is the degradable organic material and need to be thrown after operating time so that the FW would not produce any foul odor. Basically, every university has its own rule about the MSW management including FW that needed to obey by all the owner of cafeterias to prevent any environmental problems such as foul odor. On the other hand, these methods are not efficient and suitable. It is because FW is one of the renewable resources. It can be converted into valuable materials, for instance, the organic fertilizer that can be used for the plant in the university. In this study, the foods waste generation rate and the alternative method for FW management have been studied as shown in Figure 2.

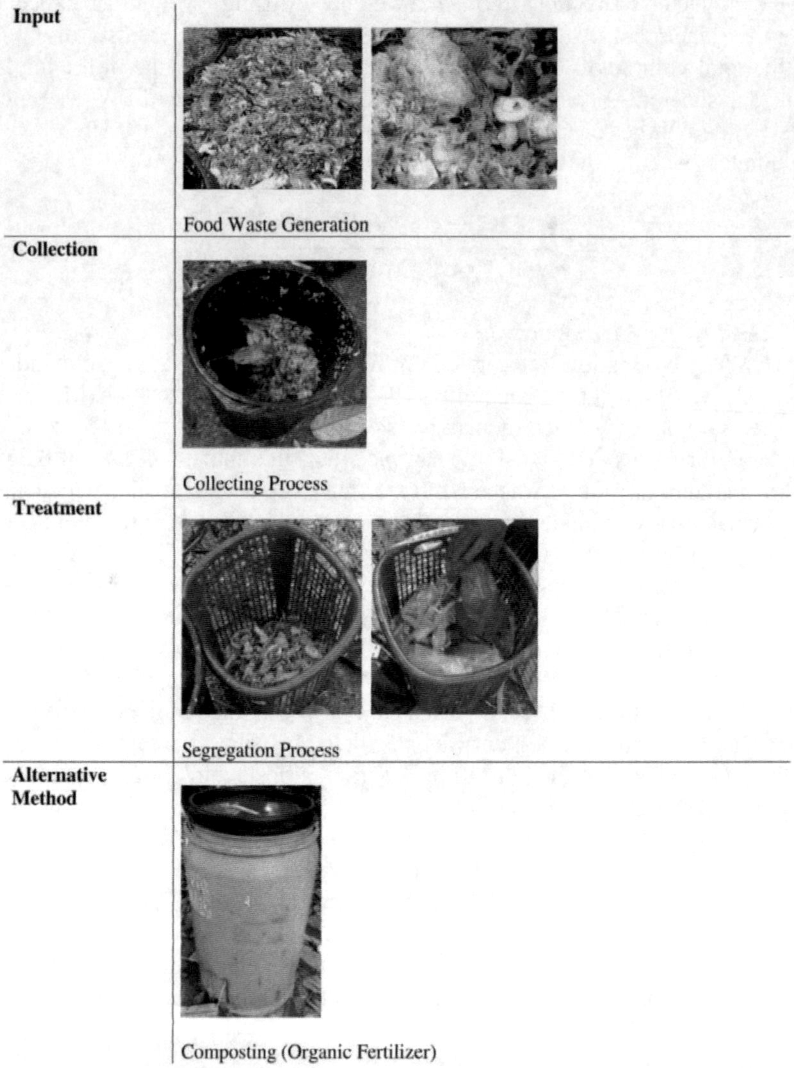

Figure 2. Alternative management system proposed for FW management in Main Campus, USM.

3.2 Food waste generation rate

Figsure 3 (a) and (b) represent the total weight of the FW generation rate that from two cafeterias within 15 days based on data shown in Appendix I. The weight of FW generated within 15 days that starting from 6 March 2018 until 23 March 2018 which from Monday to Friday that produced by two cafes in Bakti Permai, USM was totaled up for each collection day. Based on the figures, the generation rate for Café A has ranged between 41.08 g/cap/day to 112.24 g/cap/day. Meanwhile, the generation rate ranging for Cafe B was between 10.79 g/cap/day to 118.79 g/cap/day. From the analyzed figures, it could be observed that the generation rate for Café A was less than Café B. because students and staff more prefer to eat at Café A. Café A stand for Kak Ani's Café which selling rice and variety of dishes while Café B stand for Koay Teow Th'ng's Café which is selling several types of combining with soup like koay teow th'ng, noodles, and yong tau fu. So that, more students and staffs choose Café A because the food portion was bigger than Café B. There are no data on the 20th March 2018 (Monday) and 24th March 2018 (Friday) because during that day the weather is not good (raining heavily). According to the bar chart below, we can see that there was an irregular trend of WGR because of the inconsistent number of eateries.

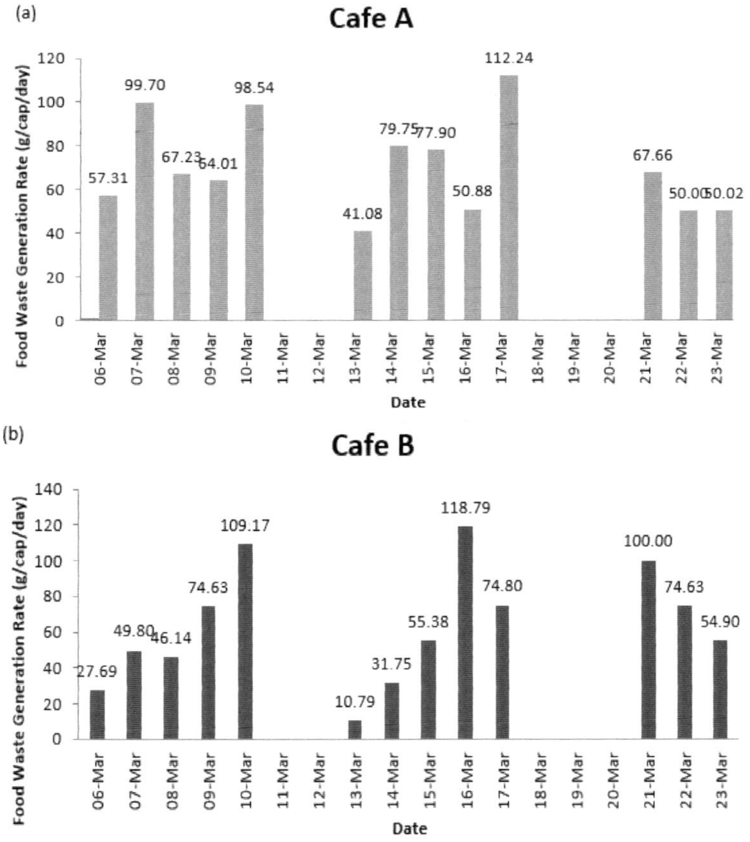

Figure 3. (a) and (b) Daily FW generation rate within 15 days for Café B.

Total weight of the FW generated in both cafes within 15 days was 332.09 kg which equals to 22.14 kg/day. This value showed a small amount of FW was generated in the Main Campus. However, there were only two cafes out of 18 cafes in the Main Campus, USM. In short, if the waste management is not properly managed, the possibility of the environmental problem of improper waste management would be plausible. Fig. 4 shows the trend of total FW generation within 15 days that was generated by two cafes based on five consecutive days. The highest value of FW generated on Tuesday, Thursday and Wednesday which was 25.79 kg, 24.01 kg and 23.67 kg compared to Monday and Friday which was 12.88 kg and 13.66 kg. The differences between the highest and a lower mean value of FW generated was 12.91 kg. It was due to the fact that there was no data recorded on Monday and Friday in the 3rd-week cause of the bad weather (raining). So, it was therefore unsuitable to do sampling because the weather would affect the data collection. Other than that, there were fewer students in the cafeteria on Monday and Friday because some of the students might free from having to attend the lecture on those days. As reported in Table 1, there was no significant between no of eateries and amount of FW where the value was 0.321. It was largely because not all the eateries mark the datasheet provided and the waste produced depends on the eating behavior

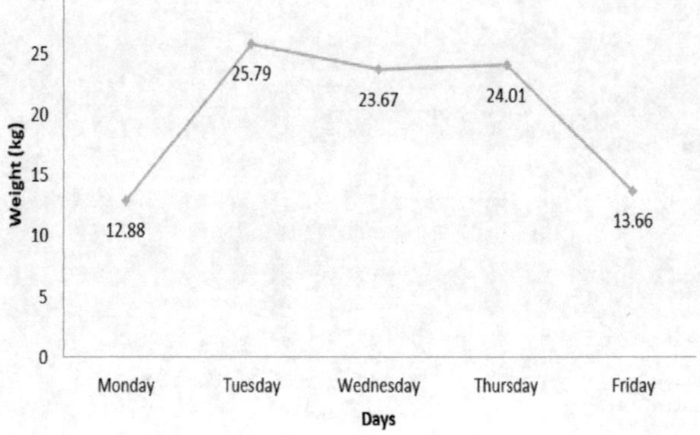

Figure 4. The trend for the amount of FW generated for both Cafes.

Table 1. Correlation between no of eateries and amount of food waste.

Correlations		FW	CS
FW	Pearson Correlation	1	.299
	Sig. (2-tailed)		.321
	N	13	13
CS	Pearson Correlation	.299	1
	Sig. (2-tailed)	.321	
	N	13	13

3.3 *Food waste composition*

Basically, Café A and Café B exerted similar food preparation process with comparable raw material used. The composition studied of FW was carried for 15 days (3 weeks). FW generated from café because of the leftover food and the food preparation activities. All the waste generated like FW, plastic rigid (food container, bottle, and others), plastic film (straw, plastic bag and so on) and paper (tissue and paper) were dumped together in the bucket. After that, the manual segregation process was carried out to obtain the percentage of waste generated. Figs. 4 (a) and (b) show the percentage of MSW generated for Café A and Café B. During segregation process, the MSW was divided into five categories which are FW, mixed paper, plastic film, plastic rigid and others (steel, aluminum, and etc.). It is because the waste composition is one of the factors that contributed to the emission of solid waste treatment.

As reported in Figs. 5 (a) and (b) there was a large amount of FW produced followed by plastic film, mixed paper, and others. The percentage of FW generate in Café A was 89.59% higher than mixed paper, plastic film, plastic rigid and others where 6.94%, 3.02%, 0.35%, and 0.10%. Meanwhile, for Café B the percentage of FW is 63.57% followed by the plastic film is 22.64%, mixed paper is 12.48%, others and plastic rigid where 0.81% and 0.50%. According to Saheri et al. [12], FW has a large percent compared to others in MSW. Thus, this study confirmed to the previous study done by Saheri et al. [12] and FW stood at 89.59% and 63.57% for Café A and Café B. So, we can proof that by doing this study where there was a higher amount of FW generated.

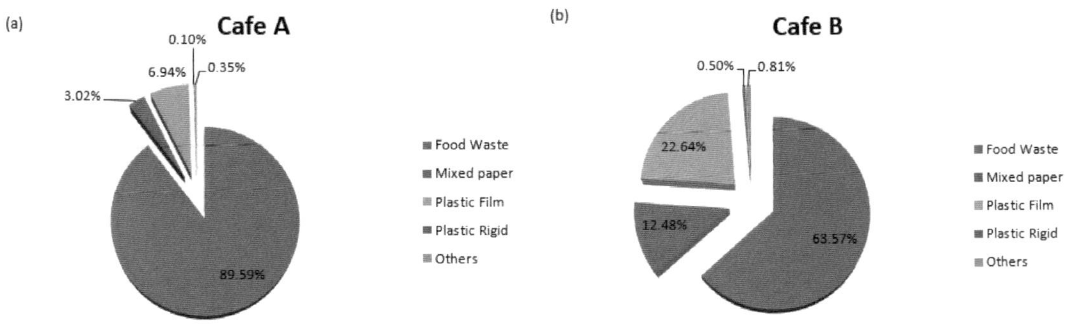

Figure 5. (a) and (b): Percentage of MSW composition generated at USM Main Campus for Café B within 15 days.

Table 2 shows the percentage of MSW and FW composition in 3 weeks based on five consecutive days for Café A and Café B. Café A generated a higher percentage of rice on Tuesday and Wednesday while a higher percentage of noodles for Café B on Tuesday and Thursday. Other than that, bones and meats waste also generated a large amount on Wednesday and for Café B a large amount of vegetable which is on Thursday. According to this study, FW composition comprised of rice, noodles, bones and meats, vegetable and shells. Rice and noodles was the main dish in Malaysia. So, the higher percentage in FW was rice and noodles followed by bones and meats, vegetable and shells as the side dish. It was predicted that the eating habit of the eateries were found to influence the composition of FW.

Table 2. Density for Café A and Café B within 15 days.

	Café A					Café B				
Composition	Mon	Tue	Wed	Thu	Fri	Mon	Tue	Wed	Thu	Fri
Mixed paper	2.78	1.85	1.58	1.64	1.24	1.45	3.35	3.35	3.18	1.16
Plastic Film	2.75	5.67	4.84	4.52	3.14	3.63	6.41	5.53	4.38	2.7
Plastic Rigid	N/A	0.14	0.11	N/A	0.06	0.17	0.1	0.11	0.09	0.03
Others	N/A	0.37	0.37	0.29	0.01	0.1	0.24	0.18	0.28	0.01
Rice	15.27	36.95	38.18	31.5	19.53	N/A	N/A	N/A	N/A	N/A
Noodles	N/A	N/A	N/A	N/A	N/A	5.53	9.6	8.14	9.41	2.7
Bones + Meat	8.4	16.7	18.21	16.39	10.1	1.1	0.49	0.99	0.73	0.37
Shell	0.5	3.49	1.23	2.63	1.61	N/A	N/A	N/A	N/A	N/A
Vegetable	9.64	13.8	7.72	11.08	6.44	2.32	4.61	5.14	7.99	4.6

3.4 Chemical characteristics of the FW

The chemical characteristics of FW obtained from the laboratory analysis are shown in Table 3. The value of pH obtained from the FW generated ranging between 6.15 to 6.17 and the average pH value is 6.16. The pH value of the FW generated can be categorized as acidic. Besides, the pH value of the FW generated from the food courts in Universiti Putra Malaysia (UPM) ranging between 4.5 to 5.5 [13]. The pH value is slightly acidic because of the mix of the citrus/juice and fruits. Other than that, the value of ash content in the FW is 25.31%. Ash content in FW represented the content of mineral. Ash content of foods is important for microbiological stability. According to the UNEP, the ash content for both cafés is higher than the value of ash content for kitchen waste in the UNEP statistics which is 15.37%.

Based on the result obtained in Table 3, the composition for carbon and nitrogen in the FW were 39.98% and 3.87% while the C/N ratio was 10.32%. C/N ratio is calculated to estimate the nutrient deficiency and ammonia inhibition. The combination of organic waste to obtain the desired value for the C/N ratio to obtain higher composting efficiency [14]. C/N ratio is a source of nutrient for compost.

The ratio of C/N ratio per gram is 6:1. The value of C/N ratio obtained is considered as low value. The lower value of C/N ratio can take a longer treatment time

Table 3. Chemical characteristics of FW.

Sample	Food Waste
pH	6.16
Ash Content	25.31%
Carbon (C)	39.92%
Nitrogen (N)	3.87%
C/N Ratio	10.32%

4. Conclusion

The field and laboratory work had indicated that the generation rate for Café A is 0.04 kg/cap/day to 0.112 kg/cap/day while for Café B is 0.01 kg/cap/day to 0.119 kg/cap/day. The density of the food waste generates is ranged between 0.035 kg/m3 to 0.077 kg/m3 for Café A and Café B within 15 days. The composition of MSW generated from both cafes consists of five categories which were food waste, paper, plastic film, plastic rigid and others (aluminum, steels). Based on the result obtained, the food waste generates a large portion compared to other components of solid waste. The composition study of FW comprises of rice, noodles, bones and meats, shells and vegetables. Rice and bones and meats are a major fraction of FW for Café A while noodles and vegetable is a large fraction of FW for Café B. There was a higher percentage of moisture content of 71.52% and slightly acidic (6.16). The value of ash content is 25.31%. Other than that, there was a higher value of CV for dried sample (4539.914 kcal/kg) compared to the wet sample (2091.3 kcal/kg). Meanwhile, the C/N ratio is 10.32%.
Further, composting is a suggested method to treat FW in this study. According to the result obtained, the higher moisture content of the FW is not convenient to used incinerator as a method in treating food waste. Besides, the higher of FW generation rate is one of the factors where landfill is not seemly to be able use in FW management. The analyze shows that the composting method is the best way for treating FW in the cafeterias. However, various issues need to be controlled before executes the composting as FW management.

Acknowledgement
Authors gratefully acknowledge the grant received from Universiti Sains Malaysia under Innovation Seed Fund Project (AUP100248).

References

[1] Kamaruddin, M.A., et al., *An overview of municipal solid waste management and landfill leachate treatment: Malaysia and Asian perspectives.* Environmental Science and Pollution Research,. **24**(35): p. 26988-27020 (2017).
[2] Kamaruddin, M., et al. *A study on the effects of different microbial inoculants on the decomposition of organic waste by using semi passive aerated reactor*. in *AIP Conference Proceedings*. AIP Publishing (2018).
[3] Kamaruddin, M., et al. *A study on the effects of carbon to nitrogen layers in semi passive aerated reactor for organic waste decomposition*. in *AIP Conference Proceedings*. AIP Publishing (2018).
[4] Saeed, M.O., M.N. Hassan, and M.A. Mujeebu, *Assessment of municipal solid waste generation and recyclable materials potential in Kuala Lumpur, Malaysia.* Waste management,. **29**(7): p. 2209-2213 (2009).
[5] Manaf, L.A., M.A.A. Samah, and N.I.M. Zukki, *Municipal solid waste management in Malaysia: Practices and challenges.* Waste management,. **29**(11): p. 2902-2906 (2009).
[6] Kamaruddin, M.A., et al., *Sustainable treatment of landfill leachate.* Applied Water Science, 2015. **5**(2): p. 113-126.

[7] Hall, K.D., et al., *The progressive increase of food waste in America and its environmental impact.* PloS one, **4**(11): p. e7940 (2009).

[8] Papargyropoulou, E., et al., *The food waste hierarchy as a framework for the management of food surplus and food waste.* Journal of Cleaner Production,. **76**: p. 106-115 (2014).

[9] Yusoff, M.S., et al., *Municipal Solid Waste Composition, Characterization And Recyclables Potential: A Case Study Evaluation In Malaysia.* The Journal of Solid Waste Technology and Management, **44**(4): p. 330-343 (2018).

[10] Adhikari, B.K., et al., *Characterization of food waste and bulking agents for composting.* Waste Management, **28**(5): p. 795-804 (2008).

[11] Kim, K.-H., et al., *Food decay and offensive odorants: A comparative analysis among three types of food.* Waste Management,. **29**(4): p. 1265-1273 (2009).

[12] Saheri, S., et al., *Solid waste management by considering composting potential in Malaysia toward a green country.* e-Bangi, **6**(1) (2017).

[13] Omar, F.N., et al., *Utilization of kitchen waste for the production of green thermoplastic polyhydroxybutyrate (PHB) by Cupriavidus necator CCGUG 52238.* African Journal of Microbiology Research,. **5**(19): p. 2873-2879 (2011).

[14] Lim, S.L., L.H. Lee, and T.Y. Wu, *Sustainability of using composting and vermicomposting technologies for organic solid waste biotransformation: recent overview, greenhouse gases emissions and economic analysis.* Journal of Cleaner Production,. **111**: p. 262-278 (2016).

IOP Conf. Series: Materials Science and Engineering **743** (2020) 012042 doi:10.1088/1757-899X/743/1/012042

Study Of The Utilization Of Carbide Waste And Fly Ash On Geopolymer Paste

Ridho Bayuaji[1], N S Putri[1], Liew Yun Ming[2] and Nurul Aida Mohd Mortar[2]

[1]Department of Civil Infrastructure Engineering, Faculty of Vocations, Institut Teknologi Sepuluh Nopember,Surabaya 60111, Indonesia
[2]Center of Excellence Geopolymer and Green Technology, School of Materials Engineering, Universiti Malaysia Perlis (UniMAP), Perlis, Malaysia

Email: bayuaji@ce.its.ac.id

Abstract. Portland cement is made of the main ingredients are limestone / lime are included in natural resource that is not renewable. So that needs holding replacement with environmentally friendly materials one of which is of geopolymer concrete. Fly Ash is the waste coal combustion products also include natural resources that can not be refurbished Based on data obtained from PT. Z, carbide waste generated in a year on average around 1978.72 m^3, it can be put to good use for waste carbide itself has a high CaO content. The composition of the material used in this study was 100% waste carbide, carbide waste 50%: 50% fly ash, as well as 100% fly ash and the activator used as NaOH and Na_2SiO_3. This study aims to determine how much influence the utilization of waste carbide and fly ash in geopolymer paste in the testing setting time, compressive strength, porosity, UPV, and permeability. The highest compressive strength besides there on the composition of a 100% fly ash is also present in 50% of waste carbide composition: 50% fly ash with activator ratio of 0.5 at the age of 56 days amounted to 39.59 Mpa. The use of fly ash in geopolymer still can not be eliminated 100%, due to the utilization of waste carbide can only be used as an additive or filler.

1. Introduction

Portland cement is an industrial product used for normal concrete building materials made of main ingredients are limestone / lime high levels of calcium and clay, or other similar materials with the composition, including into natural resources that cannot be renewed. Natural resource exploration excessive, would disrupt the environmental balance [1,2], it is necessary to the holding of a replacement with environmentally friendly materials one of which is a popular geopolymer concrete with fly ash as the main material of geopolymer concrete manufacture [6-9]. Waste fly ash is the result of burning coal in a steam power plant furnaces are pozolanik [3, 4]. Coal itself is a natural resource that cannot be updated [2] and if in continuously being explored would disrupt the balance of nature. For that we need in less utilization of fly ash, one of which is waste carbide. Waste carbide is the remainder of the evaporation of calcium carbonate powder with water from the manufacture of acetylene gas. Based on data obtained from PT. Z, carbide waste generated in a year on average around 1978.72 m^3, it can be put to good use for waste carbide itself has a high CaO content [5]. Thus, in this study utilizes carbide waste and fly ash as the main material of geopolymer paste. The main objective of this research is to prove whether the waste fly ash can replace carbide or simply to minimize the use of fly ash as filler.

Content from this work may be used under the terms of the Creative Commons Attribution 3.0 licence. Any further distribution of this work must maintain attribution to the author(s) and the title of the work, journal citation and DOI.
Published under licence by IOP Publishing Ltd

2. Methodology

The method used to answer this research is the experimental laboratory. Fly ash used in this study as the basic material for the manufacture of geopolymer from PLTU Paiton, Probolinggo. And the carbide waste is from PT. Z. To determine the chemical composition contained in fly ash and waste carbide, tests will be conducted SEM-EDX and XRF (X-Ray Fluorescence).

This study includes three series of studies, which is the study of literature, preliminary experiments and the main experiment. Activators for alkaline solution used is sodium silicate (Na2SiO3) and sodium hydroxide (NaOH) with molarity 8 M. For the manufacture of test specimens using a cylinder with a diameter of 2,5 cm x 5 cm and a cube of 15 cm x 15 cm x 5 cm. Tests include UPV and permeability with cube specimen conducted at the Laboratorium Beton Diploma Sipil, ITS Surabaya, compressive strength test using cylindrical performed in Laboratorium Struktur Teknik Sipil, ITS Surabaya and porosity test using cylinder conducted at the Laboratorium Jalan Diploma Sipil, ITS Surabaya.

3. Result and discussion

Some results from fly ash and waste carbide test are presented in the figure and the table below:

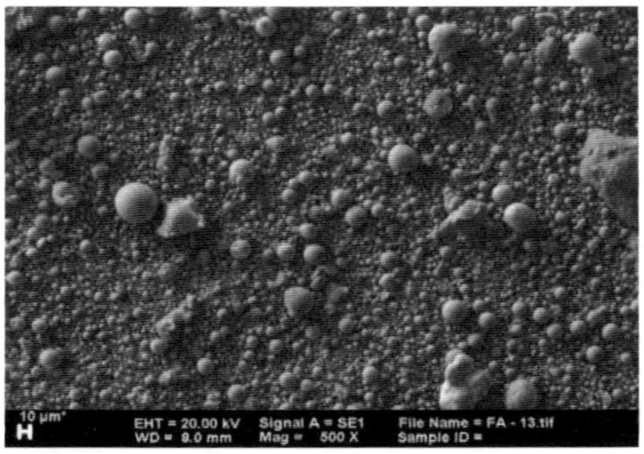

Figure 1. Fly Ash in magnification of 500x.

Figure 2. Fly Ash in magnification of 1000x.

Figure 3. Waste Carbide in magnification of 500.

Figure 4. Waste Carbide in magnification of 1000x.

SEM fly ash test result has a relatively round grain shape so it is more easily oxidized than the carbide waste that is shaped like clod. So, the faster the oxidation process, the faster the binding occurs[10].

Table 1. The XRF Test Result of Waste Carbide.

Waste Carbide (PT. Z)	
Type Compounds	Content (%)
SiO_2	3,91
Fe_2O_3	0,87
Al_2O_3	2,01
CaO	*91,48*
MgO	0,23
Na_2O	0,72
K_2O	0,07
MnO	0,02
ZnO	0,01
TiO_2	0,14
SO_3	0,27
P_2O_5	0,01

Table 2. The XRF Test Result of Fly Ash.

Fly Ash Type F (Paiton)	
Type Compounds	Content (%)
SiO_2	*47,1*
Fe_2O_3	16,07
Al_2O_3	24,25
CaO	5,83
MgO	2,62
Na_2O	0,65
K_2O	1,64
MnO	0,1
ZnO	0,29
TiO_2	1,16
SO_3	0,21
P_2O_5	0,18

From the XRF test results known that the fly ash used in this study contains a compound of Si + Al + Fe> 70%. CaO <10%. So that the fly ash used is type F. CaO content of fly ash used in this study is 5,83%, according to ASTM C 618-84 fly ash which has a CaO content of less than 10% is classified in type F.

Here are the test results setting time, compressive strength, porosity, and permeability UPV presented in Graph 1 - Graph 6. From Figure 1, note that setting the fastest time of 50% occurred in the composition of waste carbide: 50% fly ash with a time of 30 minutes 0mm in comparison with a decrease of 0.5 activator. In Figure 2 the compressive strength test, the highest yields are on a 100% fly ash composition age of 56 days with a ratio of 1.5 activator of 39.72 Mpa and 50% of waste carbide: 50% fly ash age of 56 days with a ratio of 0.5 activator 39.59 MPa. While Graph 3 test porosity, low porosity values occurred in the composition of a 100% fly ash age of 56 days with 1.5 activator ratio of 12.64%. Graph 4 UPV test, the highest yields are on a 100% fly ash composition age of 56 days with a ratio of 1.5 activator of 3556.67 m / s. Graph 5 permeability test, the result was lowest for the composition of a 100% fly ash age of 56 days with a ratio of 1.5 activator of 0.002 E-16 m2 and 50% of waste carbide: 50% fly ash age of 56 days with 0.5 activator ratio of 0.001 E-16 m2. And in Chart 6, the compressive strength of the age of 3 days with a special activator ratio of 0.5 and 1.5, the main composition of 100% Waste Carbide, 50% Waste Carbide: 50% Fly Ash and Fly Ash 100% as well as an additional 75% Waste composition carbide: 25% fly ash and 25% waste carbide: 75% Fly Ash can be seen that the percentage of waste carbide may slightly increase the compressive strength of 100%, while the percentage of waste carbide compressive strength is very weak compared with the composition of the mixture.

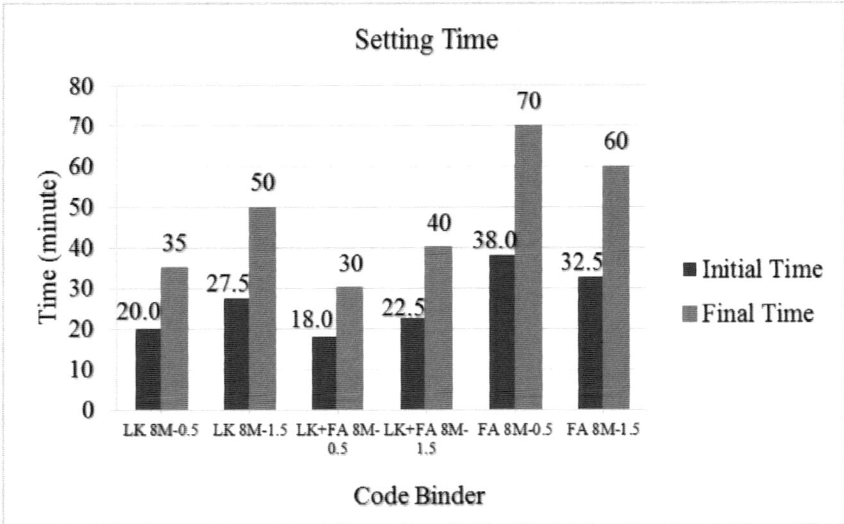

Graph 1. The Setting Time Test Result.

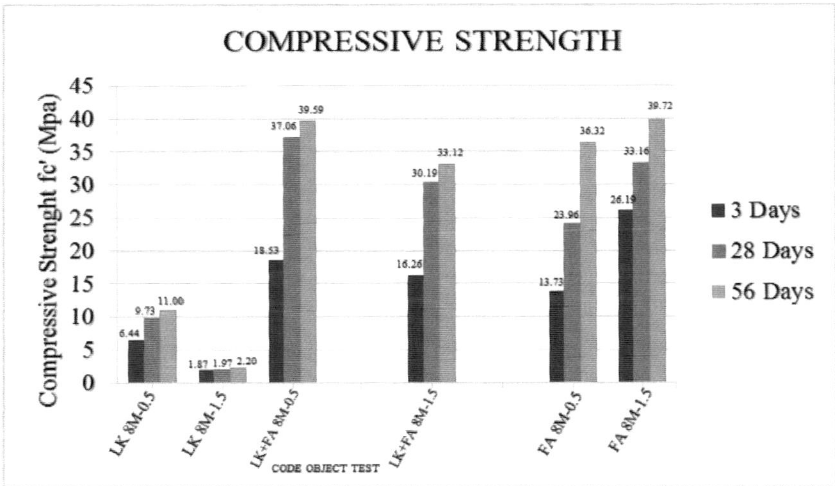

Graph 2. The Compressive Strength Test Result.

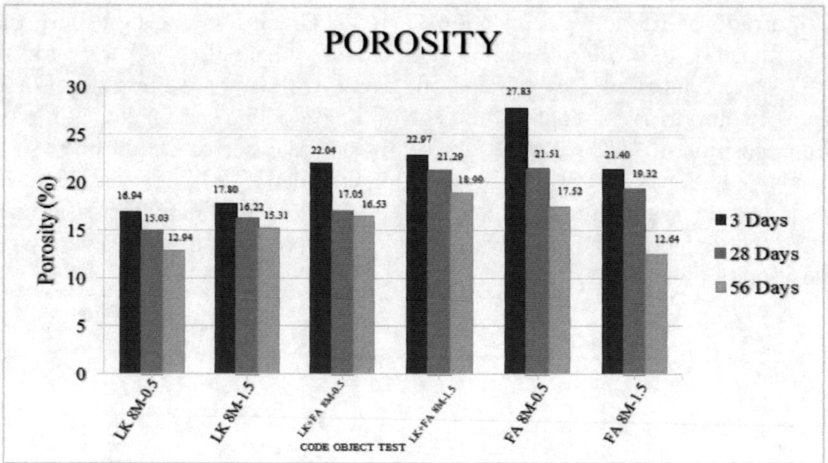

Graph 3. The Porosity Test Result.

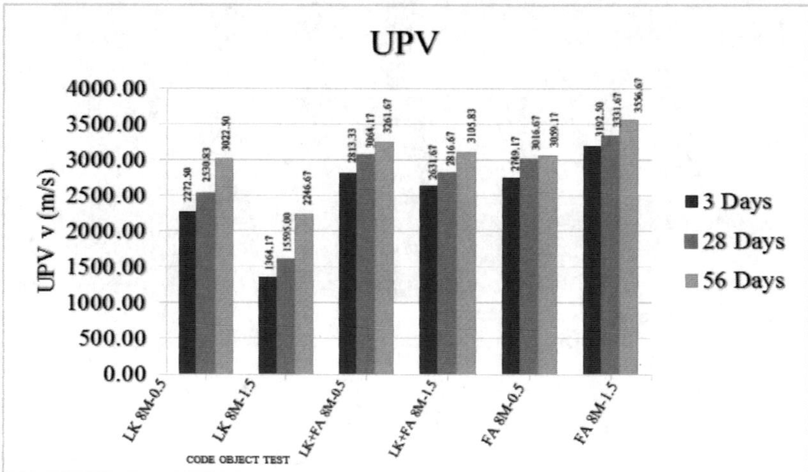

Graph 4. The Ultrasonic Pulse Velocity (UPV) Test Result.

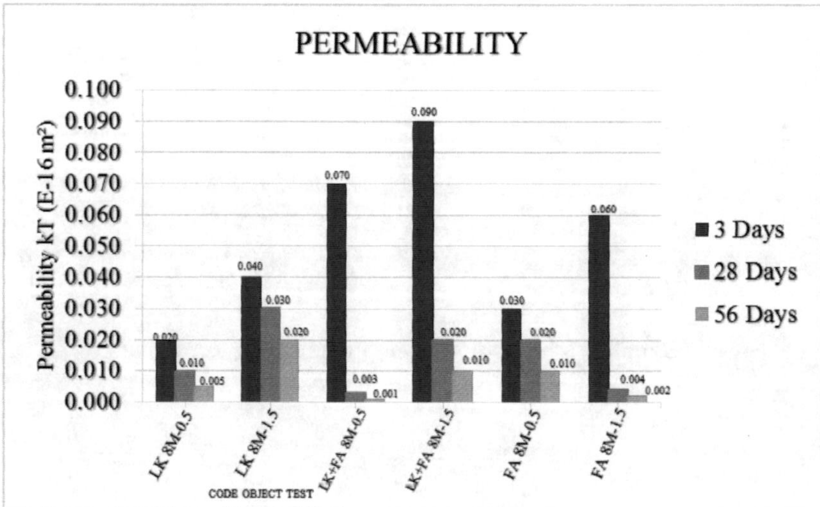

Graph 5. The Permeability Test result.

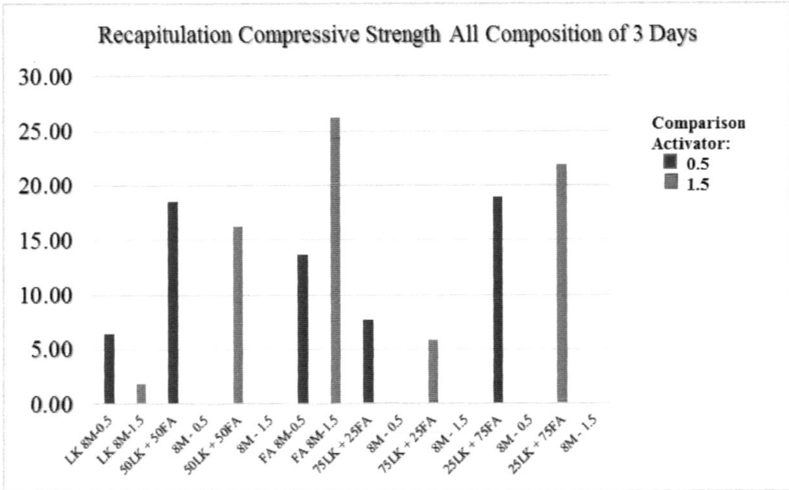

Graph 6. Recapitulation Compressive Strength 3 Days Test result.

4. Conclusion

The highest compressive strength besides there on the composition of a 100% fly ash is also present in 50% of waste carbide composition: 50% fly ash with activator ratio of 0.5 at the age of 56 days amounted to 39.59 Mpa, 0.13 Mpa only difference from the composition of 100% fly ash with a ratio of 1.5 activator at the age of 56 days of 3.72 Mpa. The use of fly ash in geopolymer still cannot be eliminated 100%, due to the compressive strength all age composition 3 days, it was found that the lower the percentage of waste carbide used the higher the compressive strength obtained. It can be concluded that the waste carbide can only be used as a filler can only minimize the use of fly ash with a maximum percentage levels of 50%.

References

[1] Davidovits,J. 1994, "*Properties of Geopolymer Cements*". Geopolymer Institute. France : Saint-Quentin.

[2] Bilvapedia, 2014. "*Coal and Natural Resources Cannot Be Updated*". Semarang.

[3] Hardjito, Djwantoro.,and Rangan, B.V. 2001, "*Development and Properties of Low-Calcium Fly Ash–Based Geopolymer Concrete*". Research Report GC I Faculty of Engineering Curtin University of Technology Perth, Australia.

[4] Lauw, 2008. "Know the Definition and Types Fly Ash". Jakarta.

[5] Natania, Dea. 2016. "*Waste Utilization Study Carbide PT.Z as Material Mixed Concrete In Light*". Surabaya: Tugas Akhir PPNS.

[6] Abdullah M M A B, Hussin K, Bnhussain M, Ismail K N, Yahya Z, Razak R A, 2012 "*Fly ash-based geopolymer lightweight concrete using foaming agent*" International Journal of Molecular Sciences, 13(6), 7186–7198.

[7] Abdullah M M A B, Jamaludin L, Kamarudin H, Binhussain M, Ruzaidi Ghazali C M, Ahmad M I, 2013. "*Study on fly ash based geopolymer for coating applications*". Advanced Materials Research, 686, 227–233.

[8] Subli M H, Omar M F, Sandu I G, Zulkepli N N, Al Bakri Abdullah M M, Sandu A V, 2014. "*Effects of hybrid fillers on the wear, Tensile and morphology properties of UHMWPE/Chitosan-ZnO composites*". Materiale Plastice, 51(4), 391–395.

[9] Cheng-Yong H, Yun-Ming L, Abdullah M M A B, Hussin K 2017. "*Thermal Resistance Variations of Fly Ash Geopolymers: Foaming Responses*". Scientific Reports, 7(November 2016), 1–11.

[10] Jaya, N.A., Al Bakri Abdullah, M.M., Ahmad, R.. 2015. "*Reviews on clay geopolymer ceramic using powder metallurgy method*", Materials Science Forum, 803, pp. 81-87.

IOP Conf. Series: Materials Science and Engineering **743** (2020) 012043 doi:10.1088/1757-899X/743/1/012043

Analysis of the connection of a nickel-chromium alloy with dental ceramics after aluminum oxide abrasive blasting

W Czepułkowska[1] and E Wołowiec-Korecka[1]

[1]Institute of Materials Science and Engineering, Lodz University of Technology, Lodz, Poland

Email: emilia.wolowiec-korecka@p.lodz.pl

Abstract. Fixed prosthetic restorations in the form of ceramic crowns or bridges are characterized by good aesthetics and durability. The most important turn out to be the mechanical attachment of the ceramic material in the unevenness of the metal, formed after abrasive blasting. The aim of this research is to investigate the impact of abrasive blasting of a nickel-chromium alloy on the covering the unevenness with ceramics. In the research was used a Ni-Cr alloy (Heraenium® NA, Hera). The metal surface was blasted with alumina of variable size (50, 110 and 250 μm) and pressure (0.4 and 0.6 MPa), which was then fired with ceramics in the form of 2 layers of IPS Classic Opaque (Ivoclar Vivadent). Samples were hot mounting and cut to obtain a cross-section of the connection. The samples was examined under a Scanning Electron Microscope. The results show the presence of air bubbles at the metal-ceramic interface and in the layers of ceramic material. The photographs show a difference in the appearance of the unevenness of the metal surface by comparing samples after blasting with an abrasive of different sizes. No significant differences in the appearance of metal roughness are observed when comparing samples after treatment with the same grain size and different pressures.

1. Introduction

Ceramic crowns and bridges on a metal substructure are one of the dental restorations made in dental prosthetics. They are characterized by good aesthetics and durability [1]. The bond strength between the metal and the ceramic material is associated with the occurrence of several mechanisms that are responsible for creating a strong connection. One of them concerns the presence of chemical bonds between materials that are formed due to the presence of an oxide layer on the metal surface. Another owes its presence to differences in coefficients of thermal expansion, where during cooling of the restoration after firing ceramics, it has compressive stress, increasing the connection strength. The last mechanism concerns the anchoring of the ceramic material in the unevenness of the metal, formed after abrasive blasting. Preparation of the surface of the metal restoration for the joint is one of the main principles aimed at ensuring the permanent use of the restoration. It is known that the shear strength of the connection between metal and ceramics depends on the development of the surface, which is why works are carried out where Co-Cr alloy [2], [3] or titanium [4] is analyzed. Sandblasting is also used in the development of zirconia surface, which is commonly used in the creation of substructures for ceramic restorations, due to its good aesthetic properties and biocompatibility. The durability of the connection is influenced by the development of the material surface, i.e. the size and shape of the unevenness created. To provide them, various treatments are used, such as chemical surface etching or mechanical roughness formation [5], [6]. The process of mechanical surface development can be performed by machines with rotary instruments or sandblasting. A combination of these treatments is

Content from this work may be used under the terms of the Creative Commons Attribution 3.0 licence. Any further distribution of this work must maintain attribution to the author(s) and the title of the work, journal citation and DOI.
Published under licence by IOP Publishing Ltd

also used to ensure the best possible surface properties [7]. Abrasive blasting directly affects the formation of the right surface roughness and the parameters used also affect its wettability or the amount of abrasive particles stuck in it. From the dental point of view, surface properties that give information about its Surfacc Free Energy and the wetting ability of the liquid may be an important aspect. Wetness of the surface with liquid ceramics is also important for the quality of the connection. When firing a ceramic material, depending on the surface properties, it may cover to varying degrees of metal unevenness leading to a reduction in joint durability. The influence of metal surface preparation for connection with dental ceramics is described in the literature, where its strength is analyzed in correlation to the condition of the substructure surface. The purpose of our research was to expand knowledge about the connection by analyzing its appearance for the surface of a Ni-Cr alloy subjected to various blasting processes.

2. Materials and methods

Six cylindrical samples made of nickel-chromium alloy (Heraenium® NA, Heraeus Kulzer), whose diameter was 8 mm and height 7 mm was used in the test. The surfaces were blasted with alumina with variable parameters, where the processes differed in the pressure used (0.4 MPa, 0.6 MPa) and the abrasive grain size (50 μm, 110 μm, 250 μm). The treatment lasted 20 seconds and the nozzle distance from the material surface was 15 mm. The angle between the nozzle and the metal for all samples was 45°. After sandblasting, the samples were cleaned in an ultrasonic cleaner (Emmi-55HC-Q, Emag) in deionized water for 8 minutes, and then dried under compressed air. Dental ceramics in several layers were fired on prepared and cleaned surfaces. The first two were the opaque IPS Classic Opaquer (Ivoclar Vivadent) and the next one was IPS Classic Dentin (Ivoclar Vivadent) in the firing parameters given in Table 1.

Table 1. Firing parameters for IPS Classic dental ceramics (Ivoclar Vivadent).

	The starting temperature [°C]	The final temperature [°C]	Heating rate [°C/min]	Heating time [min]
I opaque layer	403	980	80	1
II opaque layer	403	970	80	1
I dentin layer	403	920	60	1
II dentin layer	403	910	60	1

Then the samples were cut and included, and then metallographic specimens were created to ensure the observation of the metal-ceramics boundary. Chemical composition studies using X-ray microanalysis with EDS energy dispersion and observation of surface topography were carried out on a Scanning Electron Microscope (JSM-6610LV, JEOL).

3. Results

Point x-ray EDS microanalysis was performed in 3 places on the tested surface in order to distinguish between the present materials on the connection cross-section. At each of the analyzed points, elements characteristic of the materials used are observed (Table 2).

Table 2. Elements in the material obtained from the EDS study.

Measuring point	Current elements
Dentin	O, Si, Al, K, Na
Opaque	O, Si, Zr, K, Al, Na, Zn, Ca
Ni-Cr alloy	Ni, Cr, Mo, Si, Mn

The results of the microscopic observation of the cross-section of the Ni-Cr-ceramic alloy connection are presented in the pictures from Figure 1 to Figure 6.

IOP Conf. Series: Materials Science and Engineering **743** (2020) 012043 doi:10.1088/1757-899X/743/1/012043

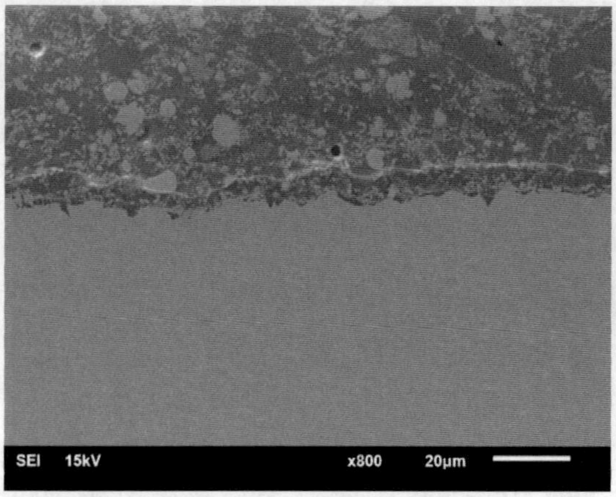

Figure 1. SEM photography of a metal-ceramic joint after treatment of a metal surface with 50 µm alumina at 0.4 MPa (approx. x800).

Figure 2. SEM photography of a metal-ceramic joint after treatment of a metal surface with 50 µm alumina at 0.6 MPa (approx. x800).

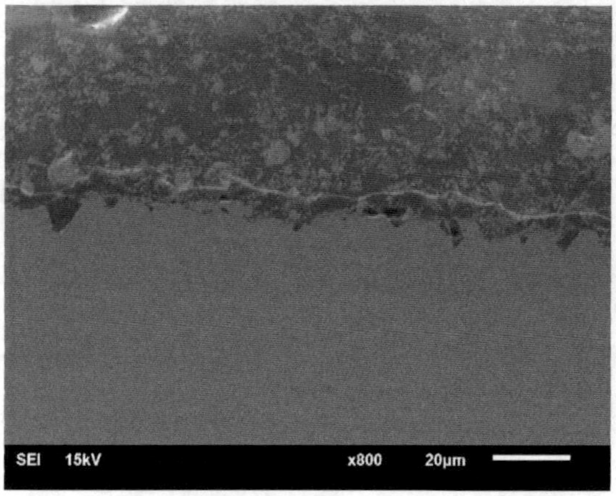

Figure 3. SEM photography of a metal-ceramic joint after treatment of a metal surface with 110 µm alumina at 0.4 MPa (approx. x800).

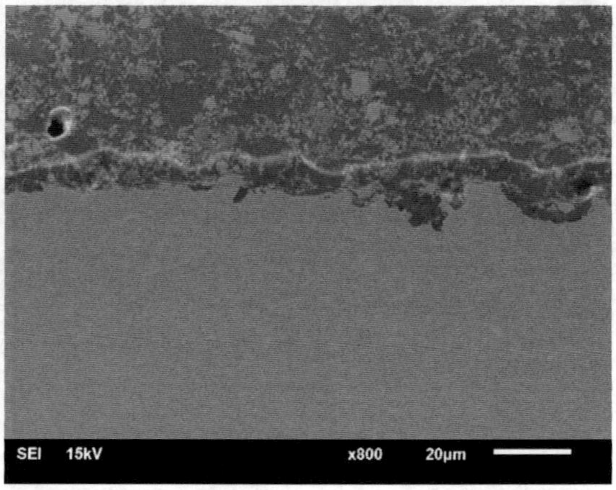

Figure 4. SEM photography of a metal-ceramic joint after treatment of a metal surface with 110 µm alumina at 0.6 MPa (approx. x800).

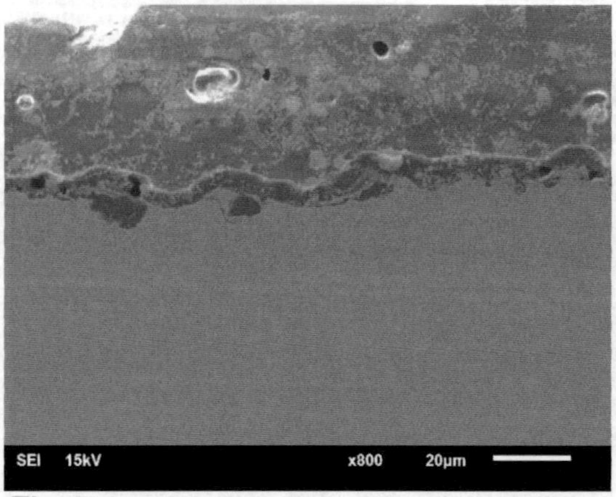

Figure 5. SEM photography of a metal-ceramic joint after treatment of a metal surface with 250 µm alumina at 0.4 MPa (approx. x800).

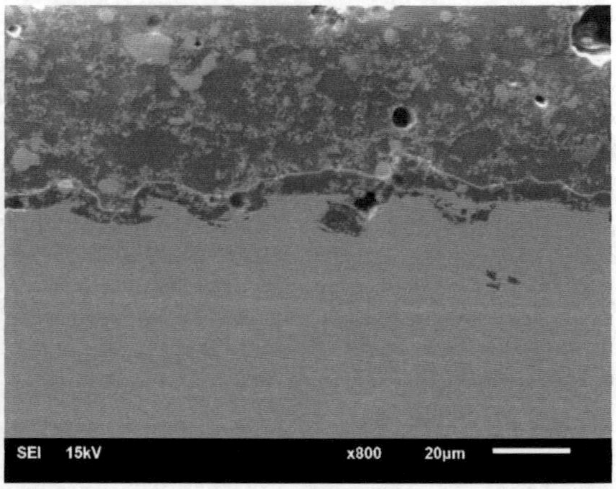

Figure 6. SEM photography of a metal-ceramic joint after treatment of a metal surface with 250 µm alumina at 0.6 MPa (approx. x800).

It is noticeable that the metal surface after blasting with alumina changes its structure due to the size of the abrasive used. For samples after sandblasting with the smallest particles (50 μm), the widths of the observed surface irregularities are much smaller, comparing them with the surface treated with grain size 110 and 250 μm. The use of higher pressure does not cause significant differences in the shape of the metal surface.

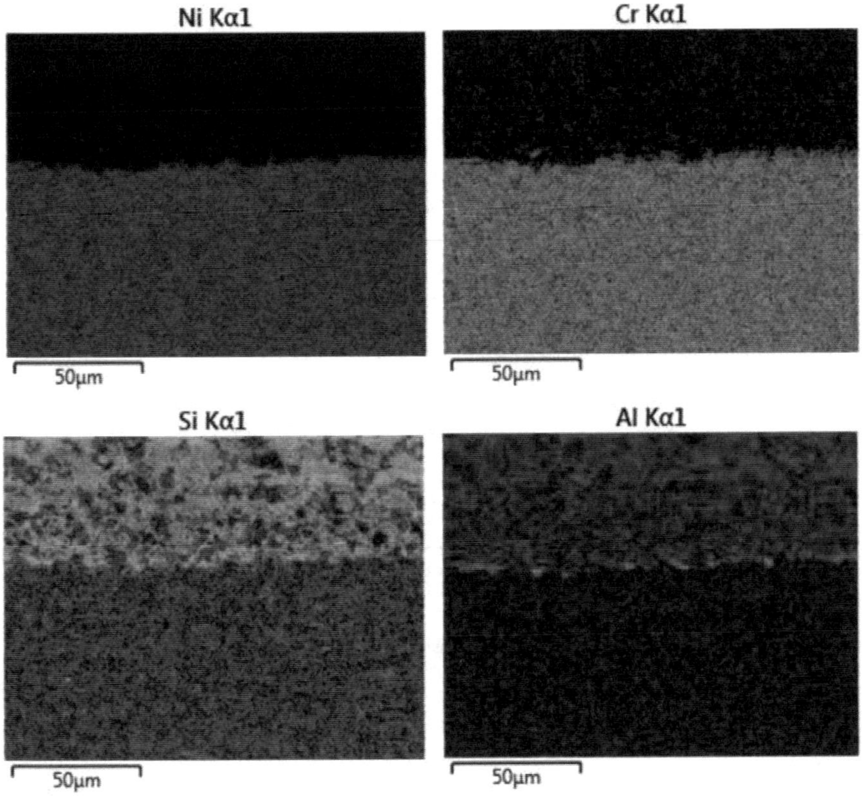

Figure 7. Surface maps of the elements on the metal-ceramic connection for sample A54.

Figure 7 contains surface maps of the elements included in the metal alloy, opaque and abrasive. On the border between metal and ceramics, an increased occurrence of aluminum is observed, which is caused by the presence of embedded abrasive particles.

4. Discussion

The study focused on the analysis of the boundary between a nickel-chromium alloy and ceramics, formed after firing a ceramic material on a metal surface. After the EDS analysis, it is possible to observe the presence of more elements contained in the first layers of ceramics and this is characteristic of this material. The opaque is a material that directly contacts the alloy, creating chemical bonds that affect the strength of the connection and covers the unevenness of the metal after abrasive blasting, creating mechanical catches [8]–[10].

Observation of the metal-ceramic connection in SEM photographs gives information on the occurrence of air bubbles in the ceramic material, where for each sample their quantity is small. The method of applying ceramic layers to the metal surface may be responsible for the main reason for their appearance in the structure. In this study, the opaque was used in the form of a paste, which is applied with a brush. This method of application is also used in creating the dentin layer. This involves the possibility of enclosing air in one of the layers (Figure 8) and this is a situation widely encountered, both in the case of research on ceramics [11]–[13] as well as in clinical cases related to failures [14].

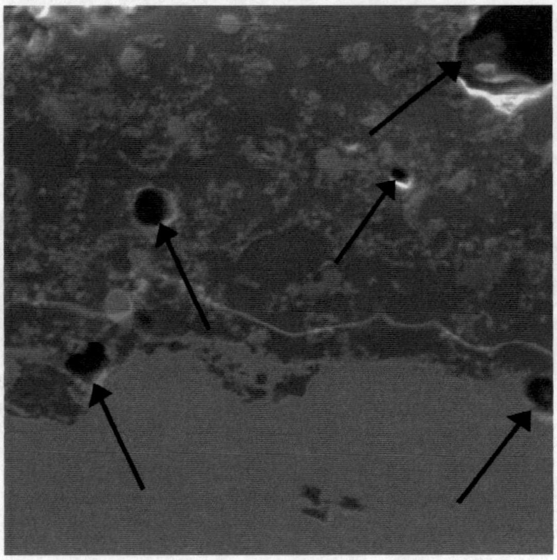

Figure 8. SEM photography of a metal-ceramic joint after 250 μm grain processing and 0.6 MPa pressure. Arrows indicate the place of occurrence of air bubbles.

The ability of the liquid to wet the alloy surface is also significant. In prosthetics, when the ceramic material is applied to the metal, it is an aqueous suspension, so its character may be more polar. When firing ceramics, by increasing the temperature, water comes out of the material and changes the character of the material to apolar, so in this case, it is important to assess the ability of ceramics to wet the metal surface [7]. The presence of roughness can also affect the cover of semi-liquid ceramics in unevenness in a different way, which is associated with a change in the viscosity of the material as well as the occurrence of various development of the metal surface. The firing temperature of the ceramics also has an impact, where better wetting properties of the ceramics are observed at higher temperatures [7]. The SEM photographs show a change in the appearance of the metal surface with a change in the used abrasive particle size. For samples whose surface has been treated with alumina with a particle size of 50 μm observed surface irregularities, which are not deep. With the increase in the particle size of the abrasive, an increase in the depth of unevenness is noted, where for the largest grain the metal surface has the largest amount of splinters. The wettability of the surface according to the Wenzel model increases as its unevenness increases [15]. Therefore, the lower the surface roughness, the less the wetting ability of the liquid should be. In addition, according to Cassie and Baxter, the rough surface consists of a solid and air, because depending on the width and depth of the unevenness, when it is wetted, the air can be enclosed in the form of bubbles at the boundary of the material. Such a mechanism can occur for roughness, whose width is small and is characterized by a large depth. In the case of wider unevenness, the material or liquid wetting the surface will not lead to air clumping at the border. Therefore, the presence of roughness and nano-roughness causes the surface to become hydrophobic. This is due to the energy state of the surface [15]. However, the SEM observation does not show a tendency to increase the occurrence of air bubbles depending on the surface roughness. They are most likely the result of the above-mentioned possible causes. The wettability of the metal may be related to the strength of the described joint due to adhesive forces. Not without significance and in this case the surface roughness turns out. Research on the impact of variable abrasive blasting on the strength of a Co-Cr alloy connection was presented by Pietnicki et.al., where durability was less for samples after 50 and 250 μm particle treatment [2]. Similar results are also observed for titanium in the studies of Gołębiowski et.al. [16]. For this material the lowest values are recorded for grains 50 and 250 μm, but also for the highest pressure used (0.6 MPa). Tests associated with the use of chemical etching surfaces in the case of titanium also results in a decrease in the strength of the metal-ceramic joint, which was presented in the study of Parchańska-Kowalik et. al. [17]. Tests showed a reduction in connection durability from 32 MPa obtained for sandblasted surfaces to approx. 16 MPa for sandblasted and chemically etched surfaces. This may be due to the removal of abrasive particles from the surface, or the reduction of unevenness on the surface. The same relationship is observed for testing the strength of the connection between titanium and zirconia [18].

Observation of the surface of the Ni-Cr alloy in our study shows that for the smallest grain treatment the irregularities are not deep, and their width is small. In this case, the ceramics are not sufficiently hooked into the roughness, which is why the joint strength turns out to be smaller than in samples after 110 μm particle treatment. In the case of samples where the metal surface has been treated with the largest grain, it may turn out that the depth of the unevenness is adequate, but their width leads to a weaker attachment of the ceramic material to them. For the Co-Cr alloy from the Pietnicki test, the optimum surface unevenness is obtained for blasting with 110 μm abrasive.

To sum up, air bubbles are observed on the border of the tested materials, which are associated with the technique of applying ceramic material to the metal. The connection cross-section shows the presence of an abrasive embedded in the surface of a Ni-Cr alloy. No significant differences in the appearance of the metal surface are observed for samples treated with the same grain size but with different pressures.

References

[1] Makhija D K *et al* 2016 *Journal of Dentistry* **55** pp 40–47/
[2] Pietnicki K, Wołowiec-Korecka E and Klimek L 2014 *Acta of Bioengineering and Biomechanics* **16** 1 pp 63–68/
[3] Külünk T, Kurt M, Ural Ç, Külünk Ş and Baba S 2011 *Journal of Dental Sciences* **6** 3 pp 140–146.
[4] Gołębiowski M, Wołowiec E and Klimek L 2015 *The Journal of Prosthetic Dentistry* **113** 5 pp 453–459.
[5] Śmielak B and Klimek L 2015 *The Journal of Prosthetic Dentistry* **113** 6 pp 596–602.
[6] Śmielak B and Klimek L 2018 *Materials* **11** 2 p. 259.
[7] Śmielak B, Klimek L, Wojciechowski R and Bąkała M 2019 *The Journal of Prosthetic Dentistry* **122** 4 p 410.
[8] Czepułkowska W, Wołowiec-Korecka E and Klimek L 2018 *Archives of Materials Science and Engineering* **92** 1 pp 5–14.
[9] Hajduga M and Danel D 2010 *Aktualne Problemy Biomechaniki* 4 pp 59–64.
[10] Messing J J, Education L M and Ray G E 2015 *Operative Dental Surgery*. Macmillan International Higher Education.
[11] Jörn D, Waddell J N and Swain M V 2010 *Journal of Dentistry* **38** pp. 143–149.
[12] Al Amri M D and Hammad I A 2012 *King Saud University Journal of Dental Sciences* **3** 2 pp 41–48.
[13] Elsaka S E, Hamouda I M, Elewady Y A, Abouelatta O B and Swain M V 2010 *Dental Materials* **26** 8 pp 793–798.
[14] Aboras M M, Muchtar A, Azhari C H and Yahaya N 2015 *6th European Conference of the International Federation for Medical and Biological Engineering* (Springer International Publishing) **45** pp 345–348.
[15] Barnat-Hunek D 2016 *Swobodna energia powierzchniowa jako czynnik kształtujący skuteczność hydrofobizacji w ochronie konstrukcji budowlanych* (Lublin: Politechnika Lubelska).
[16] Gołębiowski M, Wołowiec E and Klimek L 2015 *The Journal of Prosthetic Dentistry* **113** 5 pp 453–459.
[17] Parchańska-Kowalik M, Wołowiec-Korecka E and Klimek L 2018 *The Journal of Prosthetic Dentistry* **120** 3 pp 470–475.
[18] Śmielak B, Gołębiowski M, Klimek L and Wołowiec E 2014 *SSP* **225** pp 151–158.

IOP Conf. Series: Materials Science and Engineering **743** (2020) 012044 doi:10.1088/1757-899X/743/1/012044

Influence different amount of cellulose on the mechanical strength of dental acrylic resin

Joanna Taczała [1], Chang Fu [2], Jacek Sawicki [1] and Joanna Pietrasik [3]

[1]Lodz University of Technology, Institute of Materials Science and Engineering, Stefanowskiego St. 1/15, 90-924 Lodz, Poland
[2]Tohoku University, 2-1-1 Katahira, 980-8577 Sendai, Japan
[3]Lodz University of Technology, Institute of Polymer and Dye Technology, Stefanowskiego St. 1/15, 90-924 Lodz, Poland

Email: jacek.sawicki@p.lodz.pl

Abstract. The mechanical strength of PMMA is not sufficient when a patient applies high mastication force to the denture base plate. Many fillers to reinforce PMMA was used. However, cellulose incorporated into dental acrylic resin has never been reported. In this work, two types of modified cellulose were incorporated in the heat-cured acrylic resin, and the flexural strength was investigated. The conclusion is that the addition of properly modified cellulose could have good influence on the flexural strength of dental acrylic resin. The addition of cellulose modified by 3-Methacryloxypropylmethyldimethoxysilane has increased average flexural strength and the standard deviation was satisfactory. However, Triethoxy(octyl)silane may not appropriate to achieve the intended purpose, or the modification of the cellulose by this silane itself must still be subjected to further testing. The ability to form new chemical chains between the cellulose and acrylic surfaces had a greater impact on the strength of acrylic than facilitated even distribution of cellulose.

1. Introduction

Poly(Methyl Methacrylate) (PMMA) is the most commonly used material for fabricating denture bases since 1937 [1]. Despite its advantages, such as biocompatibility, ease of processing, and low toxicity, the mechanical strength of PMMA is not sufficient when a patient applies high mastication force to the denture base plate [2]. Therefore, mechanical reinforcement is a long-term pursuit to prolong the service time of denture bases. Up to now, plenty of studies have been conducted to solve this problem, and the results exhibit that blending with proper fillers is the most efficient way. The fillers reported previously includes: metals [3], carbon fibers [4], glass fibers [5], SiC [6], Si_3N_4 [6], ZrO_2 [6], [7], TiO2 [8], [6], SiO_2 [8], Al_2O_3 [6], [9], silver [6], graphene [10], hydroxyapatite [6], rubber [4] or polyethylene [4], [11]. However, all the methods have suffered from either limited values of enhancement or high dependence on the brand of acrylic resin.

Cellulose is known as a popular renewable and biodegradable material, exhibiting excellent mechanical properties due to its abundant hydrogen bonding and crystalline regions. It has been used to improve the mechanical properties of PMMA. Figure 1 shows how much reinforcement by properly modified cellulose can improve the tensile strength of PMMA. Figure 2 shows an increase in the Young Modulus. In both cases, modified cellulose was used 2,2,6,6-Tetramethyl 1-piperidinyloxy.

Content from this work may be used under the terms of the Creative Commons Attribution 3.0 licence. Any further distribution of this work must maintain attribution to the author(s) and the title of the work, journal citation and DOI.
Published under licence by IOP Publishing Ltd

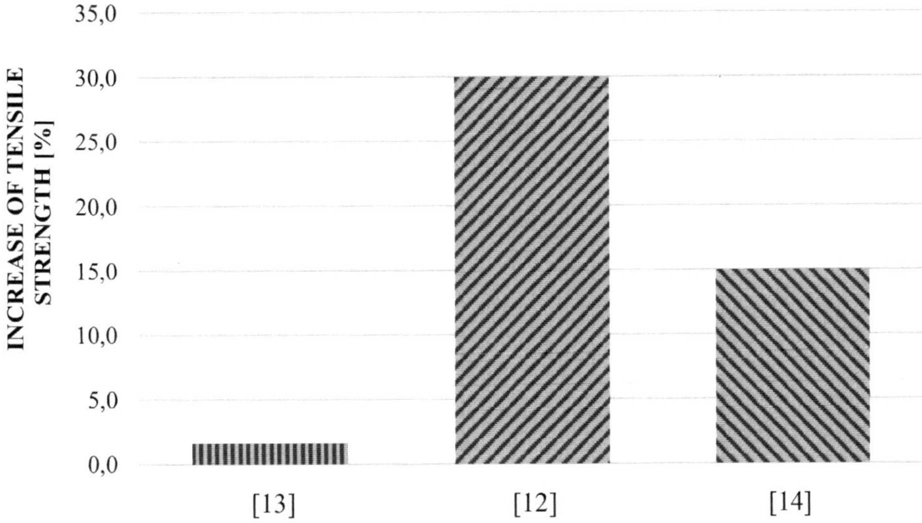

Figure 1. Graph of increase of tensile strength of PMMA reinforced with cellulose modify by 2,2,6,6-Tetramethyl-1-piperidinyloxy.

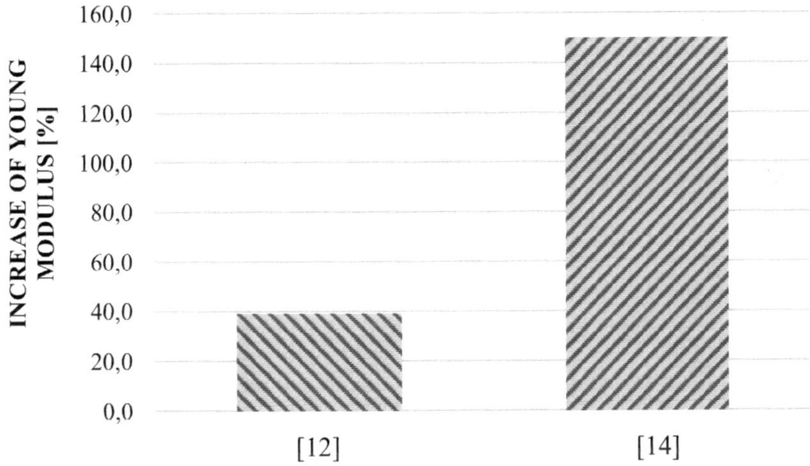

Figure 2. Graph of increase of Young modulus of PMMA reinforced with cellulose modify by 2,2,6,6-Tetramethyl-1-piperidinyloxy.

However, cellulose incorporated into dental acrylic resin has never been reported. Moreover, it is well known that the dispersion and adhesion of fillers have a great effect on the mechanical properties of composites. To achieve high compatibility between hydrophilic cellulose and hydrophobic matrix, cellulose should be modified before blending with acrylic resin.

In this work, two types of modified cellulose were incorporated in the specific heat-cured acrylic resin, and the flexural strength was investigated. That kind of acrylic resin was chosen, because this material is characterized by the highest strength of all dental acrylic resins [15] - [21]. Improvement of the strength of dental acrylic resin would reduce the thickness of the denture plate while maintaining the same strength of the material.

2. Materials and methods

2.1. Materials

Dental technicians in their laboratory use ready-made acrylic resin consist of powder and liquid. They are mixing these two products with a ratio based on manufacturer instructions. For this research, the powder stays in original form, and it was a heat-cured acrylic resin Vertex® Rapid Simplified (from Vertex Dental company). The liquid was recreated based on monomer Vertex® Rapid Simplified, and it consists of methyl methacrylate (99%) stab. (CAS number: 80-62-6, from Alfa Aesar) and ethylene glycol dimethacrylate (98%) (CAS number: 97-90-5, from Sigma-Aldrich). The second reagent was used as a cross-linking agent.

To reinforced acrylic resin was used Arbocel® UFC100 Ultrafine Cellulose for Paper and Board Coating (CAS number: 9004-34-6, from J. Rettenmaier USA LP company) after surface modification by two silanes:

- Triethoxy(octyl)silane (97%) (CAS number: 2943-75-1, from Sigma-Aldrich)
- 3-Methacryloxypropylmethyldimethoxysilane (92%) (CAS number: 14513-34-9, from ABCR company)

2.2. Methods

First, modification of cellulose was conducted to change the character of cellulose from hydrophilic to hydrophobic and to help the creation of chemical chains between methacrylic polymer and surface of cellulose. Then modified cellulose was added to the acrylic powder and mixing well. Both modified cellulose was added in three different amounts: 0,5 g, 1,0 g, and 1,5 g.

Final properly amount of acrylic liquid was added to the mixture. In the phase of hard rubber during pre-polymerization process- acrylic resin was put to the forms. Created specimens were finished with grinding machines to achieve dimensions shown in Figure 3 to the 3-point bending test. The speed test was 1 $^{mm}/_{min}$ and was conducted with a universal testing machine (from Zwick/Roell).

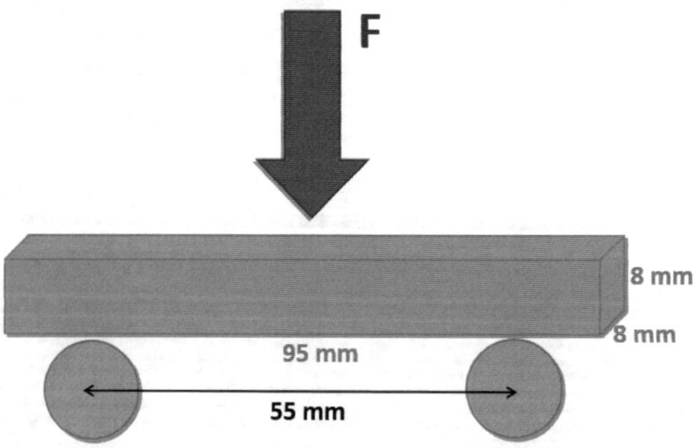

Figure 3. A diagram of force action in a 3-point bending test [source: own picture].

Also, the observation was made using SEM. This test was carried out with JSM-6610LV Scanning Electron Microscope (from JEOL USA company). To observe results specimens were covered with a thin layer of gold.

3. Results and discussion

The results of the 3-point bending test are shown in Figure 4. Prepared specimens also differed of cellulose and amount of modified cellulose. Specimen marked as "C" was pure commercial acrylic resin. Moreover, the average strength was 88,8 MPa. It is similar to information given on manufacturer's website [22], which is 85,2 MPa.

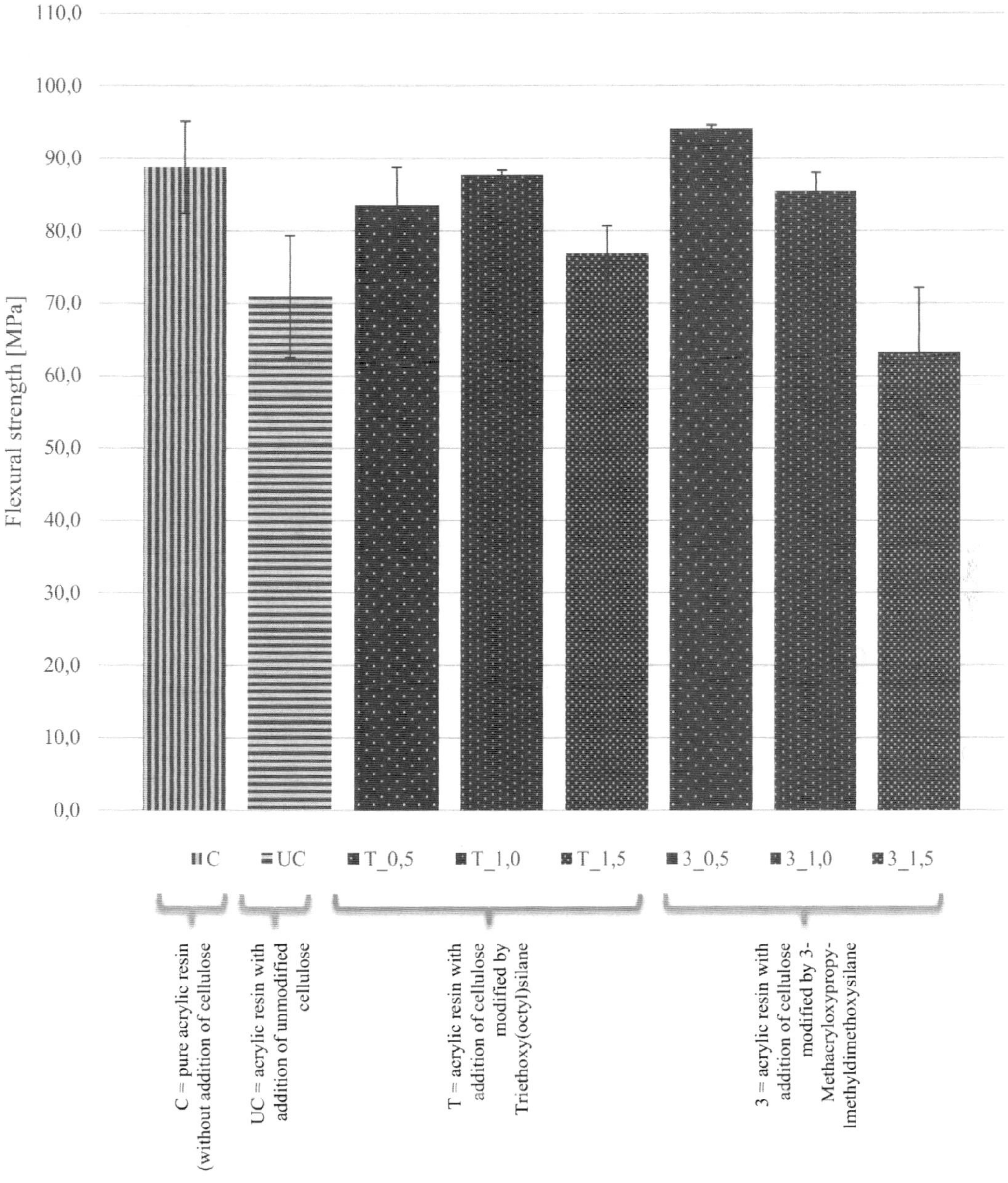

Figure 4. Diagram of average flexural strength measured by 3-point bending test.

Next is "UC" and this result is a representation of addition unmodified cellulose to commercial acrylic resin. The amount of addition was 0,5 g. Results of bigger amount are not on diagram, because agglomeration of cellulose was too extensive, that sample preparation was impossible. In this case results were not the worst, but agglomeration was visible which significantly deteriorated the strength of these samples compared to commercial acrylic. Figure 5 shows that the agglomeration of cellulose was shown. During cutting process, part of the cellulose felt off, which led to the formation of holes. This completely rejects the material for use in the creation of dental prostheses because final machine processing would be impossible.

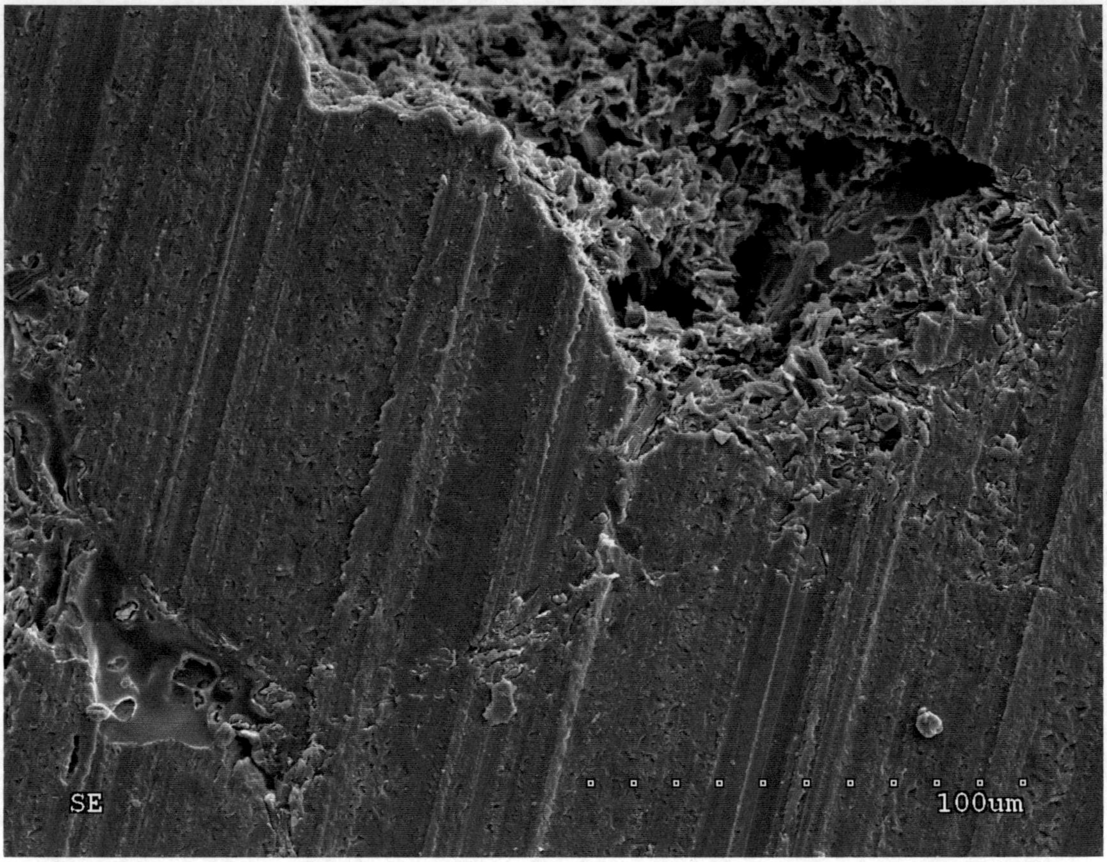

Figure 5. Image from SEM observation with a visible agglomeration of unmodified cellulose [source: own picture].

Last six results in Figure 4 are a representation of addition of cellulose modified by silanes. First set is cellulose modified by Triethoxy(octyl)silane and second- by 3-Methacryloxypropylmethyldimethoxysilane. Specimens with the addition of both modified cellulose in the amount of 0,5 g, 1,0 g, and 1,5 g were good at preparation. The addition of 2,0 g was too much, and there were difficult to mix the powder and liquid. The best results have been achieved after addition 0,5 g cellulose modified by 3-Methacryloxypropylmethyldimethoxysilane (marked on the diagram as 3_0,5). This silane was used to modify surface of cellulose to make possible to create connections between methacrylic polymer and surface of cellulose. So, addition just 0,5 g of this cellulose had positive impact on flexural strength (94,0 MPa). Both- average strength and standard deviation were satisfactory. The more this type of cellulose was added- the worse the strength and the spread of results were.

The situation is different with the addition of modified cellulose by Triethoxy(octyl)silane. This silane was used because it could change the nature of the cellulose surface from hydrophilic to hydrophobic. In this case- the addition of 1,0 g modified cellulose given the best results but still worse than in case of 3-Methacryloxypropylmethyldimethoxysilane.

A. Pure acrylic resin

B. Acrylic resin with addition of cellulose modified by Triethoxy(octyl)silane

C. Acrylic resin with addition of cellulose modified by 3-Methacryloxypropylmethyldimethoxysilane

Figure 6. Images from SEM observations [source: own picture].

SEM was observed not only to check the influence of modification on the agglomeration of cellulose but also fracture specimens. An example of fracture is shown in Figure 6. All fractures were the same, and it was a brittle type. So addition of cellulose does not have impact on type of fracture. However, during observations at higher magnification the differences were noticeable. Cellulose is visible on the picture B and C as more rough areas. The surface of pure acrylic resin is smooth. In case of cellulose modified by Triethoxy(octyl)silane cellulose fibers took up a larger area of fracture. This shows that modification with this kind of silane changed character of surface of cellulose from hydrophilic to hydrophobic and this enabled even distribution of the fibers in the PMMA core. Compare this with cellulose modified by 3-Methacryloxypropylmethyldimethoxysilane- this silane did not allow such even distribution, but the likely formation of bonds between the cellulose surface and PMMA resulted in increased acrylic strength.

4. Conclusion

Conclusion of this study is that the addition of properly modified cellulose could have good influence on the flexural strength of dental acrylic resin. Addition of cellulose modified by 3-Methacryloxypropylmethyldimethoxysilane has increased average flexural strength and standard deviation was satisfactory. However, Triethoxy(octyl)silane may not appropriate to achieve the intended purpose, or the modification of the cellulose by this silane itself must still be subjected to further testing. Improvement of modification process could have good impact on flexural strength in case of

3-Methacryloxypropylmethyldimethoxysilane. The ability to form new chemical chains between the cellulose and acrylic surfaces had a greater impact on the strength of acrylic than facilitated even distribution of cellulose. Further research can give even more visible results in this case.

References

[1] S. B. Patil, B. H. Naveen, N. P. Patil, 2006, Bonding acrylic teeth to acrylic resin denture bases: a review, *Gerodontology*, 23 (3), 131–139, https://doi.org/10.1111/j.1741-2358.2006.00129.x.

[2] A. Alhareb, Z. A. Ahmad, 2011, Effect of Al_2O_3/ZrO_2 reinforcement on the mechanical properties of PMMA denture base, *Journal of Reinforced Plastics and Composites*, 30 (1), 86–93, https://doi.org/10.1177%2F0731684410379511.

[3] S. B. Sehajpal, V. K. Sood, 1989, Effect of metal fillers on some physical properties of acrylic resin, *The Journal of Prosthetic Dentistry*, 61 (6), 746–751, https://doi.org/10.1016/S0022-3913(89)80055-1.

[4] D. C. Jagger, A. Harrison, K. D. Jandt, 1999, The reinforcement of dentures, *Journal of oral rehabilitation*, 26 (3), 185–94, https://doi.org/10.1046/j.1365-2842.1999.00375.x.

[5] F. Keyf, G. Uzun, 2000, The effect of glass fibre-reinforcement on the transverse strength, deflection and modulus of elasticity of repaired acrylic resins, *International Dental Journal*, 50 (2), 93–97, https://doi.org/10.1002/j.1875-595X.2000.tb00805.x.

[6] E. Kul, L. İ. Aladağ, R. Yesildal, 2016, Evaluation of thermal conductivity and flexural strength properties of poly(methyl methacrylate) denture base material reinforced with different fillers, *The Journal of Prosthetic Dentistry*, 116 (5), 803–810, https://doi.org/10.1016/j.prosdent.2016.03.006.

[7] V. Asopa, S. Suresh, M. Khandelwal, V. Sharma, S. S. Asopa, L. S. Kaira, 2015, A comparative evaluation of properties of zirconia reinforced high impact acrylic resin with that of high impact acrylic resin, *The Saudi Journal for Dental Research*, 6 (2), 146–151, https://doi.org/10.1016/j.sjdr.2015.02.003.

[8] A. Sodagar, A. Bahador, S. Khalil, A. S. Shahroudi, M. Z. Kassaee, 2013, The effect of TiO_2 and SiO_2 nanoparticles on flexural strength of poly (methyl methacrylate) acrylic resins, *Journal of Prosthodontic Research*, 57 (1), 15–19, https://doi.org/10.1016/j.jpor.2012.05.001.

[9] M. Vojdani, R. Bagheri, A. A. R. Khaledi, 2012, Effects of aluminum oxide addition on the flexural strength, surface hardness, and roughness of heat-polymerized acrylic resin, *Journal of Dental Sciences*, 7 (3), 238–244, https://doi.org/10.1016/j.jds.2012.05.008.

[10] F. Lin, C. Yang, Q. H. Zeng, Y. Xiang, 2018, Morphological and mechanical properties of graphene-reinforced PMMA nanocomposites using a multiscale analysis, *Computational Materials Science*, 150, 107–120, https://doi.org/10.1016/j.commatsci.2018.03.048.

[11] D. L. Dixon, L. C. Breeding, 1992, The transverse strengths of three denture base resins reinforced with polyethylene fibers, *The Journal of Prosthetic Dentistry*, 67 (3), 417–419, https://doi.org/10.1016/0022-3913(92)90261-8.

[12] T. Huang, K. Kuboyama, H. Fukuzumi, T. Ougizawa, 2018, PMMA/TEMPO-oxidized cellulose nanofiber nanocomposite with improved mechanical properties, high transparency and tunable birefringence, *Cellulose*, 25 (4), 2393–2403, https://doi.org/10.1007/s10570-018-1725-3.

[13] Y. F. Shih, M. Y. Chou, H. Y. Lian, L. R. Hsu, S. M. Chen-Wei, 2018, Highly transparent and impact-resistant PMMA nanocomposites reinforced by cellulose nanofibers of pineapple leaves modified by eco-friendly methods, *Express Polymer Letters*, 12 (9), 844–854, https://doi.org/10.3144/expresspolymlett.2018.72.

[14] M. Banerjee, S. Sain, A. Mukhopadhyay, S. Sengupta, T. Kar, D. Ray, 2014, Surface treatment of cellulose fibers with methylmethacrylate for enhanced properties of in situ polymerized PMMA/cellulose composites, *Journal of Applied Polymer Science*, 131 (2), 39808(1-9), https://doi.org/10.1002/app.39808.

[15] D. R. Radford, A. S. Juszczyk, R. K. F. Clark, 2014, Achieving a good bond in acrylic resin denture teeth, *BDJ Team*, 1, 14061(9–12), https://doi.org/10.1038/bdjteam.2014.61.

[16] Z. Raszewski, 2009, A new view at acrylic materials, *Elamed.*
[17] Y. Takahashi, J. Chai, T. Takahashi, T. Habu, 2000, Bond strength of senture teeth to senture base resins, *The International journal of prosthodontics*, 13(1), 59–65, https://www.ncbi.nlm.nih.gov/pubmed/11203611.
[18] N. S. Yadav, S. Somkuwar, S. K. Mishra, P. Hazari, R. Chitumalla, S. K. Pandey, 2015, Evaluation of bond strength of acrylic teeth to denture base using different polymerization techniques: A comparative study, *Journal of international oral health*, 7 (1), 54–6, https://www.ncbi.nlm.nih.gov/pmc/articles/PMC4516074/.
[19] J. M. S. Clancy, D. B. Boyer, 1989, Comparative bond strengths of light-cured, heat-cured, and autopolymerizing denture resins to denture teeth, *The Journal of Prosthetic Dentistry*, 61 (4), 457–462, https://doi.org/10.1016/0022-3913(89)90014-0.
[20] R. L. Schneider, E. R. Curtis, J. M. S. Clancy, 2002, Tensile bond strength of acrylic resin denture teeth to a microwave- or heat-processed denture base, *The Journal of Prosthetic Dentistry*, 88 (2), 145–50, https://doi.org/10.1067/mpr.2002.127898.
[21] N. W. Rupp, R. L. Bowen, G. C. Paffenbarger, 1971, Bonding cold-curing denture base acrylic resin to acrylic resin teeth, *The Journal of the American Dental Association*, 83 (3), 601–606, https://doi.org/10.14219/jada.archive.1971.0355.
[22] Vertex Rapid Simplified, Vertex Dental, Accessed: 04.11.2019, https://www.vertex-dental.com/en/products/18-en/26/142-vertex-rapid-simplified.

IOP Conf. Series: Materials Science and Engineering **743** (2020) 012045 doi:10.1088/1757-899X/743/1/012045

Impact of network expansions on energy losses in compressed air

Karolina Perz [1*] and Aleksandra Rewolińska

[1]Institute of Machines and Motor Vehicles, Poznan University of Technology, Poznań, Poland

Email: Karolina.Perz@put.poznan.pl

Abstract. Compressed air is a very widely used energy medium, used in various fields. Is one of the most widespread, but also one of the most expensive working media used in modern industry. According to studies available in the literature in Australia and the European Union, energy consumption by compressed air systems accounts for about 10% of total industrial energy consumption, while in the United States it accounts for 30% of total electricity consumption. In addition, these systems are one of the least efficient forms of energy - only about 10-30% of the energy generated by the compressor reaches the end point. This is primarily due to a change in the form of energy (e.g. heat), but also leaks in the network and inefficient use. During operation in the compressed air network, pressure losses (linear and local) occur as a result of pipeline construction as well as pressure losses caused by the use of devices and elements responsible for maintaining adequate air quality (local losses) in the network. The pressure drop caused by the pipeline elements will depend on the extent of the network, the diameter of the pipeline, the number of places where there is a change in the direction of air flow (elbows, venturi, etc.). However, the losses caused by the use of treatment elements will depend on the expected air quality (number of filtration stages, appropriate air dryer) as well as on the network operation time (e.g. air filter contamination). The purpose of this study is to determine the impact of compressed air network extensions on energy losses resulting from, among others friction during flow. The values of these losses depend on many parameters such as efficiency, air flow speed, diameter and length of the pipeline. Knowledge about the correlation of these parameters will allow for proper selection of compression devices, as well as for optimal design of the pneumatic network.

1. Compressed air management in industry

Efficient use of energy in industrial plants is one of the main pillars of modern energy policy in Europe. However, the world today is constantly moving steadily towards increasing primary energy consumption. Compressed air is often seen as a cheap and even free energy source. Nothing more wrong. Numerous studies confirm that the industry only uses about 30% of the compressed air produced, the rest is lost due to leaks, inadequate pressure of the prepared air, improper use, etc. Compressed air is one of the most expensive energy utilities used in production plants. Industrial pneumatics often require the highest purity classes. These requirements directly translate into the need to install special driers, filters and separators. This leads to both the increased costs of the investment itself, which is the installation of a compressed air network, but also increases the costs of maintenance and energy [1,2,3]. There is not much information available in the literature on the

Content from this work may be used under the terms of the Creative Commons Attribution 3.0 licence. Any further distribution of this work must maintain attribution to the author(s) and the title of the work, journal citation and DOI.
Published under licence by IOP Publishing Ltd

energy efficiency of compressed air networks [4,5,6]. An indicative distribution of the operating costs of the air compressor is shown in Figure 1.

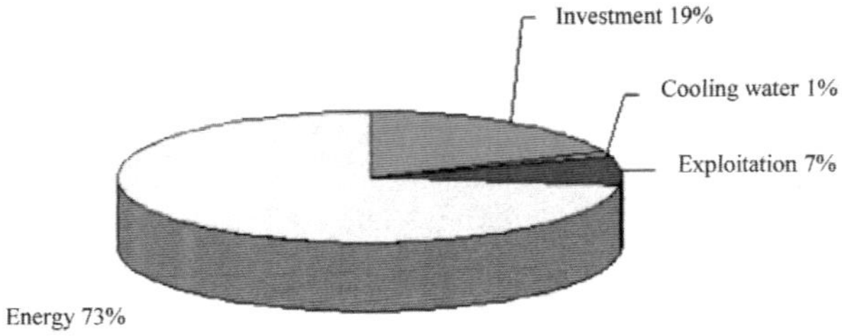

Figure 1. Cost of compressors operation [1].

Possibilities for energy savings in compressed air installations are primarily associated with reduced energy consumption for compressor drive, compressed air transmission and control of pneumatic systems.

To effectively use compressed air energy, an analysis is needed that should include the following specific tasks:

- taking an inventory of your compressed air systems,
- analysis of the efficiency of the production and use of compressed air,
- an assessment of the applicability of energy saving measures,
- adopting a cost reduction plan by improving energy efficiency,
- preparing a progress report on the implementation of the action plan.

Conducting an audit is therefore necessary in the process of determining the cumulative energy consumption of a production plant. Identifying places that can potentially have an impact on reducing energy consumption is a fundamental task in this situation.

As previously tried to demonstrate compressed air does not belong to cheap energy carriers. Such a great need for energy in the air compression system results from the fact that most of the energy used for compression is dissipated in the form of heat - the air is first heated and then it must be cooled. Often, moisture must also be removed (remove harmful water condensate from the installation), filter dust and deliver it to the place of use. If the installation is extensive, this usually means large losses due to compressed air leaks.

2. Air flow through the installation

When flowing through the network elements, the pressure of the compressed air changes. The selection of the device (usually a compressor) that is responsible for its initial value should therefore be preceded by an analysis of compressed air demand. It is carried out taking into account the required working pressure of the receivers with the admissible minimum marked, increased by pressure losses occurring in the installation connecting the individual receivers.

An integral part of the compressed air network, in addition to the elements producing (sets of compressors) and treatment (filters, driers, separators) air are the elements forming its network structure - installations connecting individual network components. Its main task is air distribution - ensuring technological flow of the factor between individual network elements. Its design (diameter of wires, number of direction changes, connections, etc.) significantly affects the operating parameters of compressed air - especially the pressure value.

In a compressed air system, pressure drops are generated by:

- length of pipelines,
- pipe diameter,
- roughness of the internal surface of the pipeline,
- operating pressure in the network - higher pressure results in higher pressure.

Branches in the compressed air network also cause additional losses, which is why each of these types of elements must be taken into account when calculating energy demand. Most often, each such element has a specific flow resistance, which is expressed as the length of the straight section. It is also disadvantageous to move the installation components too far, because the distance between the compressor and the receiver generates compressed air losses during transport. In addition, network design with remote components is disadvantageous during network upgrades or renovations. Linear and local losses are another unfavorable factor causing a pressure drop in the compressed air system. Linear losses are usually caused by internal friction. During technological processes, the working medium is transported by pipelines in which the friction effect occurs, as well as the transformation of useful work into heat.

Most linear losses are expressed by the Darcy-Weisbach equation [9]:

$$\Delta p = \lambda \cdot \frac{l}{d} \cdot \frac{\rho \cdot v^2}{2} \qquad (1)$$

where:
d - pipeline diameter,
v-flow velocity,
l-distance of measuring points,
Δp- pressure difference at points separated by l,
λ- flow resistance coefficient.

The coefficient of linear losses - friction losses Z_T is therefore equal to:

$$Z_T = \lambda \cdot \frac{l}{d} \qquad (2)$$

The degree of smoothness of the conductor is determined for the correct determination of the friction coefficient λ. The nature of the flow should also be remembered because roughness has no effect on linear losses in the stratified flow range (Re <2300) and then the friction coefficient is calculated according to the formula [9]:

$$\lambda = \frac{64}{Re} \qquad (3)$$

For turbulent flows, the coefficient of friction depends on the roughness of the pipes. Local losses also play a very important role in calculating the pressure drop. They are often called local losses and their size depends on the shape and type of additional elements used, such as: elbows, valves, orifices, diffusers and others. The values of local losses are varied and this is due to the construction of additional elements. For elbows, the amount of local losses is mainly due to the geometry of this element, because with the flow of the working medium in the corners and behind sharp angles, there are swirls that cause the dispersion of mechanical energy.

Appropriate design of the compressed air installation allows for minimizing the pressure loss of compressed air, which can significantly affect the value of the energy used to drive the compressors.

3. Impact of compressed air network design solutions on pressure losses in the compressed air network

The main task of the compressed air network is to supply air to the receiver with the appropriate parameters: pressure value and mass flow. These values will be influenced by the network span and design parameters of the fluid transport pipeline. The following study examines various design options for compressed air networks. The impact of pipeline diameter and length on air pressure losses was assessed. The amount of these losses has a direct impact on the energy costs of the compressor station. Minimizing these losses has a positive impact on the company's economic balance. The original FINCH 2.0 program was used to perform the analysis. It was created in the programming language Delphi 7 and allows the calculation of both pressure losses in compressed air installations and determines the amount of energy recovered from the air compressor in the installation. It also allows the calculation of linear and local losses in pneumatic installations, depending on additional elements used in the network. A compressed air network of 50-800 m in length (defined as equivalent length taking into account local losses resulting from the need to use pipeline structural elements, e.g. elbows, valves, etc.) and with a diameter of 0.017-0.158 m was analyzed. The next variables were the pressure behind the compressor and performance system. The study did not take into account the change in height of the network in relation to sea level, the type of flow is laminar flow. The material that was used to build the pipeline is aluminum. The results are presented in the form of the charts below.

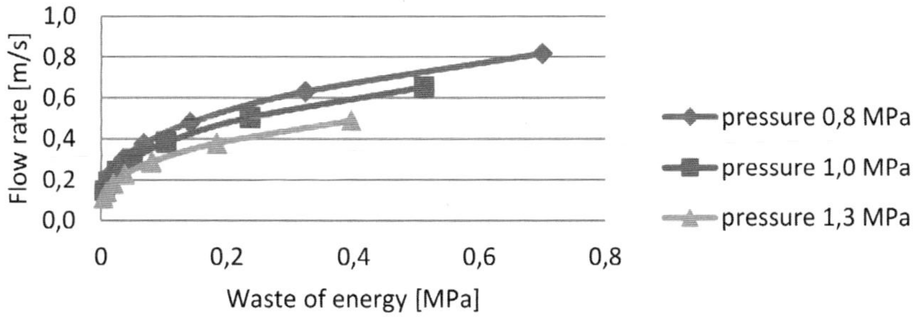

Graph 1. Dependence of energy losses on flow velocity - pipeline length 50 m.

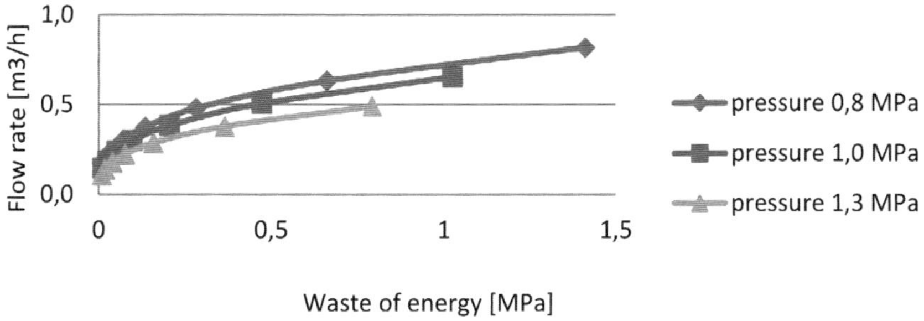

Graph 2. Dependence of energy losses on flow velocity - pipeline length 100 m.

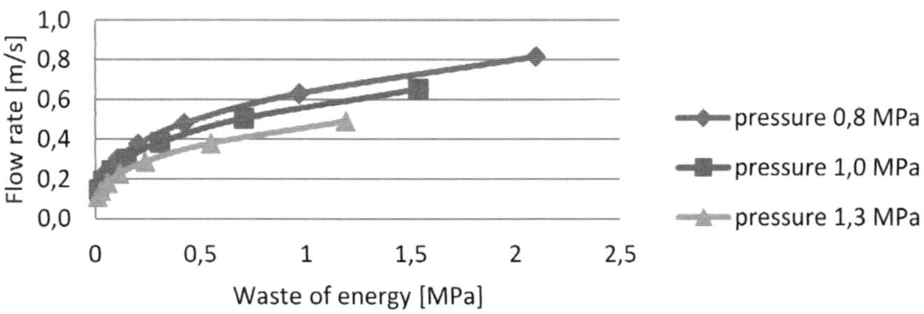

Graph 3. Dependence of energy losses on flow velocity - pipeline length 150 m.

The diagrams presented above show the dependence of energy loss generated during the air flow through the compressed air network depending on the flow speed, assuming equal network efficiency (the value of this speed depends on the diameter of the pipeline). As the flow velocity increases (and the pipeline diameter decreases), the value of air energy losses increases - the higher the velocity (smaller flow diameter), the more energy must be allocated to overcoming the resistance to movement in the pipeline resulting from min. from friction against its walls. Analyzing these graphs, it can also be seen that the energy loss increases proportionally with the length of the pipeline.

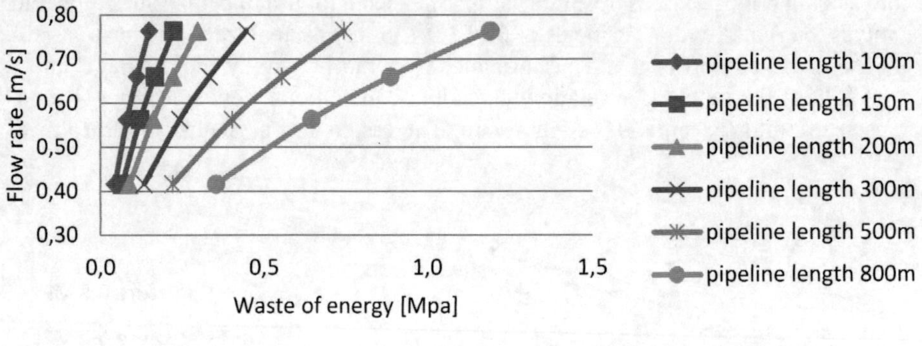

Graph 4. Dependence of energy loss on flow velocity - pipeline diameter 0.08 m.

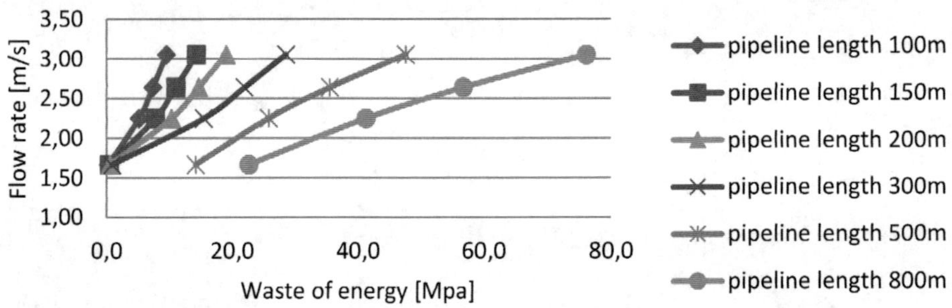

Graph 5. Dependence of energy loss on flow velocity - pipeline diameter 0.02 m.

Graph 4 and 5 show the relationship between energy loss and flow velocity for two extreme values of pipeline diameter 0.02 and 0.08m. The flow rate in this case varies due to the change in system performance. Energy losses increase with increasing flow velocity as well as with increasing pipeline length. A much larger amount of energy to overcome the resulting losses should be allocated in the case of air flow through a smaller diameter pipe (0.02m).

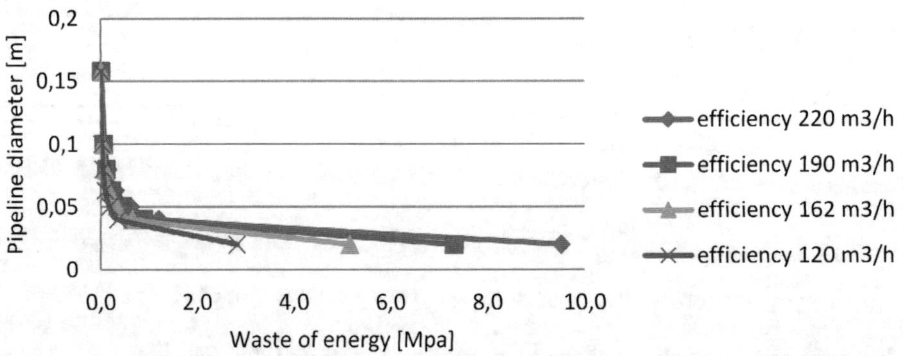

Graph 6. Dependence of energy losses on the pipeline diameter - pipeline length 100m.

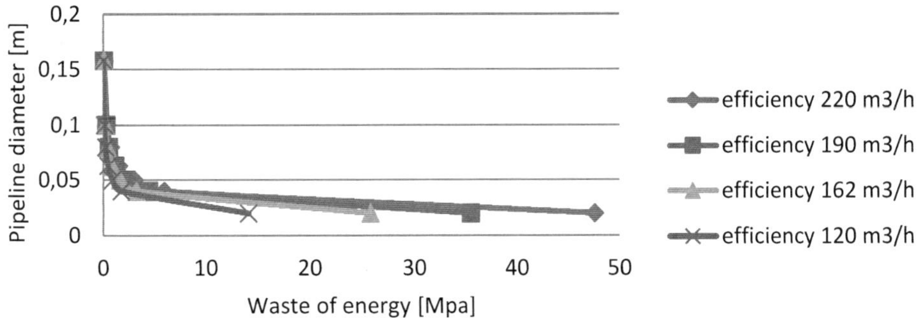

Graph 7. Dependence of energy losses on the pipeline diameter - pipeline length 500m.

The above graphs nu 6 and 7 show the dependence of energy losses on the pipeline diameter for different system performance values and for two different pipeline lengths 100 and 500 m. Based on these, it can be concluded that again the pipeline diameter value has a significant impact on the energy loss value - the smaller the diameter should be used to power the compressor to overcome frictional losses. Network performance also has an impact on this loss - the higher the performance, the greater the loss. Analyzing all the results presented above, it should be stated that both the length and diameter of the pipeline significantly affect the energy consumption of the air compressor, whose task is to create the appropriate pressure required in the technological process.

Selected variants of design solutions subjected to this analysis have been verified in real conditions. The discrepancy of the results obtained in these studies reached values compared to approx. 15% in simulation tests. In the case of energy loss analysis in real installations, component losses (filters, dryer, local starts, etc.) were taken into account, but they were not included in the value of losses. The obtained value of the discrepancy of simulation and real results allows to state that the calculation model adopted in the Finch program correctly reflects actual energy losses.

4. Conclusions

The main task of compressed air installation is to provide air to the final recipient (machine) with appropriate technological parameters. These parameters are primarily pressure, flow rate and quality of compressed air. The structure and degree of filtration of the installation itself will depend on them. Its elements influence the change of factor parameters. The value of this change must be taken into account at the design stage of the compressed air network. Important parameters during this process are the diameter and length of the compressed air network. Appropriate selection of these parameters will allow you to minimize losses, but also allow you to install a cheaper compressor (both in purchase and in operation) with lower energy consumption. Current research shows that system development is expedient, consisting of a comprehensive approach to energy savings during production (compressors, air treatment and preparation stations), distribution (installations and pipelines) and use (drives, machines, devices) of compressed air. Comprehensive control of the compressed air network can be carried out by performing an audit to show the scale of losses of this air. It can be the basis for modernizing the network in order to optimally use the energy contained in compressed air.

References

[1] Gieras M. red. 2011 Atlas Copco Technika sprężonego powietrza Atlas Copco Airpower NV ISBN 9789081535809.
[2] Perz K. 2012 Straty energii w układach sprężonego powietrza w zakładach przemysłu spożywczego *Postępy nauki i techniki* **13** pp 86-93.
[3] Perz K. 2012 Wpływ nieszczelności instalacji sprężonego powietrza na zużycie energii w przemyśle spożywczym *Problemy eksploatacji* **1** pp 141-147.
[4] Compressed air systems: Auditing and replacing air compressors 2008 *Filtration & Separation*

45 pp 41–42 doi:10.1016/s0015-1882(08)70373-2.
[5] Zhang, B. Liu M., Li Y., & Wu L. 2013 Optimization of an Industrial Air Compressor System. *Energy Engineering*, **110(6) pp** 52–64. doi:10.1080/01998595.2013.10753695.
[6] Keeney, J. W. (n.d.) 2013 Investigation of compressed air energy storage efficiency doi:10.15368/theses.2013.232.
[7] Taheri K. Gadow R. 2017 Industrial compressed air system analysis, *CIRP J. Manuf. Sci. Technol.* **18** pp 10-17 DOI:10.1016/j.cirpj.2017.04.004.
[8] Saidur R. Rahim N.A. Hasanuzzaman M. 2010 A review on compressed-air energy use and energy savings. *Renew. Sust. Energ. Rev.* **14/4** pp 1135–1153 https://doi.org/10.1016/j.rser.2009.11.013.
[9] Zasady obliczeń strat ciśnienia. Rurociągi PN-76/M-34034.

IOP Conf. Series: Materials Science and Engineering **743** (2020) 012046 doi:10.1088/1757-899X/743/1/012046

Automatic systems supporting research on modern materials and industrial bonding technologies

R Wojciechowski[1], M Bąkała[1] and A Rylski[2]

[1]Lodz University of Technology, Institute of Applied Computer Science, ul. Stefanowskiego 18/22, 90-924 Łódź, Poland
[2]Lodz University of Technology, Institute of Materials Science and Engineering, Stefanowskiego 1/15, 90-924 Łódź, Poland

Email: r.wojciechowski@kis.p.lodz.pl, m.bakala@kis.p.lodz.pl, adam.rylski@p.lodz.pl

Abstract. Industrial bonding technologies, including soldering and brazing, operate on solid-liquid phase boundaries. Commonly applied in many areas of industry (such as electronics, metallurgy or aviation), these bonding processes rely on the wetting of joining surfaces using a liquid bonding material. Understanding the wetting process and quantitative description of its parameters enables the design of new materials and joining technologies, as well as the optimization of existing methods. From the primary interfacial impact parameters, i.e. the contact angle and surface tension, the dynamics of the wetting process can be determined, including adhesive tension. This article investigates two independent measurement systems for wetting parameters in industrial brazing technologies: the sessile drop method and the Wilhelmy plate method. Both methods can be applied effectively in research on wetting processes. The measurement systems are compared in an example study.

1. Introduction

Phenomena occurring on interphase surfaces can be described using primary capillarity equations. These formulate the relationships between phase surface energies and capillary pressure, wettability, interphase adhesion and adsorption. The wetting process of joining surfaces using a liquid bonding material is affected by primary interfacial impact parameters, i.e. the contact angle and surface tension. Knowledge of these parameters enables technological aspects of the dynamics of the wetting process to be determined, including adhesive tension. Surface tension denotes a force tending to minimize the area of the surface. It is caused by asymmetries in the intermolecular forces between surface molecules. In the equilibrium state, the tension forces (liquid σ_{LV}, solid σ_{SV} and interphase σ_{SL}) at each point of contact between the three phases, between the liquid drop and the flat solid surface, are balanced. When a liquid comes into contact with a solid in a bulk, gaseous phase, an interface between the surface of the liquid and the outline of the contact surface is created. This is known as the contact angle. When the value of the contact angle is less than 90°, the solid is wetted with the fluid. When it is greater than 90°, the fluid does not wet the substrate surface (figure. 1). When one phase is wetted by another, the surface separation disappears and energy is released, due to the respective surface tensions. The relationship between the wetting angle θ and the surface energies ω is described by the Young equation [1, 2, 3, 4, 5, 6].

The contact angle and surface tension are common parameters of wettability. However, knowledge of adhesive tension enables more adequate analysis of the technological processes involved. Adhesive tension describes the difference between the surface energies when a fluid in a bulk, gaseous phase

Content from this work may be used under the terms of the Creative Commons Attribution 3.0 licence. Any further distribution of this work must maintain attribution to the author(s) and the title of the work, journal citation and DOI.
Published under licence by IOP Publishing Ltd

spreads over a solid surface and becomes a solid-fluid surface. In industrial bonding, the dynamics of the wetting process can be changed to customize the bonding process by modifying adhesive tension over time.

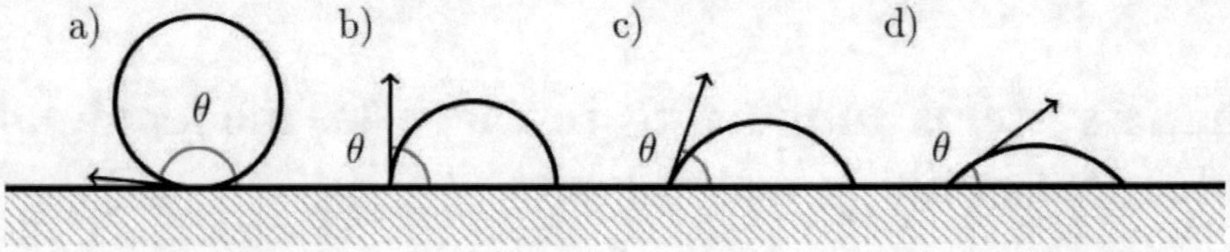

Figure 1. Liquid drop shapes in contact with: a), b) non-wettable (wetting angle $\theta \geq 90°$) and c), d) wettable (wetting angle $\theta < 90°$) substrates.

2. Measurement of wetting parameters

The sessile drop method is a general measurement procedure, which may be used to determine common parameters of wettability, including surface tension and contact angles, on the interface between liquid and solid surfaces. This method relies on geometrical traits analysis of a stationary liquid drop on a substrate. Drop shape analysis and geometrical properties extraction are performed using image processing algorithms. The accuracy of the method depends primarily on the quality of the images grabbed during the experimental process. The theoretical geometry of a liquid drop on a non-wettable substrate is presented in figure 2 a. The spreading of a liquid material along a wettable substrate is shown in figure 2 b [7, 8, 9, 10, 11].

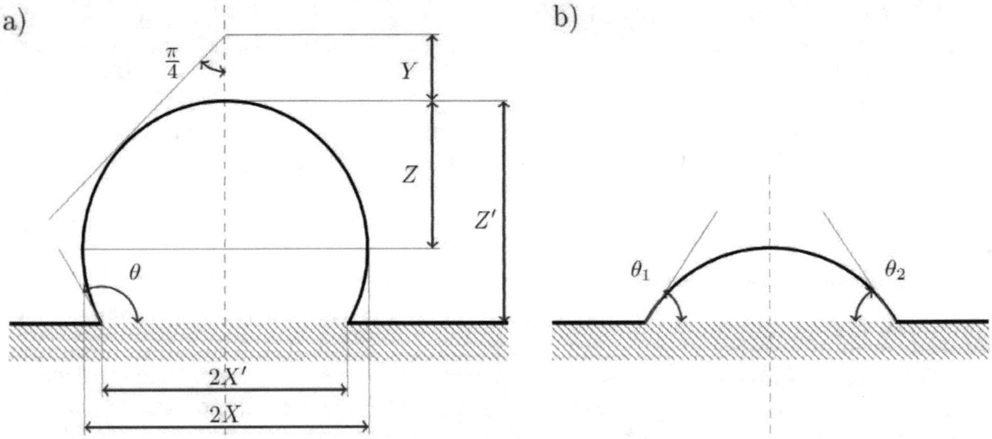

Figure 2. Theoretical geometry of a liquid drop: a) on a non-wettable substrate, b) spreading along a wettable substrate.

Surface tension can be described based on the Bashforth-Adams and Dorsey equations:

$$\sigma = g \cdot \Delta\rho \cdot \alpha^2 \quad (1)$$

where σ is surface tension, g the gravity acceleration, $\Delta\rho$ the difference in density between the metallic liquid phase and gaseous bulk phase and α is an empirical parameter calculated using Dorsey's or Porter's formulas. For small drop measurements, the preferred method of estimating α is Porter's formula, which for the theoretical geometry of a liquid drop on a non-wettable substrate (figure 2 a) is formulated as follows [7, 8, 9, 10, 11]:

$$\frac{\alpha^2}{X^2} = \left(\frac{Z}{X}\right)^2 - 0.66 \cdot \left(\frac{Z}{X}\right)^3 \cdot \left[1 - 4.05 \cdot \left(\frac{Z}{X}\right)^2\right] \tag{2}$$

The braze spread on a wettable substrate can be measured to determine the extreme wetting angles θ_1 and θ_2 (figure 2 b). In terms of technological analysis, verification and optimization of the process can be achieved based on knowledge of the adhesive tension and its variability with regard to process duration. A strict correlation exists between the parameter of adhesive tension and common parameters of wettability. Adhesive tension depends on the surface tension of the braze material and the current wetting angle, which converges over time with the extreme wetting angles. The Wilhelmy plate method is a measurement procedure which enables determination of adhesive tension. This method requires real-time observation of the forces acting on the specimen during its immersion in a fluid braze bath (gravity, capillary and buoyancy forces according to the experiment phase). In the case of immersion in a liquid braze, the adhesive tension is denoted by the formula [2, 12, 13, 14, 15]:

$$\sigma_A = \sigma \cdot \cos\theta \tag{3}$$

where σ is the surface tension of the liquid braze and θ is the current wetting angle. Including the distribution of forces,

$$\sigma_A = \frac{1}{P_S} \cdot \left(F_{m2} - F_{m1} + C_S \cdot \rho \cdot g \cdot h\right) \tag{4}$$

where F_{m1}, F_{m2} are the resultant forces measured before and after immersion of the specimen in the fluid braze, ρ is the density of the specimen, g is gravity acceleration, h is immersion depth, P_S is the specimen perimeter and C_S is the cross-section of the specimen.

3. Laboratory stands for measuring wetting parameters

ThermoWet is an automatic laboratory stand, which enables accurate measurement of surface tension and wetting angles using the sessile drop method, at temperatures of up to 1800° in the presence of a protective gas atmosphere. The main part of the measurement system is a custom-build resistance furnace, in which the specimen and a solid substrate are placed during the experiment. The temperature is measured and controlled precisely using a Eurotherm process PID controller. This enables temperature control accuracy of 0.5 K [18]. Images of the specimen are registered during the experiment by a vision system composed of a Watec camera with set of vision protection filters. They are then submitted to further analysis using dedicated image processing algorithms in an external application, a component of the ThermoWet system [16, 19]. A PC computer communicates with all of the components of the device using USB and serial ports, allowing planning of the experiment (temperature set points, camera settings), operation of the PID temperature controller, monitoring of system behaviour and registering of the camera images. The system produces good quality images, features an automatic measurement cycle and provides accurate temperature control. Figure 4 presents an overview of the ThermoWet measurement system, with close-ups of the vision- and specimen loading subsystems. Figure 5 shows the system architecture [7, 16].

Figure 3. Overview of Thermo-Wet measurement system with close-ups of the vision- and specimen loading subsystems.

The measurement procedure is based on the sessile drop method. While the specimen is heated up and melting, the measurement system registers images of the specimen. The grabbed images are analyzed, with the main goal of extracting the geometrical properties of the drop against the base surface. The geometrical dimensions of the drop of material are determined analytically, based on approximation of its shape (polynomial, elliptical). The contact angles are computed at points of contact between the drop and the substrate base for each of the three phases (solid-, gas- and liquid). The accuracy of the analysis depends primarily on image quality. The measurement system should enable precise determination of the borders between the drop, the surface of the base and the background. Its accuracy can be reduced, however, by phenomena associated with high-temperatures, such as glare or auras. Using vision protection filters can improve the results [7, 16].

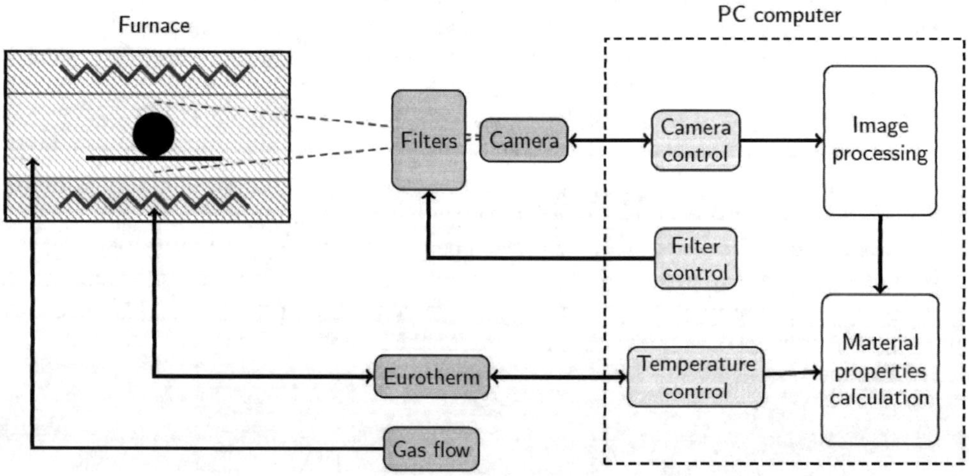

Figure 4. Architecture of the ThermoWet measurement system [7, 16].

WAST is an autonomous measurement laboratory system, which enables complex research into the dynamic properties of brazing processes. Adhesive tension can be analyzed at temperatures of up to 1000 °C under various technological gas atmospheres. This measurement system is principally composed of a movable cylindrical resistance furnace, in which the specimen immersed in a braze bath is placed, connected by scales. Figure 5 provides an overview of the WAST measurement system, with close-ups of the PLC control- MFCs- and driver sub-systems and of the entrance to the furnace, with optical sensors and a gas protection atmosphere gate. The measurement procedure is based on the Wilhelmy plate method. In the main part of the experiment, the specimen is immersed and remains in a fluid braze bath. The weighing module registers resultant forces acting on the specimen. A state diagram of the measurement experiment is presented in figure 6.

Figure 5. Overview of the WAST measurement system, with close-ups of the PLC control-, MFCs and driver sub-systems and the furnace entrance with gas protection atmosphere gate and optical sensors.

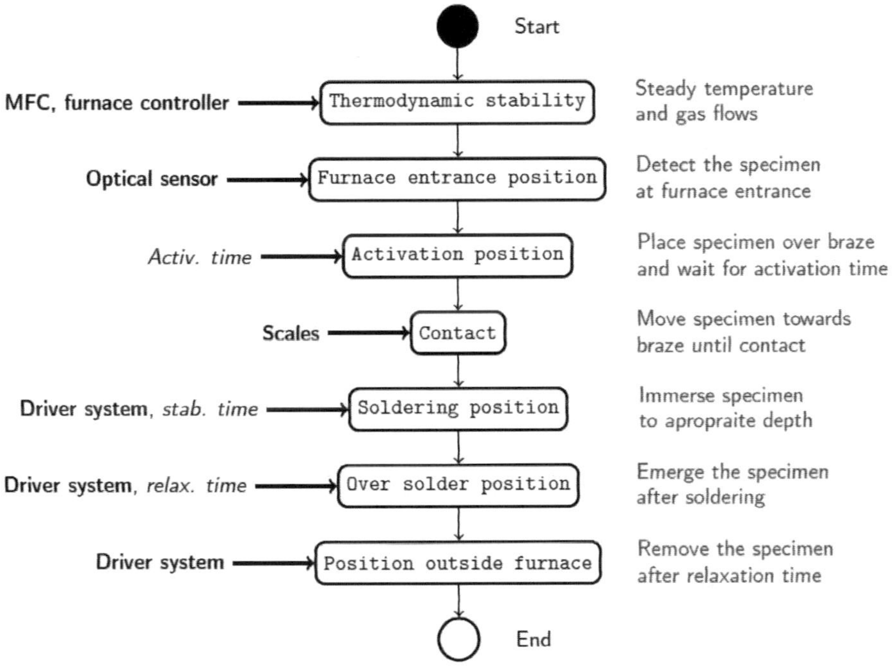

Figure 6. State diagram of measurement experiment.

Knowledge of the weight, geometrical specimen parameters and braze surface tension enables determination of the technological parameters – wetting force, adhesive tension or wetting angle. These in turn permit analysis of the wetting dynamics (e.g. advancing and receding angles, observation of the wetting changes as the specimen remains static in the fluid braze). To understand the results, this research should be supplemented by additional studies, such as analysis of the microstructure of the specimen and its resistance against oxides, or of specific phases generated in the process [12, 17].

4. Example analysis

As an example analysis, an investigation was carried out on a sample of L-Ag5P (DIN EN 1044-CP104 composed of 89% Cu, 6% P, 5% Ag) braze. The surface tension was determined using the sessile drop method, by melting the braze on a non-wettable Al_2O_3 base substrate. The extreme contact angles were calculated using a spreading test, based on the same methodology. The wetting dynamics for the L-Ag5P braze and Cu specimen were also analyzed based on Wilhelmy plate immersion tests. The main aim of the example analysis was to verify and compare the two measurement methods. The parameters of the experiments are shown in table 1.

Table 1. Parameters of experiments using the sessile drop method and Wilhelmy plate tests.

Laying drop experiment		Wilhelmy plate experiment	
Parameter	Value	Parameter	Value
Temperature range [°C]	750	Temperature [°C]	750
Gas reduct. atmosphere	H_2	Immersion depth [mm]	4
Gas protect. atmosphere	Ar	Immersion speed [mm s^{-1}]	1
		Activation time [s]	30
		Stabilization time [s]	100
		Cooling time [s]	30
		Gas reduct. atmosphere	H_2 + Ar
		Gas protect. atmosphere	N_2

Figure 7 a shows an image registered by the ThermoWet device of a L-Ag5P drop on a non-wettable Al_2O_3 substrate. The calculated mean value of the contact angle between the braze drop and base substrate is around 155° and the surface tension of the L-Ag5P braze $\sigma = 0.946$ [N/m]. A spreading test was performed for the same braze material on the copper substrate. The resulting images grabbed during this experiment, illustrating changes in the contact angles over time during the melting / spreading process, are presented in figure 7 b. The extreme contact angle value was approximately 12.5°. Based on (4) and the forces registered during the Wilhelmy plate experiment, characteristic changes in adhesive tension were calculated. From (3) and the value of surface tension determined in the ThermoWet stand, the contact angle dynamics over time were also calculated (figure. 8) At point A, contact between the hanging specimen face and the braze surface was detected by the weighing module. Next, the specimen was immersed in the fluid braze until the specified position of 4 mm depth was reached (points B, B'). During the immersion process, buoyancy and wetting forces began to act on the specimen. At point C,

IOP Conf. Series: Materials Science and Engineering **743** (2020) 012046 doi:10.1088/1757-899X/743/1/012046

the wetting angle reached 90º, which implies that the wetting force factor was zero, as the weighing module registers only gravity and buoyancy forces. After point C, the wetting angle reached an acute angle, causing the adhesive tension to rise. At point D, the adhesive tension reached 90% of its maximum value, which is technologically optimal for creating a solid, high quality join. Point E (E') represents the start of the process of specimen emergence. At point E, the contact angle was approximately 11.8º, which is similar to the value determined in the spreading test.

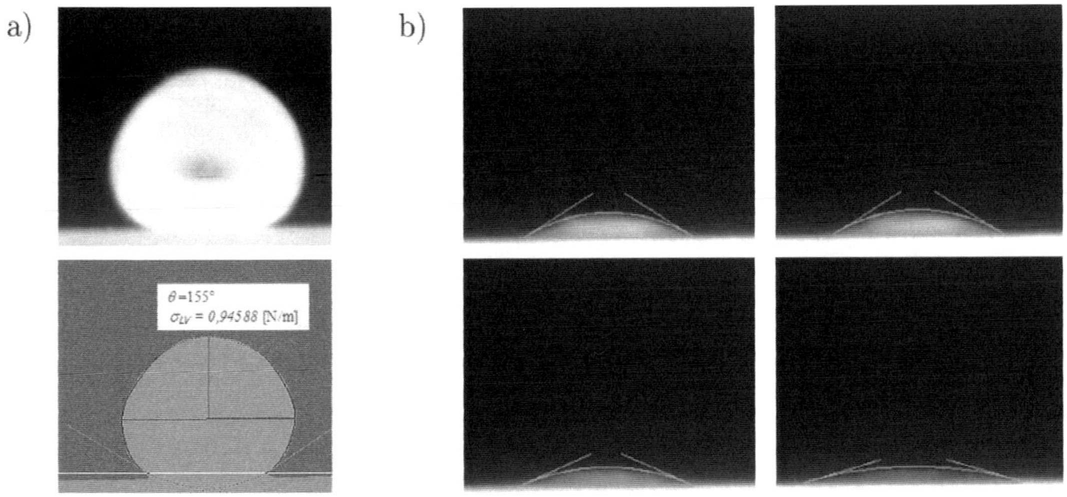

Figure 7. The laying drop experiment: a) determination of surface tension for L-Ag5P braze on Al_2O_3 base substrate, b) spreading test for L-Ag5P braze on copper substrate.

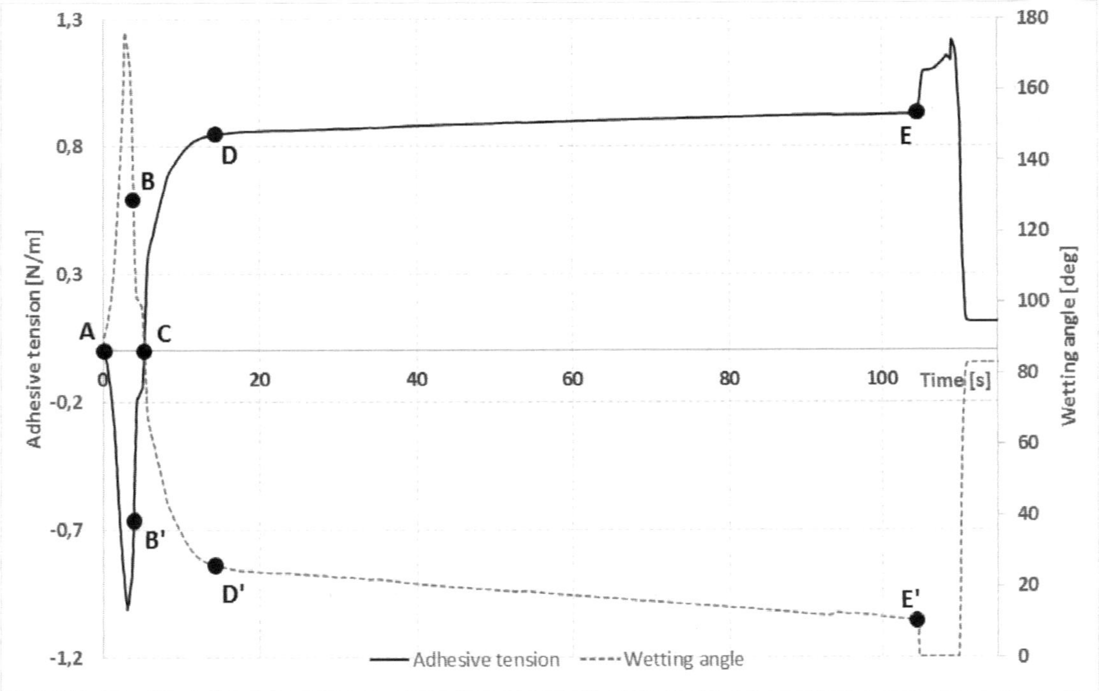

Figure 8. Changes in adhesive tension and contact angle for L-Ag5P braze and copper specimen material at 750 ºC.

IOP Conf. Series: Materials Science and Engineering **743** (2020) 012046 doi:10.1088/1757-899X/743/1/012046

5. Conclusions

This article has presented and compared two independent measurement methods: the sessile drop method and the Wilhelmy plate method. The sessile drop method is based on approximation of the drop shape and determination of its geometrical traits, using ThermoWet equipment to grab and process the images. The Wilhelmy plate method consists of analogue measurement of the forces acting on a specimen during immersion in a fluid bath. It employs the WAST laboratory system. In sessile drop experiments, image processing algorithms enable accurate measurement, but this depends primarily on the quality and resolution of the images (allowing effective approximation of the edges of the specimen and the substrate). Phenomena commonly observed in high-temperature processes (such as glare) reduce the quality of the results, but their effect can be limited through use of different vision filters. The experiment is time consuming, but it is important to analyze the interphase reactions accurately and to obtain high-resolution results. The second method described here, the Wilhelmy plate method, provides a relatively rapid method for verifying the suitability of materials to be joined, in different technological environments and using different criteria (conductivity, strength of connection, etc.). Implemented in the WAST system, measurements based on the plate method yield information on changes in adhesive tension over time at a certain temperature. To describe changes in the wetting angle over time, knowledge of surface tension is also needed. For well-known materials, the values for surface tension can be obtained from reference tables in the literature. For new materials, however, additional experiments, such as using ThermoWet, are needed. The energy state of the prepared surface, as well as the assumed process criteria (temperature, gas protection/reduction atmosphere, interphase reaction time, etc.) should be the same for both experimental processes. To verify each of the methods, a comparative analysis was conducted, focusing on the same technological aspects. Based on the convergence of the results, both methods can be used effectively to design new materials and joining technologies, as well as for optimizing already existing materials and technologies. The optimization of production processes using different technological parameters based on knowledge obtained using various measurement methods can significantly reduce overall economic costs and improve the quality of the final product.

References

[1] Missol W. (1975). *Energia powierzchni rozdziału faz w metalach*. Wydawnictwo Śląsk, Katowice, Poland.

[2] Eustanthopoulos N., Nicholas M. G., Drevet B. (1999). *Wettability at High Temperatures*. Pergamon Press.

[3] Adamson A. W., Gas A. P. (1997). *Physical Chemistry of Surfaces*. John Wiley and Sons.

[4] Kwok D. Y., Neumann A. W. (2000). Contact angle interpretation in terms of solid surface tension. Colloids and Surfaces A Physicochem. and Eng. Aspects, 161(1), 31–48.

[5] Takamichi I., Guthrie R. (2015). *The Thermophysical Properties of Metallic Liquids. Volume 1. Fundamentals*. Oxford University Press.

[6] Schwartz M. (1995). *Brazing: For the Engineering Technologist*. Chapman and hall.

[7] Sankowski D., Mosorov V., Strzecha K. (2011). *Przetwarzanie i analiza obrazów w systemach pzemysłowych. Wybrane zastosowania*. Wydawnictwo Naukowe PWN, Poland.

[8] Lahooti S., del Rio O. L., Cheng P., Neumann A. W. (1996). *Axisymmetric dropshape analysis*. AppliedSurface Thermodynamics. 509–556.

[9] Lam C. N. C., Lu J. Y., Neumann A. W. (2001). *Measuring contact angle*. Handbook of Applied Surface and Colloid Chemistry, Part 5: Analysis and Characterization in Surface Chemistry.

[10] Siwiec G., Botor J., Machulec B. (2003). *The surface tension determination through the estimation of the parameters of the sessile drop equation*. Archives of Metallurgy, 48(2). 209–221.

[11] Spelt J. K., Rotenberg Y., Absolom D. R., Neumann A. W. (1978). *Sessile Drop Contact Angle Measurements Using Axisymmetric Drop Shape Analysis (ADSA)*. Colloids.

[12] Bąkała, M., Wojciechowski, R., Sankowski, D., Rylski, A. (2017). *Semi-automatic apparatus for measuring wetting properties at high temperatures*. Metrol. Meas. Syst., 24(1), 175–184.
[13] Aniello M., Morrow N. R. (1995). *Point-by-point method of determining contact angles form dynamic Wilhelmy plate data for oil/brine/solid systems*. Journal of Colloid and Interface Science, 172.
[14] James L., Jellison D. R., Hosking M. (1976). *Statical interpretation of meniscograph solderability test*. IEEE Transaction on Parts, Hybrids, and Packaging, 12.
[15] Ning W., Jialin D., Fortunato J. (1999). *Dynamic surface tension measurement with a dynamic Wilhelmy plate technique*. J. of Colloid and Interface Science, 215.
[16] ThermoWet. Information about the laboratory stand. *http://www.iis.p.lodz.pl*. (Apr. 2018).
[17] WAST. Information about the laboratory stand. *http://www.iis.p.lodz.pl*. (Apr. 2018).
[18] Eurotherm. Temperature and process controllers manufacturer. *http://www.eurotherm.com*. (Apr. 2018).
[19] Watec. Cameras manufacturer. *http://www.wateccameras.com*. (Apr. 2018).

IOP Conf. Series: Materials Science and Engineering **743** (2020) 012047 doi:10.1088/1757-899X/743/1/012047

Numerical simulation of phase transformation during gas quenching after low pressure carburizing

Krzysztof Krupanek[1], Jacek Sawicki[1] and Victoria Buzalski[2]

[1]Institute of Material Science and Engineering, Lodz University of Technology, Stefanowskiego 1/15, 90-924 Lodz, Poland
[2]State University of Western Parana, 645 Rua da Faculdade, 85903-000 Toledo, Brazil

Email: jacek.sawicki@p.lodz.pl

Abstract. Low pressure carburizing followed by high pressure gas quenching is often applied to steel elements in order to increase their strength and durability, but undesirable distortions are also formed due to this process. In order to minimize these distortions, the Vacuum UCM furnace manufactured by SECO/WARWICK conducted by single piece flow method is proposed. A refined simulation that describes the quenching phenomena with enough accuracy can reduce the project and optimization costs as it replaces the trial and error method. The goal of this study was to create a computer model that can predict the behaviour of the phase transformations that occurs at 4D quenching chamber after low pressure carburizing in single piece flow technology. First, an investigation about the simulations that has been developed in this area was done. Based on physical phenomena involved in this process, a numerical scheme was created; The numerical scheme was applied to the real steel element that was submitted to the single piece flow technology process; The numerical simulation was compared with the experimental data. As a result, a numerical model able to simulate phase transformation after complex chemical-heat treatment in the single peace flow technology was obtained.

1. Introduction

Thermal surface treatments as low pressure carburizing (LPC) followed by high pressure gas quenching (HPGQ) are often applied to steel elements in order to increase their strength and durability. LPC started being applied in industries two decades ago, since it gives better results in terms of carburizing precision and evenness in comparison to the traditional carburizing in atmospheres [1], [2], [3]. The HPGQ improved significantly the result of the quenching process by decreasing the deformation rate when compared with the traditional oil quenching [2], [4], [5], [6]. The goal of these treatments is to provide carbon diffusion into the material and then the phase transformation from austenite to martensite. As a result, the hardness in the surface of the material are improved, promoting greater durability [7].

Undesirable distortions are caused in the steel during these processes. They are mainly formed by the change of the steel structure volume that occurs due the phase transformation and non-uniform parameters during the quenching [6]. The correction of these distortions is one of the most expensive processes [4]. For this reason, despite they are inevitable [3], the distortions need to be avoided as much as possible.

The Vacuum UCM furnace manufactured by SECO/WARWICK (Figure 1) proposes a thermochemical treatment being conducted by single piece flow technology that comes to replace the traditional batch method for LPC/HPGC in order to minimize distortions. As in the batch method more than one element is submitted to the chamber, the physical and metallurgical proprieties varies over the

Content from this work may be used under the terms of the Creative Commons Attribution 3.0 licence. Any further distribution of this work must maintain attribution to the author(s) and the title of the work, journal citation and DOI.
Published under licence by IOP Publishing Ltd

placement which the element occupies [8], [2]. On the other hand, the single piece flow model proposes that every single piece goes through the exact same position and process conditions as every other piece, furthermore the carburization in this method produces a favourable, i.e. compressible distribution of residual stress in the technological surface layer [4], [9]. In the 4D quenching chamber, the device provides the inflow of cooling gas from all sides and the rotation of parts to support uniform cooling effect [10]. Through becoming more uniform the parameters, especially during the quenching, it is possible to obtain better outcomes.

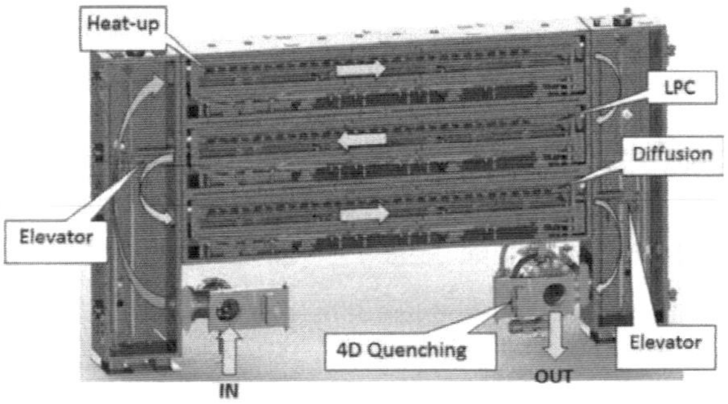

Figure 1. Vacuum UCM furnace SECO/WARWICK [1].

A refined simulation that describes the quenching phenomena with enough accuracy can reduce the project and optimization costs. The computer model allows better comprehension about the process, the device and how the design influences the results [11]. Understanding the parameters that affect the intensity of the process is crucial to accomplish the desired performance [2]. Therefore, a computer model replaces the trial and error method that is costly in terms of time and resources, affording the process optimization.

However, currently, there is no commercial software that simulates the influence of the complex fluid flow field in the phase transformation that occurs in high pressure gas quenching (HPGQ). In recent researches, including the works developed by Ferguson et al. [12], Li [3] and Sims et al. [6], [11], [13], phase transformation simulations were done at DANTE modelling software in order to predict the stress formation during HPGQ. In all these cases, the heat transfer coefficient was considered constant and the combination of more than one modelling program was required. The heat transfer coefficient has significant influence on the quenching results, including distortion, residual stresses and hardness distribution [2]. Hence, an algorithm scheme that takes in account the complex fluid flow and the heat transfer coefficient as function of the space and time is needed.

2. Numerical Algorithm Scheme

The numerical simulation of phase transformation that occurs in the steel element in 4D quenching chamber in the single piece flow technology is based on a Multiphysics Field. It takes into account the carbon diffusion, heat and mass balances, heat transport and the kinetics of phase transformation.

The goal of the algorithm scheme is to obtain the carbon concentration, cooling speed and temperature field which are parameters that determine the transformation to martensite at the steel. These solved variables are used to evaluate the phase transformation based on data generated in JMatPro.

The system consists of the fluid domain in the quenching chamber and the solid steel element. The computer model was simulated in Ansys CFX solver for computational fluid dynamics. Hexagonal elements of the grid was created, whose regular shapes assist in obtaining more accurate results in numerical simulation [14].

Numerical Algorithm Scheme is illustrated in Figure 2 . With the results obtained, it is possible to analyse if some change in geometry or input data would be profitable in view of optimization. If so, the simulation process restarts.

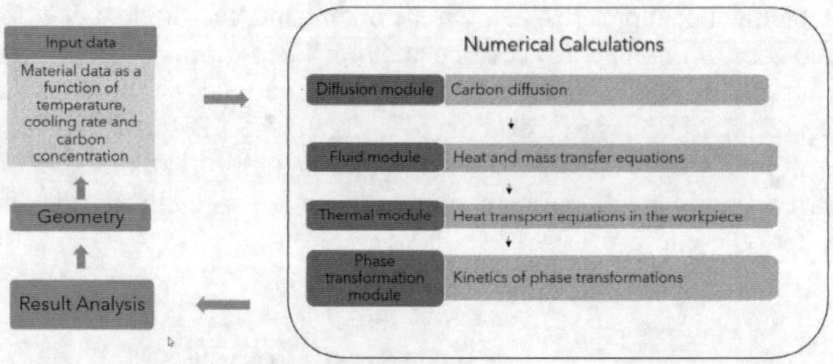

Figure 2. Numerical Algorithm Scheme.

2.1. Diffusion module
In the Low Pressure Carburizing chamber, the steel is submitted to the boost and diffusion sequence procedure, where the surface is saturated with carbon. After it is transferred and indexed through the chamber where the diffusion occurs [1]. The carbon diffusion is governed by Fick's Laws.

- Fick's Law:

$$J = -\mathrm{D}\nabla\emptyset \tag{1}$$

$$\frac{\partial\emptyset}{\partial t} = -\mathrm{D}\nabla^2\emptyset \tag{2}$$

Where J is the flux, D the diffusion coefficient, $\emptyset$ the carbon concentration, t the time. The equation (1) is used as boundary condition and the equation (2) describes the carbon concentration distribution in the element.

2.2. Fluid module
The flow field is described by momentum equation, continuity equation and energy equation [15].

- Momentum:

$$\frac{\mathrm{D}U}{\mathrm{D}t}\rho = \rho\vec{g} - \nabla\mathrm{P} + \mu\nabla^2 U \tag{3}$$

- Continuity:

$$\frac{\partial\rho}{\partial t} + \mathrm{div}\,\rho U = 0 \tag{4}$$

- Energy:

$$\frac{\partial(\rho h)}{\partial t} + \nabla(\rho U h) = \nabla(\lambda\nabla T) + \tau{:}\,\nabla U + \mathrm{S_E} \tag{5}$$

Where ρ is the specific mass, h is the enthalpy, U is the velocity, T is the temperature, λ is the conduction coefficient, $\tau{:}\,\nabla U$ is the viscous dissipation and $\mathrm{S_E}$ the source term.

2.2.1. Turbulence Model. The turbulence model is done according to k-ω based SST (Shear Stress Transport) models available in Ansys CFX, that is a combination of k- ω and k- ε models. In the boundary layer, k-ω model is applied. As k-ω model is replaced by k- ε in the layers located further away from the wall because it models turbulence in free flow very well and it is less sensitive to inlet conditions for values that describe the turbulence. The desired characteristics of both models are joined into one due the fact that the standard k- ε model can be transformed into equations on k and ω, as the

ω is the dissipation of the turbulence kinetic energy, therefore, $\omega = \frac{\varepsilon}{k}$. The equations of this model are then multiplied by the function whose value is 1 in the free flow and 0 by the wall, and the standard k-ω model equations are multiplied by F_1 function [14]. These two-equation models are the most prominent in this area, supported by a large number of validation studies [16]. The SST model is proposed for accurate boundary layer simulation [15].

The applied model equation are as follows:

- Turbulent Kinetic energy:

$$\frac{\partial(\rho k)}{\partial t} + \frac{\partial(\rho u_j k)}{\partial x_j} = -P - \beta^* \rho\omega k + \frac{\partial}{\partial x_j}\left[\frac{(\mu + \sigma_k \mu_t)\,\partial k}{\partial x_j}\right] \quad (6)$$

- Dissipation of the turbulent kinetic energy:

$$\frac{\partial(\rho\omega)}{\partial t} + \frac{\partial(\rho u_j \omega)}{\partial x_j} = -\frac{\gamma}{v_t}P - \beta^* \rho\omega^2 + \frac{\partial}{\partial x_j}\left[(\mu + \sigma_\omega \mu_t)\frac{\partial\omega}{\partial x_j}\right] + 2(1 - F_1)\frac{\rho\sigma_{\omega 2}}{\partial\omega}\frac{\partial k}{\partial x_j}\frac{\partial\omega}{\partial x_j} \quad (7)$$

Where:

- Turbulent viscosity:

$$v_t = \frac{a_{1k}}{\max(a_1\omega, SF_2)} \quad (8)$$

- Production term:

$$P = \mu_t SS \quad (9)$$

Where:

$$S = \left(\frac{1}{2}\left(\frac{\partial u_i}{\partial x_j} + \frac{\partial u_j}{\partial x_i}\right)^2\right)^{\frac{1}{2}} \quad (10)$$

- Functions F_1 and F_2:

$$F_1 = tanh(arg_1^4), arg_1 = min(max\left(\frac{k^{\frac{1}{2}}}{0.09\omega y}, \frac{500v}{\omega y^2}\right), \frac{4\rho\sigma\omega^2 k}{CDk\omega y^2} \quad (11)$$

$$F_2 = tanh(arg_2^2), arg_2 = max\left(2\frac{k^{\frac{1}{2}}}{0.009\omega y}, \frac{500v}{\omega y^2}\right) \quad (12)$$

- Fixed coefficients:

$$a_{k1} = 0.85, \sigma_{\omega 1} = 0.5, \beta_1 = 0.075, \beta^* = 0.09, \kappa = 0.41, \gamma_1 = \frac{\beta_1}{\beta^*} - \sigma_{\omega 1}\kappa^2(\beta^*)^{\frac{1}{2}}, \sigma_{k2} = 1.0,$$

$$\sigma_{\omega 2} = 0.856, \beta_2 = 0.0828, \beta^* = 0.09, \kappa = 0.41, \gamma_2 = \frac{\beta_2}{\beta^*} - \sigma_{\omega 2}\kappa^2(\beta^*)^{\frac{1}{2}}.$$

2.2.2. Modification of the production term. The SST k-ω model tends to over produce turbulence in the pile-up areas because of the high S values generated in these regions. Kato and Launder proposes to replace the tangential stresses S in the turbulence production equation with rotation Ω, then:

$$P = \mu_t S\Omega \quad (13)$$

IOP Conf. Series: Materials Science and Engineering **743** (2020) 012047 doi:10.1088/1757-899X/743/1/012047

Where:

$$\Omega = \left(\frac{1}{2}\left(\frac{\partial \mu_i}{\partial x_j} - \frac{\partial \mu_j}{\partial x_i}\right)^2\right)^{\frac{1}{2}} \quad (14)$$

Therefore, the model enables to introduce a term that limits over-production of the turbulence kinetic energy in the areas of high-pressure gradients[14].

2.3. Thermal module

Calculation of heat transport based on preset boundary and initial conditions. The equation (15) is dependent of material's characteristics, as conduction coefficient k_t, specific heat C_p and density ρ. But these proprieties varies with temperature, cooling speed and carbon concentration. Therefore an iterative procedure is required.

$$\rho C_p \dot{T} = \nabla \cdot (k_t \nabla T) \quad (15)$$

2.4. Phase transformation module

Due the conditions, during the quenching process occurs an extensive rearrangement of atoms. The phase transformations can be classified as diffusional or martensitic. The diffusional phase transformation involves diffusion of carbon and other solute atoms over the time. The martensitic phase transformation is diffusionless and occurs very rapidly. During the quenching these two phase transformations can occurs simultaneously at different locations [17]. The temperature field, cooling speed and carbon concentration are used to evaluate the phase transformation based on data generated in JMatPro.

3. Results

Based on the numerical scheme described above (Figure 2), computational simulation where carried out. The element which was simulated is a flat ring made of 20MnCr5 steel. The material data varying with temperature, cooling speed and carbon concertation was created and implemented to every module of numerical scheme.

3.1. Diffusion

First, the carbon distribution was obtained after heating to 860°C and low pressure carburizing (Figure 3). The numerical results where compared with experimental data by plotting the value of carbon concertation in the middle of top surface (Figure 4). The carbon concentration in a layer was measured with a glow discharge optical emission spectrometer – LECO GDS850A. The comparison shows a good match between both.

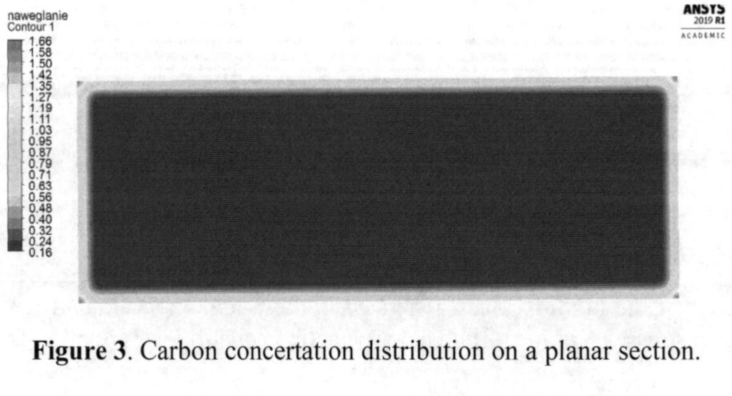

Figure 3. Carbon concertation distribution on a planar section.

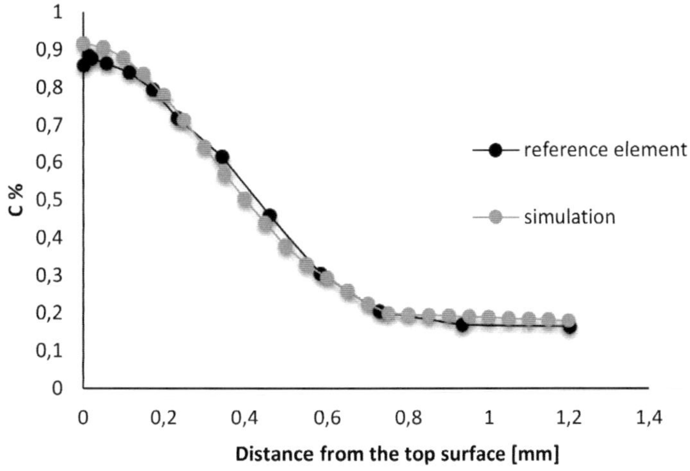

Figure 4. Carbon concertation.

3.2. Fluid module

Secondly, the CFD summations where carried out to obtain complex flow of cooling gas (Figure 5 and Figure 6) which has the biggest influence on wall heat transfer distribution on the element surface during high pressure gas cooling (Figure 7).

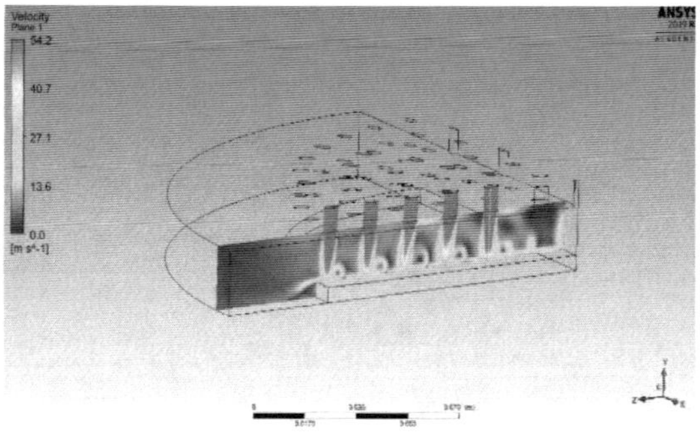

Figure 5. Velocity distribution on a plane perpendicular to the top cooling surface.

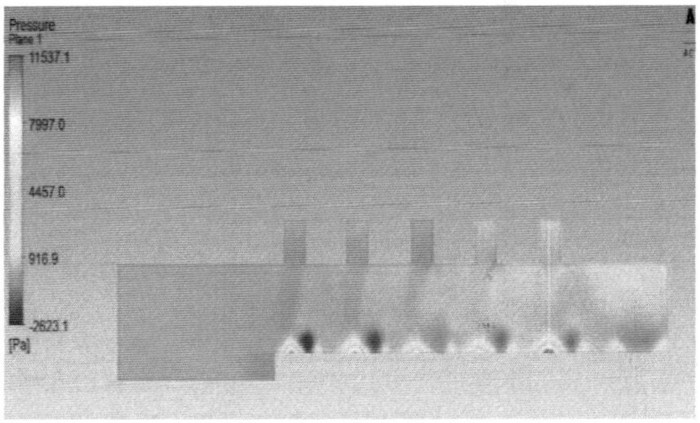

Figure 6. Velocity distribution on a plane perpendicular to the top cooling surface.

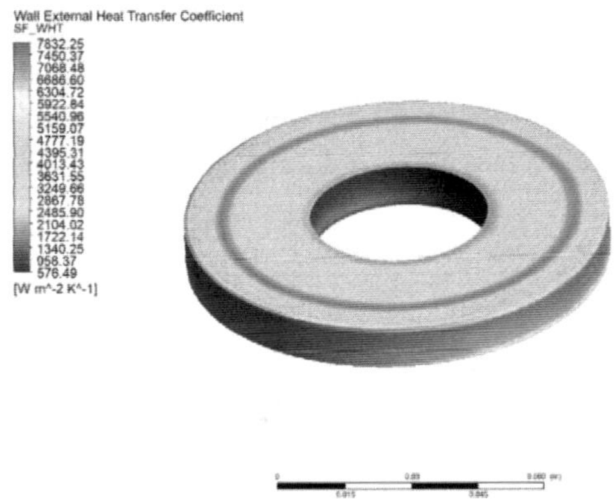

Figure 7. Heat transfer coefficient distribution in the wall of the element.

IOP Conf. Series: Materials Science and Engineering **743** (2020) 012047 doi:10.1088/1757-899X/743/1/012047

3.3. Thermal module

Then, after obtaining the wall heat transfer coefficient distribution over the surface of the element and material data, the cooling process could be simulated. Simulation shows that uneven wall heat transfer coefficient affects the temperature distribution over time (Figure 8).

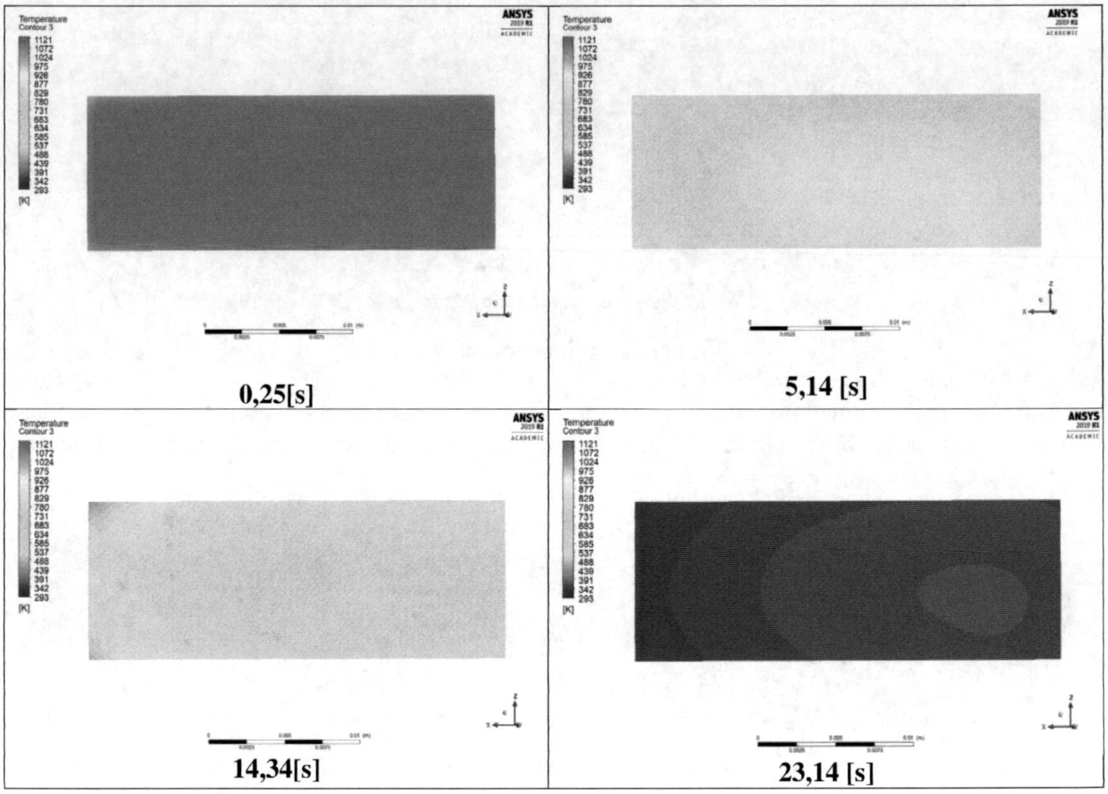

Figure 8. Temperature distribution over time on cross section.

3.4. Phase transformation module

Finally based on generated material data depended on the calculated cooling speed, transient temperature distribution and carbon concertation the numerical scheme is able to calculate the initiation of martensitic transformation. As the Figure 9 shows the martensitic transformation occurs differently depending on location. This phenomena is due to uneven heat transfer coefficient distribution on the element surfaces which may lead to an increase of undesirable quenching deformations.

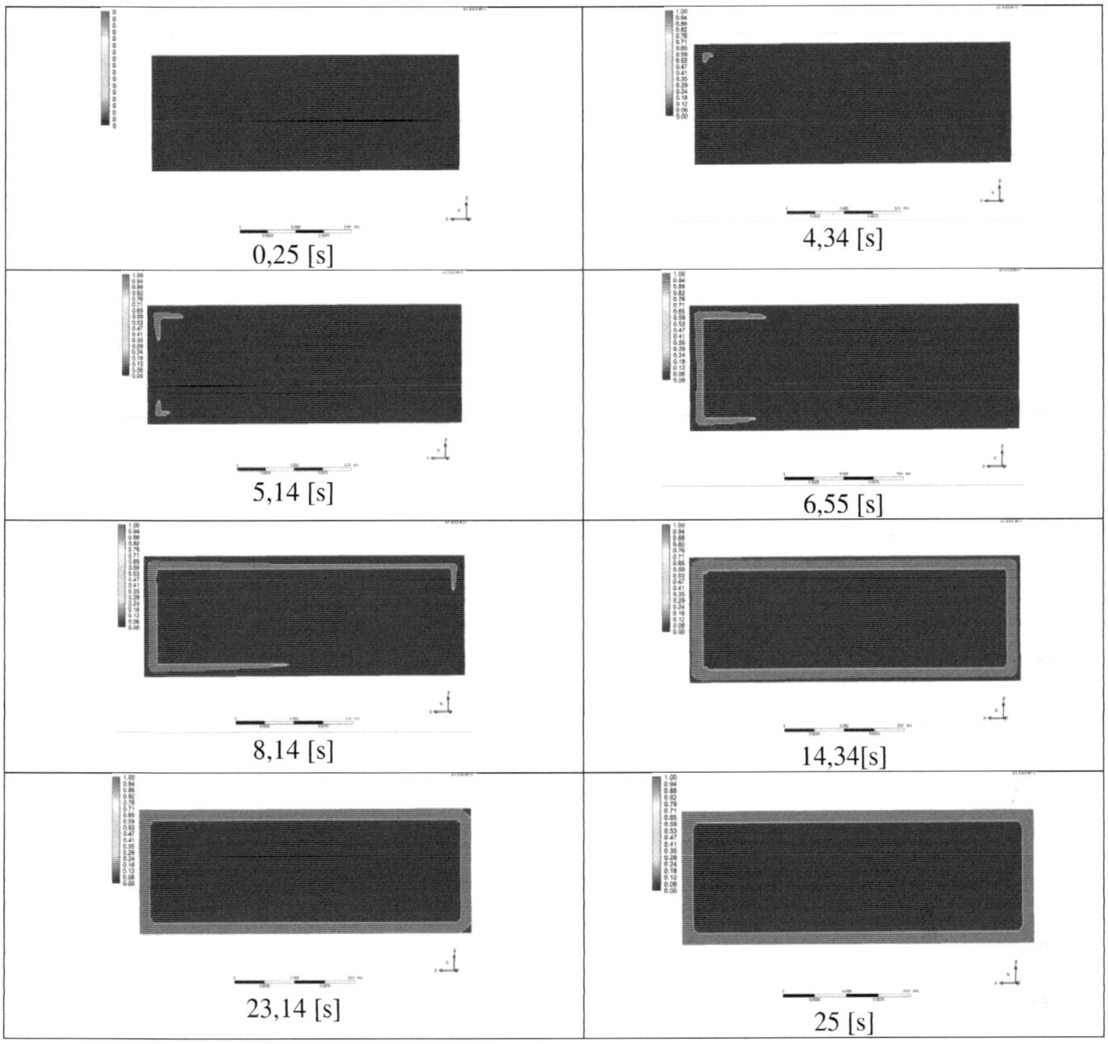

Figure 9. Martensite initiation on cross section.

4. Conclusion

The phase transformation during the gas quenching is a very critical part of the process in term of deformation generation. Therefore, it is advantageous to invest in simulations that describes properly the phenomena in sense of having a good device that operates with improved parameters. The simulation will provide us optimized process, avoiding unnecessary and costly. The presented numerical algorithm scheme was verified positively by the experiment.

Acknowledgments

This research was funded by the Polish National Centre for Research and Development, grant number POIR.04.01.04-00-0087/15.

References

[1] Korecki M, Wołowiec-Korecka E, Sut M, Brewka A, Stachurski W and Zgórniak P 2017 Precision case hardening by low pressure carburizing (LPC) for high volume production *HTM J. Heat Treatm. Mat.* **72/3** 175-183.

[2] Wołowiec-Korecka E 2018 Modeling methods for gas quenching, low-pressure carburizing and

low-pressure nitriding *Elsevier* **177** 489-505.

[3] Li Z C 2015 Heat treatment response of steel fatigue sample during vacuum, *Proceedings of the International Manufacturing Science and Engineering Conference (ASME 2015), Charlotte, North Carolina, USA, June 8-12.*

[4] Korecki M, Wołowiec-Korecka E and Glenn D 2015 Single-Piece, High-Volume, and Low-Distortion Case Hardening of Gears *AGMA Technical Paper* **15FTM04.**

[5] Dybowski K, Sawicki J, Kula P, Januszewicz B, Atraszkiewicz R and Lipa S 2016 The effect of the quenching method on the deformations size of gear wheels after vacuum carburizing *Arch. Metall. Mater.* **61/2B** 1057-1062.

[6] Sims J, Li Z C, and Ferguson B L 2019 Causes of Distortion during High Pressure Gas Quenching Process of Steel Parts, *Proceedings of the 30th ASM Heat Treating Society Conference*, *Detroit, Michigan, October 15-17.*

[7] Wołowiec-Korecka E, Korecki M, Stachurski W, Zgórniak P, Sawicki J, Brewka A, Sut M and Bazel M 2016 System of single-piece flow case hardening for high volume production *Arch. Mater. Sci. Eng.* **79(1)** 37-44.

[8] MacKenzie D S, Kumar A, Metwally H, Paingankar S, Li Z, and Ferguson B L 2009 Prediction of distortion of automotive pinion gears during quenching using CFD and FEA *J. ASTM Int.* **6(1)** 1-10.

[9] Stachurski W, Krupanek K, Januszewicz B, Rosik R and Wójcik R 2018 An effect of grinding on microhardness and residual stress in 20mncr5 following single-piece flow low-pressure carburizing *J. Mach. Eng.* **18/4** 73-85 .

[10] Stachurski W, Zgórniak P, Sawicki J and Przybysz M 2017 Hardening-related deformations of gear wheels after vacuum carburising and quenching in a 4D quenching chamber *Adv. Sci. Technol. Res. J.* **11(1)** 237-245.

[11] Habean S, Sims J, Li Z C, and Ferguson B L 2019 Quenching Process Improvement Through Modeling, *Proceedings of the 30th ASM Heat Treating Society Conference*, *Detroit, Michigan, October 15-17.*

[12] Ferguson B L, Li Z C, and Freborg A 2014 Simulating the manufacturing process chain, *Proceedings of the 21st Congress IFHTSE combined wih European Conference, Munich, Germany, May 12-15.*

[13] Li Z C, Sims J, Fetty J, and Baker T 2018 Minimizing Distortion During High Pressure Gas Quenching Processes, *Conference Proceedings of the AHS International 74th Annual Forum and Technology Display, Phoenix, Arizona, USA, May 14-17.*

[14] Stachurski W, Sawicki J, Krupanek K, and Nadolny K 2019 Numerical analysis of coolant flow in the grinding zone *Int. J. Adv. Manuf. Technol.* **104(5-8)** 1999-2012.

[15] ANSYS Inc. Release 2019 R3, ANSYS CFX - Solver Modeling Guide.

[16] Bardina J E, Huang P G, Coakley T J, 1997 Turbulence Modeling Validation, Testing, and Development *Nasa Tech. Memo.* **110446** 8-20.

[17] Li Z, Grandhi R V, Srinivasan R 2006 Distortion minimization during gas quenching process *J. Mater. Process. Technol.* **172(2)** 249-257.

IOP Conf. Series: Materials Science and Engineering **743** (2020) 012048 doi:10.1088/1757-899X/743/1/012048

Influence of quenching temperature of the individual quenching method on the geometrical dimensions of the elements

E Wołowiec-Korecka[1], P Zgórniak[2], A Brewka[1,3] and K Krupanek[1]

[1]Institute of Materials Science and Engineering, Lodz University of Technology, Lodz, Poland
[2]Institute of Machine Tools and Production Engineering, Lodz University of Technology, Lodz, Poland
[3]Seco/Warwick S.A, Świebodzin, Poland

Email: emilia.wolowiec@p.lodz.pl

Abstract. The quenching process is related to the internal-stress phenomena, resulting in geometric changes (distortions). In this paper, the impact of hardening temperature on the quenching distortions occurring during low-pressure carburizing with gas quenching using the individual quenching method was analyzed. The reference elements were subjected to carburizing at 980°C, followed by gas quenching at temperatures of 860°C, 920°C and 980°C. The geometrical measurements of the elements were made before and after the chemical treatment and the size of the quenching distortions of their geometrical parameters was determined: external and internal diameters as well as ring thickness. The study examined whether there is a statistically significant difference between the geometrical dimensions of the elements before and after the thermo-chemical treatment for the elements in three temperature groups. It was shown that for the ring thickness parameter there are no significant differences between the groups of elements before and after heat treatment for the temperature 980°C, while for the temperature 860°C and 920°C these changes are significant. However, for external diameter bottom, there is always a difference regardless of temperature. For the single piece-flow vacuum carburizing with gas quenching using the individual quenching method, it is advisable to use the highest possible quenching temperatures to reduce quenching distortions.

1. Introduction

The problem of quenching distortions is inextricably linked to machine parts production process. The purpose of quenching is to introduce martensite in hardened steel in order to ensure a high level of hardness. The final value of martensitic structure depends on cooling time and the amount of carbon contained in the structure. In practice, only steel grades that contain more than 0.3% C are quenched. Machine parts are often carburized to obtain the desired carbon content in the material before quenching.

Many researchers highlight that an important cause of quenching distortions (geometric changes) are improper cooling conditions, especially the selection of unsuitable coolant type, cooling temperature and the temperature while rinsing the element with coolant during quenching, which causes too slow or uneven heat absorption from the element. Dybowski et al. indicate that low pressure carburizing followed by high pressure quenching causes less distortion than low pressure carburizing with oil quenching [1]. According to Dybowski et al., in case of oil quenching, the mean radial run-out is lower

Content from this work may be used under the terms of the Creative Commons Attribution 3.0 licence. Any further distribution of this work must maintain attribution to the author(s) and the title of the work, journal citation and DOI.
Published under licence by IOP Publishing Ltd

than in case of gas quenching [2,3]. Stachurski et al. report that the use of a higher cooling gas pressure results in a smaller bore diameter compared to the use of lower cooling gas pressure [4]. They also emphasize that in order to minimize distortion after quenching, the structural and operating parameters of the quenching chamber should be optimized. Atraszkiewicz et al. indicate that the type of coolant is also an important factor for the process of forming distortions [5]. They demonstrate that the use of helium instead of nitrogen in the quenching process reduces unevenness and distortion. For gas quenching, the nature of the distortions is uniform, unlike in the case of oil quenching. In addition, in the case of oil quenching, partial shrinkage and partial increase of the pitch diameter of the wheel can occur. Additionally, phase transitions, phase number changes and thermal shrinkages during heat treatment also play a key role during quenching. In particular, it is known that the plasticity of transformation attributed to permanent distortion caused by phase transition is the reason for unexpected dimensional instability and difficulties in process optimization [6–8].

In recent years, several attempts have been made to design machines and their instrumentation with the purpose of reducing the distortions obtained and achieving uniformity [9–12]. However, there is no conclusive information on the effect of quenching temperature on distortions.

2. Materials and Methods

30 bushings were made of 20MnCr5 steel. Their outer and inner diameter amounted to 120 mm and 95 mm, and their height was 35 mm. All the elements were subjected to low pressure carburizing using the stream method [13] in a temperature of 980°C (Table 1), and then to gas quenching at a temperature of 860°C (group A), 920°C (group B) and 980°C (group C). A rotation device for individual quenching was used for carrying out quenching [14]. The elements were individually inserted into a chamber where the cooling nozzles were arranged around them and the base with the coolant was rotated thus ensuring identical quenching conditions for each of the elements [9,15]. Geometrical measurements of elements were made before and after chemical treatment, providing the basis for determining the extent of quenching distortions of their geometrical parameters: internal diameter, external diameter, ring thickness. The measurement results were analyzed in order to verify whether any statistically significant differences exist in the average values of geometrical parameters between groups 860°C, 920°C and 980°C. Depending on the assumptions fulfilled, the individual research hypotheses were tested using the t-Student test or a Wilcoxon test, with statistical significance assumed at a level of $\alpha=0.05$

Table 1. Thermal and chemical treatment process parameters.

Group	LPC temperature [°C]	HPGQ temperature [°C]
1	980	860
2	980	920
3	980	980

3. Results

An increase in the value of the geometrical parameters: external diameter, internal diameter, ring thickness was noticed in the temperature range examined. Analyzing the measurement results, a decrease in internal and external diameters of the bushings was observed. The average difference in the inner diameter was -260±5 μm for temperature 860°C, -259±6 μm for temperature 920°C and -262±7 μm for temperature 980°C (Figure 1, Figure 2, Figure 3). The average difference in the external diameter was -134±6 μm for temperature 860°C, -121±6 μm for temperature 920°C and -116±4 μm for temperature 980°C (Figure 4, Figure 5, Figure 6). The average difference in the ring thickness was -7.8±3.4 μm for temperature 860°C, -6.6±5.9 μm for temperature 920°C and 0.5±5.3 μm for temperature 980°C (Figure 7, Figure 8, Figure 9). These differences are statistically significant for all parameters, in all temperature groups ($p<0.05$) except for ring thickness, where for 980°C there were no differences between the values before and after heat treatment.

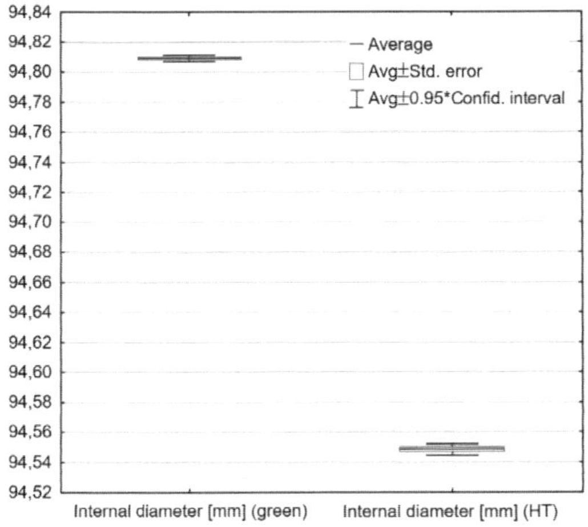

Figure 1. Internal diameters of elements before and after thermo-chemical treatment at 860°C.

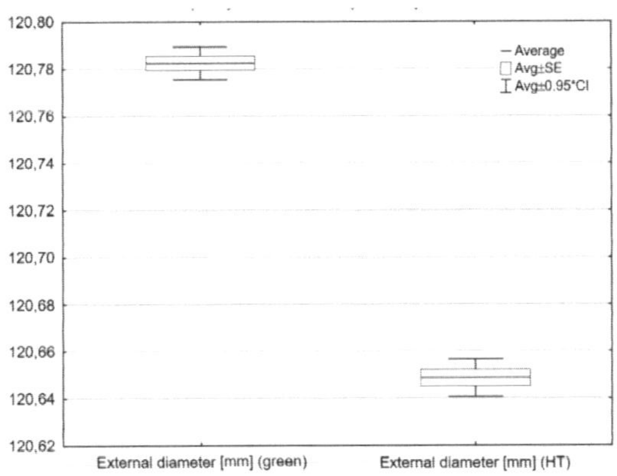

Figure 2. External diameters of elements before and after thermo-chemical treatment at 860°C.

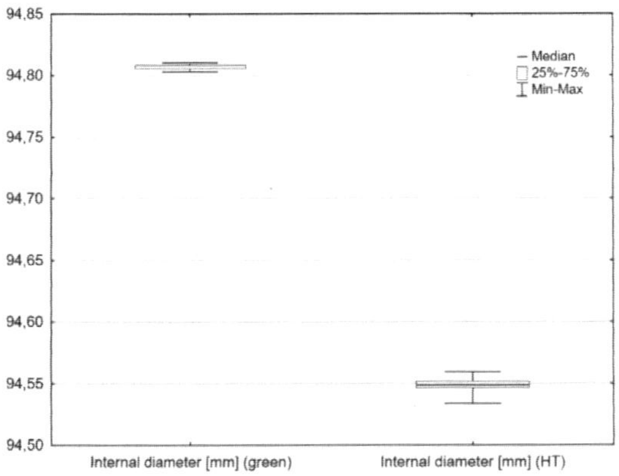

Figure 3. Internal diameters of elements before and after thermo-chemical treatment at 920°C.

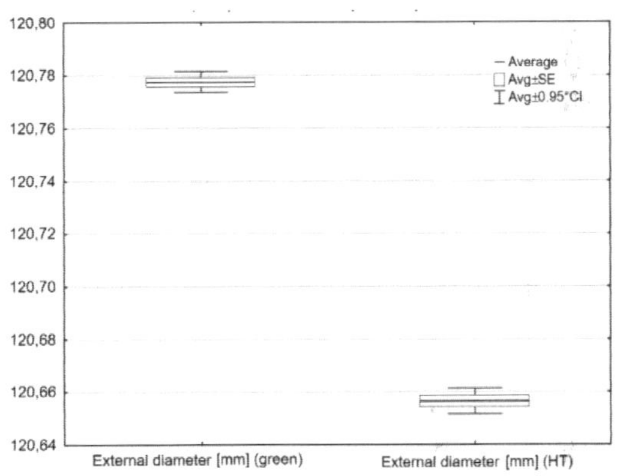

Figure 4. External diameters of elements before and after thermo-chemical treatment at 920°C.

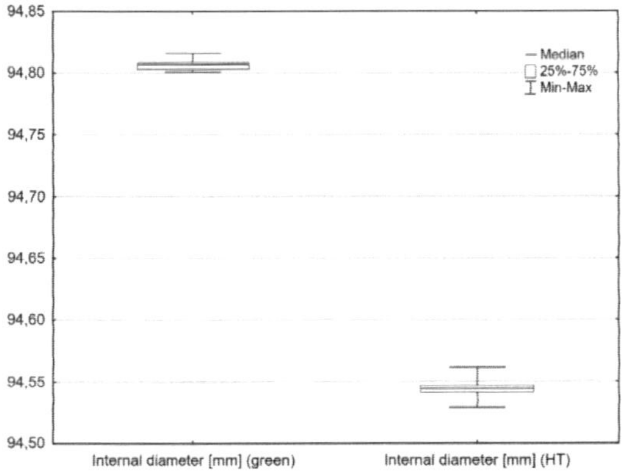

Figure 5. Internal diameters of elements before and after thermo-chemical treatment at 980°C.

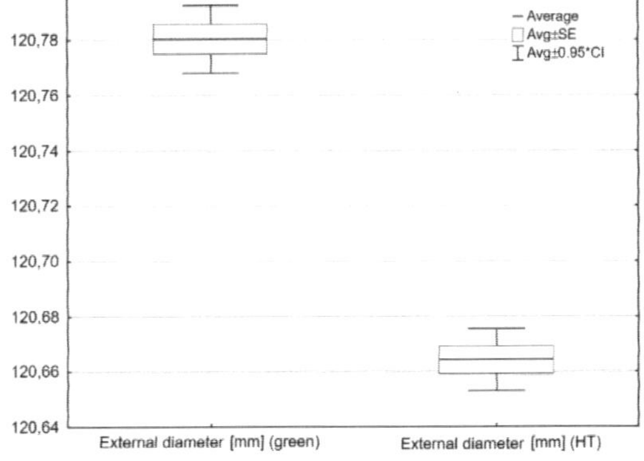

Figure 6. External diameters of elements before and after thermo-chemical treatment at 980°C.

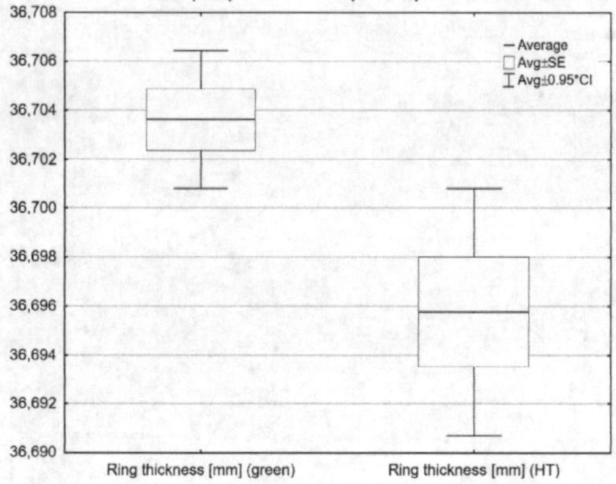

Figure 7. Ring thickness before and after thermo-chemical treatment at 920°C.

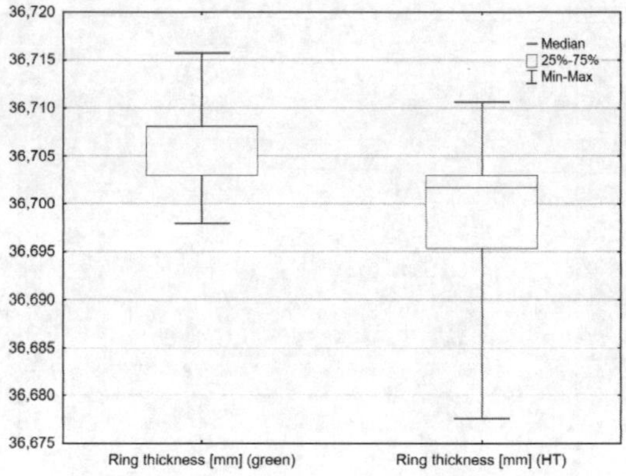

Figure 8. Ring thickness before and after thermo-chemical treatment at 920°C.

Furthermore, using the Tukey test it was demonstrated that the differences in internal diameters are significantly greater for 860°C ($p<0.05$) (Figure 10) and using the Kruskal-Wallis test it was shown that ring thickness was significantly smaller for 980°C, compared to other temperatures ($p< 0.05$) (Figure 11).

4. Discussion

Statistical analysis of the measurements made indicates that statistically significant differences exist between temperature groups for the following parameters: outer diameter and ring thickness. The lower quenching temperature (860°C) results in unfavorable changes in the outer diameter and ring thickness parameters. A high temperature of quenching (980°C) reduces the degree of distortion of the outer diameter and ring thickness parameters. In the opinion of the authors, quenching from high temperatures (usually from quenching temperatures) results in the immediate occurrence of martensitic transition and a "freezing" of the geometrical dimensions on the outer side of the element. Consequently, the "deformation front" is pushed deep inside the element and the inevitable changes in volume, related to transition of the material from austenite to martensite, take place in the inner parts of the element. Quenching from low temperatures (i.e. after the stage of precooling for quenching) delays the moment of martensitic transition in the outer parts in relation to the inner parts of elements. It seems that quenching from lower temperatures is a more favorable process.

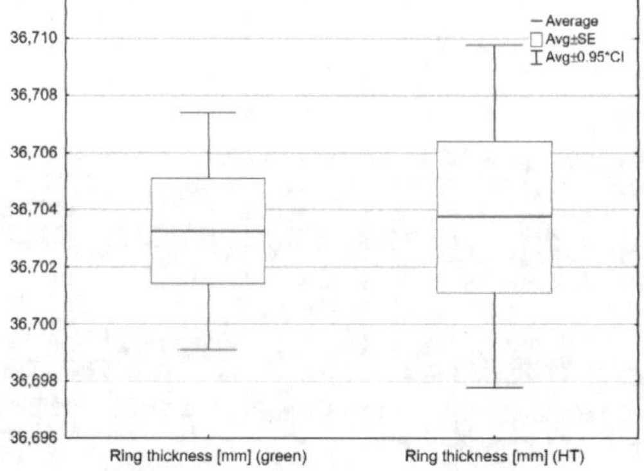

Figure 9. Ring thickness before and after thermo-chemical treatment at 980°C.

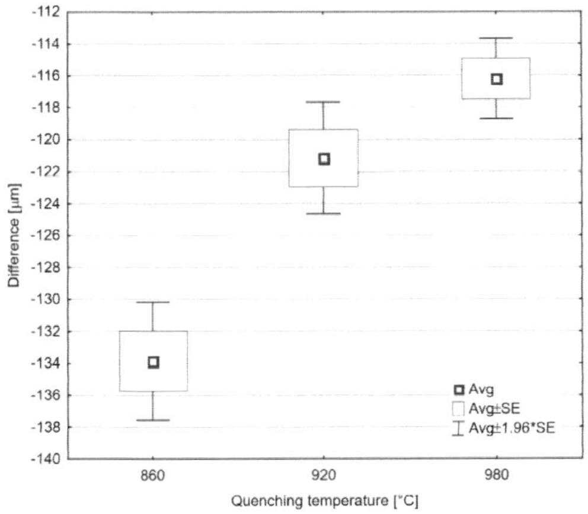

Figure 10. Differences in ring external diameter before and after thermo-chemical treatment depending on quenching temperature.

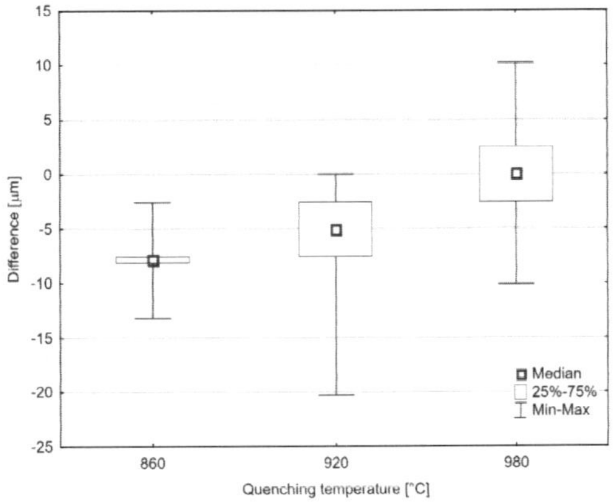

Figure 11. Differences in ring thickness before and after thermo-chemical treatment depending on quenching temperature.

5. Conclusions

Based on the results of the research, the following conclusions were reached:

1. A low temperature of quenching has an unfavorable effect on changing the external diameter and ring thickness parameters.
2. High temperature quenching reduces quenching distortions of external diameter and ring thickness.
3. at the same time, reduces the size of distortions of the outer parameters (outer diameters, length of elements).
4. High temperature quenching does not change the ring thickness parameter before and after heat treatment.

Acknowledgments

This research was funded by the Polish National Centre for Research and Development, grant number POIR.04.01.04-00-0087/15.

References

[1] Dybowski K, Sawicki J, Kula P, Januszewicz B, Atraszkiewicz R and Lipa S 2016 The effect of the quenching method on the deformations size of gear wheels after vacuum carburizing, *Archives of Metallurgy and Materials* **61** 1057–1062.

[2] Dybowski K, Kula P, Atraszkiewicz R, Januszewicz B, Rzepkowski A and Jakubowski K 2014 Wpływ sposobu hartowania na wielkość odkształceń detali wykonanych ze stali 16MnCr5 nawęglanych niskociśnieniowo *Inżynieria Materiałowa* **35** 367–369.

[3] Dybowski K, Niewiedzielski R 2017 Distortion of 16MnCr5 steel parts during low-pressure carburizing *Advances in Science and Technology Research Journal* **11** 201–207.

[4] Stachurski W, Zgórniak P, Sawicki J and Przybysz M 2017 Hardening-related deformations of gear wheels after vacuum carburising and quenching in gas *Advances in Science & Technology - Research Journal* **11** 237–245.

[5] Atraszkiewicz R, Januszewicz B, Kaczmarek Ł, Stachurski W, Dybowski K and Rzepkowski A 2012 High pressure gas quenching: Distortion analysis in gears after heat treatment *Materials Science and Engineering: A* **558** 550–557.

[6] Kim D, Cho H, Lee W, Cho K, Cho Y, Kim S and Han H 2016 A finite element simulation for carburizing heat treatment of automotive gear ring incorporating transformation plasticity *Materials & Design* **99** 243–253.

IOP Conf. Series: Materials Science and Engineering **743** (2020) 012048 doi:10.1088/1757-899X/743/1/012048

[7] Sawicki J, Kruszyński B and Wójcik R 2017 The influence of grinding conditions on the distribution of residual stress in the surface layer of 17CrNi6-6 steel after carburizing *Advances in Science and Technology Research Journal* **11** 17–22.
[8] Lipa S, Sawicki J, Wołowiec-Korecka R, Dybowski K and Kula P 2016 Method of determining the strain hardening of carburized elements in Ansys environment *Solid State Phenomena* **240** 74–80.
[9] Korecki M, Wołowiec-Korecka E, Sut M, Brewka A, Stachurski W and Zgórniak P 2017 Precision case hardening by low pressure carburizing (LPC) for high volume production *HTM Journal of Heat Treatment and Materials* **72** 175–183.
[10] Korecki M, Wołowiec-Korecka E and Brewka A 2016 Unicase Master-In-line, high-volume, low-distortion, precision case hardening for automotive, transmission and bearing industry, in: IFHTSE Conference, Prague.
[11] Wołowiec-Korecka E, Korecki M, Stachurski W, Sawicki J, Brewka A, Sut M and Bazel M 2016 System of single-piece flow case hardening for high volume production *Archives of Materials Science and Engineering*. **79** 37–44.
[12] Krupanek K, Staszczyk A, Sawicki J and Byczkowska P 2018 The impact of nozzle configuration on the heat transfer coefficient *Archives of Materials Science and Engineering.* **90** 16–24.
[13] Korecki M, Fujak W, Olejnik J, Stankiewicz M and Wołowiec-Korecka E 2018 Multi-chamber furnace from vacuum carburizing and quenching of gears, shafts, rings and similar workpieces *US 9,989,311.*
[14] Korecki M, Fujak W, Olejnik J, Stankiewicz M and Wołowiec-Korecka E 2018 Device for individual quench hardening of technical equipment components *US 10,072,315.*
[15] Korecki M, Wołowiec-Korecka E and Glenn D 2016 Single-piece, high-volume, and low-distortion case hardening of gears *Thermal Processing for Gear Solutions* **5** 32–39.

IOP Conf. Series: Materials Science and Engineering **743** (2020) 012049 doi:10.1088/1757-899X/743/1/012049

The influence of grinding wheel type on microhardness and residual stresses in vacuum-carburised 20MnCr5 steel using the single-piece flow method

P Zgórniak[1], E Wołowiec-Korecka[2], B Januszewicz[2], M Sikora[1] and R Rosik[1]

[1]Institute of Machine Tools and Production Engineering, Lodz University of Technology, Lodz, Poland
[2]Institute of Materials Science and Engineering, Lodz University of Technology, Lodz, Poland

Email: piotr.zgorniak@p.lodz.pl

Abstract. The aim of the experiment described in the paper was to determine the effect of selected conditions of abrasive machining on the size and distribution of microhardness and residual stresses developed in the technological surface layer of flat specimens made of 20MnCr5 steel. The specimens were subjected to single-piece flow low-pressure carburising (LPC) and high-pressure gas quenching (HPGQ) in a 4D Quenching chamber in order to achieve an effective case depth of ECD=0.4 mm. This was followed by grinding the specimens with Quantum and Vortex alumina grinding wheels. Cooling and lubricating fluid were supplied to the grinding zone using the flood (WET) method. The samples were ground in one pass of the grinding wheel using one grinding depth (ae = 0.01 mm). The measurements for each specimen were made twice - after the thermo-chemical treatment and after the grinding. Microhardness and residual stress were measured using the X-ray method sin2Ψ. The final part of the article provides an analysis of the measurement results and presents conclusions and recommendations for further studies.

1. Introduction

The properties of the top layer formed during grinding have a direct impact on functional properties of the item processed such as fatigue strength, resistance to abrasion and corrosion etc. [1]. The important parameters describing the condition of the technological top layer include microhardness and residual stresses. The course of microhardness and residual stresses is influenced by, among other things, the type of heat treatment that precedes grinding and the characteristics of the grinding wheel.

One of the most frequently used methods of surface heat treatment is carburising with subsequent quenching [2]. The low-pressure variety of carburising [3,4] outperforms conventional carburising [5-9] in terms of efficiency and is characterised by a number of advantages such as a lack of internal oxidation and higher uniformity of the layers obtained.

Residual stresses occurring in the heat and chemical treatment procedures exist in both the substrate and the surface layer [10]. The analysis of the condition of residual stresses is of major importance in view of their impact on fatigue strength, tribological wear, corrosion, brittle fracture and contact fatigue [11-14]. In the event of taking mechanical properties into consideration, this impact may be beneficial but may also lead to damage to the item or the whole device, depending on the type of stresses and their superposition with operation stresses origination from external excitations. According to the literature

Content from this work may be used under the terms of the Creative Commons Attribution 3.0 licence. Any further distribution of this work must maintain attribution to the author(s) and the title of the work, journal citation and DOI.
Published under licence by IOP Publishing Ltd

review, creating compressive stresses in the top layer (TL) that are compensated with tensile stresses in the core may contribute to increased fatigue strength [15-19].

In the case of grinding with the application of grinding wheels with Al_2O_3 abrasive grains, the grinding power is increased as a result of increased efficiency, which leads to an increased grinding temperature in the item processed [1]. Its course is the main cause of changes in microhardness and residual stresses in comparison to the material after heat treatment. The increased heat load of the top layer causes the occurrence of unfavourable residual tensile stresses that reduce the fatigue strength of dynamically loaded parts of machines and also a reduction of microhardness deep in the technological top layer. At the same time, it has to be underlined that the risk of unfavourable heat impact on the top layer is lower in the case of improving the efficiency of the cooling lubricant reaching the zone of contact between the active abrasive grains and the surface ground [20,21].

The objective of the experimental studies described in this article was establishing the impact of changes in grinding conditions, such as the grinding wheel type, on the value and distribution of microhardness and residual stresses occurring in the technological top layer of flat samples made of 20MnCr5 steel. To this end, the samples were first subjected to low- pressure carburising (LPC) using the single-piece flow method and then to high-pressure gas quenching (HPGQ). Finally, the samples were ground with aloxite grinding wheels of the Quantum and Vortex type. During grinding, the liquid cooling lubricant – in the form of oil emulsion – was supplied to the processing zone using the conventional submersion method.

2. Single-piece flow low-pressure carburising and quenching

For the purposes of conducting heat and chemical treatment, a UCM low-pressure furnace of the SECO/WARWICK Company (Poland), shown in figure 1, was used.

a) b)

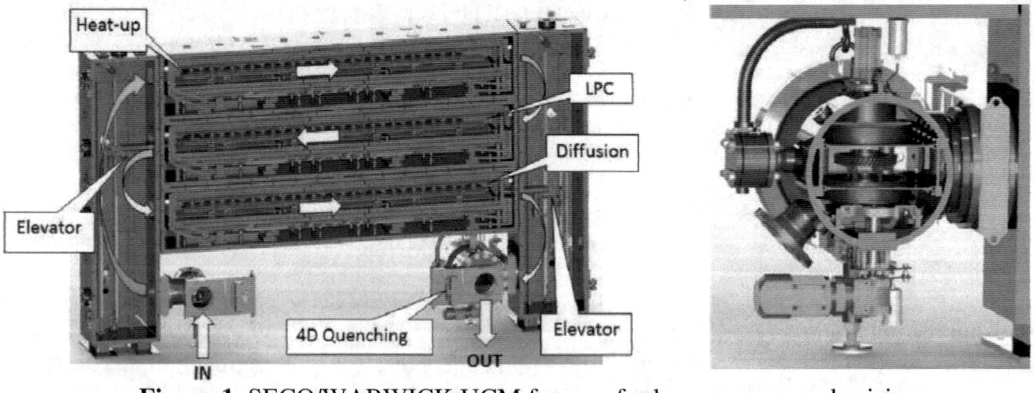

Figure 1. SECO/WARWICK UCM furnace for low-pressure carburising: a) general view, b) quenching chamber [22].

It is an innovative device in which the heat and chemical treatment is carried out using the single-piece flow method. In this method, each piece passes individually through identical positions and process conditions prevalent in the furnace [22-24]. As a result, carburising of this type is characterised by very high precision and reproducibility in comparison to conventional methods. Additionally, the application of high-pressure gas quenching (HPGQ) – in a quenching chamber of the 4D Quenching type for individual gas – cooling of each piece enables free shaping of the colling curve and achieving optimal microstructure and properties of steel. An important characteristic of this solution is the application of a system of cooling nozzles surrounding the piece and ensuring a uniform inflow of cooling gas from all directions (3D). At the same time, the uniformity of cooling is supported by a table that rotates together with the piece (4D). Such a cooling system enables the obtainment of a cooling intensity comparable to that of oil systems without having to use helium (He).

For the purposes of experimental studies, four flat samples with dimensions of 100x100x10 made of 20MnCr5 steel were selected. The samples' dimensions resulted from the structure of parts of the mechanism transporting them within the UCM furnace. The samples were carburised at a temperature of 920°C, achieving the effective layer thickness of ECD=0.4 mm. Next, the samples were quenched in a quenching chamber at the pressure of 7 bar and then tempered at a temperature of 190°C for 3 hours.

3. Grinding

Further examinations of samples previously subject to the heat and chemical treatment process using the single-piece flow method in a UCM furnace were carried out in the process of circumferential grinding of planes. To that end, a type SPD-30B conventional grinder for flat surfaces of the Jotes SA Company (Poland) was used. During the studies, two samples made of 20MnCr5 steel (52±1 HRC) were being ground. Two grinding wheels of the Norton Company (Poland) were used as the tool – 2NQ60JVS3 and IPA60EH20VTX. The 2NQ60JVS3 grinding wheel of the Quantum type is a grinding wheel with a ceramic binder – the volume of which consists of 20% Norton Quantum abrasive grains and 80% alundum grains. It is a soft grinding wheel (J) with a closed structure. The IPA60EH20VTX grinding wheel is a Vortex type grinding wheel made of aloxite abrasive grains with a ceramic binder (VTX). It is a hard grinding wheel (EH) with an open structure and increased porosity (also referred to as a large-pore grinding wheel). Both grinding wheels feature the same size of abrasive grains, the grain number of which is 60. The grinding wheels were conditioned prior to every test with the use of a single-point diamond dresser of the M1020 type.

The machining allowance was removed in a single work cycle consisting of an overtravel and return travel (the same and opposite rotational direction) using a constant grinding depth of a_e=0.01 mm. For the purposes of tests, a constant value of grinding wheel peripheral speed of v_s=25.6 m/s and machined item speed of v_w=18 m/min were also assumed. Table 1 features a comprehensive summary of the machining conditions applied during grinding.

Table 1. Grinding conditions.

Grinding mode	Single-pass longitudinal circumferential surface grinding
Grinding machine	Flat-surface grinder SPD-30B by Jotes Co. Ltd. (Poland)
Workpiece material	20MnCr5, carburized and hardened with 61±1 HRC
Grinding wheels	2NQ60JVS3 and IPA60EH20VTX
Grinding wheel rotational speed	n_s = 1400 rpm
Grinding wheel peripheral speed	v_s = 25.6 m/s
Workpiece peripheral speed	v_w = 18 m/min
Working engagement	a_e = 0.01 mm
Dresser	Single grain diamond dresser type M1020
Dresser weight	Q_d = 2.0 kt
Grinding wheel peripheral speed while dressing	v_{sd} = 10 m/s
Dressing allowance	a_d = 0.01 mm
Axial table feed speed while dressing	v_{fd} = 5.0 mm/min
Number of dressing passes	i_d = 4
Conventional grinding fluid (GF)	Emulgol ES-12 in a 5% concentration
Conventional GF flow rate	Q_{GF} = 4 l/min

The samples were ground with the use of a cooling lubricant supplied using the submersion method. An oil-in-water emulsion employing the Emulgol ES-12 (5%) oil was used as a conventional machining liquid in the submersion method and was supplied to the grinding zone with an expenditure of Q_{GF} = 4 l/min.

4. Measurements of microhardness and residual stress by the X-ray method

The measurements of sample surface microhardness after grinding carried out using the KB10BVZ-FA microhardness meter of the KB Prüftechnik GmbH Company (Germany). Microhardness was determined using the Vickers scale at a load of 0.9807 N, in compliance with the PN-EN ISO 6507 standard. The measurements were carried out on cuts perpendicular to the surface ground, down to a depth of 0.3 mm. Three microhardness measurements were performed for each sample. The mean measurement results obtained were subjected to interpolation using cubic B-spline functions.

The measurements of residual stresses in the samples ground were carried out using the sin2ψ X-ray method in ω geometry using a PROTO iXRD apparatus fitted with two position-sensitive semiconductor detectors. The source of X-radiation was a Cr anode tube emitting characteristic X-radiation with a wavelength of λ=2.29 A. The change in position reflection (211) of iron, positioned at an angle of 2θ=156.4°, was examined. For calculation purposes, there were adopted X-ray elastic constants of ½ S2=5.92 1/TPa and S1=-1.27 1/TPa. The measurement was carried out for an area limited by a collimator with a diameter of φ=2 mm. The exposure time was 1s. In order to obtain stress distributions deep in the substrates examined, electrochemical point etching by means of an 8818-V3 electropolishing unit of the PROTO Company was used. A stress measurement was performed after each etching.

5. Results and discussion

5.1. Microhardness

The microhardness tests conducted after grinding showed that the smallest changes in it (by about 340 HV near the surface) in comparison to microhardness of the material prior to grinding are obtained during machining with a grinding wheel of the Vortex type (figure 2a). In case of grinding with a Quantum type grinding wheel (figure 2b) the reduction of microhardness at the surface, in comparison to the material prior to grinding, was greater and amounted to approximately 420 HV.

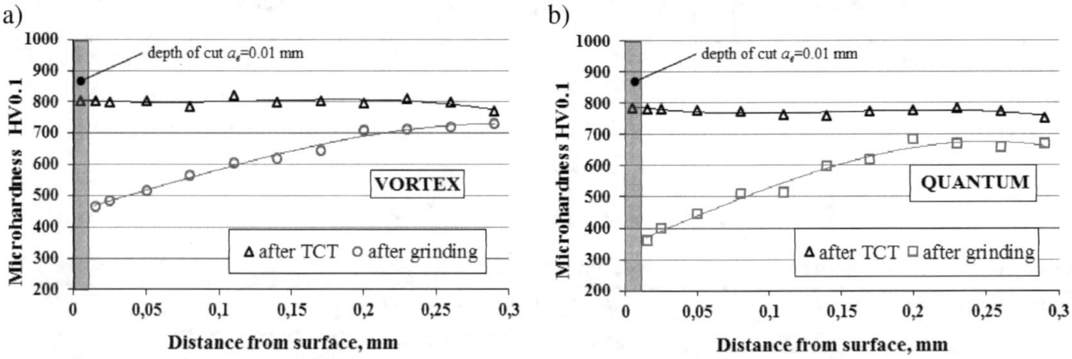

Figure 2. Distribution of microhardness in 20MnCr5 steel ground with aloxite grinding wheels: a) IPA60EH20VTX (Vortex) b) 2NQ60JVS3 (Quantum).

It is worth noting that, as shown in figure 3, the microhardness at the surface obtained for the sample ground with the Quantum type grinding wheels differs by about 105 HV from the value obtained for the Quantum grinding wheel. Additionally, the comparison of microhardness distributions for samples ground with these two grinding wheels indicates that similar values of microhardness were obtained in a distance of about 0.15 mm from the surface.

5.2. Residual stress

Figure 4 presents the results which are the mean of 3 measurements of residual stresses in samples after the thermo-chemical treatment (TCT) process (prior to grinding). It can be noticed that the value of residual stresses on the surface of samples carburised in low pressure amounted to -257 MPa. These stresses were increasing monotonically with distance from the surface, achieving a value of -498 MPa at a depth of 0.3 mm.

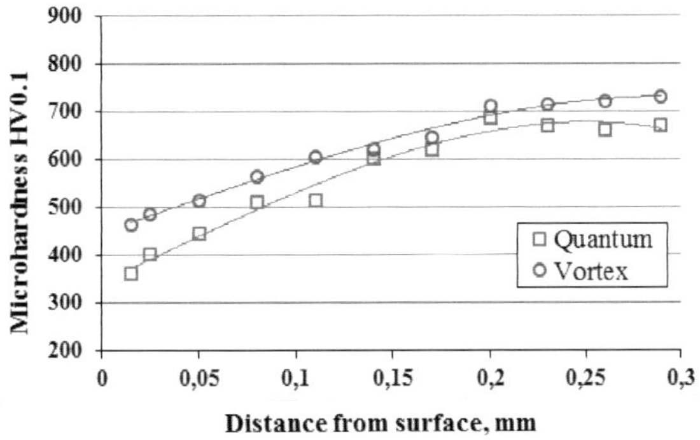

Figure 3. Comparison of microhardness distribution in 20MnCr5 steel ground with aloxite grinding wheels: 2NQ60JVS3 (Quantum) and IPA60EH20VTX (Vortex).

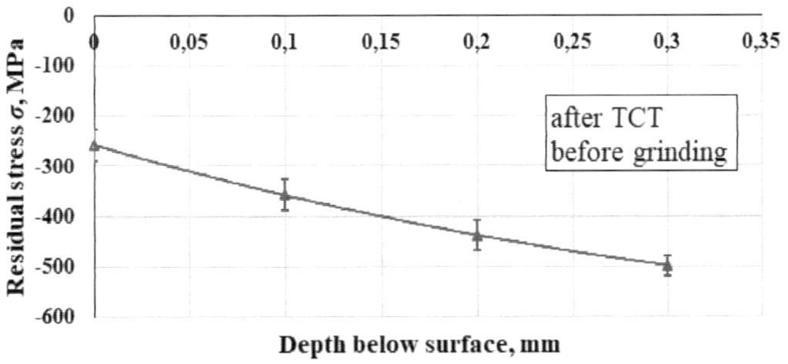

Figure 4. Residual stresses in 20MnCr5 steel – after TCT, prior to grinding.

Figure 5 presents the distribution of residual stresses for samples carburised in low pressure and then ground with an aloxite grinding wheel of the Vortex type and an aloxite grinding wheel of the Quantum type.

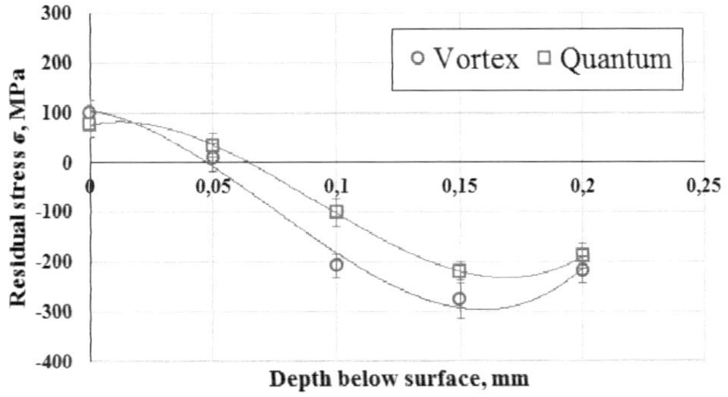

Figure 5. Residual stresses in 20MnCr5 steel after grinding with aloxite grinding wheels: 2NQ60JVS3 (Quantum) and IPA60EH20VTX (Vortex).

After comparing figure 4 to figure 5, it is clear that grinding with aloxite grinding wheels results in a worsened condition of residual stresses in comparison to the material after heat and chemical treatment (prior to grinding). In both cases, unfavourable tensioning residual stresses were obtained right under the surface. This is caused by a large amount of heat flowing to the item and by relatively high grinding temperatures, which cause unfavourable structural changes (among other things, the steel tempering process).

The most favourable condition of residual stresses was obtained in the sample ground with the IPA60EH20VTX grinding wheel of the Vortex type. In this case, the residual stresses change from tensioning to beneficial compressing ones at about 0.05 mm from the surface, while for the Quantum grinding wheel this distance is about 0.065 mm. It has to be noted that for the Vortex grinding wheel the residual stresses obtained right under the surface were greater than in the case of the Quantum grinding wheel. However, this difference is little and amounts to about 25 HV units. In the opinion of the authors, the above property results from the increased porosity of the grinding wheel structure, characterised by large intergrain spaces in comparison to the Quantum grinding wheel. As a result of that, the zone of contact between the active abrasive grains and the material processed received a larger amount of lubricant, which reduces friction and leads to a reduction in grinding temperature. The reduction in temperature is also caused by a lower number of active abrasive grains in comparison to the Quantum grinding wheel and, in consequence, by lower friction in the grinding zone. As a result, a lower grinding temperature has a more beneficial impact on the condition of stresses caused in the top layer of the steel ground.

6. Conclusion

On the basis of the results obtained within the scope of test conditions applied, it can be stated that:

- the low pressure carburising process conducted using the ‘single-piece flow’ method enables achieving a beneficial (i.e. compressing) distribution of residual stresses in the technological top layer
- the process of grinding with aloxite grinding wheels contributes to worsening of the condition of residual stresses in comparison to the material after low-pressure carburising treatment by causing adverse residual tensioning stresses near the surface in the technological top layer
- the best effect of grinding in relation to the distribution of microhardness and residual stresses are obtained while grinding with open structure grinding wheel characterised by increased porosity (IPA60EH20VTX grinding wheel of the Vortex type)

Further research will focus on the selection of treatment conditions with the aim of increasing its efficiency while retaining the desired properties of the top layer. The authors also plan to expand the research by performing measurements of grinding force and surface roughness.

Acknowledgments

This research was funded by the Polish National Centre for Research and Development, grant number POIR.04.01.04-00-0087/15.

References

[1] Kruszyński B W and Wójcik R 2001 Residual stress in grinding *J. Mater. Process. Technol.* **109** 254-257.
[2] Korecki M, Wołowiec-Korecka E, Sut M, Brewka A, Stachurski W and Zgórniak P 2017 Precision case hardening by low pressure carburizing (LPC) for high volume production *HTM J. Heat Treatm. Mat.* **72/3** 175-183.
[3] Gräfen W and Edenhofer B 2005 New developments in thermo-chemical diffusion processes *Surf. Coat. Technol.* **200/5-6** 1830-1836.
[4] Kula P, Dybowski K, Wołowiec E and Pietrasik R 2014 “Boost-diffusion” vacuum carburising – Process optimisation *Vacuum* **99** 175-179.
[5] Atraszkiewicz R, Januszewicz B, Kaczmarek Ł, Stachurski W, Dybowski K and Rzepkowski A 2012 High pressure gas quenching: Distortion analysis in gears after heat treatment *Mater. Sci. Eng. A* **558** 550-557.

[6] Dybowski K, Sawicki J, Kula P, Januszewicz B, Atraszkiewicz R and Lipa S 2016 The effect of the quenching method on the deformations size of gear wheels after vacuum carburizing *Arch. Metall. Mater.* **61/2B** 1057-1062.
[7] Gawroński Z, Malasiński A and Sawicki J 2010 Elimination of galvanic copper plating process used in hardening of conventionally carburized gear wheels *Int. J. Automot. Technol.* **11/1** 127-131.
[8] Gawroński Z, Malasiński A and Sawicki J 2010 A selection of the protective atmosphere eliminating the inter-operational copper plating step in the processing of gear wheels *Arch. Mater. Sci. Eng.* **44/1** 51-57.
[9] Sawicki J, Kruszyński B and Wójcik R 2017 The influence of grinding conditions on the distribution of residual stress in the surface layer of 17CrNi6-6 steel after carburizing *Adv. Sci. Technol. Res. J.* **11(2)** 17-22.
[10] Stachurski W, Krupanek K, Januszewicz B, Rosik R and Wójcik R 2018 An effect of grinding on microhardness and residual stress in 20MnCr5 following single-piece flow low-pressure carburizing *J. Mach. Eng.* **18/4** 73-85.
[11] Kruszyński B, Gawroński Z, Sawicki J and Zgórniak P 2008 Enhancement of gears fatigue properties by modern termo-chemical treatment and grinding processes *Mech. Mech. Eng.* **12/4** 387-395.
[12] Lipa S, Sawicki J, Dybowski K, Pietrasik R and Januszewicz B 2018 The effect of non-metallic inclusions formed as a result of deoxidation on the fatigue strength of 15CrNi6 and 16MnCr5 steel *Arch. Metall. Mater.* **63/3** 1345-1350.
[13] Sawicki J, Górecki M, Kaczmarek Ł, Gawroński Z, Dybowski K, Pietrasik R and Pawlak W 2013 Increasing the durability of pressure dies by modern surface treatment methods *Chiang Mai J. Sci.* **40(5)** 886-897.
[14] Webster G A 2000 Role of residual stress in engineering applications *Mater. Sci. Forum* **347-349** 1-11.
[15] Gawroński Z and Sawicki J 2006 Technological surface layer selection for small module pitches of gear wheels working under cyclic contact loads *Mater. Sci. Forum* **513** 69-74.
[16] Lipa S, Sawicki J, Wołowiec E, Dybowski K and Kula P 2016 Method of determining the strain hardening of carburized elements in Ansys environment *Solid State Phenomena* **240** 74-80.
[17] Gawroński Z and Sawicki J 2000 Toothed wheel optimization by means of the finite element analysis *Mech. Mech. Eng.* **4(2)** 183-189.
[18] Sawicki J, Siedlaczek P and Staszczyk A 2018 Finite-element analysis of residual stresses generated under nitriding process: a three-dimensional model *Met. Sci. Heat Treat.* **59(11-12)** 799-804.
[19] Sawicki J, Siedlaczek P and Staszczyk A 2018 *F*atigue life predicting for nitrided steel – finite element analysis *Arch. Metall. Mater.* **63/2** 917-923.
[20] Stachurski W, Sawicki J, Wójcik R and Nadolny K 2018 Influence of application of hybrid MQL-CCA method of applying coolant during hob cutter sharpening on cutting blade surface condition *J. Clean. Prod.* **171** 892-910.
[21] Stachurski W and Nadolny K 2018 Influence of the condition of the surface layer of a hob cutter sharpened using the MQL-CCA hybrid method of coolant provision on its operational wear *Int. J. Adv. Manuf. Technol.* **98** 2185-2200.
[22] Korecki M, Wołowiec-Korecka E and Glenn D 2015 Single-Piece, High-Volume, Low-Distortion Case Hardening of Gears *AGMA Technical Paper* **15FTM04.**
[23] Wołowiec-Korecka E, Korecki M, Stachurski W, Zgórniak P, Sawicki J, Brewka A, Sut M and Bazel M 2016 System of single-piece flow case hardening for high volume production *Arch. Mater. Sci. Eng.* **79(1)** 37-44.
[24] Stachurski W, Zgórniak P, Sawicki J and Przybysz M 2017 Hardening-related deformations of gear wheels after vacuum carburising and quenching in a 4D quenching chamber *Adv. Sci. Technol. Res. J.* **11(1)** 237-245.

The influence of grinding conditions on surface roughness parameters of 20MnCr5 steel vacuum-carburised by single-piece flow method

M Sikora[1], P Zgórniak[1] and E Wołowiec-Korecka[2]

[1]Institute of Machine Tools and Production Engineering, Lodz University of Technology, Lodz, Poland
[2]Institute of Materials Science and Engineering, Lodz University of Technology, Lodz, Poland

Email: malgorzata.sikora@p.lodz.pl

Abstract. The purpose of the experimental tests described in the article was to determine the effect of selected abrasive machining conditions on the value of surface roughness of 20MnCr5 steel samples. The samples were subjected to vacuum carburising (LPC) using the 'single-piece flow' method and high-pressure gas quenching (HPGQ) in a 4D Quenching chamber to obtain an effective case depth (ECD) of 0.4 mm. Next, the samples were ground with a Norton Vortex corundum grinding wheel. Cooling and lubricating fluid was delivered to the grinding zone using the flooding method (WET) and with the minimum quantity lubrication method (MQL). The samples were ground in one pass of the grinding wheel using two grinding depths – a_e = 0.01 mm and a_e = 0.02 mm. For each grinding test, surface roughness measurements were made using a stationary laboratory profilometer T8000 from Hommel Werke. These measurements enabled the determination of the 2D and 3D roughness parameters. Analysis of the measurement results showed that for the adopted test conditions, the use of the MQL method resulted in a decrease in surface roughness compared to the surface ground using the WET method. This property applies to both tested grinding depths.

1. Introduction
An important factor that has an impact on the resulting parameters of the grinding process is the grinding temperature. The most efficient method of lowering it is by applying a liquid cooling lubricant (GF) [1,2]. From among many methods of supplying it to the grinding zone, the most popular is the submersion method (WET) employing oil-in-water emulsions [3,4]. This method is still considered the most effective way of reducing high temperature in the grinding zone by reducing friction as a result of lubricating and cooling the zone of contact between the grinding wheel and the material processed. Another advantage of this method is flushing the shavings from the grinding zone as well as wetting and cleaning the grinding wheel.

The main downside of the submersion method is the high expenditures of cooling lubricant (GF) and the fact that only a small portion of its volume reaches the zone of contact between the grinding wheel and the material processed. It is estimated [5,6] that the cost of applying cooling lubricants constitutes up to 17% of the total production costs and are markedly higher than the tool costs (2-4%). Thus, endeavours are made to reduce cooling lubricant (GF) expenditure or eliminate it totally due to economic reasons. Additionally, such limiting is also motivated by environmental issues and the

Content from this work may be used under the terms of the Creative Commons Attribution 3.0 licence. Any further distribution of this work must maintain attribution to the author(s) and the title of the work, journal citation and DOI.
Published under licence by IOP Publishing Ltd

necessity to adjust to increasingly rigorous regulations related to the protection of the environment and OHS [7].

Taking into account the above-listed environmental and technical obstacles related to the application of conventional cooling lubricants applied using the submersion method, attempts are made at eliminating them by applying alternative methods of lubrication and cooling.

The most frequently employed method for significantly reducing the amount of cooling lubricant supplied to the processing zone and ensuring effective lubrication of the machining zone is the application of the minimum quantity lubrication (MQL) method [7-10]. The MQL method consists of the continuous production of oil mist and delivering it directly to the grinding zone, usually onto the grinding wheel active surface (GWAS). The flow of the lubricant takes place with the participation of a transporting agent – a stream of compressed air, also performing the function of a coolant. The data in the literature indicated the in the MQL method the lubricating agent is supplied in the amount of 20-100 ml/h [1,7,8,11]. To compare, oil-in-water emulsion in the submersion method is used in amounts of over 120,000 ml/h whereas during grinding this amount varies from 300,000 to 1,200,000 ml/h, depending on the process variety. The cooling lubricants applied most often are synthetic esters or fatty alcohols [9]. Plant oils have also been used for some time as cooling lubricants in the MQL method [1,12].

An important aspect of the application of the MQL method during grinding is the method of positioning of the oil mist spraying nozzle, which has an impact on the effectiveness of this method. It has to be underlined here that a similar relationship exists with regard to the positioning of the nozzle in the submersion method (WET) [13]. The best results, similar to those provided by the submersion method, were achieved when the nozzle was positioned as close as possible to the zone of contact between the grinding wheel and the item processed [14]. However, the method of angular positioning of the nozzle in relation to the grinding wheel should be considered individually for each case of grinding [8].

The aim of the experimental studies described in this article was to establish the impact of the employed method of supplying cooling lubricant to the grinding zone on the value of surface roughness after grinding. Two methods were selected for comparison – the submersion method (WET) and the minimum quantity lubrication (MQL) method. Two machining depths were used during the research. The samples ground were previously subjected to low-pressure carburising (LPC) using the single-piece flow method and then to high-pressure gas quenching (HPGQ). It is necessary to underline here that carburising with subsequent quenching is one of the most frequently used methods of surface heat treatment [15]. The low-pressure variety of carburising [16,17] outperforms conventional carburising [11,18-21] in terms of efficiency and is characterised by a number of advantages such as, among other things, a lack of internal oxidation and higher uniformity of the layers obtained.

2. Single-piece flow low-pressure carburising and quenching

For the purposes of conducting heat and chemical treatment, a UCM low-pressure furnace from the SECO/WARWICK Company (Poland) was used [22]. It is an innovative device in which the heat and chemical treatment is carried out using the single-piece flow method. In this method, each piece passes individually through identical positions and process conditions prevalent in the furnace [15,23-25]. As a result, carburising of this type is characterised by very high precision and reproducibility in comparison to conventional methods. Additionally, the application of high-pressure gas quenching (HPGQ) in a quenching chamber of the 4D Quenching type for individual gas cooling of each piece enables free shaping of the cooling curve and achieving optimal microstructure and properties of steel. An important characteristic of this solution is the application of a system of cooling nozzles surrounding the piece and ensuring a uniform inflow of cooling gas from all directions (3D). At the same time, the uniformity of cooling is supported by a table that rotates together with the piece (4D). Such a cooling system enables a cooling intensity comparable to that of oil systems to be obtained without having to use helium (He).

For the purposes of experimental studies, four flat samples with dimensions of 100x100x10, made of 20MnCr5 steel, were selected. The samples' dimensions resulted from the structure of parts of the mechanism transporting them within the UCM furnace. The samples were carburised at a temperature of 920°C, achieving the effective layer thickness of ECD=0.4 mm. Next, the samples were quenched in a quenching chamber at a pressure of 7 bar and then tempered at a temperature of 190ºC for 3 hours. Table 1 summarises the parameter of heat and chemical treatment.

Table 1. Thermo-chemical treatment parameters.

Number of samples	Parameters of treatment		
V_WET-1 V_WET-2 V_MQL-1 V_MQL-2	Vacuum carburizing at 920°C	Quenching in nitrogen at 7 bar and at a temperature of 850 ºC	Tempering at 190 ºC for 180 min

3. Grinding

Further examinations of samples previously subjected to the heat and chemical treatment process using the single-piece flow method in a UCM furnace were carried out in the process of circumferential grinding of planes. To that end, a type SPD-30B conventional grinder for flat surfaces from the Jotes SA Company (Poland) was used. During the studies, four samples made of 20MnCr5 steel (52±1 HRC) were ground. An IPA60EH20VTX grinding wheel from the Norton Company (Poland) was used as the tool. The IPA60EH20VTX grinding wheel is a Vortex type of grinding wheel made of aloxite abrasive grains with a ceramic binder (VTX). It is a hard grinding wheel (EH) with an open structure, increased porosity (also referred to as large-pore grinding wheel) and abrasive grain granulation with a grain size number of 60. The grinding wheel was conditioned prior to every test with the use of a single-point diamond dresser of the M1020 type.

The machining allowance was removed in ten work cycles. A single work cycle consisted of an overtravel and return travel. Two grinding depths were used for removing the material during overtravel: a_{e1}=0.01 mm and a_{e2}=0.02 mm. For the purposes of tests, a constant value of grinding wheel peripheral speed of v_s=25.6 m/s and machined item speed of v_w=18 m/min were also assumed. Table 2 features a comprehensive summary of the machining conditions applied during grinding.

Table 2. Grinding conditions.

Grinding mode	Longitudinal circumferential surface grinding
Grinding machine	Flat-surface grinder SPD-30B by Jotes Co. Ltd. (Poland)
Workpiece material	20MnCr5, carburized and hardened with 61±1 HRC
Grinding wheels	IPA60EH20VTX (Vortex type)
Grinding wheel rotational speed	n_s = 1400 rpm
Grinding wheel peripheral speed	v_s = 25.6 m/s
Workpiece peripheral speed	v_w = 18 m/min
Working engagement	a_{e1} = 0.01 mm; a_{e2} = 0.02 mm
Dresser	Single grain diamond dresser type M1020 (2.0 kt)
Dressing allowance	a_d = 0.01 mm
Environments	WET – conventional fluid MQL – minimum quantity lubrication
Conventional grinding fluid (GF)	Emulgol ES-12 in a 5% concentration
Conventional GF flow rate	Q_{GF} = 4 l/min
MQL system	Ecolubric MQL Booster - oil-mist generator with single external nozzle
MQL fluid	Ecolubric E200L – cold-pressed rapeseed oil without additives
MQL flow rate	Q_{MQL} = 100 ml/h
MQL supply air pressure	P = 0.6 MPa

The samples were ground with the use of a cooling lubricant supplied using the submersion method (WET) and the minimum quantity lubrication (MQL) method. An oil-in-water emulsion employing the Emulgol ES-12 (5%) oil was used as a conventional machining liquid in the submersion method and was supplied to the grinding zone with an expenditure of Q_{GF} = 4 l/min. For creating the oil mist in the MQL method, an external Ecolubric MQL Booster device from the Accu-Svenska AB Company (Sweden) was used. Ecolubric E200L rapeseed oil supplied at an expenditure rate of Q_{MQL} = 100 ml/h was used as the cooling lubricant in the MQL method. During the studies, a single spray nozzle positioned tangentially in relation to the active grinding wheel surface was used.

Table 3 shows the set of variable processing conditions employed in the studies described.

Table 3. Variable conditions of grinding employed in the studies.

Number of samples	Environments	Working engagement
V_WET-1	WET	a_{e1} = 0.01 mm
V_MQL-1	MQL	
V_WET-2	WET	a_{e2} = 0.02 mm

4. Surface roughness measurements

The sample surface roughness measurements after grinding were carried out using a Hommel Tester T8000 profilometer from the Hommelwerke Company (Germany). The measurement conditions were established on the basis of PN-EN ISO 3274:2011E and PN-EN ISO 4288:2011E standards and summarised in Table 4. A 2D parameter, determined from the roughness profile, and a 3D parameter – obtained as a result of surface topography measurements – were used for describing the roughness of the surface ground. The value of the maximum height of roughness profile *Rz* was employed as a 2D parameter (according to PN-EN ISO 4287:1999). The roughness of the surface of a single sample was determined as an arithmetic mean from three measurements. The amplitude parameter *Sz* was used as a 3D parameter.

Table 4. Surface roughness – measuring conditions.

Profilometer	Hommel Tester T8000 by Hommelwerke company (Germany)
Stylus type	TKU 300
Stylus tip radius	r_{tip} = 2 μm
Traverse width	l = 4 mm
Traverse length	lt = 4.8 mm
Evaluation length	ln = 4.0 mm
Sampling length	lr = 0.8 mm
Traverse speed	v_t = 0.05 mm/s
Cut-off	λc = 0.8 mm
Measuring ranges	±80 μm

5. Results and discussion

Figure 1 presents the results of surface roughness measurements for samples after grinding with the use of two different liquid cooling lubricant supply methods – WET and MQL.

On the basis of *Rz* and *Sz* roughness parameter graphs presented in figure 1, it can be stated that, for both methods of cooling and lubricating, the roughness increases together with increasing grinding depth, which results from the geometrical correlation between the size of a single abrasive grain and the grinding depth.

The lowest values of the *Rz* parameter (figure 1a) were measured on the surface of the sample ground with Emulgol ES-12 emulsion supplied using the submersion method (WET) at the grinding

IOP Conf. Series: Materials Science and Engineering **743** (2020) 012050 doi:10.1088/1757-899X/743/1/012050

depth of a_{e1} = 0.01 mm. In the case of the application of the MQL method, the surface roughness obtained for the same grinding depth was higher by about 7% in comparison to the WET method. A similar dependence between the *Rz* parameter value and the liquid cooling lubricant supply method was noticed for the grinding depth of a_{e2} = 0.02 mm. In this case, the value of *Rz* for the sample ground using the MQL method is only about 5% higher in comparison to that obtained for the submersion method (WET).

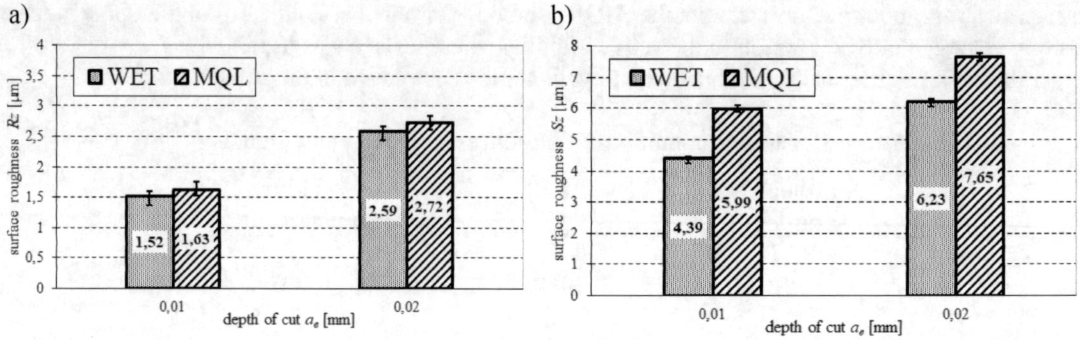

Figure 1. Surface roughness parameters: a) *Rz*, b) *Sz*.

The above observations can be explained by the worse efficiency of the liquid cooling lubricant supply to the grinding zone in the MQL method. The increased friction in the grinding zone, combined with insufficient cooling thereof characteristic for the MQL method, resulted in a temperature in the top layer of the item processed, resulting from heat accumulated in the item processed after subsequent passes of the grinding wheel, tempering the surface of that item. As a result, the phenomenon of fissure formation, the appearance of side burrs with significant height and rubbing of the tempered surface with grinding products that were not removed from the grinding zone. An additional probable cause of an increase in surface roughness could be the appearance of the grinding wheel agglutination phenomenon. Additionally, due to technical limitations in spray nozzle positioning in the MQL method, the oil mist stream does not enable the correct removal of shavings and other products of grinding from the processing zone. Shavings stick to the surface tempered, which leads to surface rubbing that increases the roughness of the ground sample surface. It is also necessary to remember that the effectiveness of lubrication depends, to a large extent on – among other things – the distance of the MQL nozzle from the item processed. As a result of the nozzle position applied, a part of the air-oil aerosol stream could undergo dispersion and end beyond the zone of contact between the grinding wheel and the material processed thus reducing lubrication effectiveness, which leads to increased friction in the grinding zone and worsening the shavings removal effectiveness. It has to be underlined here that the outlet of the MQL nozzle was located as close as possible to the grinding zone and further reduction of this distance was impossible due to technical considerations.

Regarding the *Sz* parameter (figure 1b), which due to lack of vulnerability to impact of individual accidental elevations and cavities has high generalising capabilities, its lowest value was noticed, similarly to the case of the *Rz* parameter, for the surface ground with the use of submersion method (WET) of supplying liquid cooling lubricant during grinding. In the case of grinding a sample to the depth of a_{e1} = 0.01 mm and with the use of the MQL method, the *Sz* value was 36% higher in comparison to the WET method. For the depth of a_{e1} = 0.01 mm this difference amounted to 23%.

When comparing the values of *Rz* and *Sz* parameters, it is noticeable that the values of the *Sz* parameter for all the four surfaces ground using various methods of cooling lubricant supply are several times greater than the values of the *Rz* parameter. The cause of such a situation can be found in differing methods of determining the value of individual parameters. In the case of the *Rz* parameter, it is described as the sum of the height of the highest-profile elevation and the depth of the deepest cavity of the profile inside the elementary section whereas the *Sz* parameter is the mean value of absolute height of five largest elevations and five deepest cavities within the sampling area. Another cause of the occurrence of differences in values of the *Rz* and *Sz* parameters can be the different reference bases in relation to which the individual parameters are calculated. For the profile, it is the mean line and, for the 3D measurements, the surface. Additionally, the value of the surface topography parameters is influenced by issues related to, for example, maintaining an accurate height reference basis between passages or maintaining accurate sampling spacing on each track in relation to other ones.

6. Conclusion

On the basis of the results obtained within the scope of the test conditions applied, it can be stated that:

- Within the scope of each of the grinding depths considered, the surface roughness obtained during grinding using the MQL method is similar to the value obtained for the submersion method (WET).
- The minor differences in the *Rz* parameter between results obtained for the MQL and WET methods most likely result from the temperature difference between individual tests. The higher temperature taking place during grinding with the application of the minimum quantity lubrication (MQL) method may result from insufficient lubrication and cooling of the grinding zone in comparison to the submersion method (WET). As a result of the impact of higher temperature, in the tempered and softer material, the plastic deformations increase when the blade is cutting in, manifesting in the formation of side burrs of significant height and occurrence of the fissuring phenomenon. This results in increased roughness of the surface ground.
- Another cause of increased roughness expressed by the *Rz* parameter in case of grinding with the use of the MQL method may be the fact that the stream of oil mist does not enable the appropriate removal of shavings and other products of grinding from the processing zone. As a result of this, the shavings stick to the surface tempered, leading to surface rubbing that increases the roughness of the ground sample surface.

Further research will focus on the selection of treatment conditions with the aim of lowering the surfaces' roughness obtained after grinding. The authors plan to conduct studies on improving the effectiveness of supplying oil mist to the grinding zone through: 1) changes in the positioning of the spraying nozzle, 2) increasing oil expenditure as a result of using additional nozzles spraying the oil most. Additionally, the authors plan to conduct measurements of the grinding temperature.

Acknowledgments

This research was funded by the Polish National Centre for Research and Development, grant number POIR.04.01.04-00-0087/15.

References

[1] Sharma V S, Singh G and Sørby K 2015 A review on minimum quantity lubrication for machining processes *Mater. Manuf. Process.* **30/8** 935-953.

[2] Weinert K, Inasaki I, Sutherland J W and Wakabayashi T 2004 Dry machining and minimum quantity lubrication *CIRP Ann.-Manuf. Techn.* **53/2** 511-537.

[3] Debnath S, Reddy M M and Yi Q S 2014 Environmental friendly cutting fluids and cooling techniques in machining: A review *J. Clean. Prod.* **83** 33-47.

[4] Kruszyński B, Gawroński Z, Sawicki J and Zgórniak P 2008 Enhancement of gears fatigue properties by modern termo-chemical treatment and grinding processes *Mech. Mech. Eng.* **12/4** 387-395.

[5] Fratila D 2010 Macro-level environmental comparison of near-dry machining and flood machining *J. Clean. Prod.* **18** 1031-1039.

[6] Pusavec F, Kramar D, Krajnik P and Kopac J 2010 Transitioning to sustainable production – Part II: evaluation of sustainable machining technologies *J. Clean. Prod.* **18** 1211-1221.

[7] Madanchi N, Kurle D, Winter M, Thiede S and Herrmann CH 2015 Energy efficient process chain: The impact of cutting fluid strategies *Proc. CIRP* **29** 360-365.

[8] Tawakoli T, Hadad M J, Sadeghi M H, Daneshi A and Sadeghi B 2011 Minimum quantity lubrication in grinding: effects of abrasive and coolant-lubricant types *J. Clean. Prod.* **19** 2088-2099.

[9] Stachurski W, Sawicki J, Wójcik R and Nadolny K 2018 Influence of application of hybrid MQL-CCA method of applying coolant during hob cutter sharpening on cutting blade surface condition *J. Clean. Prod.* **171** 892-910.

[10] Stachurski W and Nadolny K 2018 Influence of the condition of the surface layer of a hob cutter sharpened using the MQL-CCA hybrid method of coolant provision on its operational wear *Int. J. Adv. Manuf. Technol.* **98** 2185-2200.

[11] Sawicki J, Kruszyński B and Wójcik R 2017 The influence of grinding conditions on the distribution of residual stress in the surface layer of 17CrNi6-6 steel after carburizing *Adv. Sci. Technol. Res. J.* **11(2)** 17-22.

[12] Lawal S, Choudhury I and Nukman Y 2013 A critical assessment of lubrication techniques in machining processes: a case for minimum quantity lubrication using vegetable oil-based lubricant *J. Clean. Prod.* **41** 201-221.

[13] Stachurski W, Sawicki J, Krupanek K and Nadolny K 2019 Numerical analysis of coolant flow in the grinding zone *Int. J. Adv. Manuf. Technol.* **104** 1999-2012.

[14] Brinksmeier E, Heinzel C and Wittmann M 1999 Friction, cooling and lubrication in grinding *CIRP Ann.-Manuf. Techn.* **48/2** 581-598.

[15] Korecki M, Wołowiec-Korecka E, Sut M, Brewka A, Stachurski W and Zgórniak P 2017 Precision case hardening by low pressure carburizing (LPC) for high volume production *HTM J. Heat Treatm. Mat.* **72/3** 175-183.

[16] Gräfen W and Edenhofer B 2005 New developments in thermo-chemical diffusion processes *Surf. Coat. Technol.* **200/5-6** 1830-1836.

[17] Kula P, Dybowski K, Wołowiec E and Pietrasik R 2014 "Boost-diffusion" vacuum carburising – Process optimisation *Vacuum* **99** 175-179.

[18] Atraszkiewicz R, Januszewicz B, Kaczmarek Ł, Stachurski W, Dybowski K and Rzepkowski A 2012 High pressure gas quenching: Distortion analysis in gears after heat treatment *Mater. Sci. Eng. A* **558** 550-557.

[19] Dybowski K, Sawicki J, Kula P, Januszewicz B, Atraszkiewicz R and Lipa S 2016 The effect of the quenching method on the deformations size of gear wheels after vacuum carburizing *Arch. Metall. Mater.* **61/2B** 1057-1062.

[20] Gawroński Z, Malasiński A and Sawicki J 2010 Elimination of galvanic copper plating process used in hardening of conventionally carburized gear wheels *Int. J. Automot. Technol.* **11/1** 127-131.

[21] Gawroński Z, Malasiński A and Sawicki J 2010 A selection of the protective atmosphere eliminating the inter-operational copper plating step in the processing of gear wheels *Arch. Mater. Sci. Eng.* **44/1** 51-57.

[22] Korecki M, Wołowiec-Korecka E and Glenn D 2015 Single-Piece, High-Volume, Low-Distortion Case Hardening of Gears *AGMA Technical Paper* **15FTM04.**

[23] Wołowiec-Korecka E, Korecki M, Stachurski W, Zgórniak P, Sawicki J, Brewka A, Sut M and Bazel M 2016 System of single-piece flow case hardening for high volume production *Arch. Mater. Sci. Eng.* **79(1)** 37-44.

[24] Stachurski W, Zgórniak P, Sawicki J and Przybysz M 2017 Hardening-related deformations of gear wheels after vacuum carburising and quenching in a 4D quenching chamber *Adv. Sci. Technol. Res. J.* **11(1)** 237-245.

[25] Stachurski W, Krupanek K, Januszewicz B, Rosik R and Wójcik R 2018 An effect of grinding on microhardness and residual stress in 20MnCr5 following single-piece flow low-pressure carburizing *J. Mach. Eng.* **18/4** 73-85.

IOP Conf. Series: Materials Science and Engineering **743** (2020) 012051 doi:10.1088/1757-899X/743/1/012051

Frictional behaviour of TiN coatings grown by atomic layer deposition

Wojciech Grodzicki[1] and Lukasz Kolodziejczyk[1*]

[1]Institute of Materials Science and Engineering, Lodz University of Technology, Stefanowskiego 1/15, 90-924 Lodz, Poland

Email: * lukasz.kolodziejczyk@p.lodz.pl

Abstract. The main idea of the presented work is the evaluation of frictional properties of the TiN coatings deposited by atomic layer deposition technology using atomic force microscopy approach. Ogletree method was employed in order to calibrate the lateral force and estimate the average friction coefficient of TiN coatings. Obtained titanium nitride coatings were also characterized in terms of morphology, surface topography and mechanical properties. The evaluation has been done with reference to the commercially available TiN produced by commonly known magnetron sputtering method.

1. Introduction

Nowadays, material scientists are focused on manufacturing and testing advanced materials (i.e. composites) with excellent and various properties (i.e. high temperature resistance, electrical conductivity). One of the most important property is a resistance to mechanical factors (i.e. friction, fatigue, torsion). Ceramics (i.e. titanium nitride) are known as materials with the best mechanical resistance. Such materials can be used in the space industry, biomedical engineering and wherever structures are exposed to strong mechanical impact. They have also wide application in the production of coatings.

Atomic Layer Deposition (ALD) is a thin-film deposition technique with atomic scale precision growth. During the ALD process occurs saturated chemisorption of gases (precursors) on the surface of substrates. A monoatomic layer is formed in one ALD cycle deposited material. The thickness of the layer depends on the number of deposition cycles. The resulting layers are stoichiometric and show excellent homogeneity over the entire surface.

There are several factors responsible for the differences in friction properties in the micro- and nanoscale. These may include particles that are products of wear and impurities, the transition from elastic to plastic deformation regime as well as the effect of the meniscus formation. One of the extension of atomic force microscopy, enabling measurement of normal and friction forces with pico-Newton resolution is so-called lateral force microscopy (LFM). This technique allows study the tribological properties, including coefficient of friction and wear in the nanometre scale. In this method cantilever with a tip moves on a sample's surface in the vertical direction which causes a torsion of the cantilever.

Titanium nitride is the one of the hardest ceramics material. TiN is often used as a coating on titanium alloys, steel, carbide and aluminium. It is extremely durable, thermally stable and has a very high level of a resistant to chemical etching. TiN has a very high hardness and modulus of elasticity. This material has also outstanding optical properties, it perfectly reflects infrared radiation. The

Content from this work may be used under the terms of the Creative Commons Attribution 3.0 licence. Any further distribution of this work must maintain attribution to the author(s) and the title of the work, journal citation and DOI.
Published under licence by IOP Publishing Ltd

IOP Conf. Series: Materials Science and Engineering **743** (2020) 012051 doi:10.1088/1757-899X/743/1/012051

material has a wide application due to these properties. Titanium Nitride is an example of advanced material which can be use in space industry and biomedical engineering. TiN can be applied into fuselages of spacecraft and can play a role as a protection against damages and scratching in implants. Additionally, it becomes superconducting at very low temperatures.

2. Materials and methods

2.1. Preparation of coatings

The coatings were synthesized using atomic layer deposition technique. Two different precursors combinations were used for TiN synthesis: i) titanium tetrachloride, $TiCl_4$ and monomethylhydrazine, MMH (later in this work abbreviated as $TiCl_4$/MMH) or ii) $TiCl_4$ and dimethylhydrazine DMH (abbreviated as $TiCl_4$/DMH). As a substrate material a native oxide silicon wafer was used. For the purpose of achieving higher deposition rates the ALD semi-industrial machine, the Beneq TFS 200 was used. Electron diffraction (EDX) analysis revealed that compositions of obtained TiN films were close to stoichiometry: Ti (40–45.5 at.%) and N (43.5–49 at.%) with low Cl contamination (0.5–1 at.%). Two groups of samples with different thickness (resulted from different number of ALD deposition cycles) and grown at varying temperatures have been selected for further investigation, according to Table 1.

Table 1. Parameters of deposition processes.

Sample	Growth temperature, °C	Number of deposition cycles	Pulses duration, s	Thickness, nm
$TiCl_4$/MMH@300°C	300°C	600	0.3	10-30
$TiCl_4$/MMH@400°C	400°C			
$TiCl_4$/MMH@500°C	500°C			
$TiCl_4$/DMH@350°C	350°C			
$TiCl_4$/DMH@400°C	400°C			
$TiCl_4$/DMH@450°C	450°C			
$TiCl_4$/DMH@500°C	500°C			
$TiCl_4$/MMH@420°C	420°C	7000	0.3	80

Thickness measurements were performed using X-ray reflectometry (XRR) technique and by depth profiling on nanoindenter. As a reference sample 80 nm thick commercial TiN layer synthesized by magnetron sputtering technique (MS) was used.

2.2. Characterization

Surface morphology, topography and lateral force measurements were performed under ambient conditions using Multimode 5 atomic force microscope equipped with Nanoscope V controller (Bruker Corporation). Topography measurements were made in tapping mode and the size of the images were 1×1 µm. Commercial silicon cantilevers type HQ:NSC15 (MicroMasch) with nominal tip radius ~8 nm, cantilever spring constant of 40 N/m and resonant frequency of 325 kHz were used. Image acquisition was performed with use of Nanoscope 7.3 software and further image processing was done using Nanoscope Analysis 1.9 (Bruker Corporation) and MountainsMap Premium 5.0 (Digital Surf) software.

For friction measurements the areas of 200×200 nm were used at increasing loads up to ~80 nN (10 normal loads ramping equally over the single image) at 1 Hz scan rate according to the procedure described already in our previous papers [1,2]. Commercial silicon cantilevers type HQ:CSC37 (MicroMasch) with nominal tip radius of 8 nm and nominal cantilever spring constant of 0.8 N/m was used. A commercial TGG1 silicon calibration grid (NT-MDT) with 1-D array of triangular steps was used for lateral force calibration [3]. All LFM tests were conducted above adhesion-only regime.

IOP Conf. Series: Materials Science and Engineering **743** (2020) 012051 doi:10.1088/1757-899X/743/1/012051

Mechanical properties, hardness and elastic modulus, of the coatings were measured using nanoindentation technique on Nano Indenter G200 system (KLA Corporation). For nanoindentation a diamond Berkovich tip (Micro Star Technologies) and the continuous stiffness measurement mode were used. The tip shape was calibrated by conducting experiments on a fused silica standard and data were analysed using the Oliver and Pharr [4] approach. Nine experiments were performed on each sample at a strain rate of 0.05 s^{-1}.

Nano-wear tests were conducted in reciprocating sliding mode with the use of the nanoindenter equipped with a diamond conical tip with 1 μm radius and apex angle of 90 deg. The tests were carried out at low normal force of 6 mN for 100 cycles, sliding distance and frequency of 50 μm and 1 Hz, respectively. All of the tests were performed under ambient conditions. Five experiments under same conditions were performed on each sample. Cross profile measurements in the location of the wear tests allowed wear coefficient calculation, based on Archard equation [5].

3. Results and discussion

3.1. Surface topography

In Fig. 1 AFM surface topography scans of all investigated coatings are presented while the roughness parameters are gathered in Fig. 2. Commonly known geometric surface structure parameters of the samples, average roughness (Ra) and root-mean square roughness (RMS), were defined as average values taken from 512 surface profiles. The error was calculated as the standard deviation among all surface profiles.

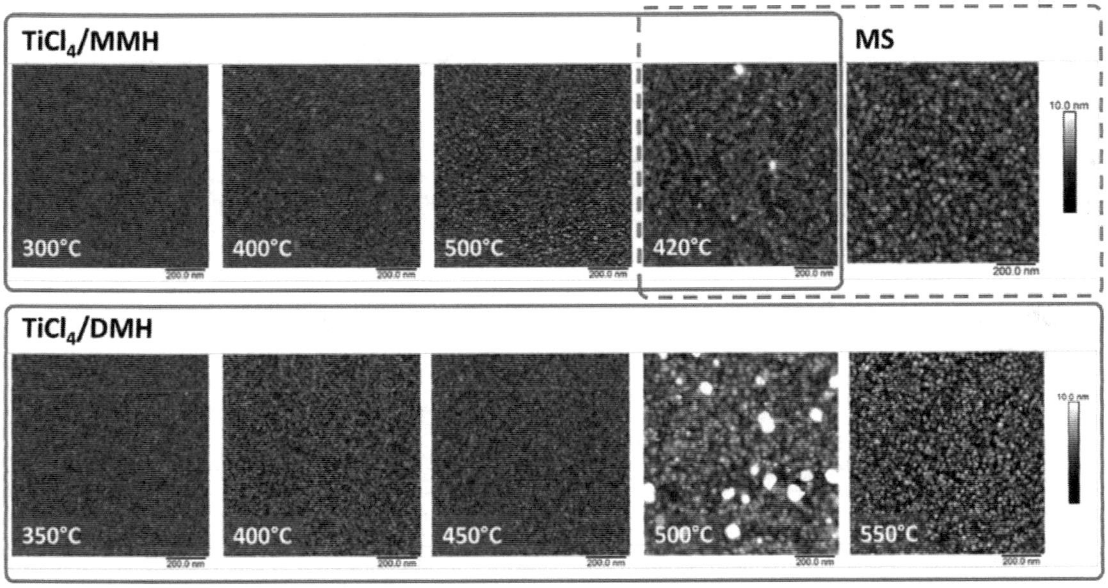

Figure 1. AFM topography images of the analysed coatings (scale bar 200 nm). Red dot line area represents images of thicker (80 nm) coatings.

In the case of TiN coatings produced in the presence of $TiCl_4$/MMH precursors a clear proportional relationship between the synthesis temperature and the geometric surface structure can be noticed. The values of Ra and RMS increase with linear fashion from 0.29 to 0.70 nm and from 0.35 to 0.87 nm, respectively.

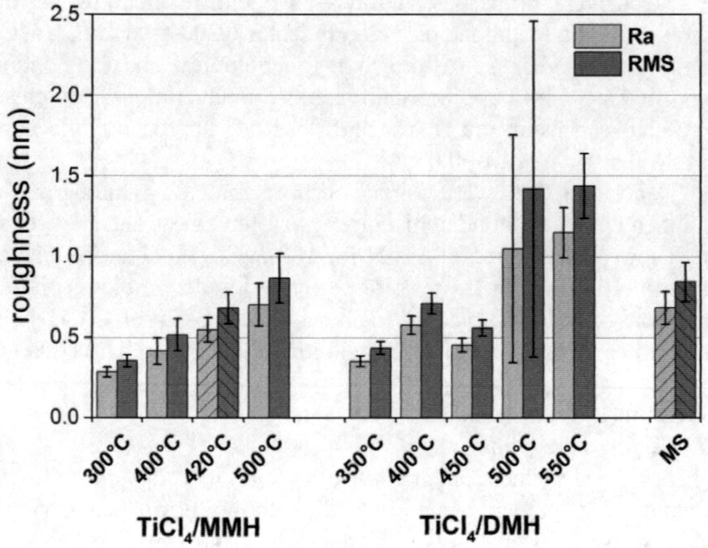

Figure 2. Selected roughness parameters (Ra and RMS) of analysed coatings acquired from AFM images. Sparse stipes bars refer to thicker coatings.

For the coatings grown in the presence of $TiCl_4$/DMH precursors a sharp increase in roughness parameters after reaching the temperature of 500°C is observed (Ra and RMS higher than 1.0 nm and 1.4 nm, respectively). Large values of standard deviation of roughness parameters for the coating grown at 500°C is directly linked to the presence of conglomerates of considerable size on the surface, what can be seen on Fig. 1. For the same precursor combination the coatings grown at temperatures below 500°C show roughness parameters within the range 0.35 – 0.58 nm (Ra) and 0.43 – 0.71 nm (RMS). The roughness of magnetron sputtered TiN coating is similar to that of ALD-derived TiN using $TiCl_4$/MMH precursors at temperature 500°C, namely 0.68 nm (Ra) and 0.84 nm (RMS).

3.2. Mechanical properties

The samples of the first group of coatings (with thickness in the range 10 – 30 nm) are too thin to have directly reliable results by nanoindentation, thus the further analysis of those samples mainly focused on LFM investigation (Section 3.3). From the nanoindentation theory, which says that the depth of penetration should not exceed 10% of the coating thickness, it appears that in the case of indenters whose radius of rounding is less than 20 nm, the coating should have a thickness of approx. 200 nm.

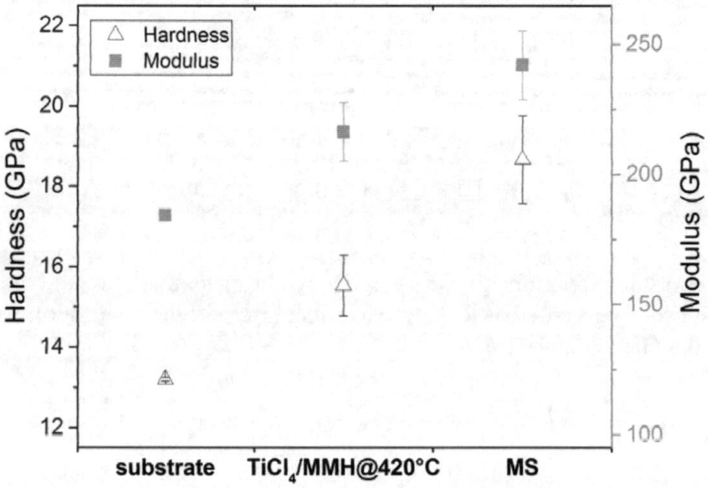

Figure 3. Hardness and elastic modulus values of 80 nm thick TiN coatings with reference to the substrate.

The fulfilment of this condition is necessary and allows to receive real values of mechanical properties with minimized substrate effect on the result [6,7]. In practice, the mechanical properties of coatings of much smaller thickness are successfully determined and the generally accepted limit is approx. 50 nm [8]. The thinner the coating, the smaller the convergence of the obtained result with the actual value. It depends on many factors, including one of the most important relationships between the mechanical properties of the substrate-coating and the superposition of these properties. Due to the imperfect geometry of the used indenters it is recommended to perform measurements on coatings with a thickness of approx. 100 nm [7]. Even after adopting the smallest limiting thickness found in the majority of publications referring to the measurement of ultra-thin coatings, the provided coatings are too thin to be able to determine even comparably changes in their properties. Nevertheless a careful work was done on both, $TiCl_4$/MMH@420°C and reference magnetron sputtered TiN samples, in order to extrapolate hardness (H) and elastic modulus (E) values. Obtained values (Fig. 3) were supplemented with the results calculated for bare substrate. The ALD-derived TiN exhibited H and E values of 15.5±0.8 GPa and 216±11 GPa, respectively. In case of magnetron sputtered TiN mechanical properties were higher than those of ALD-synthesized TiN and not too different to those found in literature, for TiN produced by other techniques (H=18.7±1.1 GPa, E=242±13 GPa). The smallest values of H and E were achieved for the bare substrate (13.2±0.1 GPa and 184±1 GPa, respectively).

3.3. Tribological properties

In Fig. 4 average values of friction coefficient for all investigated coatings are presented.

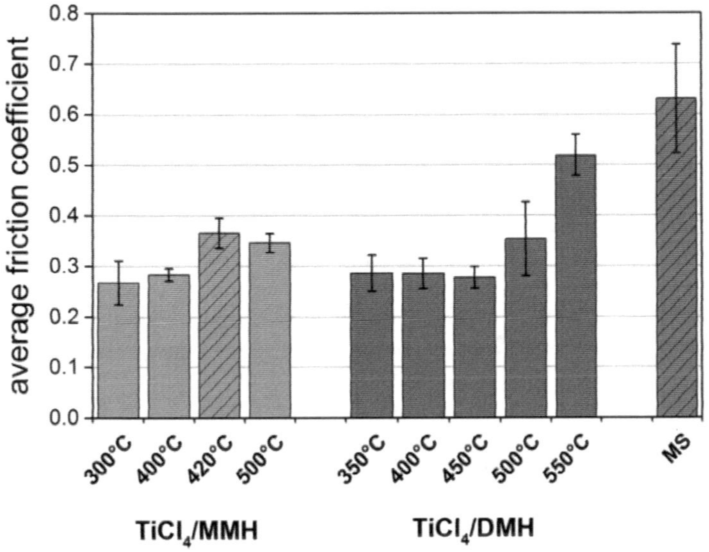

Figure 4. Mean values of friction coefficient for all investigated coatings obtained during LFM experiment. Sparse stipes bars refer to thicker coatings.

From the graph it can be noticed that for ALD-derived samples the values of friction coefficient maintain at similar level (ranging from 0.3 to 0.4) up to annealing temperature of 500°C, independently on precursors used. Increase of the annealing temperature above 500°C results in friction coefficient increase above 0.5. However when compared to the results of reference TiN sample it is clearly seen that magnetron sputtered TiN film is characterized by the highest friction coefficient value (>0.6) amongst all investigated samples. Moreover, taking into account that roughness of the MS-derived TiN is similar to majority of ALD-synthesized TiN coatings, no general correlation of this parameter with the friction coefficient could be found. Thus, it can be suggested that besides topography other factors were present and governed the frictional characteristics of these coatings.

Measurement of H and E (Section 3.2) allowed determination of elastic strain to failure (which is related to H/E) and resistance to the plastic deformation (often associated with H^3/E^2) translating directly into the behaviour of coating during wear and scratch tests (assuming uniformity of the coating) [9–13]. Wear behaviour and scratch adhesion of the studied coatings is presented in Table 2, supplemented with results of H/E and H^3/E^2 ratios.

Table 2. Wear coefficient and scratch critical load of the studied coatings (80 nm thick) supplemented with corresponding results of H/E and H^3/E^2.

	Wear coefficient K mm³/Nm	**Critical load Lc mN**	**H/E**	**H^3/E^2 GPa**
$TiCl_4$/MMH@420°C	$(1.2\pm0.9)\cdot10^{-4}$	10.5±0.3	0.072	0.080
MS	$(9.2\pm1.3)\cdot10^{-7}$	14.6±0.2	0.077	0.111

Wear behaviour analysis showed that in case of ALD-derived TiN film the wear coefficient K is more than 2 orders of magnitude higher than for magnetron sputtering sample. A measure of adhesion strength of the coating to the substrate is the load at which total peeling-off of the coating from the substrate surface occurs, denoted here as Lc parameter. The critical loads obtained in nanoscratch tests, for ALD and MS-derived TiN films were 10.5 mN and 14.6 mN, respectively. Comparing the results of wear and scratch resistance with calculated values of H/E and H^3/E^2 it can be noticed that presented correlations are consistent with literature. Most of the failure mechanisms begins with the plastic deformation thus the coating should be more resistant to wear when it has a high resistance to plastic deformation and, therefore, it reaches a high value of H^3/E^2 (H/E).

4. Conclusions

This paper describes few experimental results of frictional properties of the TiN coatings deposited by atomic layer deposition technology using lateral force microscopy. LFM results revealed that for ALD-derived samples grown at temperature below 500°C the values of friction coefficient maintain at similar and relatively low level (ranging from 0.3 to 0.4) independently on precursors used. Wear and scratch resistance tests results showed correlation with H/E and H^3/E^2 ratios to be consistent with literature. Obtained results will constitute basis for the study of tribological properties of MAX phases and MXenes based on nitrides (i.e. TiN) and carbides – novel materials synthesized by the tailored bottom-up growth by means of atomic layer deposition on substrates with different nature with a fine control of the chemical composition and microstructure.

Acknowledgments

The study is financed by the National Science Centre, project acronym ALD4MAX, project title: "Atomic Layer Deposition for tailored bottom-top growth of MAX and MXene films", agreement number 2016/22/Z/ST5/00693.

References

[1] Martínez-Martínez D, Kolodziejczyk L, Sánchez-López J C and Fernández A 2009 Tribological carbon-based coatings: An AFM and LFM study *Surf. Sci.* **603** 973–9.

[2] Kolodziejczyk L, Kula P, Szymanski W, Atraszkiewicz R, Dybowski K and Pietrasik R 2016 Frictional behaviour of polycrystalline graphene grown on liquid metallic matrix *Tribol. Int.* **93** 628–39.

[3] Ogletree D F, Carpick R W and Salmeron M 1996 Calibration of frictional forces in atomic force microscopy *Rev. Sci. Instrum.* **67** 3298.

[4] Pharr G M, Oliver W C and Brotzen F R 1992 On the generality of the relationship among contact stiffness, contact area, and elastic modulus during indentation *J. Mater. Res.* **7** 613–7

[5] Archard J F 1953 Contact and Rubbing of Flat Surfaces *J. Appl. Phys.* **24** 981.

[6] Li X and Bhushan B 2002 A review of nanoindentation continuous stiffness measurement technique and its applications *Mater. Charact.* **48** 11–36.

[7] Mocko W, Szymanska M, Smietana M and Kalisz M 2014 Simulation of nanoindentation experiments of single-layer and double-layer thin films using finite element method *Surf. Interface Anal.* **46** 1071–6.

[8] Geng K, Yang F, Druffel T and Grulke E A 2005 Nanoindentation behavior of ultrathin polymeric films *Polymer (Guildf).* **46** 11768–72.

[9] Leyland A and Matthews A 2000 On the significance of the H/ E ratio in wear control: a nanocomposite coating approach to optimised tribological behaviour *Wear* **246** 1–11.

[10] Musil J, Kunc F, Zeman H and Polakova H 2002 Relationships between hardness, Young's modulus and elastic recovery in hard nanocomposite coatings *Surf. Coatings Technol.* **154** 304–13.

[11] Musil J 2000 Hard and superhard nanocomposite coatings *Surf. Coatings Technol.* **125** 322–30

[12] Voevodin A A and Zabinski J S 1998 Load-adaptive crystalline-amorphous nanocomposites *J. Mater. Sci.* **33** 319–27.

[13] Beake B D, Vishnyakov V M and Harris A J 2011 Relationship between mechanical properties of thin nitride-based films and their behaviour in nano-scratch tests *Tribol. Int.* **44** 468–75.

IOP Conf. Series: Materials Science and Engineering **743** (2020) 012052 doi:10.1088/1757-899X/743/1/012052

Comparison of mechanical behaviour of microstructures of 2024 aluminium alloy containing precipitates of different morphologies

Anna Staszczyk[1] and Jacek Sawicki[1]

[1]Institute of Materials Science and Engineering, Lodz University of Technology, Stefanowskiego 1/15, 90-924 Lodz, Poland

Email: jacek.sawicki@p.lodz.pl

Abstract. The phenomena occurring in the microstructure of the material are responsible for its macroscopic properties, such as strength, hardness, and toughness. Strengthening phases with the appropriate composition and characteristics increase the strength of the alloy. However, excess brittle intermetallic phases may cause problems with cracking and plasticity of the material. The amount and morphology of those particles are related to the chemical composition of the alloy. Simulation of tensile loading of the microstructure of three different 2024 aluminium alloys shows how the distribution of intermetallic phases affects stress states in the material. The results show that variations in the amount of the alloying additions in the range described by the standard can influence the deformation mechanisms in the material.

1. Introduction

Aluminium alloys of Al-Cu-Mg system are popular because of their good mechanical properties and low density [1]. Their strength and hardness depend strongly on the microstructure being the product of multi-stage heat treatment, typically solution treatment followed by artificial aging. There are several aspects in the microstructure of an alloy that has to be taken into consideration while predicting its mechanical properties. There are many different types of particles that can form during heat treatment; while some of them have a strengthening effect, the others can contribute to brittle fracture of the material.

Several works focus on the complexity of the microstructure of 2024 alloys. Multiple alloying additions contribute to the formation of many intermetallic phases with varying chemical compositions. Silicon and iron have poor solubility in the alpha solution of an Al-Cu-Mg alloy. They form intermetallic phases with copper, magnesium, and manganese during solidification of the cast. It is not possible to dissolve them entirely during solution treatment, and they do not change during further heat treatments in elevated temperatures. Boag et al. described presence of multiphase, composite particles in 2024-T3 material [2]. Mrówka-Nowotnik and Sieniawski examined precipitates of phases present in the microstructure just after casting [3]. Shabestari et al. examined the formation of those intermatllic compounds in 2024 alloy during solidification with different cooling rates [4].

The intermetallic particles are usually much harder and more brittle than the matrix or S phase, as measured by Radutoiu et al. [5]. They also act as the initiators of localised corrosion as proved by DeRose et al. [6]. Multiphase precipitates in 2024 alloy with morphology resembling core and shell were subject of works by Kaczmarek [7][8] and Lipa [9].

Content from this work may be used under the terms of the Creative Commons Attribution 3.0 licence. Any further distribution of this work must maintain attribution to the author(s) and the title of the work, journal citation and DOI.
Published under licence by IOP Publishing Ltd

In this work, the authors expand the existing method for determining the tensile behaviour of a heterogeneous alloy microstructure. The numerical algorithm has been previously described in other papers [10][11] [12] and is based on a combination of the cellular automata method with digital material representation.

2. Materials and methods

The materials used for this study were rods of three 2024 alloys obtained from different suppliers with varying chemical compositions (shown in Table 1). The considerable difference can be observed in the amounts of silicon and iron present.

The specimens cut from the rods were subjected to the heat treatment processes, solution treatment at 500°C for 4h, and artificial aging at 180°C for 10h. The process parameters were designed so that they follow the most popular procedures applied in industrial settings. For reference, the Vickers hardness was measured for all three alloys on Innovatest Verzus 700AS (Table 3). All of the alloys had similar hardness.

Table 1. Chemical composition of the examined alloys.

Alloy	Cu	Mg	Mn	Si	Fe	Zn	Ti	Al
A	4.76	1.36	0.79	0.16	0.12	0.04	0.02	Rest
B	4.15	1.51	0.64	0.24	0.25	0.09	0.06	Rest
C	4.39	1.62	0.74	0.41	0.43	0.04	0.05	Rest

The samples were examined on the Scanning Electron Microscope, Jeol JSM-6610LV. The images of microstructures were all taken with 1000x magnification. The observed phases were additionally examined with Energy Dispersive Spectroscopy to identify their chemical composition. Two main types of precipitates were observed; round ones and those of more irregular shapes. The round dispersoids were only found in the alloy A and had only Al and Cu or additionally Mg in them. The rest of the intermetallic particles contained Cu, Mn, Fe, Si, and Mg in different proportions.

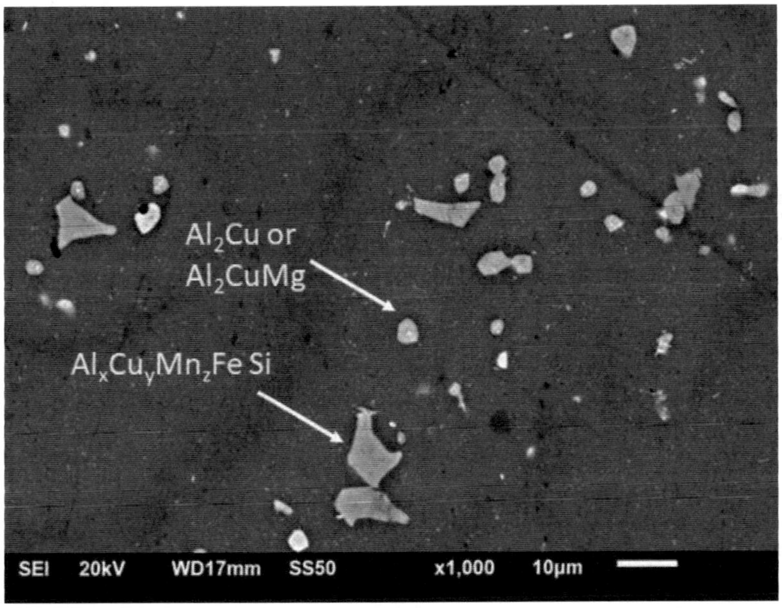

Figure 1. Example of the phase identification in the alloy A.

Mechanical properties (hardness and elastic moduli) of observed phases forming the precipitates as well as the particle-free zones were measured with the use of G200 nanoindenter (MTS Nano

Instruments). The Young's modulus of the matrix (measured at random places between precipitates) was the same for all of the specimens and was around 90 GPa. The round precipitates with Cu and Mg had the average modulus close to 120 GPa, while the big irregular particles had the average modulus of 160 GPa. Those mean values were later used in the simulation.

Table 2. Elastic modulus of examined phases in the microstructure of alloys.

	Matrix	Round precipitates	Other precipitates
Young's modulus [GPa]	90	120	160

Table 3. Vickers hardness of examined alloys.

Alloy	Hardness in T6 state
A	138 ± 2.7 HV
B	134 ± 1.9 HV
C	135 ± 3.2 HV

For the calculational purposes, the images were subjected to post-processing. The goal was to feed the algorithm with a domain where the phases of one type are one colour and distinguishable from the matrix. The algorithm than associates the colour of the pixels with a certain value of the Young's modulus measured from the nanoindentation test. The program simulates the tensile loading of the two-dimensional domain when it is subjected to 0.05% deformation. The outcome of the simulation is a colour map showing the von Mises Equivalent stress in the microstructure.

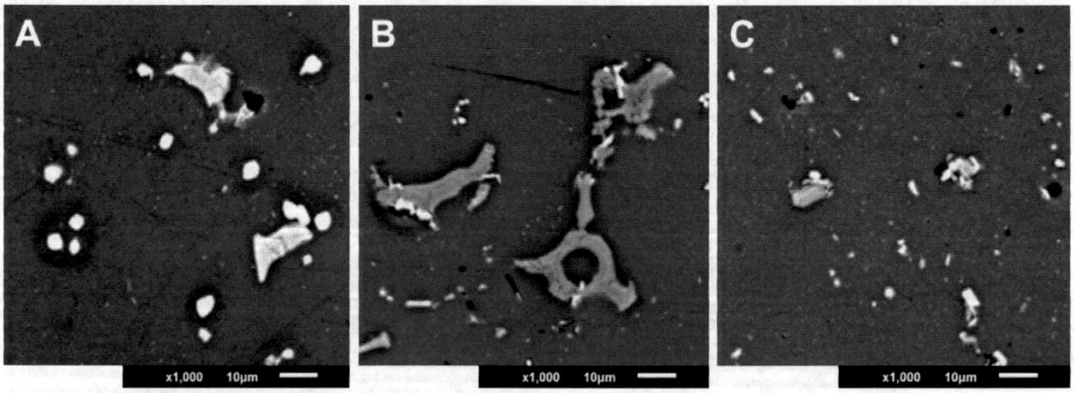

Figure 2. The microstructure images of the alloys A, B and C used in the simulation.

3. Results and Discussion

The microstructure of the alloy A contains two main types of precipitation exhibiting visual distinction. Some of them are round in shape and have been identified with EDS as θ or S phases (Al2Cu or Al2CuMg). Precipitates of such stoichiometry are usually encountered on smaller length scales and are believed to be responsible for strengthening mechanisms. Here, observed particles are notably larger, with dimensions up to 5µm. It is important to note that such precipitates were not observed in two other alloys. Nanoindentation results show that they are slightly less hard than the rest. That confirms a similar conclusion obtained by Radutoiu et al. [13]. The microstructure fragment chosen for simulation contains the same amount of both particle types

The particles in alloy B are much larger and have rather irregular complicated shapes. It is common for them to form circles surrounding small areas of the matrix. Such precipitates are often referred to as the “Chinese script”. Their sizes can go up to 50μm. Since they are large and brittle, they crack easily, which can be observed in the examined specimens as a result of grinding and polishing of the surface as the decomposition of the precipitate into smaller fragments. The microstructure chosen for simulation contains the typical particles, one of them being visibly cracked and one of them having a “hole” in the centre. It is assumed that those are all precipitates of the same phase.

The microstructure of the alloy C is much more dispersed and consists of many smaller precipitates of the hard intermetallic phase with lengths up to 10μm. They were identified to be Al(CuMnFe) phases, occasionally containing Si, which is similar to the bigger phases in alloys A and B.

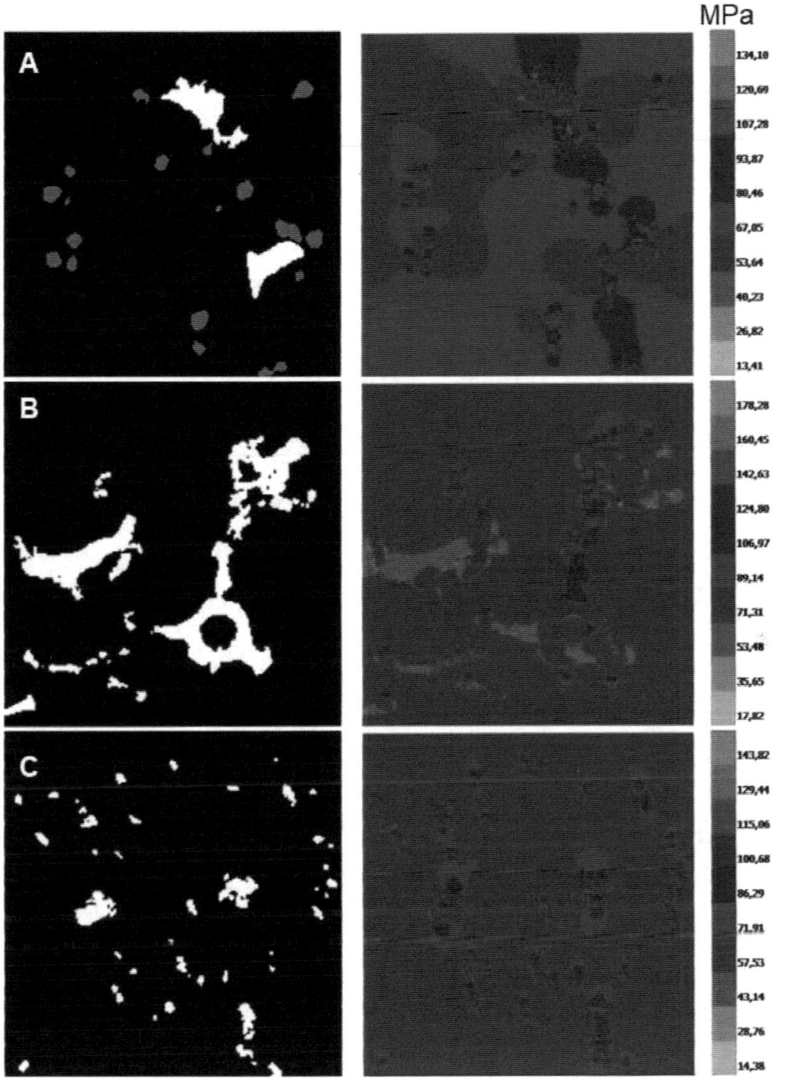

Figure 3. Results of the simulation of the tensile loading of the microstructure. The equivalent stress states are shown for alloys A, B, and C.

The overall mechanical properties of an aluminium alloy depend on the combination of effects caused both by large dispersoids and small second-phase particles present in the microstructure. The amount of them, their composition, and morphology change depending on the amount of the alloying elements in the material. In this case, the overall amount of the intermetallic phases was similar for all of the three alloys. However, there was a considerable difference in dispersion of the particles that were observed in SEM photos. The simulation results show that the highest stresses concentrate on the edges of precipitates. The highest stress was obtained in one of the large precipitates of the alloy B. It shows that the narrow parts of the hard precipitate are subjected to large stresses, which in real life may result in the crack initiation. The lowest maximum equivalent stress was obtained for the alloy A. However, the distribution shows the concentration points on the verges of the harder precipitates again. In this case, the round precipitates have less influence on the stress state than the others. The most uniform distribution is shown for the alloy C. It correlates with the generally more uniform distribution of precipitates and their dispersion.

In conclusion, the numerical simulation of the tensile loading of the microstructure of the Al 2024 alloy showed considerable differences in stress states between alloys with different chemical compositions. The simulation was based on the experimental measurements of mechanical properties of intermetallic phases. The results showed that amount of alloying elements has an influence of size and dispersion of intermetallic particles, and in some cases, on their type. More dispersed distribution results in more uniform equivalent stress state in the material. On the other hand, when particles are large and very hard they tend to concentrate high-stress values on their edges, which may result in cracking.

References

[1] Ł. Kaczmarek, P. Kula, J. Sawicki, S. Armand, T. Castro, P. Kruszyński, A. Rochel, Arch. Metall. Mater. 54, (2009) 25–30.

[2] A. Boag, A.E. Hughes, N.C. Wilson, A. Torpy, C.M. MacRae, A.M. Glenn, T.H. Muster, How complex is the microstructure of AA2024-T3?, Corros. Sci. 51 (2009) 1565–1568.

[3] G. Mrówka -Nowotnik, J. Sieniawski, Analysis of intermetallic phases in 2024 aluminium alloy, (2013).

[4] S.G. Shabestari, M.H. Ghoncheh, H. Momeni, Evaluation of formation of intermetallic compounds in Al2024 alloy using thermal analysis technique, Thermochim. Acta. 589 (2014) 174–182.

[5] N. Radutoiu, J. Alexis, L. Lacroix, J.-A. Petit, M. Abrudeanu, V. Rizea, S. Vulpe, Effect of the over-ageing treatment on the mechanical properties of AA2024 aluminum alloy, Rev. Chim. (Chemistry Mag. 63 (2012) 1042–1045.

[6] J.A. DeRose, T. Suter, A. Balkowiec, J. Michalski, K.J. Kurzydlowski, P. Schmutz, Localised corrosion initiation and microstructural characterisation of an Al 2024 alloy with a higher Cu to Mg ratio, Corros. Sci. 55 (2012) 313–325.

[7] L. Kaczmarek, M. Steglinski, J. Sawicki, J. Swiniarski, D. Batory, K. Kyziol, L. Kolodziejczyk, W. Szymanski, P. Zawadzki, D. Kottfer, Optimization of the heat treatment and tribological properties of 2024 and 7075 aluminium alloys, Arch. Metall. Mater. 58 (2013) 535–540.

[8] L. Kaczmarek, M. Stegliski, H. Radziszewska, L. Kolodziejczyk, J. Sawicki, W. Szymanski, R. Atraszkiewicz, J. Swiniarski, Effect of double-phase segregations formed due to two-stage aging on the strength properties of alloy PN-EN 2024, Met. Sci. Heat Treat. 54 (2013) 477–482.

[9] S. Lipa, L. Kaczmarek, M. Steglinski, H. Radziszewska, K. Kyziol, D. Kottfer, Effect of core/shell precipitations on fatigue strength of 2024-T6I6 alloy, Int. J. Fatigue. (2019).

[10] A. Staszczyk, M. Debyser, J. Sawicki, Modelling and Simulation of Core–Shell Precipitates under Tensial Loading Using the Cellular Automata Method, Mech. Mech. Eng. 21 (2017) 443–452.

[11] A. Staszczyk, J. Sawicki, Application of an equivalent truss model for determining the stress state in multi-phase materials with Cellular Automata method, Adv. Sci. Technol. Res. J. 11 (2017) 51–57.

[12] A. Staszczyk, J. Sawicki, .M. Stegliński: Modeling of mechanical behavior of double phase precipitates in 2024 aluminum alloy. AIP Conference Proceedings 2078, 020056 (2019); https://doi.org/10.1063/1.5092059.

[13] N. Radutoiu, J. Alexis, L. Lacroix, M. Abrudeanu, J.-A. Petit, Study of the influence of the artificial ageing temperature on the AA2024 alloy microstructure, Key Eng. Mater. 550 (2013) 115-125.

AUTHOR INDEX

Abdullah, M A H 161, 235
Abdullah, M M A B 43
Abdullah, Mohd Mustafa Al Bakri 8, 30, 36, 71, 77, 86, 148, 189
Adam, Hussaini 103, 133
Adam, Hussain 97
Adam, Tijjani 103
Ahmad, Muhammad Naim 174
Ahmad, Nurul Atiqah 263
Ahmad, Romisuhani 86, 93, 148, 155, 189
Alrozi, Rasyidah 281
Amini, Hazim Mohamad 16
Awang, Arita Hanim 174
Ayob, A. 179
Ayub, M. A. R. M. 59
Azani, A 43
Azhari, AW 43
Azli, Muhammad Amirul Aziem Noor 189
Bahrin, M. A. F. 124, 214
Bakar, B. H. Abu 179
Bakar, Mohamad Bashree Abu 268
Bayuaji, Ridho 30, 290
Brewka, A 335
Buzalski, Victoria 326
Bąkała, M 317
Chobpattana, V 43
Czepułkowska, W 296
Derman, M N 1
Derman, M. N. 22
Fu, Chang 302
Ghazali, M. J. 22
Gopinath, C. B. Subash 97, 103
Grodzicki, Wojciech 355
Halin, D S Che 43
Hashim, N H 168
Hashim, Noratikah 86
Hashim, U. 115, 133, 141, 207, 214
Hashim, Uda 97, 103, 109
Hashim, U 124
Hui, Beh Hui 241
Hui-Teng, Ng 36
Hussin, Kamarudin 30, 36
Ibrahim, Fazdliel Aswad 227
Ibrahim, Nur Hulwani 97, 109
Ibrahim, Wan Mastura Wan 86, 93, 155, 189
Isa, N F 168
Ishak, Muhammad Ikman 49, 54, 59, 65
Ishak, Noorhidayah 59
Ismail, C N 54
Ismail, Haris 65
Ismail, Muhd Nazmi 275
Ismail, Nur Syakimah 263
Jaih, M K M 221
Jambek, Asral Bahari 115, 124, 133, 141, 207, 214
Jantira, Nurul Nadia 281
Januszewicz, B 341
Janvekar, Ayub Ahmed 268
Jaya, Nur Ain 30
Jin, Lai Shan 197
Jing, Leow Yee 77
Junaidi, Subaer 71
Kamal, Syazni Zainul 197
Kamaruddin, Mohamad Anuar 275, 281
Keat, Lim Chun 227
Khor, C Y 54
Khor, C. Y. 49, 59, 65
Kolodziejczyk, Lukasz 355
Krupanek, Krzysztof 326
Krupanek, K 335
Lazim, Azwan Mat 268
Li, Long-Yuan 77
Mahmed, N 43
Ming, Liew Yun 290
Mohamad, Khairunnisa 174
Mohamed, Mazlan 268
Mortar, Nurul Aida Mohd 36, 71, 86, 148, 155, 189, 290
Muhamad, Khairunnisa 71
Muhamad, K 168, 221
Muhammad, Noorhafiza 8
Naamandadin, N A 168
Naamandadin, Nurul Akmam 174
Nawi, M A M 54
Nawi, M. A. M. 49, 59, 65
Nazri, F. Mohamed 179
Noor, S N A M 221
Noor, S N F M 8
Noor, Siti Nur Aishah Mohd 174
Omar, M F 161, 235
Omar, Mardhiah Farhana 227
Othman, M K 221
Parmin, N. A. 97, 109, 133, 207
Perz, Karolina 310
Pietrasik, Joanna 302
Putri, N S 290
Rahim, Ruslinda A 263

Rahman, Syakir Amir Abdul 174
Ramli, M M 43
Rani, A L Abdul 161, 235
Rashid, N A 161, 235
Razak, K A 43
Razak, R A 168
Razak, R. Abdul 179
Razak, Rafiza Abd 71, 77
Razak, S M S 221
Rewolińska, Aleksandra 310
Rezek, Bohuslav 263
Rosik, R 341
Roslan, Nurul Syiffa Husna Che 254
Rosli, M U 54
Rosli, M. U. 49, 59, 65
Rozyanty, R 16
Rylski, A 317
Said, R Mohd 1
Salimi, M. N. 109
Salimi, Midhat Nabil 197, 241, 254
Salleh, M A A M 43
Salleh, MAA Mohd 1
Salleh, Mohd Arif Anuar Mohd 148
Sandu, Andrei Victor 36, 86, 148
Saud, N 1
Sauffi, Ahmad Syauqi 155
Saufi, Ahmad Syauqi 93
Sawicki, Jacek 302, 326, 362
Shaharuddin, S. N. A. 133, 207
Shahidan, Nurzarihah 268
Sikora, M 341, 348
Soberi, Nur Suhaili Mohd 16
Staszczyk, Anna 362
Sukeri, N S Mohamad 43
Taczała, Joanna 302
Tahir, Muhammad Faheem Mohd 30, 93, 189
Teik, Lim Ewe 275
Termizi, Siti Nor Azreen Ahmad 59
Tham, P Y 168
Tusimin, Z 168
Uda, M. N. A. 97, 103, 109, 115, 124, 133, 141, 207, 214
Uda, M. N. Afnan 97, 103, 109, 115, 124, 133, 141, 207, 214
Wahab, J. A. 22
Wei, Z. X. 49
Wojciechowski, R 317
Wołowiec-Korecka, E 296, 335, 341, 348
Yahya, N 168, 221
Yahya, Zarina 71, 77, 93
Yong, Heah Cheng 30, 36
Yong-Sing, Ng 36
Yun-Ming, Liew 30, 36
Zahid, M. Z. A. Mohd 179
Zaidi, Fakhryna Ahmad 155
Zaidi, Fakhryna Hannanee Ahmad 93
Zailani, Warid Wazien Ahmad 86, 155
Zainol, Norrazman Zaiha 174, 227
Zainuddin, Firuz 16
Zakaria, Nurul Husna 8
Zgórniak, P 335, 341, 348
Zhe, Adeline Choo Khang 71
Zulkifli, Abuzaar Mohd 227
Zulkifli, Nadiah 'Izzati 148